Current Topics in

Bone Biology

Current Topics in
Bone Biology

Editors

Hong-wen Deng

Creighton University, USA
Xi'an Jiaotong University, P R China &
Hunan Normal University, P R China

Yao-zhong Liu

Creighton University, USA

Associate Editors

Chun-yuan Guo

P&G Pharmaceuticals, USA

Di Chen

University of Rochester Medical Center, USA &
Nankai University Medical College, China

World Scientific

NEW JERSEY • LONDON • SINGAPORE • BEIJING • SHANGHAI • HONG KONG • TAIPEI • CHENNAI

Published by

World Scientific Publishing Co. Pte. Ltd.

5 Toh Tuck Link, Singapore 596224

USA office: 27 Warren Street, Suite 401-402, Hackensack, NJ 07601

UK office: 57 Shelton Street, Covent Garden, London WC2H 9HE

British Library Cataloguing-in-Publication Data
A catalogue record for this book is available from the British Library.

CURRENT TOPICS IN BONE BIOLOGY

ISBN 981-256-209-5

Printed in Singapore by World Scientific Printers (S) Pte Ltd

CONTENTS

PREFACE

The volume entitled **Current Topics in Bone Biology** is the first to be generated by the members and friends of International Chinese Hard Tissue Society (ICHTS), which underscores the Society's commitment to promoting excellence in bone and mineral research and facilitating the translation of advances of bone biology to health care and clinical practice. The publication of the volume will be translated into Chinese and published by a prestigious publishing house in China, Higher Education Press, to accelerate our support of ICHTS' main mission "to promote scientific and profession excellence and enhance communication among scientists of Chinese heritage and other internationals in the field of hard tissue research and related areas".

The International Chinese Hard Tissue Society is indebted to the editors and authors who made this book possible. I did not think it was possible to recruit so many in this endeavor because of the problem that these authors encountered. Traditionally, scholars writing chapters for books are given credit and encouraged by their institution. However, government research administration in European countries and some academic institutions have changed this policy. Now the career evaluation system for scholars have downgraded the value of writing chapters and gives no credit for this activity. Publication credit in many academic institutions is based on impact, a measure of circulation and prestige of the journal in which the publication occurs. Thus, my hat is off to each of the dedicated scholars who took time to help further to fill in the void to our understanding of physiology and pathology and to "connect the dots" in our understanding to generate new paradigms even though it may not advance their careers.

To implement publication of this volume, the International Chinese Hard Tissue Society is indebted to the tremendous efforts of Professor Hong-Wen Deng, Editor-in-Chief, Dr. Yaozhong Liu Co-Editor-in-Chief, and Associate Editors Dr. Chun-Yuan Gao and Dr. Di Chen. They successfully coaxed and cajoled authors of 20 chapters to complete their tasks on schedule. In addition, this monumental effort succeeded only with the continual support of the President, Dr. Darren Ji and the help of Dr. Ying Lu who introduced us to the publisher, World Scientific Publishing Co.

This book is intended for students, teachers, practitioners and investigators of the skeletal system. Basic scientists and clinical investigators interested in bone and their adjacent soft tissues will find this "Current Topics in Bone Biology" useful at all levels of inquiry, including molecular biology, cell biology, biochemistry, physiology, genetics, pathology and biomechanics. The volume is interesting reading. There are many chapters in this volume that describe skeletal effects of genetic and environmental factors from which the reader can formulate their own opinions and paradigms.

I recommend this volume strongly, especially to principal investigators, research associates, post-doctoral fellows, graduate students, as well as established investigators and clinicians. Digesting it will pave the way for all to fill in the blanks and develop new paradigms for skeletal physiology and pathology.

Webster S.S. Jee, Ph.D.
Professor of Anatomy & Radiobiology-
University of Utah School of Medicine;
Co-Editor-in-Chief-
Journal of Musculoskeletal & Neuronal Interaction;
Founding Member & Chairman of the Board-
International Chinese Hard Tissue Society

CHAPTER 1

INTERNATIONAL CHINESE HARD TISSUE SOCIETY THE POWER THAT CONNECTS THE WORLD OF SCIENCE AND CULTURE

Darren X. Ji, MD, PhD, President
Webster SS Jee, PhD, Chairman, Board of Directors
International Chinese Hard Tissue Society (ICHTS)

It is a delight to see the fruition of the published book on bone biology and osteoporosis compiled by ICHTS in its 10th anniversary. ICHTS has come a long way from a small group of scientists gathering at the Sun Valley International Hard Tissue Workshop to a professional organization with over 700 members around the world.
The mission of ICHTS is to promote scientific and professional excellence and to enhance communication among scientists of Chinese heritage and other international scholars in the field of hard tissue research and related areas. As it has grown in the past ten years, ICHTS has gone through excitements, challenges and inevitable growing pains and has mature into a cohesive and strong organization. ICHTS has participated, sponsored, and organized various scientific events to promote information exchanges and research collaborations around the world. In this short overview, we invite you to join us on the growing path of ICHTS and its impact on its members and the overall scientific community.

1. Sun Valley, Idaho – the birthplace of ICHTS

Ten years ago (1994), a small group of Chinese scientists attended the 24th International Sun Valley Hard Tissue Workshop. The workshop was an annual event since 1965, organized by Dr. Webster Jee, a professor at the University of Utah and a pioneer of bone biology and bone histomorphometry. These scientists - Web Jee, Mei-Shu Shih, Jian

Li, David Ke, Liya Tang, Mei Li, Pancras Wong, Linda Ma, and Baijin Han, *et al* had known each other and/or been working together, over the years. Coming from the same cultural background, they shared similar interests and recognized the potential synergy of a cohesive supporting group to facilitate scientific exchanges and to assist career growth of new comers. During this meeting, the seed was planted to establish a professional group. It was named the North American Chinese Hard Tissue Society. In about two years as the membership significantly expanded with scientists joining from Europe and other parts of world, the Society was renamed as International Chinese Hard Tissue Society (ICHTS). Dr. Mei-Shu Shih served as a coordinator and then the first president of the Society. Dr. Shih charged the establishment of the society mission, bylaw and charters, and newsletters. This was a critical step that laid a professional foundation for the future growth of the Society. The Society members started to communicate frequently and convened whenever they could to discuss science and life. Upon Dr. Webster Jee started his term as the 2nd president, Jian Li (the 3rd president) chartered the development of the society logo and the Chinese heading as we are still using today (Fig. 1). It was also during this period of time that Jian Li launched the first website of ICHTS and secured the official web domain: www.ichts.org.

One of the first expansions of membership was to those Chinese researchers attending the American Society for Bone and Mineral Research (ASBMR), since it attracts most of the attendees around the world in the research field of bone and mineral. With more members joining in, the membership meetings started to take place in conjunction with the annual ASBMR meeting. The format was a semi-formal gathering, mostly a gathering in a Chinese restaurant. The members got to hear the society update, participated in scientific discussions, and enjoyed great Chinese food. It remained to be the single most effective networking event that new and old members got to know each other quickly.

In 2000 when leadership team recognized that dinner and society business meeting were not a good match since activities at the dinner table took much attention away from the society business. A renovation took place and for the first time that the ICHTS annual meeting was held

國際華人骨研學會

Fig. 1. The society logo and the Chinese heading, developed by Jian Li.

in a formal meeting room at the Crowne Plaza Toronto Center on a Sunday evening during the ASBMR meeting in 2000. Only light refreshments were provided. Much to the delight of the organizers, such a meeting format was well received and the society business discussion became much more productive. This meeting format continued for the following years until today.

In the early years and continuing, the Society began to invite distinguished honorary members to ICHTS of all nationalities. To name a few here: Douglas Axelrod, Charlie Bleau, David Burr, David Dempster, Harold Frost, Juerg Gasser, Harry Genant, George Jaworski, Chris Jerome, Conrad Johnston, Stephan Krane, David Lacey, Robert Lindsay, T. Jack Martin, Les Matthews, Gregory Mundy, Carol Pibeam, Robert Recker, Gideon Rodan, Tom Sanchez, Masahiko Sato, Mitchell Schaffler, Hans Schiessl, William Sietsma, Steve Teitelbaum, David Thompson, Tom Wronski. Over the years, these distinguished scientists and business leaders served as mentors and sponsors and contributed toward the nurturing of ICHTS or it's members. Some became close personal and family friends.

When Dr. David Ke became the president in 2000, efforts began to focus on formulating the society development strategy and establishing effective operating structures. One of the important working principles set during this time was that ICHTS would collaborate primarily with organizations whose main purpose was to promote scientific exchanges, not financial gains. Such principles set a tone in the later decision-makings on the society positioning in dealing with potential collaborators. It was also during this time that executive offices were not only established but also put into effective operation. For example, establishment of audit committee independent from Treasurer and the President, and setting up expense limits that can be authorized by society President and Chairman ensured the just use of the society budget and the transparency of accounting.

In May 2003, ICHTS proudly re-launched its website incorporating the latest web technologies and a professional look, through a friendship contribution of Steve Surman, a web professional at Procter & Gamble. The site is subsequently maintained by Ming-qing Chen, with Darren Ji being the primary content owner. In May 2004, ICHTS launched its Chinese mirror site in Beijing (www.ichts.org.cn), developed by Dong Li, providing much easier access to the growing members from China and Asia.

In ten years, the society has gone through transitions of five presidents, with as much joys as the growing pains. Nevertheless, when the five past and current presidents got together in Minneapolis in September 2003 (Fig. 2), they shared one smiling content – ICHTS has come a long way, and is becoming an irreplaceable powerhouse to nurture the growth of its members and the scientific community.

2. The Consummate Union of the Chinese Philosophy and Western Logics

The over 700 plus ICHTS members today consist of diverse backgrounds connected by the common bond of Chinese heritage and the shared interest of promoting ICHTS' mission.. Many of its members and honorary members are from different nationalities and ethnicities, sharing a same enthusiasm on what ICHTS can bring into the hard tissue field.

Members of ICHTS are consistently yielding outstanding contributions to the research community through scientific publications and presentations. Many have achieved world-class recognition, becoming leaders in academia, pharmaceutical and biotechnology

Fig. 2. It has been a joyful journey overall –five ICHTS presidents got together in September 2003, Minneapolis, MN, *from left to right*:
1st President: Mei-Shu Shih, DVM, PhD (1994 - 1996)
2nd President: Webster SS Jee, PhD (1996 - 1997)
3rd President: Xiao-Jian Li, MD (1997 - 2000)
4th President: Hua-Zhu David Ke, MD (2000 - 2003)
5th President: Darren Xiaohui Ji, MD, PhD (2003 - 2005).

industries. In a world of constantly changing rules, ICHTS members have started to realize that more than ever, networking and collaboration have become critical components of one's sustained success, in science and in professional career. More and more intra- and inter-continental collaboration and partnership among the ICHTS members have been established. Energized by the profound economic and scientific potential of China, many ICHTS members have ongoing working relationships or hold joint appointments in Chinese academic institutions. Some of the Society members have been recognized with prestigious awards, such as "Yangtze River Scholars" of the Chinese government and the "Hundred Scientists Plan" which honors outstanding contributions in the advancement of science and research in China.

Anytime we look at these outstanding members either as a group or an individual, we could easily identify some common traits that are deeply rooted in the consummate union of the Chinese philosophy and Western logics. The Chinese brain nurtured patience, perseverance and being contented with achievement of baby steps, while the Western teaching provides the perpetual optimism and confidence of being oneself.

It is under such nicely married cultures that ICHTS becomes an exceptional source of scientific expertise and cultural strength, promoting communication and synergy and cultivating the professional and scientific excellence.

3. A Cultivator of Technical and Career Growth

Over the years, ICHTS continued to introduce programs to help the growth of its members. Each year, ICHTS grants a number of travel awards to the outstanding young scientists to attend scientific conferences such as ASBMR, ORS (Orthopedic Research Society) and CSOS (Chinese Speaking Orthopedic Society) (Fig. 3). These awards not only provide recognition of outstanding scientific contributions, but also give the opportunity for the young members to be exposed to and get connected with the mainstream development of each field. In 2004, the leadership team decided to re-name the ICHTS travel award as ICHTS Webster Jee Young Investigator Award, in honor of the pioneering contributions of Dr. Web Jee to the hard tissue research and to the development of ICHTS.

The Career Development Workshop and Grant Writing Workshop have proved very useful for the junior members and faculties. The mentorship program was also developed to provide advice and guidance in technical and career issues.

Everyone needs a hand from time to time – this is the principle that many of the senior ICHTS members make themselves available in helping the junior members with the problems such as job seeking, reference letters and guidance in personal and technical growth.

4. A Model of Alliance and Collaboration for Enhanced Scientific Exchanges

Uniquely positioned to provide technical expertise and cultural bridge, ICHTS has been collaborating vigorously with various societies and organizations around the world to promote science and cultural exchanges. The primary means of such promotions is through alliance to co-organize or co-sponsor scientific meetings or workshops. During his presidency between 1997 – 2000, Dr. Jian Li initiated a number of collaborations with many prominent colleagues in China including Dr. Meng Xunwu of Chinese Medical Association, Dr. Liu Zhonghou of Chinese Osteoporosis Foundation, Dr. Zhao Yanling of HOMA Symposium.

Since 1999, ICHTS started to make more strategic choices and co-sponsored the Third International Osteoporosis Conference in Xian, China (1999), the SIROT meeting in Egypt (2002), and International Bone Research Instructional Course and Hands-on Workshop (2002). Most significantly, ICHTS co-organized with the Chinese Medical Association (CMA) the 1st International Conference of Osteoporosis and Bone Research (ICOBR) in Beijing in 2003. This was the first time that ICHTS acted in China as a co-organizer of an

Fig. 3. Dr. Di Chen (right), Vice-President of ICHTS, is conferring a travel award to Ms. Lanjuan Zhao, a graduate student from Creighton University, during the 2003 annual meeting, Minneapolis, MN.

international conference. The meeting was a recognized success in the quality of scientific programs and the level of audience participations. Based on this experience, the two parties agreed to continue the collaborating in every two years to co-organize ICOBR in China. The next one is to be held in October 20-24, 2005.

In 2004, ICHTS formed another significant alliance with Orthopaedic Research Society (ORS) to co-promote ICHTS travel awards to the ORS meeting. ICHTS had its formal entry into the ORS arena and recruited over 80 new members in two months.

Recognizing the growing potential of the research capabilities in China, ICHTS has developed working relationships with the Chinese embassy and consulate in the US to seek opportunities for joint development. ICHTS has also reached tentative agreements with a few Chinese academic institutions to establish ICHTS research centers in China.

5. Sponsorship and Friendships

ICHTS's achievements could not be parted from the continuous sponsorships from our friends and corporate partners including (in alphabetical order):

- Amgen
- Hologic
- Merck, Inc.
- Norland
- OrthoLogic
- OsteoMetrics
- Pfizer, Inc
- Pfizer Foundation
- Procter & Gamble Pharmaceuticals
- Scanco
- SkeleTech
- Wyeth

Over the years ICHTS has developed strong friendships with these individuals and coorporations, generating synergies to benefit the growth of all parties.

6. An Exciting Journey and a Bright future

Looking into the future, ICHTS is committed to becoming an energizing ground for our members around the world to unite and help each other. ICHTS will continue its strong involvement in the scientific exchanges, and will make special effort to promote our members' career growth. For example, periodic topics and discussions will be facilitated in the area of general skills such as communication and leadership. The mentorship program will be strengthened to allow junior members to receive adequate guidance and advice when needed. We will keep improving communications and connections with our colleagues and members in China to help with education, and adequate representation of their interest in the international communities.

There are a number of goals ICHTS strives to achieve in the next five years. We plan to break into the other countries in Asia than China for sustained membership growth. We will expand our presence to the scientific meetings such as those of American Dental Association and Bioengineering society for enhanced networking and collaboration opportunities. We will attract more clinicians to broaden the scope of our expertise in the hard tissue field. We will provide training fellowships and grants to help junior members to jumpstart their careers. We will promote the exploration of Chinese herbs for standardized therapeutic uses. Most of all, we seek to establish an ICHTS forum to establish scientific and cultural synergies among the scientists from the East and the West.

We strongly encourage our members to bring new ideas to help the society to grow. We cordially invite those who are not yet members to join this exciting and dynamic society. To join the society, please visit the ICHTS website (www.ichts.org).

CHAPTER 2

INTEGRATED BONE TISSUE ANATOMY AND PHYSIOLOGY

Xiao Jian Li, M.D., Principal Scientist

Wyeth Research
200 Cambridge Park Drive
Cambridge, MA 02140
E-mail: jli@wyeth.com

Webster S. S. Jee, Ph.D., Professor

Division of Radiobiolog, University of Utah
729 Arapeen Drive, Suite 2338
Salt Lake City, UT 84108-1218
E-mail: webster.jee@hsc.utah.edu

1. Introduction

The skeletal system, a collective of many individual bones joined by connective tissue, provides both biomechanical support and metabolic supply for the entire body. Bone tissue formed by inorganic salts embedded in organic matrix, has great rigidity and hardness compared to other connective tissues. The skeleton system primarily provides basic biomechanical functions to 1) maintain the shape of the body; 2) protect the soft tissues of the cranial, thoracic, and pelvic cavities; 3) provide the framework for the bone marrow; and 4) transmit the force of muscular contraction from one part of the body to another to produce movements. The skeletal system also serves as a mineral ion bank, contributing to the regulation of extracellular fluid composition. The skeletal system constantly renews its structural material, adapts its mass, shape, and properties to the changing mechanical environment, and endures voluntary physical activity. In this chapter we will introduce the skeletal elements at both the organ level (Skeletal Anatomy) and the tissue level (Skeletal Histology). We will also describe major skeletal biologic mechanisms and explain how they work together in modifying the skeletal tissue to adapt to the mechanical and non-mechanical environments.

2. Skeletal Anatomy

A typical adult long bone consists of a central cylindrical diaphysis, and an epiphysis at each end. The metaphysis is the conical region connecting the diaphysis with the epiphysis (Fig. 1a). The joint is the connecting point between two bones. It is responsible for transferring the load from one bone to the other. The joint is covered by articular cartilage to minimize friction and wear between the two bony ends during movement.

Subchondral bone as a part of the epiphysis supports articular cartilage. The epiphysis is composed of a trabecular network with a thin peripheral cortical shell and a subchondral bony top (Fig. 1b).

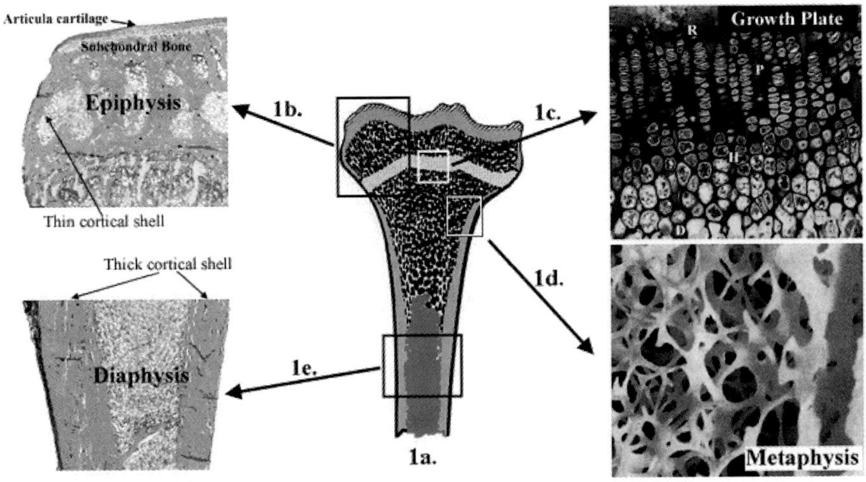

Figure 1. Microphotograph of a proximal tibia showing the gross profile of a typical long bone (See text for details)

Beneath the epiphysis there is a growth plate complex (Fig. 1c). It is composed of a number of chondrocyte columns, which are surrounded by the hyaline cartilage columns. Each chondrocyte column is divided into 4 functional zones: resting (R), proliferating (P), hypertrophic (H) and degenerating (D) zones. The hyaline cartilage in hypertrophic and degenerating zones is mineralized to become calcified cartilage, which later becomes the core structure for the metaphyseal primary spongiosa. After skeletal maturity a thin layer of bone replaces the growth plate complex, which is referred to as the closure or the fusion of the growth plate.

The metaphysis is composed of a sponge-like network of interconnected trabecular plates and spicules (Fig. 1d), generated by the growth plate complex, which is divided into primary and secondary spongiosa. The primary spongiosa has a calcified cartilage core that is surrounded by the woven bone matrix. In the secondary spongiosa the trabecular network is primarily composed of lamellar bone, with few remnants of calcified cartilage core and woven bone matrix.

Extending from the metaphysis, the diaphysis is a thick cylindrical cortical bone shaft (Fig. 1e). This cortical shell contains a central marrow space that is occupied primarily by the hematopoietic and/or fatty tissue with minimal or no cancellous bone.

Articular cartilage has a much weaker biomechanical property and very different function than bone. In order for articular cartilage to endure the forces transmitted during load bearing and locomotion, its load bearing surface area has to be much larger. The metaphysis is a funnel shape trabecular network structure that transmits mechanical loads from large articular cartilage surface to the small diaphyseal cortical bone. From the upper metaphysis toward the diaphysis, the thin cortical sheller is gradually and significantly thickened, while the dense trabecular network gradually becomes sparse and is eventually phased away. The total cross sectional area of the long bone is also gradually decreased from the growth plate toward the diaphysis [1-3].

2.1. *Axial and appendicular bones*

Axial and appendicular bones are the two subdivisions of the skeleton. Axial bones, such as vertebral bodies in the spinal column, have a thin cortical shell, a rich cancellous network, and are located adjacent to the viscera. They contain hematopoietic marrow and turnover at a high rate. The appendicular bones, such as femurs and tibiae, have a thick cortical shell, with cancellous bone in their epiphyseal and metaphyseal regions, and are surrounded by muscles. They contain mainly fatty marrow and turnover at a low rate [1-2].

2.2. *Woven and lamellar bone*

Woven and lamellar bones are the two sub-tissue types in the skeletal system based on matrix organization. In general, woven bone is an immature bone while lamellar bone is a mature one. The immature woven bone is formed rapidly during embryonic skeletal development, longitudinal bone growth under the growth-plate-complex, early fracture healing, or the osteosarcoma formation. Because of the rapid formation, the interwoven coarse collagen fibers are arranged in a random fashion. The distribution of osteocytes generally follows that of collagen fiber and is therefore also in a random fashion. When a fracture occurs, a large woven bony callus is rapidly formed, yielding a temporary functional structure to partially restore mechanical properties, such as stiffness. This allows the return of functional loading on the healing bone. The loading condition initiates secondary bone remodeling, replacing the woven bone callus with lamellar bone. In this situation, the functions of this temporary woven bone callus include: 1) rapid restoration of biomechanical properties to enable physical activity; and 2) providing a scaffold for secondary lamellar bone replacement. In all cases, woven bone is considered to be an interim material that is eventually resorbed and replaced by lamellar bone. By nature, woven bone is less organized and shorter-lived than lamellar bone. By 3 years of age, woven bone produced during human fetal bone development is completely replaced by lamellar bone. The mechanical strength of woven bone is weak due to its randomly orientated and loosely bundled collagen fibers and low mineral deposition (Figs. 2a & 2b) [1-3].

In contrast, lamellar bone is formed at a much slower pace. The collagen fibers produced during bone formation are laid on to the existing bone surfaces in an orderly fashion. Collagen fibers made by osteoblasts are laid down layer by layer with strict organization. For each layer, collagen fibers are laid parallel to each other, forming a bone matrix sheet, which is called a lamella. In the next layer, the direction of the collagen fibers is perpendicular to that of the previous layer. As such, the histological appearance of this lamination under polarized light is alternating light–dark layers, representing the cross-sectional and longitudinal orientated collagen fibers on the histologic section (Figs. 2c & 2d). The mechanical strength of lamellar bone is strong due to its orderly orientated and stably bundled collagen fibers and high mineral deposition [1-3].

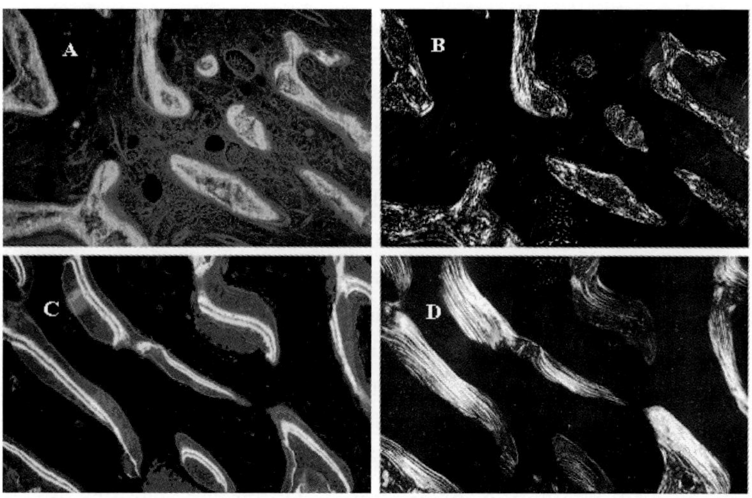

Figure 2a-2d. Microphotograph showing detail histologic characteristics of cancellous woven trabeculae diffusely labeled with tetracycline viewed under UV light (A, upper left), and its collagen fiber randomly orientated viewed under polarized light (B, upper right). Cancellous lamellar trabeculae are linearly double-labeled viewed under UV light (C, lower left), and their collagen fiber orderly orientated as the lamination viewed under polarized light (D, lower right).

Three distinct lamellar patterns can be found in the adult cortical bone: (a) Haversian system or osteon - concentric layers of lamellar sheets arc formed surrounding a longitudinally vascular channel; (b) circumferential lamellae - multiple layers of lamellar sheets are formed uninterruptedly around part or all of either periosteal or endocortical surfaces (see Bone Surfaces section); and (c) interstitial lamellae - fragments of old lamellae are observed in regions between Haversian systems. Similar to cortical bone, cancellous bone contains two major lamellar patterns within the trabecular spicules. They are trabecular packets (hemiosteons shaped like shallow crescents) and interstitial lamellae [1-9].

2.3. Cortical and Cancellous Bones

The skeletal system can also be divided into cancellous bone and cortical bone. While cancellous bone is a sponge-like trabecular network structure that occupies the inner region of the epiphysis and metaphysis (Figs. 1b & 1d), cortical bone is a semi-solid shell that covers the entire bone. This cortical shell is thin in the epiphyseal and metaphyseal regions, but is thick in the diaphyseal region [1-3].

Cortical bone is the primary tissue type of the skeletal system as it contributes 80% of the entire (1,400,000 mm^3) adult skeletal mass in humans. Porosity in cortical bone is due to Haversian canals, Volkmann's canals, and resorption cavities, containing primarily nervous tissue and blood vessels. The surface area of the cortical bone is relatively small as it contributes to only 33% of the total bone surface. Therefore, its surface to volume ratio is only about 2.5. This small surface adjacent to marrow results in a low turnover rate in cortical bone. The main function of cortical bone is to provide biomechanical, supportive, and protective properties. Haversian canals run nearly parallel (at a 11° angle) with the major axis of bone. They are interconnected with Volkmann's canals, which are oriented perpendicular to the skeletal loading axis and run horizontally from periosteal (outer) surface to the endocortical (inner) surface of the cortical bone. As such, a three dimensional network of canals exist throughout cortical bone. Along with it is the network of circulatory vessels and nerves, as well as an extracellular fluid path, which allows exchanges of nutrition, nerve signals, and metabolites between cortical bone and its neighboring environment (outside tissues or inner marrow). Cortical bone is constantly renewing/remodeling itself in response to altered mechanical and nonmechanical environmental signals, as well as microdamage. Since cortical bone is a semi-solid material, its renewal requires initiating a complex process called remodeling, in which the removal of existing intracortical bone is followed by the generation of new osteons. During this process, one or more pre-existing osteons are partially removed to yield space for the newly generated osteons. Their remnant is then left in between new osteons as interstitial bone. In cortical bone more than 60% of the total bone volume is occupied by osteonal lamellae, while the remaining 40% is occupied by the interstitial or sub-periosteal/endosteal-circumferential lamellae (Fig. 3) [1-3].

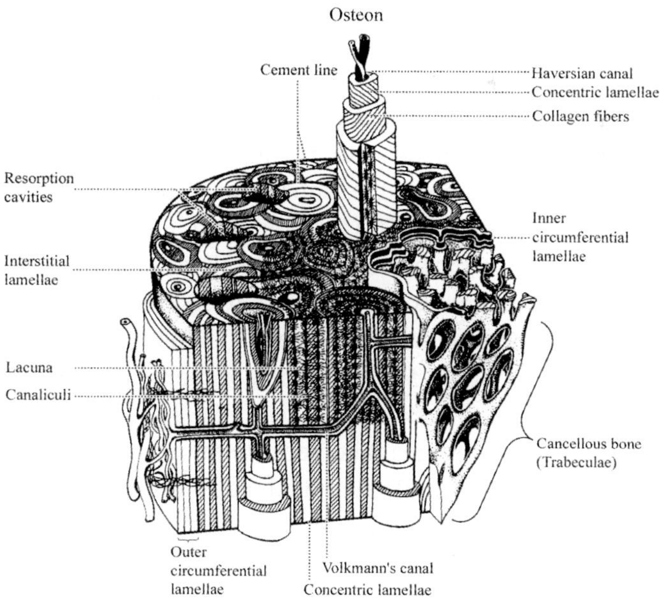

Figure 3. Three dimensional schematic view of histological details of cortical bone. (From Weiss, L., Ed., Cell and Tissue Biology, A Textbook of Histology, Urban and Schwarzenberg, Baltimore, 1988. [1].)

In contrast, cancellous bone contributes only 20% of the total bone mass (350,000 mm^3). Cancellous bone is composed of a large number of rod or plate shaped trabeculae, forming a sponge-like trabecular network. The primary composition in cancellous bone marrow is hematopoietic. The surface area of the cancellous bone is large, contributing to 75% of the total surface of the skeletal system. Therefore its surface to volume ratio is approximately 20, 7-fold higher than that of cortical bone. This large surface adjacent to marrow results in a high turnover rate in cancellous bone. The main function of the cancellous bone is to provide biomechanical support and fulfill homeostatic demands [1-3].

The proportion of each tissue type varies greatly between different bone sites. The proportion of cortical bone to total bone can be as high as 92% in the ulna, while it is 62% in a typical vertebra [1-3].

2.4. Bone Surfaces

Compared to other tissues, bone surfaces have unique importance as they are the only region available for bone cellular activities (modeling and remodeling) in response to mechanical and nonmechanical demands. Bone surfaces are categorized into periosteal and endosteal surfaces. The periosteal surface covers the entire outer perimeter of bone. It contributes to 4.4% (roughly 500,000 mm^2) of the total bone surface. The endosteal surface is further divided into: 1) intracortical surface (the surface of Haversian canals), which contributes to 30.4% (roughly 3,500,000 mm^2) of the total bone surface; 2) endocortical surface, which contributes to 4.4% (roughly 500,000 mm^2) of the total bone surface; and 3) trabecular surfaces, which contribute to 60.8% (roughly 7,000,000 mm^2) of the total bone surface. The periosteum is a fibrous sheet with a deep cambium layer of undifferentiated cells covering the periosteal surface. During growth, the periosteum actively initiates circumferential radial bone growth by adding new bone onto the outer surface. This consequently enlarges the cross-sectional area of the long bone. During adulthood periosteal bone formation is minimized to a negligible level, although it may be more active in advanced age. When fractured, the periosteum actively participates in the bone repair process. The endosteum is a cellular layer covering the endocortical surface and outlining the marrow cavity of all individual bones [1, 3, 10, 11].

The skeletal system constantly renews itself on the endosteal surface, through remodeling, to maintain its biomechanical strength and meet metabolic needs. The process of bone remodeling includes 1) resorption, 2) formation, and 3) quiescence. Therefore, at any specific time a given bone surfaces is in one of the above 3 functional stages. At any given time, there is 0.6% cortical and 1.2% cancellous bone surfaces undergoing bone resorption; and 3% cortical and 6% cancellous bone surfaces undergoing bone formation. The most commonly seen stage in a bone sample is the quiescent or resting state. In both cortical and cancellous bone, greater than 93% of total bone surfaces is in the quiescent state [1-4, 11].

2.5. Composition of Bone

Components of bone include organic matrix (20-40%), inorganic mineral (50-70%), cellular elements (5-10%), and lipids (3%).

The organic bone matrix consists predominately of type I collagen with a small quantity of types III, V and X collagen [11]. Many collagen molecules (tropocollagens) are bundled together to form collagen fibers, which are further aligned parallel to each other to form a lamella sheet. Between the ends of tropocollagens, interfibrillar cross-links are formed by crosslinking between tri-valent pyrodinolines and pyrroles thus stabilizing the matrix. Tropocollagens are aligned in a quarter-staggered end-overlap fashion, forming a collagen fiber with numerous gap regions, into which hydroxyapatite crystals are deposited as mineralization occurs (Fig. 4). Trace amounts of type III, V and X collagen may regulate the diameter of collagen fibrils during certain stages of bone matrix formation.

Hydroxyapatite is the predominate molecule of bone mineral. It exists in the form of needle-, plate- or rod-shaped crystals, which are deposited into the gap regions of collagen fibers. It provides bone matrix with mechanical rigidity and load bearing capacity. Hydroxyapatite contains many impurities, such as carbonate, citrate, magnesium, fluoride and strontium that are either incorporated into the crystal lattice or

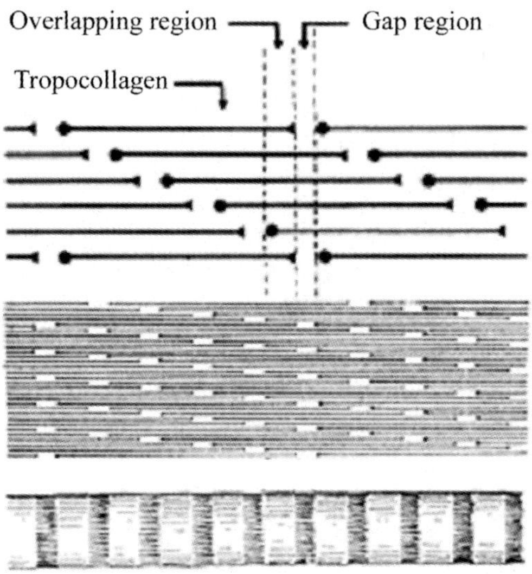

Figure 4. Special arrangement of tropocollagen to form gap regions for mineral crystal deposition during bone matrix mineralization.

absorbed onto the crystal surface. The imperfect crystals are more soluble than geologic apatite, enabling bone to be re-solubilized and release its calcium, phosphate or magnesium ions into the extracellular fluid as needed. As such, bone serves as a mineral ion bank to meet homeostatic demands of the whole body. Bone seeking substances that have high bone affinity can also be incorporated into the bone matrix as mineralization occurs. These substances include tetracyclines, polyphosphates, bisphosphonates, and bone-seeking radionuclides.

In bone, various non-collagenous proteins (NCPs) make up 10% of its total organic matrix. Their specific roles are unclear, but they are considered important in the calcification process, the fixation of the hydroxyapatite crystals to the collagen, as well as the control of osteoblastic and osteoclastic metabolism. Among the NCPs in bone, the most abundant ones are osteocalcin, osteonectin, osteopontin and bone sialoprotein. Only bone sialoprotein and osteocalcin are unique (specific) to the skeleton. Some NCPs, including hydroxyproline and collagen cross-links, are released into the circulation during the breakdown of bone matrix. Levels of these substances reflect bone turnover and thus act as biomarkers in blood and urine samples [13]. Therefore, evaluating urinary and plasma/serum levels of these unique NCP's to monitor the status of bone turnover has been clinically useful to diagnose, detect progression, and determine treatment efficacy of skeletal metabolic diseases [12-17].

3. Skeletal Histology

3.1. Bone cellularity

The major cellular elements of bone include osteoclasts, osteoblasts, osteocytes, bone-lining cells, along with the precursors of these specialized cells, and cells of the marrow compartment and the immune regulatory system [1-4, 6-11, 18-27]. In this section, cellularity descriptions are focused on the four specialized bone cell types and their precursors.

The osteoclasts

As bone resorbing cells, osteoclasts are multinucleated giant cells with a diameter ranging from 20 to over 100 microns. Osteoclasts have acidophilic cytoplasm containing numerous vesicles.

Osteoclastogenesis originates from the granulocyte–macrophage colony-forming unit (GM-CFU). The GM-CFU is the mononuclear/ phagocytic lineage from either the local hematopoietic marrow or systemic circulation. Although early promonocytes are the primary cell source from which the osteoclasts differentiate, monocytes and macrophages are also capable of osteoclastic differentiation under special circumstances. Functional osteoclasts may live for up to 7 weeks with a half-life around 6-10 days. An osteoclastic nucleus has approximately a 10-day lifespan. As osteoclasts complete their resorptive function, they migrate into adjacent marrow space, where they undergo apoptosis.

Although the details are unclear, it is a general view that osteoblastic lineage cells initiate osteoclastogenesis. The activation signal induces the ingrowth of blood vessels, from which the osteoclastic precursor migrates into the adjacent marrow. The activation signal also stimulates the bone-lining cells to contract, by which they release factors that digest the underlying osteoid layer, exposing the mineralized surface for osteoclastic resorption. Bone-lining cells may lift from the bone surface and migrate into the adjacent marrow, where they directly contact osteoclast precursors. During this cell-cell contact, RANKL on bone-lining cell can bind to its receptor RANK on osteoclastic precursor, thus initiating the osteoclastogenic process. As such, multiple committed mononucleated precursor cells fuse together to form mature multinucleated osteoclasts, which are subsequently activated through polarization to form ruffled borders.

After activation, osteoclasts adhere their membrane perimeter to mineralized matrix through a structure called the clear zone, which lacks organelles but is rich in active filaments and integrin receptors. This clear zone serves as a circulating permeable wall between the apical surface of the osteoclasts and bone surface, which forms a small chamber that is sealed from extracellular fluid. As such, a suitable microenvironment for bone resorption can be created and maintained within the chamber. The bottom side of the osteoclastic membrane is extensively enfolded, forming a striated ruffled border that secretes products into the chamber that leads to bone demineralization and degradation (Fig. 5) [1, 2, 6, 10, 19-25]. The cytoplasm and plasma

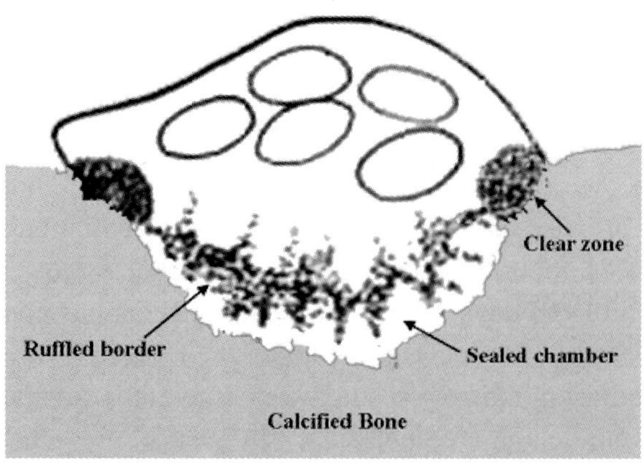

Figure 5. Schematic representation of an activated osteoclast is conducting bone resorption.

membrane of the ruffled border contains chloride channels, encased in Na+, K+ ATPase, HCO_3/CL exchanges, Na+/H+ exchanges, lysosomal proteins, and RANK [6]. RANK, calcitonin and vitronectin (integrin $\alpha_V\beta3$) receptors are markers specifically expressed by osteoclasts [6, 23].

Hormonal osteoclastic regulators include calcitonin, parathyroid hormone (PTH), and $1,25(OH)_2$ vitamin D_3. Receptors for all of the above regulatory hormones are found in osteoclasts except for PTH (Fig. 6). The absence of PTH receptor in osteoclasts has long triggered the hypothesis that cells of the osteoblastic lineage mediate the osteoclastic response to PTH. The recent discovery of the RANK-RANKL pathway validates this hypothesis. Non-hormonal osteoclastic regulators include many local factors and cytokines, such as IL1 and IL6.

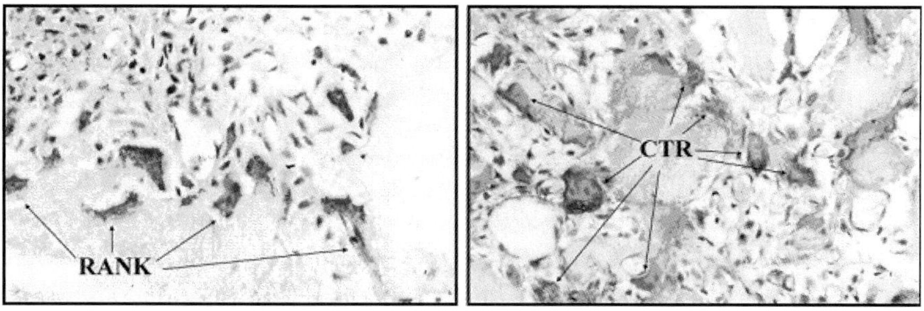

Figure 6. Immunohistochemical staining of osteoclasts with antibodies of RANK (left) and calcitonin receptor (CTR).

The osteoblasts

As bone-forming cells, osteoblasts are involved in the entire bone formation process. They synthesize and secrete collagen to form unmineralized bone matrix (osteoid) and participate in osteoid calcification by regulating the flux of calcium and phosphate in and out of bone. Typical osteoblasts are 15-30 microns cuboidal-shaped cells lined up side-by-side to form a cellular sheet that covers the entire formation surface. The osteoblast has a large nucleus localized in the bottom half of its cytoplasm. The osteoblast also has many cellular processes, abundant endoplasmic reticulum, enlarged Golgi, and collagen-containing secretory vesicles. Among osteoblastic synthetic products, osteocalcin and bone sialoprotein are the only two bone-specific proteins. They serve as biomarkers for osteoblastic identification and functional evaluation. Functional characteristics of active osteoblasts also include the appearance of intensive alkaline phosphatase, the secretion of type I collagen and the synthesis of noncollagenous protein in response to specific mechanical and nonmechanical stimuli [1-4, 6, 11, 14, 18-19, 25, 27].

Osteoblasts are originated in part from stroma located in bone marrow adjacent to the endosteum or in the periosteum. Mesenchymal progenitors in bone marrow or connective tissue are the primary cell source from which osteoblasts are derived. There is strong evidence suggesting that pericytes and endothelial cells are also osteoblastic precursors. The possible fates of an active osteoblast are to become a bone lining cell, an osteocyte, or apoptosis [28-34].

Osteoblasts contain receptors for parathyroid hormone (PTH), parathyroid hormone-related protein (PTHrP), prostaglandins, vitamin D metabolites, bone morphogenetic proteins (BMPs), gonadal and adrenal steroids, certain cytokines, lymphokines, colony-stimulating factor (CSF-1), RANKL. Through these receptors, regulatory factors may activate, enhance, or inhibit the differentiation, proliferation, vigor and apoptosis of osteoblasts and their precursors. For example, recent studies have suggested that PTH and prostaglandin E_2 inhibit osteoblastic apoptosis [34-37]. Some regulatory factors also play a role in balancing activities between various cell types. RANKL produced by osteoblasts, initiates osteoclastogenesis. Osteoprotegerin, a decoy RANK receptor, which is also secreted by the osteoblast, inhibits osteoclastic formation (Fig. 7).

Bone formation occurs in two distinct stages: matrix formation and mineralization (Fig. 8). Osteoblasts synthesize and secrete collagen fibers, laying them on an existing surface in an orderly fashion as multi-layered lamellae. During matrix formation some osteoblasts lag behind and become embedded in the newly formed matrix. They differentiate into osteocytes and extend out processes to communicate with neighboring osteocytes, surface osteoblasts, or lining cells.

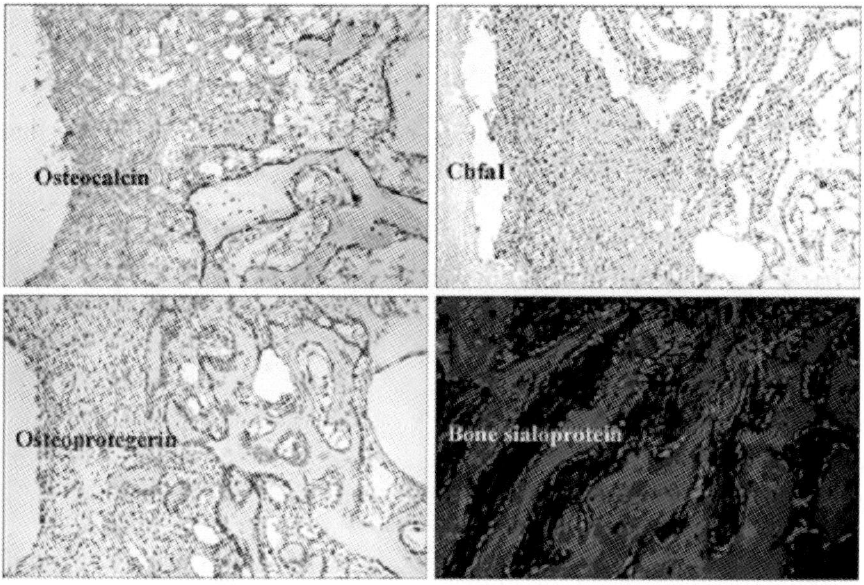

Figure 7. Immunohistochemical staining of osteoblasts with antibodies of Osteocalcin (upper left) and Cbfa1 (upper right), Osteoprotegerin (lower left) and bone sialoprotein (lower right).

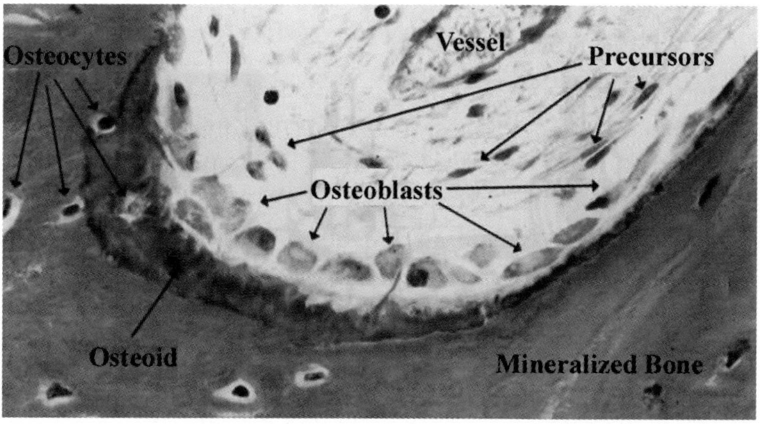

Figure 8. Microphotograph showing detail histologic characteristics of osteoid, osteoblasts and osteocyte.

Corresponding to their vigor, the height of osteoblasts is reduced gradually from an average of 7 microns at the beginning to 1 micron at the end of the bone formation period. As soon as the bone matrix is produced, tropocollagens start to reorganize. Tropocollagens align themselves in a quarter-staggered end-overlap manner, forming numerous gap regions within. Cross-links are formed between tri-valent pyrodinolines and pyrroles to stabilize the collagen fibrous structure (Fig. 4). The process of gap region formation and interfibrillar cross-linking is called bone matrix (osteoid) maturation, by which the osteoid prepares itself for the subsequent mineralization.

As osteoid matures, osteocytes regulate an influx of mineral ions from extracellular fluid to form hydroxyapatite molecule crystals, which are then deposited into the stabilized gap regions of the osteoid. Mineralization occurs at the interface between mineralized bone and unmineralized osteoid, which is called the mineralization front. A 7-10-micron thick layer of immature osteoid, which is referred to as the osteoid seam, is frequently observed at the active bone forming sites, between the mineralization front and the surface osteoblast layer [1-3, 5, 10-11, 18, 38-42].

The bone-lining cells

Bone-lining cells are a layer of elongated flat cells. They are interconnected as a cellular sheet that covers the quiescent bone surfaces. They are also known as resting osteoblasts [1-4, 19, 28, 34-36]. Bone-lining cells are about 1 micron thick with a 12-micron diameter. They have a thin, flat nucleus with an attenuated cytoplasm (Fig. 9). These cells extend their processes to communicate with adjacent bone-lining cells, as well as osteocytes through gap junctions (Fig. 10). In adult dogs, the cellular density of trabecular or endosteal surfaces is approximately 19/mm bone surface in fatty marrow, and it is greater in the hematopoietic marrow. The bone-lining cell density decreases with age [28, 34].

Bone-lining cells are derived from surface osteoblasts when they have completed their historical role as bone forming cells. During the quiescent period, bone lining cells, together with a one-micron underneath layer of osteoid, serve as a barrier to protect bone surfaces from inappropriate resorption by osteoclasts or other inflammatory cells (Fig. 10). The ultimate fate of the bone-lining cells is not known. They may return to the pool of stem cells or pre-osteoblasts, undergo

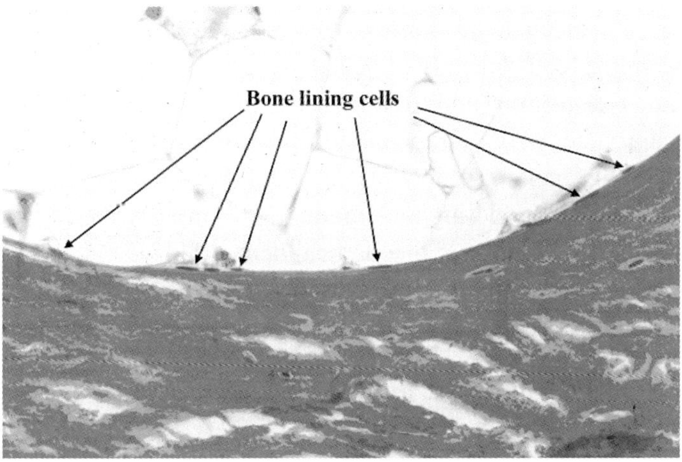

Figure 9. Microphotograph showing bone lining cells cover the resting endosteal surfaces.

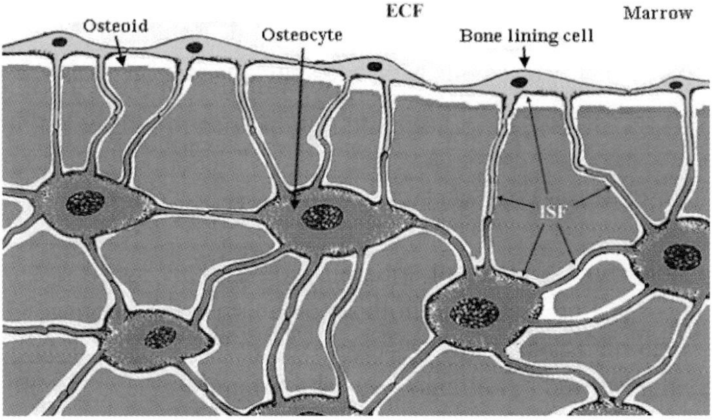

Figure 10. Diagrammatic representation of the interstitial fluid compartment (ISF) and the extracellular fluid compartment (ECF), as well as their composition elements: bone, osteocyte, bone-lining cell, as well as the underlying osteoid.

apoptosis, or may be reactivated into osteoblasts, which can form bone when a new modeling cycle is initiated by bone anabolic agents [28, 34-36].

Bone-lining cells, together with the osteocytic lacuno-canalicular 3-dimensional network system, are considered as the most likely candidates for mechanical sensors in Frost's Mechanostat hypothesis. Since the lining cell sheet is attached tightly to most of the endosteal surface, any distortion and deformation of bone caused by external mechanical impact will be transmitted to the lining cells. As a result, the disturbed lining cells produce specific signals to initiate biological mechanisms, either modeling or remodeling, to modify the skeletal shape and size to meet the changed mechanical demand [43].

Bone-lining cells are interconnected as a cellular sheet that covers almost the entire endosteal surface. This cellular sheet separates the extracellular fluid from the interstitial fluid percolating through the osteocytic lacunar-canalicular system, thus serving as an ion barrier. The cellular barrier may have a role in maintaining a suitable microenvironment for the growth of bone crystals, as well as in regulating the influx and efflux of calcium and phosphate for mineral homeostasis. Bone-lining cells are also involved in the initiation of bone modeling and remodeling cycles. After receiving an activation signal, lining cells self-contract and secrete neutral proteases to digest the surface osteoid. The bone surface barrier (lining cell-osteoid) is therefore removed and mineralized matrix is then exposed for osteoclastic resorption (Figs 9 & 10). Thus, the bone surface enters the initial stage of bone remodeling [19, 24-25].

The osteocytes

The osteocyte is the most abundant cell type in bone tissue. In mature bone about 95% of total bone cells are osteocytes. Osteocytes are the only cell type to be embedded within the bone matrix. A young osteocyte has a smaller cell size and fewer protein-synthetic organelles than its predecessor – the osteoblast. The cell size and organelles are further reduced as the osteocyte ages. The life span of osteocytes depends largely on the rate of bone turnover, during which osteocytes are removed. In slow bone turnover sites, osteocytes have an average half-life of 25 years according to Frost (87-89).

During bone formation some osteoblasts are left behind and housed in small chambers known as lacunae, which are embedded in the newly formed osteoid as bone formation advances. During this period the embedded osteoblasts differentiate into osteocytes by loss of most of their organelles and extension of many long and slender processes that are encased in the canaliculi. By connecting their canaliculi with that of other osteocytes, surface osteoblasts or lining cells via gap-junctions, a three-dimensional osteocytic lacuno-canalicular network is formed throughout the entire bone (Figs. 3, 10 & 11) [1-4, 6-9, 12, 14, 44-49].

The size of a canalicula is very small. This tiny tubular canalicula contains an osteocytic process and the surrounding interstitial fluid. The periosteocytic space is the space between the osteocytic cytoplasmic membrane and the bony wall of lacuno-canalicular network (Figs. 10 & 11). The periosteocytic space is filled with 1.0-1.5 liters of interstitial fluid. The total surface area of the lacuno-canalicular wall in the skeletal

Figure 11. Microphotograph of a cross sectional osteon showing the densely distributed lacuno-canalicular network.

system ranges between 1,000-5,000 m^2. The special structural configuration of this osteocytic lacuno-canalicular network allows the external mechanical force to be transformed into an internal mechanical signal that is detectable by the sensor cells. Any distortion/deformation of bone matrix will compress the osteocytic lacuno-canalicular periosteocytic space and cause the rapid movement of the interstitial fluid. The rapid fluid flows produce shear stresses on the cytoplasmic membrane of the osteocytic process. Consequently, this stimulates osteocytes, the mechanosensors, to synthesize biochemical signals, such as prostaglandins or nitric oxide, and transmit them to the effector cells through the 3D network [1-4, 43].

The physiological function of the osteocyte is not well understood, but some scientists have proposed the following hypotheses. With its 3D network throughout the bone, osteocytes work with lining cells to regulate the exchange of mineral ions between interstitial fluid and extracellular fluid. Therefore, it maintains an appropriate local mineral ionic milieu that is suitable for bone matrix mineralization. Based on its unique position within bone matrix, osteocytes are responsible for detecting microdamages and for initiating the repair process. Osteocytes (together with lining cells) may also serve as mechanosensors of the Mechanostat system. External mechanical bone matrix distortion and deformation can cause rapid movement of interstitial fluid within the periosteocytic space. This rapid fluid flow produces shear stress on osteocytic cytoplasmic membrane, generating electric and biochemical signals. As a result, biological activities such as modeling or remodeling may be initiated to modify the skeletal mass, shape and size to meet mechanical demands [43].

Most osteocytes eventually undergo apoptosis. Aging, unloading, chronic glucocorticoid administration and loss of estrogen are factors known to increase osteocyte apoptosis and subsequent bone loss [31-33, 48-50]. Treatment with estrogen inhibits osteocyte apoptosis in ovariectomized rats [32]. An increase in loading inhibits the osteocyte apoptosis in rat cortical bone [33]. In aged human and dogs, absence of osteocytes is consistently observed in the hypermineralized lacunae of cortical bone [48]. Loss of osteocytes in aged bone tissue may contribute to osteonecrosis or osteoporosis in patients suffering from vascular disease with circulative vessel degeneration. Studies have shown that treatment with PTH and bisphosphonates also prevent osteocyte apoptosis [37].

3.2. Bone structural unit

In the skeletal tissue different cell types do not work independently because they are regulated by the same governing system -- the mechanical and non-mechanical environments. In fact, during the process of adaptation to the mechanical and non-mechanical environment, these cells work together in a highly synchronized fashion as individual components of the same biologic mechanisms. These processes are characterized by the coupling between resorption and formation in bone remodeling, as well as the synchronized coordination between formation drift and resorption drift in bone modeling.

Accordingly, two terms were created by Frost to define such a highly synchronized collective work accomplished by the teamwork of these various bone cells. Basic multicellular unit (BMU) is a functional term defining a unit of new bone created by the remodeling process, in which all cellular elements (progenitors, lining cells, osteoclasts, osteoblasts, osteocytes… etc,) are involved. Bone structural unit (BSU) is a histologic term, which focuses on defining the existence of the end product of bone remodeling. In cortical bone, the BSU is an osteon or

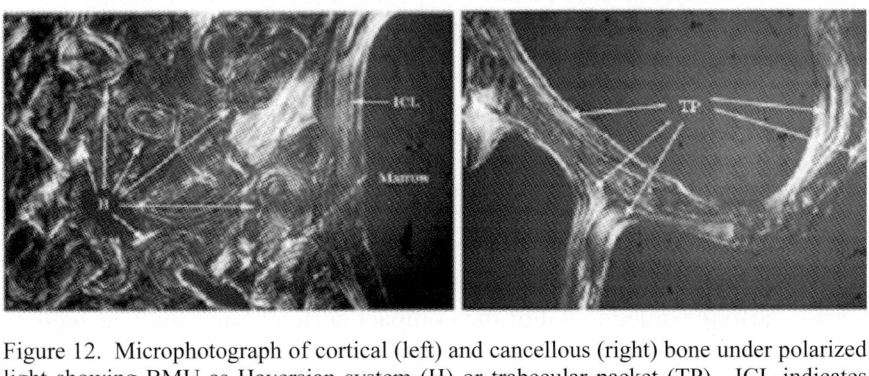

Figure 12. Microphotograph of cortical (left) and cancellous (right) bone under polarized light showing BMU as Haversian system (H) or trabecular packet (TP). ICL indicates inner circumferential lamellae.

a Haversian system (Fig. 12). In cancellous bone, the BSU is a hemiosteon or a trabecular packet (Fig. 12) [1-4, 10-18].

The Haversian system or osteon is formed by 20 to 30 concentric lamellar sheets, which surround a central canal (Haversian canal) containing nerves, circulatory vessels (blood vessels and lymphatics) and loose connective tissue. The wall of an osteon is approximately 70 to 100 microns thick. The diameter of an osteon ranges from 200 to 250 microns. On a cross section of cortical bone, an irregular or scalloped reversal cement line highlights the outer border of each osteon. This cement line is a 1 to 2 micron thick layer of mineralized matrix lacking in collagen fibers. The length of an osteon is approximately 2.5 mm. Osteons are oriented nearly parallel to the bone's loading axis, running vertically from the proximal end to the distal end.

The trabecular packet is also known as a hemiosteon (Fig. 12). It is a shallow crescent with a wall about 50 microns thick and 1 mm long. In cancellous bone, scalloped cement lines hold the basic structure units - trabecular packets - together with interstitial lamellar bone, forming many rod or plate-shaped individual trabeculae that connect to each other to further construct a trabecular spongiosa.

The differences in BMUs between cortical and cancellous bone are significant. In the entire skeletal system, there are 21×10^6 osteons and 14×10^6 trabecular packets. The density of BMU is $15/mm^3$ in cortical bone and $40/mm^3$ in cancellous bone. The time required for bone remodeling (remodeling period) is longer in cortical bone (148 days) than in cancellous bone (112 days). During a one-year period only 3% of cortical bone is renewed, while 26% cancellous bone is renewed.

4. Osteogenesis

4.1. Intramembranous ossification

Intramembranous ossification, also known as membrane bone formation, is an osteogenic process through which bone is formed directly without a prior cartilage anlage. All of the compact cortical bone shell, as well as certain bones of the cranium, are formed by this mechanism. Intramembranous ossification is initiated by condensing the well-vascularized fetal mesenchymal and epithelial tissue. As the number of cells and fibers increase to a certain degree, the mesenchymal cells differentiate into osteoblasts. Osteoblasts form many thin primary woven trabecular spicules, which thicken and become connected by continued osteoblastic formation, eventually forming a spongy-like woven cancellous network. Primary osteons will eventually fill all porosity within the cancellous network to form primary cortical bone. After birth, Haversian systems gradually replace primary compacta by the remodeling process [1-3, 6, 10].

4.2. Endochondral ossification

Endochondral ossification, also known as cartilaginous bone formation, is the formation of bone matrix on a calcified cartilaginous template. It is an osteogenic process through which cancellous bone is formed. Initially, chondrocytes proliferate and deposit cartilage matrix to form a cartilaginous model of the future bone. As the chondrocytes mature the cartilage matrix calcifies. These calcified cartilage cores serve as a scaffold, on which osteoblasts produce and deposit woven bone matrix to form primary spongiosa. The primary spongiosa models and remodels itself into the secondary spongiosa by removing packets of woven bone and replacing them with lamellar trabecular packets [1-3, 6, 10].

5. Skeletal Biological Mechanisms

Growth (longitudinal and radial growth), modeling and remodeling are the three major distinct biological mechanisms that modify bone mass and structure of the skeletal system for its adaptation to the mechanical and non-mechanical environments. The distinct differences among the three mechanisms include the involvement of cellular components, the cellular working sequence and the cellular working location (Table 1). Longitudinal growth is mainly responsible for increasing bone length (Fig. 13A upper panel). Radial growth is mainly responsible for enlarging bone cross-sectional area (Fig. 13A lower panel). Modeling is mainly responsible for maintaining bone shape or profile (Fig. 13A, B & C). Remodeling is mainly responsible for converting the woven spongiosa into the lamellar spongiosa during growth and maintaining bone integrity after skeletal maturity. However, all three mechanisms work together in a highly synchronized fashion during growth, to enlarge the bone length and diameter, and to maintain the appropriate profile for all individual bones. As such, their teamwork produces a proportional gain in bone mass, structure and strength that is adequate to adapt to the 20-fold gain of mechanical impact between birth and skeletal maturity. During aging, remodeling becomes the dominant mechanism to maintain the integrity of the established mass-structure-strength configuration, by replacing parts that are damaged by aging or parts that are unfit to the altered mechanical usage and non-mechanical agents.

Table 1 Comparison of Skeletal Biologic Mechanisms

			Growth	
Parameters	**Remodeling**	**Modeling**	**Longitudinal**	**Radial**
Cellular component	Osteoclast/Osteoblast and their precursors	Osteoclast/Osteoblast and their precursors	Chrondrocyte/Osteoclast Osteoblast and their precursors	Osteoblast and precursors
Location	Spatially related	Different surfaces	Growth plate at bone ends	Periosteal
Coupling	A ⟶ R ⟶ F	A ⟶ F; A ⟶ R	Cell death ⟶ R ⟶ F	A ⟶ F
Timing	Cyclical	Continuous	Continuous	Continuous
Extent	Small (< 20%)*	Large (> 90%)	Large	Large
Apposition rate	Slow (0.3-1.0 μm/d)	Fast (2-10 μm/d)	Fast woven bone formation	Fast (2-10 μm/d in growth) Slow (< 1 μm/d in adults)
Cement line	Scalloped	Smooth	None	Smooth
Balance	No change or net loss	Net gain	Net gain	Net gain, cross sectional area
Occurrence	Throughout life span	Prominent in growth; ineffective in adults	During growth only	Rapid during growth; slow to none in adults
MES threshold**	< 200 microstain	> 1,000 microstain	Genetically defined	Genetically defined

*Of available surface; **MES–minimum effective strain; A-activation; R-resorption; F-formation

5.1. Bone growth

During the growing period, bone growth is accomplished by increasing length and cross-sectional area, respectively, through the endochondral ossification osteogenic pathway and formation-drift of bone modeling mechanism. The former is called longitudinal growth and the later is referred to as radial growth (Fig. 13A) [1-3].

Longitudinal growth

The growth plate complex is the center of longitudinal growth. It is composed of a large number of chondrocytes and hyaline cartilage matrix. The hyaline cartilage appears as many matrix columns, each having multiple shelves that stack chondrocytes together into vertical chondrocyte columns. In each chondrocyte column, the chondrocyte evolves itself from a flat precursor to a large mature functional cell. The resting zone contains layers of undifferentiated, self-renewable flat precursors, providing a constant source of cells for the longitudinal growth. These precursors differentiate to chondrocytes and proliferate to maximize their volume and produce a large quantity of cartilage matrix, which is subsequently mineralized to become calcified cartilage. Finally, chondrocytes undergo cellular apoptosis, in which cells breakdown to release their entire contents, and ultimately result in cell death. This gives way to the concurrent vascular ingrowth and osteoclastic resorption, which is followed by osteoblastic woven bone formation.

Chondrocyte apoptosis results in empty lacunae, which exposes the calcified cartilage columns (spicules) in between. After osteoclastic matrix cleanup, these calcified cartilage spicules become the core scaffold, on which osteoblasts form woven bone, building a densely connected trabecular network called the primary spongiosa. As bone elongation progresses, primary woven bone is replaced by lamellar bone in a trabecular network, also known as the secondary spongiosa.

This process adds a new metaphyseal region at the epiphyseal end and removes the old cancellous bone at the diaphseal end, which consequently increases the length of the entire bone (Fig. 13A upper panel).

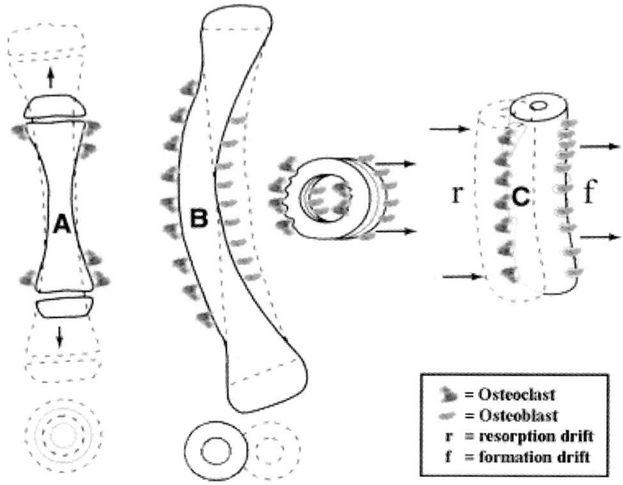

Figure 13. Diagrammatic representation of the longitudinal bone growth (A, upper), radial growth (A, lower), and modeling (B & C; resorption drift and formation drift).

Radial growth

As the whole bone elongates longitudinally, its cross-sectional area also proportionally enlarges by radial growth. During growth, the periosteum is constantly stimulated by increased mechanical demands, activating the modeling – formation drift resulting in radial bone growth. The formation drift adds new lamellar bone onto the outer cortical surface, which consequentially enlarges the cross-sectional area of the long bone (Fig. 13A, lower panel).

Bone modeling

Bone modeling is one of the predominant biological mechanisms that governs the enlargement of each individual bone during growth. It is minimized in adults [1-3, 6, 10, 51]. It is so named because of its similarity with the home modeling, in that either a new component is added as an addition but not a replacement, or an old component is removed without a subsequent replacement. Bone modeling is divided into bone formation drift and bone resorption drift that involve osteoblastic bone formation and osteoclastic bone resorption, respectively. Although the two processes are activated and conducted separately on different bone surfaces, they work together in a synchronized fashion at the organ level to enlarge the size of each individual bone and shape it into an appropriate profile. These series of events are actively ongoing throughout the entire growth period. After skeletal maturity, the resorption drift apparently completely ceases while the formation drift is minimized on the periosteal surface [1-3, 7, 10, 51-59].

Bone remodeling

Bone remodeling is a biological mechanism that adapts bone to mechanical and non-mechanical stimuli throughout the lifespan in humans. It is so named because of its similarity with the home remodeling, in that an old component is firstly removed, and subsequently replaced by a new component at the same location. During the first 3 years of childhood, the remodeling process is mainly responsible for replacing immature woven bone with the more biomechanically and metabolically competent lamellar bone. During growth, the remodeling process is responsible for converting the primary spongiosa woven bone into the secondary spongiosa lamellar bone, which enables bone elongation. During adulthood, the remodeling process is the biologic mechanism responsible for replacing aged bones that are damaged, or that are mechanically unfit, therefore maintaining the skeletal mechanical capacity.

Through remodeling, all cancellous bone is completely renewed every 1 to 4 years while all cortical bone is completely renewed every 20 years. This periodic bone replacement is referred to as bone turnover. After age 30, the coupling between bone formation and resorption becomes delinquent. Consequently, a focal bone deficit or a negative bone balance is created in each remodeling cycle for all remodeling BMUs. The overall deficit leads to an age-related bone loss at the rate of 1% total bone mass per year. The magnitude of total bone loss is primarily determined by activation frequency (Acf). Activation frequency is an index for the total number of remodeling BMU per day.

Assuming the deficit is a constant for all remodeling BMUs, the magnitude of the total bone loss equals Acf multiplied by the deficit constant. The higher the Acf is the greater magnitude thc total bone loss will be.

Under normal conditions, approximately 10% of endosteal surfaces are undergoing either resorption or formation due to remodeling. This surface resorption creates Howship's lacunae, which will not completely fill prior to the end of the bone formation period. The total volume of newly creased and yet to be fully filled lacunae ranges from 2 to 8% of the total bone volume. This is referred to as remodeling space. The remodeling space is determined by the total number of remodeling BMUs (Acf). In ovariectomized rat skeleton, estrogen deficiency-induced increase of Acf results in an enlarged remodeling space equivalent to 20% of the total bone volume. In contrast, bisphosphonate treatment-induced reduction of Acf results in a reduced remodeling space of less than 1% of the total bone volume [1-4, 6, 10-11, 18, 24, 60-63].

Different from modeling, osteoclastic bone resorption and osteoblastic bone formation in the remodeling process occur at the same surface as serial sequential events. These events include activation, resorption, reversal, formation and quiescence as detailed below (Fig. 14) [1, 3, 6, 10-11, 18, 58, 60-63].

Activation

This is a process through which the bone surface is converted from the quiescent stage to the active resorption stage. The factor that initiates this process is unknown, but activation is believed to occur partly in response to local structural, metabolic, mechanical and nonmechanical requirements. When mechanosensors, the osteocytes, detect mechanically unfit bone matrix, damaged bone matrix, or increased demand for mineral ions, they produce an activation signal to initiate the bone remodeling process. This signal induces the ingrowth of capillaries, from which the mononcleated precursors of osteoclasts migrate to the adjacent marrow region. The signal also stimulates the bone-lining cells to contract themselves and release factors that digest the underlying osteoid layer. This exposes the mineralized surface for osteoclastic resorption. Bone-lining cells may lift from the bone surface and migrate into the adjacent marrow, where they directly communicate with osteoclastic precursors through cell-cell contact. As such, RANKL on bone-lining cells binds to its receptor RANK on osteoclastic precursors, initiating the osteoclastogenic process. Thereafter multinucleated osteoclasts are formed by the fusion of several mononucleated precursor cells. These mature osteoclasts then migrate and attach to the exposed mineralized surface. In the meantime, they become activated, polarized, and form ruffled borders.

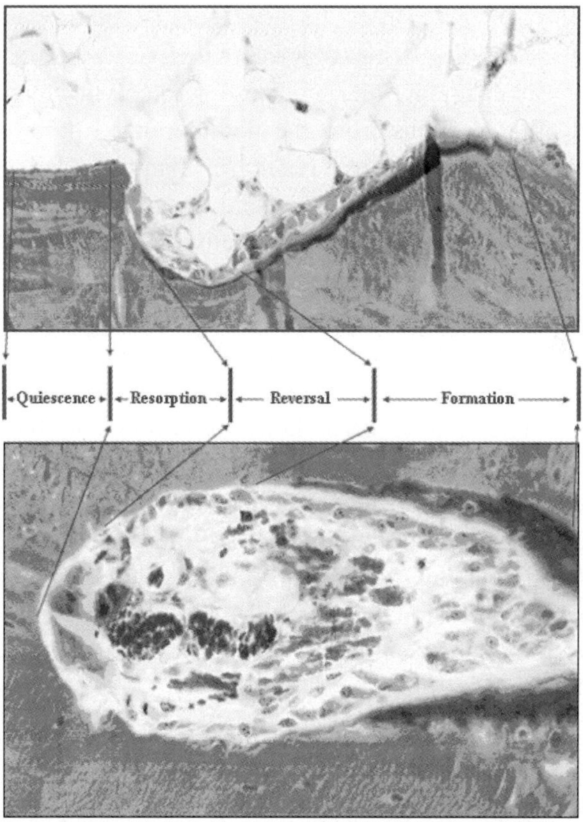

Figure 14. Diagrammatic representation of the trabecular packet (upper) and cortical osteon (lower) remodeling process: activation, resorption, reversal, formation and quiescent phases.

Resorption

Osteoclasts attach their perimeter membrane to the bone matrix through the clear zone. This clear zone becomes a circulated wall that connects the osteoclastic apical membrane and the covered bone surface, forming a small chamber that is sealed from extracellular fluid. Within the chamber, the bottom side of the osteoclastic membrane is extensively enfolded to form a striated ruffled border. The ruffled border secretes H^+ ions into the sealed chamber, which acidifies the microenvironment to a pH of 3.5. This acidic environment dissolves minerals from the bone matrix and releases them into the extracellular fluid. As it progresses, the ruffled border further secretes numerous proteolytic enzymes, including metalloproteinases, cystiene-proteinases, phosphatase, cathepsin K, etc. These proteolytic enzymes digest collagen and other proteins from the demineralized bone matrix. The removal of demineralized matrix results in many resorptive cavities on trabecular and cortical bone. In special cases, trabecular perforation occurs when the enhanced osteoclastic activities increase the resorption depth over the normal trabecular thickness, or when the resorbing trabeculae are thinner than normal resorption depth. The trabecular perforation eliminates the trabecular elements along with their surface, thus decreasing the total surface area available for subsequent bone formation. In this unique case, bone-formation and bone-resorption become uncoupled [6, 19-25].

In cortical bone, the osteoclasts in the cutting cone travel in a direction roughly parallel to the long axis of the bone at a speed of about 20 to 40 microns per day and radially about 5 to 10 microns per day. The mean depth of erosion is about 60 microns in trabecular bone and about 100 microns in cortical bone. This phase takes about 1 to 3 weeks [11, 18, 62, 63].

Reversal (coupling)

The reversal period is the 1- to 2-week period during which the bone surface is reversed from resorption phase to the formation phase. The reversal period has a defined histologic appearance within an ongoing remodeling unit. A group of mononuclear cells are observed between the point where the osteoclasts fade away and the point where the osteoid starts to appear (Fig. 14). The physiological function of these mononuclear cells is yet to be identified. Some suggest that monocytes close to the osteoclasts are capable of digesting and clearing the demineralized matrix. The breakdown products of this matrix removal process, such as IGF I and II, TGFβ and FGFs, are known to stimulate osteoblastic activity for new bone formation. They are therefore postulated as "coupling factors" that are capable of repopulating the osteoblasts into the resorption foci, as well as determining their vigor and life span. Consequently, the upcoming osteoblasts will have the full capacity to produce sufficient new bone matrix to completely fill Howship's lacunae. The mononuclear cells close to the osteoid are considered to be preosteoblasts. They migrate to the reversal surface by the chemotactic effects of the "coupling factors". Thereafter preosteoblasts differentiate into mature functional osteoblasts that initiate the formation phase. The coupling phenomenon is delinquent in aging patients or those who suffer from skeletal metabolic diseases due to the diminished osteoblast vigor and/or shortened osteoblastic life span. In these cases, osteoblastic formation either does not occur after bone resorption, or produces insufficient bone matrix that results in incompletely filled Howship's lacunae.

Formation

Bone formation occurs in two distinct stages: bone matrix formation and bone matrix mineralization.

Bone matrix formation is a process in which a large number of mature osteoblasts synchronize their work as a team to synthesize and secrete type I collagen fibers. The collagen fibers are oriented in an orderly fashion to form multilayer lamellar sheets. The orientation of the collagen fibers is parallel to each other in the same lamellar sheet, and perpendicular to each other between neighboring lamellar sheets. As the layers of lamellar sheets accumulate, a series of changes occurs in the osteoblast. While most osteoblasts remain on the top layer lamellar sheet, few lag behind and are housed in lacunae with their processes extending out circumferentially within canaliculi. These buried osteoblasts differentiate into osteocytes and communicate to their neighboring osteocytes through gap-junctions. As such, a three dimensional osteocytic lacunae-canaliculi network is established throughout the entire bone. Corresponding to reducing vigor of the surface osteoblasts, their height is reduced as bone formation progresses, from an average of 7 microns to 1 micron throughout this process. After bone formation, osteoblasts gradually differentiate into bone-lining cells. When bone matrix is produced, the maturation process is initiated to form the tropocollagen array and to establish cross-links between tropocollagens. As such, the gap-regions for mineral crystal deposition are formed and the matrix is stabilized.

The mineralization process occurs only in the mature bone matrix in which the gap regions are ready for hydroxyapatite deposition. As osteoid matures, osteocytes regulate an influx of mineral ions from extracellular fluid, to form hydroxyapatite molecule crystals at the local mineralization sites. These crystals are then deposited into the stabilized gap regions of mature osteoid. The process of mineralization starts at the mineralization front, an interface between mineralized bone and unmineralized osteoid, where the matrix is mature. The process then advances toward the upper lamellar layer as their maturation completes. The ten-day period required for matrix maturation is called mineralization lag time. This mineralization lag time results in the existence of a 7-10-micron layer of immature osteoid called the osteoid seam. Upon completion of bone formation, a one-micron layer of unmineralized matrix remains on the bone surface.

The progress of bone formation can be labeled by administering calcium-seeking substances with fluorochrome properties, such as tetracycline. If a tetracycline is administered during the mineralization period, it is incorporated into the hydroxyapatite crystal lattice at the mineralization front. In the histologic section viewed through an epi-florescent microscope, a sharp band of fluorochrome label is clearly visible. This technique is widely utilized to determine the rate of bone formation. Two injections of tetracycline are separately administered with a prescribed time interval, creating two bands of florescent label on the formation site of the undecalcified bone section (Fig. 15). The

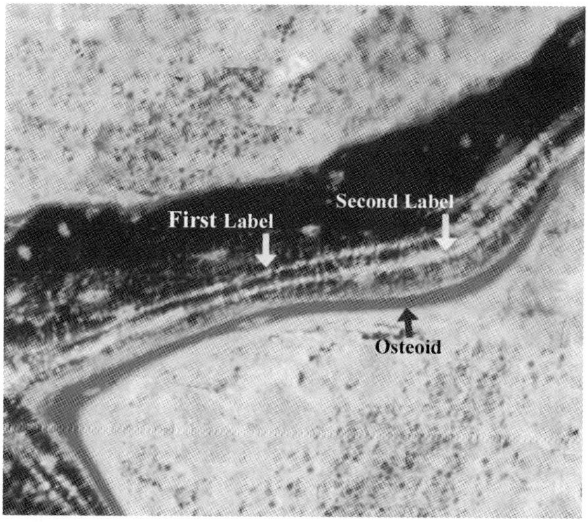

Figure 15. Microphotograph showing a double-labeled trabecula.

distance between the two fluorochrome-bands (Ir.L.Wi) represents the length for which the bone formation (mineralization) had advanced during the time interval. The rate of mineral apposition can be obtained by dividing Ir.L.Wi by the interval [1-3, 5, 10-11, 18, 38-42].

Resting (quiescence)

Normally, about 80% of the cancellous and cortical endosteal surfaces and about 95% of the intracortical surfaces are in the resting stage. They are covered by bone-lining cells that may function as osteogenic precursor cells and an endosteal membrane, a 0.1 micron layer of unmineralized connective tissue with fewer collagen fibers and less amorphous ground substance than found in bone [38].

6. Lifetime Skeletal Changes

6.1. Fetal skeleton

During the early fetal period, mesenchymal condensations gradually ossify and develop into individual bones, according to their genetically designated morphology, through either endochondral or intramembranous ossification. Membrane (dermal) bones are developed through the intramembranous ossification path. They include parts of the scapula, clavicle and mandible, most bones in the skull vault and facial region, as well as many bones in the sensory organs. Cartilaginous bones are developed through the endochondral ossification path. They include most of the bones at the base of the skull, the vertebral column, the pelvis and the extremities. However, the bone forming processes of these two distinct osteogenic paths are similar in that they both begin with an increased number of cells and fibrous extracellular matrix. The proliferated cells are differentiated into osteoblasts. These mature osteoblasts lay down an unmineralized woven matrix, which is mineralized almost immediately. At birth, all individual bones are completely developed into their final shape and size, with poorly mineralized woven bone tissue. Without the presence of sufficient mechanical loading, embryonic osteogenesis involves only woven bone formation. The fine-bundled lamellar bone that replaces woven bone through the remodeling process is not observed until the post-natal period, when an effective mechanical load is present [1, 2].

6.2. Growing skeleton

During the first 3 years of childhood, the woven bone in each individual bone is replaced with biomechanically and metabolically competent lamellar bone, through the bone remodeling process.

During the growing period, the hyaline cartilage, produced by the chondrocytes in the growth plate complex, is calcified and converted to mineralized bone through the endochondral ossification process. This process results in formation of a new metaphyseal region at the epiphyseal end and removal of the old cancellous bone at the diaphseal end, which consequently elongates the entire bone. The cumulative elongation therefore increases the length of the growing long bone. As the whole bone elongates, the diaphyseal cortical bone deformation magnitude and its direction are changed. These changes activate the periosteum and initiate radial bone growth by modeling – formation drift. The formation drift adds new lamellar bone onto the outer cortical surface, which consequentially enlarges the cross-sectional area of the long bone. In the meantime, due to the changed loading pattern, some periosteal or endosteal surfaces are misaligned from the loading axis and are therefore subject to disuse. This activates the modeling - resorption drift to remove the non-weight-bearing bone from these surfaces, which consequentially realigns the diaphyseal cortical bone into the center of the loading axis with an appropriate profile.

During growth the two predominate tissue-level biological mechanisms are longitudinal bone growth and modeling. These two mechanisms work together in a highly coordinated fashion, to proportionally increase both length and diameter of each individual bone. Bone formation drift is involved to induce circumferential bone growth. Bone resorption drift is involved to change shape until the bone is aligned with the predominant loading axis, and to shape the bone into the appropriate profile. During growth, bone remodeling is also involved as the third biological mechanism, to replace primary spongiosa by secondary spongiosa in metaphyseal cancellous bone, and to replace primary osteons with secondary osteons in intracortical bone. Overall, bone balance during growth is a positive one. In this period the positive bone balance accumulates bone mass to its peak level by the age of early twenties, when the skeleton matures. Between birth and maturity, mechanical forces on bones increases about 20 times. In the meantime, the three above biologic mechanisms work together in a highly synchronized fashion, producing a proportional gain in bone mass-architecture-strength configuration that is sufficient to offset the 20-fold gain of mechanical loads. This fact is in agreement with Frost's mechanostat hypothesis. Bone mass is maintained at the peak level until the age of late twenties. Thereafter it starts to decline [1-3, 7, 10, 51, 53-59].

6.3. Adult and aging skeleton

Following closure of the growth plate complex, both longitudinal bone growth and modeling bone-resorption-drift cease completely, while the modeling bone-formation-drift is minimized on the periosteal surface. The only predominate tissue-level biological mechanism during adulthood is bone remodeling. Normally 3.6% of intracortical surfaces and 7.2% of cancellous surfaces are at various remodeling stages at a given time. When these surfaces undergo remodeling, a quantum amount of bone is first removed by osteoclastic resorption to create a resorption cavity, which is refilled by subsequent osteoblastic bone formation. As a result, the resorption cavity is filled with a new osteon in the intracortical bone, and with a packet or hemiosteon in cancellous bone. In ones twenties, the processes are fully coupled as the osteoblasts are capable of filling the resorption cavity completely. This coupling phenomenon of bone remodeling allows the total bone mass to be maintained at its peak level for almost a decade in this age group. After age 30, the coupling becomes unbalanced or delinquent. This creates a focal bone deficit or a negative bone balance in all remodeling BMUs. The overall deficit leads to an age-related bone loss at the rate of 1% total bone mass per year. Because of the net bone resorption at the BMU-level, activation frequency (Acf) is a major determinant of the magnitude of the total bone loss, and increases with age. In the female menopausal population, a sudden loss of estrogen stirs a large birth of remodeling BMUs. This large increase in Acf leads to a much higher rate of total bone loss. As bone loss accumulates, residual bone mass declines to the level where fracture risk is high. The combination of low bone mass and the deteriorated microstructure significantly increase fracture risk in postmenopausal women. At age 70, the cumulative loss of bone mass can be more than 30% relative to young adult bone mass [1-3, 26, 65-75].

As a universal phenomenon, age-related bone loss has the following common characteristics: 1) Osteopenia is reversible before skeletal maturity but it is irreversible thereafter; 2) Women begin losing bone 10 years earlier than men. Women also lose bone at a faster (2x) rate then men in the postmenopausal period; 3) In adulthood, although being minimized, the periosteal formation drift of bone modeling is important because it efficiently increases or maintains bone strength. In men the small bone gain at the periosteal surface can partially offsets weakened mechanical properties caused by age-related bone loss on endosteal surfaces. In contrast, age-related bone loss in women is so profound that the periosteal bone gain is insufficient to offset the lost mechanical property [8, 112]; 4) The amount of bone replaced at the trabecular-endosteal remodeling sites appears to decrease with advancing age. This uncoupling phenomino is due mainly to the diminished osteoblast vigor and/or shortened osteoblastic life span. The sequence of A → R → F can also be uncoupled when trabecular plates are perforated or lost. This perforation reduces the available bone surface for the subsequent bone formation. The net effect is further loss of bone mass and architecture.

7. Skeletal Adaptation to Mechanical and Non-Mechanical Stimuli

It is widely accepted that mechanical usage plays an important role in skeletal development and maintenance. However, the specific mechanisms driving the regulation of bone mass and its architecture by mechanical usage and the nonmechanical environment are not well understood. In the last century, many scientific pioneers have attempted to relate the lifetime bone changes to the external mechanical environment. Among them Frost has introduced the Mechanostat hypothesis. In this section we attempt to discuss components of the Mechanostat and how it may regulate bone biology [2, 3, 10, 58-59].

7.1. The mechanostat hypothesis

Mechanostat is a name for the bone mass control system, in which a variety of biologic mechanisms are working together in a synchronized fashion to modify skeletal mass and architecture to fit normal physical activities. It is so named because its working mechanism is similar to that of a **thermostat**, the temperature control system in a house. It is well accepted that bone mass adapts to mechanical usage (MU) in a special way: The MU constantly activates skeletal biologic mechanisms to correct the unfit portion of bone mass/architecture to meet mechanical demands. As such, this mechanism must behave like a home thermostat, in that the responding biologic mechanisms turns "on" in response to an error signal sent from the sensor mechanism, to correct the bone error that is disproportionate to MU. Thereafter, the feedback loop mechanism allows the sensor mechanism to detect the correction, thus stopping the production of the error signal. With the absence of an error signal, the responding mechanism turns off (Relation 1).

Accordingly, the mechanostat has to consist of the following biologic adaptive mechanisms: 1) A threshold mechanism that can set the minimum effective strain (MES) and determine whether the current bone mass/architecture configuration fits the MU; 2) A sensor mechanism that is capable of the following functions: a. converting the mechanical forces into local biomechanical signals (e.g. fluid shear stresses), b. determining whether the mechanical signals are above or below the threshold, c. transforming the biomechanical signal to biochemical ones (prostaglandin and nitric oxide); and d. transmitting the biochemical signals from sensor cells to the responding mechanisms; 3) Responding mechanisms that are capable of modifying bone mass and its architecture to adapt to the MU changes; 4) A feedback loop mechanism that allows the sensor mechanism to detect the correction, thus stop the signal production and turn off the responding mechanisms.

7.2. The mechanothreshold mechanism

This is a set-point system in bone tissue called minimal effective strain (MES), which is capable of determining whether the external MU is an excessive or a moderate one, under the current bone mass/architecture configuration as described in Fig. 16. The set point defines a range of strain values produced by external MU. Within the set point range of the Adaptation Window (AW) no biologic response is activated. If the external MU produces a MES that is either below or above the AW, an appropriate response will occur to alter muscle, and bone mass, and thus strength accordingly. MES less than the AW (< 200 microstrains) will turn on the remodeling mechanism that causes bone loss adjacent to marrow. MES exceeding the AW (> 1000 microstrains), approaching the Moderate Overload Window (MOW), will turn on the modeling mechanism that increases lamellar bone mass (Fig. 16). MES exceeding the MOW (> 3000 microstrains), approaching the Pathologic Overload Window (POW), will turn on the modeling mechanism that increases woven bone formation. If there is a persistent underload, such as long-term bed-rest or hypogravity, muscle and bone mass will diminish to a level corresponding to the lower load. In contrast, if there is a persistent overload, such as in extensive weight-lifting training, muscle and bone mass will increase to a level corresponding to the heavy load.

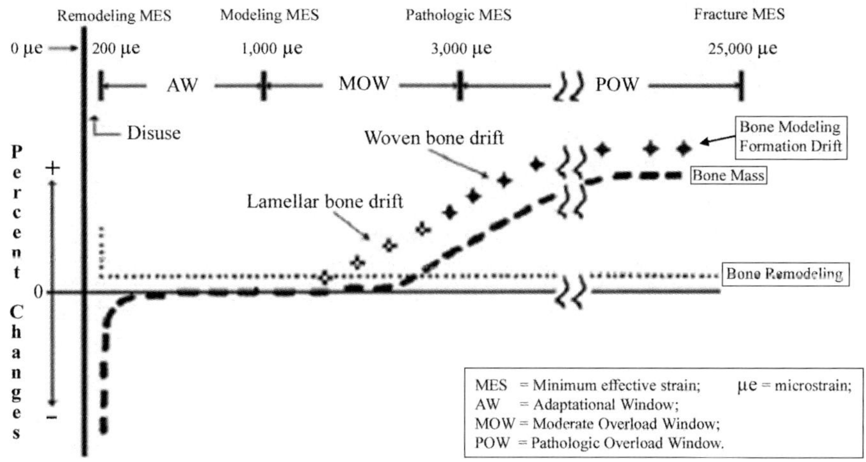

Figure 16. Diagrammatic representation of the mechanical usage window or set-point range as the threshold mechanism. **με** microstrain.

7.3. The mechanosensor mechanism

The best candidate for the sensor mechanism is the osteocytic lacuno-canalicular network system (together with bone-lining cells). Because the network distributes throughout the entire bone (cortical and cancellous, woven and lamellar), any bone deformity or distortion caused by external mechanical force will compress the interstitial fluid to move rapidly around the osteocytic cytoplasmic membrane. The mechanical deformation is therefore converted to shear stress as a local biomechanical signal. The osteocyte detects and transforms the biomechanical signal into a biochemical signal, which may include prostaglandin production and nitric oxide release. These biochemical signals are transmitted via the gap-junctions to the lining cells on the endosteal surface, or precursor cells in the adjacent marrow, where the responding mechanisms are initiated to alter the bone mass accordingly.

Osteocytic lacuno-canalicular network can also sense the abnormal MU and bone microdamage by other means. Since the network is densely distributed throughout the bonc matrix, the production of a microcrack may interrupt canaliculi and stimulate the osteocyte to produce signals. A persistent absence of mechanical stress has been reported to cause osteocytic cell death. Osteocytic apoptosis alone is sufficient to initiate the responding mechanism – the bone remodeling process.

7.4. The mechanoresponding mechanism

The two major mechanoresponding mechanisms are bone modeling and bone remodeling. They are capable of modifying the bone mass and its architecture to adapt to the MU changes, after receiving a specific signal from the sensor mechanism. Modeling is turned on to add more bone when the sensor mechanism detects a MES exceeding 3000 microstrain. In contrast, remodeling is turned on to remove the excessive bone mass when the sensor mechanism detects a MES less than 200 microstrain.

7.5. The mechanofeedback loop mechanism

A feedback loop mechanism is a mechanism that governs responding mechanisms that avoid the over correction of a bone error. It detects whether a bone error correction is appropriately conducted and sending the signal accordingly to control the responding mechanisms. If it detects that the error is corrected appropriately, a signal is sent to turn off the responding mechanism. If it detects that the error is over corrected, a signal is sent to turn on the opposite responding mechanism to re-correct the error. As such, the feedback loop mechanism protects the bone mass from being inappropriately corrected. During long-term bed rest, although the external mechanical load is minimal, bone loss will plateau at a certain level due to the protective feedback loop mechanism. Treatment with anabolic agents induces a large bone gain in rat skeleton. However, withdrawal of the treatment will cause a bone loss so the bone mass returns to the pre-treatment level, also due to the feedback loop mechanism.

7.6. Utah Paradigm for bone physiology

Recently a more comprehensive vision has constructively built into Frost's mechanostat theory, by adding non-mechanical factors, as showed in Relation 2 [2, 31, 58-59, 75, 77, 90-92]. First, mechanical factor plays a major role in controlling the biologic responding mechanisms, which are capable of modifying postnatal load-bearing bone and mass. Second, osteoblasts and osteoclasts do not perform their work independently, nor are they regulated independently. Instead they both belong to different parts of the same biological mechanisms, either remodeling (resorption coupled with formation) or modeling (coordinated formation and resorption drifts); and they are both under the regulation of the same controlling mechanism – mechanical usage and non-mechanical agents. Modeling usually increases bone mass and strength. Remodeling repairs microdamage and conserves or removes unneeded bone adjacent to marrow. Third, nonmechanical stimuli, such as growth hormones, parathyroid hormone, androgens, calcium and vitamin D, can directly influence biologic mechanisms, because their effects on bone cells can alter the rate of bone formation or bone resorption, or the set point of the mechanostat. They can also indirectly influence biologic responding mechanisms, because their effects on muscles can alter the mechanical loading and strain (Relation 2).

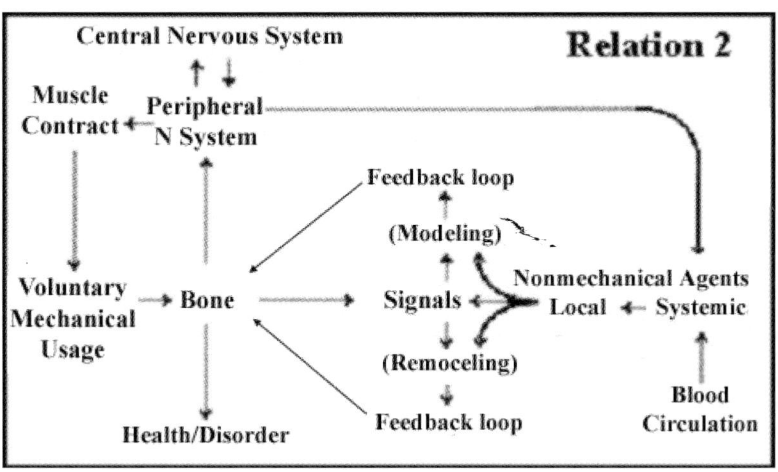

This broadened vision of bone biology has raised four interesting points: (1) The nonmechanical environment is essential to insure that effector cells work properly so biologic mechanisms can sufficiently maintain skeletal health and prevent skeletal disease; (2) Mechanical factors determine the time and location in which biologic mechanisms are onset; (3) After birth, biologic mechanisms are under the control of neuromotor physiology and anatomy; and (4) Most nonmechanical factors can directly or indirectly enhance or hinder the mechanical control.

As indicated in Relation 2, nonmechanical stimuli can directly or indirectly influence every step of the mechanical regulation. They can directly influence bone cells and precursors, to either enhance or inhibit their recruitment and/or vigor. They can also directly influence osteocytes to alter their sensitivity to detect mechanical signals, such as the set-point alteration suggested by Frost [57]. A similar view has also been suggested by Carter et al., who consider that non-mechanical factors alter the ØAS, an attractor stress stimulus that regulates the mechanically related bone remodeling stimulus [76-77]. For example, in the ovariectomy-induced estrogen deficient rat skeleton, the threshold for bone remodeling is raised to a point that a previously normal MU (producing no mechanical signal because the MES is within set-point range) now becomes abnormal and produces a disuse signal (MES is below the raised set-point range). This signal indicates that there is "excessive" bone, thus turns on the remodeling biologic mechanism to remove it. Therefore, it causes a significant bone loss. The relationship between mechanical and non-mechanical stimuli can be summarized as such that while the mechanical factors dominate bone regulation, the degree of such regulation can be effectively altered by non-mechanical factors. Recently it has been shown that both mechanical and non-mechanical factors are important. The effect of mechanical stimulus depends on the non-mechanical environment and their combination generates an additive or synergistic anabolic bone effect [36, 78-86].

Dedication

This chapter is dedicated to the late Harold M. Frost, M.D., Dr.Sc. (hon) whose awesome contribution changed the paradigm of skeletal physiology and pathophysiology. His works inspired us to generate the chapter to celebrate his highly original contributions and their influence on future development to bone biology. We will miss him dearly.

We also wish to dedicate the chapter to Katie Z. Li and Becky Setterberg. Without them, this chapter would have not been possible.

References

1. Jee, WSS, in: *Cell and tissue biology, a textbook of histology, sixth edition,* Ed. Weiss, L, (Urban and Schwarzenberg, Baltimore, 1988), p. 211.
2. Jee, WSS, in: *Orthopaedics, principles of basic and clinical science,* Ed. Bronner, F and Worrell, RV, (CRC Press, Boca Raton, 1999) p. 3.
3. Frost, H M, *Introduction to a new skeletal physiology,* Volumes I and II, (Pajaro Group, Pueblo, 1995).
4. Schenk, RK, Felix, R and Hofstetter, W, in: *Connective tissue and its heritable disorders. Molecular, genetic and medical aspects,* Ed. Royce, PM and Steinmann, B, (Wiley-Liss, New York, 1993), p. 85.
5. Harris, WH, *Nature,*188 (1960).
6. Baron, R, in: *Primer on the metabolic bone disease and disorder of mineral metabolism,* fifth edition, Ed. Favus, MJ, (American Society for Bone and Mineral Research, Wash. D.C., 2003), p. 1.
7. Martin, RB and Burr, DB, in: *Structure, function and adaptation of compact bone,* (Raven Press, New York, 1989), pp. 18, 85, 186 and 214.
8. Marotti, G, Palazzini, S, Palumbo, D, Ferretti, M, *Bone*, 19 (1996).
9. Donahue, HJ, *Calcif Tissue Int,* 62 (1998).
10. Jee, WSS, in: *Bone mechanics handbook*, second edition, Ed. Cowin, SC, (CRC Press, Boca Raton, 2001), p. 5, 33, 38.
11. Parfitt, A M, in: *Bone histomorphometry: techniques and interpretation,* Ed. Recker, R R, (CRC Press, Boca Raton, 1983), p. 143.
12. Lian, JB, Stein, GS, Canalis, E, Robey PG and Boskey Al, in: *Primer on the metabolic bone diseases and disorder of mineral metabolism,* fourth edition, E., Favus, MJ, (Lippincott Williams and Wilkins, Philadelphia, 1999), p. 21.
13. Khosla, S and Kleerekoper, M, in: *Primer on the metabolic bone diseases and disorders of mineral metabolism,* fifth edition, Ed. Favus, MJ, (American Society for Bone and Mineral Research, Wash. D.C., 2003), p. 166.
14. Lian, JB, Stein, GS, Canalis, E, Gehron-Robey, P and Boskey, AL, in: *Primer on the metabolic bone diseases and disorders of mineral metabolism,* fifth edition, Ed. Favus, MJ, (American Society for Bone and Mineral Research, Wash. D.C., 2003), p. 13.

15. Gehron-Robey, P and Boskey, AL, in: *Osteoporosis,* Ed. Marcus, F, Feldman, D, Kelsey, J, (Academic Press, San Diego, 1996), p. 95.
16. Gorski, JP, Critical Review of Oral Biology, 9 (1998).
17. Watts, NB, in: *Primer on the metabolic bone diseases and disorders of mineral metabolism,* fifth edition, Ed. Favus, MJ, (American Society for Bone and Mineral Research, Wash. D.C., 2003), p. 336.
18. Ericksen, EF, Axelrod, DW and Melsen, F, *Bone histomorphology,* (Raven Press, 1994).
19. Rodan, GA and Rodan, SB, in: *Osteoporosis: etiolology, diagnosis and management,* second edition, Ed. Riggs, BL. and Melton III, LJ, (Lippincott-Raven, Philadelphia, 1995), p. 3.
20. Teitelbaum, SL, Tondravi, MM and Ross, FP, in: *Osteoporosis,* Ed. Marcus, R, Feldman, D and Kelsey, J, (Academic Press, San Diego, 1996), p. 61.
21. Suda, T, Takahashi, N and Martin, TJ, *Endocrinol Rev,* 13 (1992).
22. Roodman, GD, *Endocrinol Rev,* 17 (1996).
23. Baron, R, Chakraborty, M, Chatterjee, D, Horne, W, Lomri, A and Ravesloot, J-H, in: *Physiology and pharmacology of bone,* Ed. Mundy, GR and Martin TJ, (Springer-Verlag, New York, 1993), p. 111.
24. Mundy, FR, in: *Primer on the metabolic bone diseases and disorders of mineral metabolism,* fifth edition, Ed. Favus, MJ, (American Society for Bone and Mineral Research, Wash. D.C., 2003) p. 46.
25. Puzas, EJ and Lewis, GD, in: *Orthopaedics. Principles of basic and clinical science,* Ed. Bronner, F and Worrell, RV, (CRC Press, Boca Raton, 1999), p. 45.
26. Pacifici, R, Aging and cytokine production, *Calcif Tissue Int,* 65 (1999).
27. Lian, JB and Stein, GS, in: *Osteoporosis,* Ed. Marcus, R, Feldman, D and Kelsey, J, (Academic Press, San Diego, 1996), p. 23.
28. Miller, SC and Jee, WSS, *Calcif Tissue Int,* 41 (1992).
29. Noble, B, Stevens, H, Loveridge, N and Reeve, J, *Bone,* 20 (1997).
30. Tomkinson, A, Reeve, J, Shaw, RW and Noble, BS, *J Clin Endocrinol Metab,* 82 (1997).
31. Weinstein, RS, Jilka, RL, Parfitt, AM and Manolagas, SC, *J Clin Invest,* 102 (1998).
32. Tomkinson, A, Bevers, EF, Wit, JM, Reeve, J and Noble, BS, *J Bone Miner Res,* 13 (1998).
33. Noble, BS, Stevens, H, Mosley, JR, Pitsillides, AA, Reeve, J and Lanyon, L, *J Bone Miner Res,* 12 (1997).
34. Miller, SC, Bowman, BM, Smith, JM and Jee, WSS, *Anat Rec,* 198 (1980).
35. Dobnig, H and Turner, R, *Endocrinology,* 136 (1995).
36. Jee, WSS and Ma, YF, *Bone,* 21 (1997).
37. Jilka, RL, Weinstein, RS, Bellido, T, Roberson, P, Parfitt, AM and Manolagas, S, *J Clin Invest,* 104 (1999).
38. Harris, WH, Haywood, EA, Lavorgna, J and Hamblen, DL, *J Bone Jt Surg,* 50A (1968).
39. Frost, HM, *Calcif Tissue Res*, 3 (1969).
40. Merz, W and Schenk, R, *Acta Anatomica,* 76 (1970).

41. Recker, RR, Ed: Bone histomorphometry: techniques and interpretation, (CRC Press, Boca Raton, 1983).
42. Parfitt, AM, Drezner, MK, Glorieux, FH, Kanis, JA, Malluche, H, Meunier, PJ, Ott, SM and Recker, RR, *J Bone Miner Res,* 2 (1987).
43. Burger, EH and Lein Nulend, J, in: *Principles of bone biology,* Ed. Bilezikian, JP, Raisz, LG and Rodan, GA, (Academic Press, San Diego, 2002) p. 101.
44. Marotti, G Ferretti, M, Muglia, MA, Palumbo, C and Palazzini, S, *Bone,* 13 (1992).
45. Marotti, GI, Ferretti, M, Remaggi R and Palumbo C, *Bone,* 16 (1995).
46. Marotti, G, *Ital J Anat Embryol,* 101 (1996).
47. Martin, RB, *Bone,* 26 (2000).
48. Metz, LN, Martin, RB and Turner, AS, *Bone,* 33 (2003).
49. Frost, HM, *J Bone Jt Sur,* 42A (1960).
50. Frost, HM, *J Bone Jt Surg,* 42A (1960).
51. Frost, HM, in: *Paediatric osteology,* Ed. Schönau, E, (Elsevier, Amsterdam, 1996), p. 3.
52. Frost, HM, *Bone modeling and skeletal modeling errors,*(C. C. Thomas, Springfield, 1973).
53. Frost, HM, *Anat Rec,* 219 (1987).
54. Frost, HM, *Bone and Mineral,* 2 (1987).
55. Frost, HM, *Clin Orthop Rel Res,* 175 (1983).
56. Frost, HM, *Anat Rec,* 226 (1990).
57. Frost, HM, *J Bone Miner Res,* 7 (1992).
58. Frost, HM, Osteoporoses: new concepts and some implications for future diagnosis, treatment and research (based on insights from the Utah paradigm), (Ernst Schering Research Foundation AG, Berlin, 1998).
59. Frost, HM, *Anat Rec,* 275 (2003).
60. Parfitt, AM, in: *Osteoporosis,* Ed. Marcus, R, Feldman, D and Kelsey, J, (Academic Press, San Diego, 1996), p. 315.
61. Frost, HM, *Anat Rec,* 226 (1990).
62. Ericksen, EF, Gundersen, HJG, Melsen, F and Mosekilde, L, *Metab Bone Dis Rel Res,* 5 243, (1984).
63. Ericksen, EF, Mosekilde, L, Melsen F,. *Bone,* 6, (1985).
64. Garn, SM, Rohmann, CG and Wagner, B, *Federated Proceedings,* 26 (1967).
65. Garn, SM, Wagner, B, Rohmann, CG and Ascoli, W, *Amer J Physiol and Anthro,* 28 (1968).
66. Sherman, S, Heaney, RP, Parfitt, AM, Hadley, EC and Dutta, C, Eds. *Calcif Tissue Int,* 53 Suppl 1 (1993).
67. Lindsay, R, in: *Osteoporosis, etiology, diagnosis and management, second edition* Eds. Riggs, BL. and Melton III, LJ, (Lippincott-Raven, Philadelphia 1995), p. 133.
68. Blumshon, A and Eastell, R, in: *Osteoporosis, etiology, diagnosis and management,* second edition, Ed. Riggs, BL and Melton III, LJ, (Lippincott-Raven, Philadelphia, 1995), p. 161.
69. Marcus, R, Feldman, D and Kelsey J, Eds, *J Osteoporosis,* (Academic Press, San Diego, 1996).
70. Frost, HM, *J Bone Mineral Metab,* 17 (1999).

71. Frost, HM, *J Bone Mineral Res,* 14 (1999).
72. Rosen, CJ and Kiel, DP, in: *Primer on the metabolic bone diseases and disorders of mineral metabolism,* fifth edition, Ed. Favus, MJ, (American Society for Bone and Mineral Research, 2003), p. 57.
73. Cooper, C, in: *Primer on the metabolic bone diseases and disorders of mineral metabolism,* fifth edition, Ed. Favus, MJ, (American Society for Bone and Mineral Research, 2003), p. 307.
74. Reid, IR, in: *Primer on the metabolic bone diseases and disorders of mineral metabolism,* fifth edition, Ed. Favus, MJ, (American Society for Bone and Mineral Research, 2003), p. 86.
75. Frost, HM, *J Bone Miner Res,* 12 (1997).
76. Carter, DR, Harris, WH, Vasu, R and Caler, WE, in: *Mechanical properties of bone,* Ed. Cowin, SC, (ASME, New York, 1981), chapter 6.
77. Carter, DR, Critical Rev Biomed Eng, 87 (1982).
78. Jee, WSS, Ma, YF, Li, M, Liang, XQ, Lin, BY, Li, XJ, Ke, HZ. Mori, S, Setterberg, RB and Kimmel, DB, in: Sex steroids and bone, Ed. Ziegler, R, Pfeilschifter, J, Brautigam, M, (Springer, Berlin, 1994), p. 119.
79. Yeh, JK, Aloia, JF and Chen, M, Calcif Tissue Int, 54 (1994).
80. Tam, CS, Akhter, MP, Johnston, E, Covey, MA, Pearse, A, Ringer, L, Recker, RR, *Bone*, 23 (1998).
81. Tang, LY, Raab-Cullen, DM, Yee, JA, Jee, WSS and Kimmel, DB, *J Bone Miner Res*, 12 (1997).
82. Chow, JWM, Fox, SW, Jagger CJ and Chambers, TJ, *Amer J Physiol*, 274 (1998).
83. Mosekilde, L, Thomsen, JS, Orhii, PB, McCarter, RJ, Meya, W and Kalu, DN, *Bone*, 244 (1999).
84. Gasser, JA, J Japanese Soc Bone Morphometry, 7 (1997).
85. Gasser, JA, in: *Musculoskeletal interactions*, Volume 2, Ed. Lyritis, GP, (Hylonome Editions, Athens, 1999), p. 77.
86. Jee, WSS, Zhou, H, Yao, W, Cui, and Ma, YF, in: *Osteoporosis update 1999* (Proceedings, Third International Congress on Osteoporosis, Xi'an, P.R. China, Beijing, China, 1999), p. 78.
87. Frost HM. Introduction to Biomechanics. Charles C. Thomas, Springfield, (1963).
88. Frost HM. Laws of Bone structure. Charles C. Thomas, Springfield, (1964).
89. Frost HM. Bone Dynamics in Osteoporosis and Osteomalacia. Charles C. Thomas, Springfield. (1966).
90. Frost HM. Bone mechanostat: A 2003 Update. Anat Rec 275A:1081-1101, (2003).
91. Frost HM. A 2003 update of bone physiology and Wolff's law for clinicians. Angle Orthodontist 74:3-71, (2004).
92. Frost HM. The Utah Paradigm of Skeletal Physiology, Vol I. Bone and Bones (and Associated Problems). International Society of Musculoskeletal and Neuronal Interactions, Athens, Greece, (2004).

CHAPTER 3

SKELETAL STEM CELLS

Martin Connolly and Gang Li

Department of Trauma and Orthopaedic Surgery, School of Medicine, Queen's University Belfast, Musgrave Park Hospital, Belfast, BT9 7JB, UK

Mesenchymal stem cells (MSCs) are known to have the capability to differentiate into many cell types of skeletal tissues. MSCs were first identified in the bone marrow but since then they have been isolated and identified in many other tissues. Many possible therapeutic uses of MSCs require a careful review of their sources. Bone marrow, adipose tissue, periosteum, skeletal muscle, adult peripheral blood, umbilical cord blood, vascular pericytes, bone tissue, amniotic fluid, spleen, and dermis are sources of MSCs. Bone marrow is the most established source which has been investigated most, understood best and its use *in vivo* is promising. Adipose tissue, skeletal muscle and periosteum have also been proven to contain MSCs and may have possible future uses. Recently, MSCs have also been found in adult peripheral blood at low numbers. More research is needed to develop adult peripheral blood as a viable option for MSCs. If it were, the future use of MSCs would be greatly facilitated with the ease of its collection. Evidence of MSCs in the other tissues, such as umbilical cord blood, vascular pericytes, bone tissue, amniotic fluid, spleen, and dermis also existed but they are of limited use.

1. Introduction

In an adult, the production of new cells usually involves a chain of processes that begins with cell proliferation and involves migration, differentiation and maturation. The first cell in this chain is termed a stem cell, which has clonogenic and self-renewing capabilities and can differentiate into multiple cell lineages. The bone marrow is the site of two separate and distinct stem cell lineages – the haemopoietic stem lineage, from which blood cells and osteoclasts develop, and the stromal/ Osteoblast (OB) cell lineage from which OBs develop. The name most commonly attributed to stromal stem cell in current literature is "mesenchymal stem cell" (MSC). Other names include connective tissue stem cells, stromal stem cells, and stromal fibroblastic stem cells.[i] Contrary to what the name "MSC" may suggest, it does not give rise to all tissues derived from the embryonic mesenchyme. The mesenchyme is a specialized tissue in the embryo that gives rise not only to the muscle, bone, and other connective tissue, but also to the blood and other cells. MSCs are so named because they develop from the mesenchyme.[1] Until recently, it was assumed that adult stem cells were committed to differentiate into the tissues in which the stem cell resides. However, recent investigations have shown that this assumption was incorrect. For example, haemopoietic stem cells have been differentiated into hepatocytes,[2, 3] and neural stem cells have been observed differentiating into blood cells.[4] Similarly, MSCs not only have the ability to differentiate into OBs, but also adipocytes,[5] chondroblasts,[6] myoblasts,[7] astrocytes,[8] and fibroblasts.[9]

Recognizing that MSCs may differentiate into a number of skeletal cells, it is interesting to investigate if MSCs are present in a number of skeletal tissues, i.e. adipose tissues or muscle tissues etc. Indeed one would hypothesize that MSCs may travel in the peripheral blood, in which way enable them to spread in a wide range of tissues. A critical review of potential sources of MSCs is worthwhile due to the promising therapeutic potentials of MSCs.

2. MSC Differentiation in the Osteoblast Lineage

The differentiation of a MSC into an OB has four main stages and five main cell types in the lineage. This is summarized in Figure 1.

There are many factors, which control this differentiation process. Runx2 (previously named Cbfa1) is crucial for OB development.[10] In mice, the deletion of Runx2 leads to animals which have a skeleton comprising only of chondrocytes and cartilage; OBs and bone are not evident.[11] In addition to Runx2, Indian Hedgehog (a secreted growth

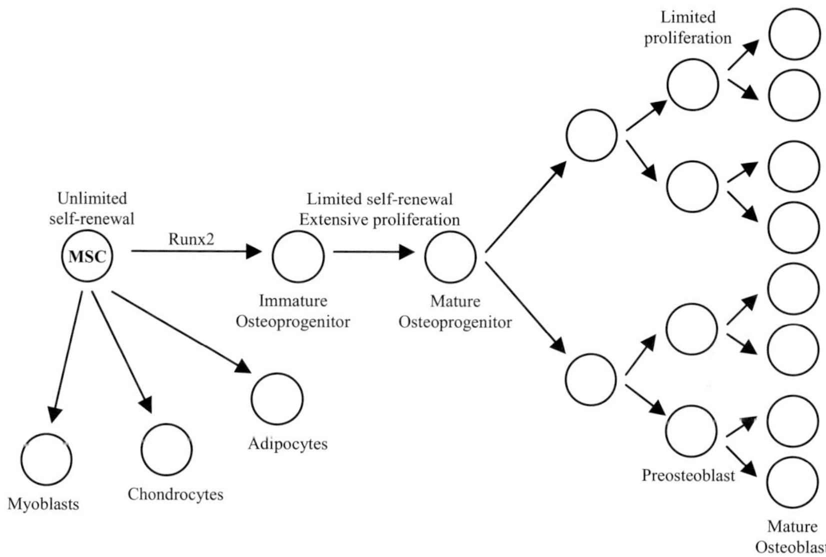

Figure 1. The differentiation pathways of MSCs. Adapted from: Aubin JE, Triffitt JT. Mesenchymal stem cells and osteoblast differentiation. In: Bilezikian JP, Raisz LG, Rodan GA, eds. Principles of Bone Biology. 3rd Edn. San Deigo: Academic Press 2002

factor) is required for the differentiation of MSCs to OBs in endochondral bone, but not in intramembranous bone.[12] Bone morphogenetic proteins (BMPs) are also important regulators of OB development. They tend to promote OB differentiation and bone matrix formation in the more mature OB cell lineage but having inhibitory effects in the earlier differentiation process.[10]

2.1. Markers of osteoblast

There are a large number of markers for MSCs and more specific markers for more differentiated cells along the OB lineage.

Surface markers: There are a variety of surface markers which are detectable in very immature osteoprogenitors but which are also detectable in MSCs. Antibodies which react with such surface markers include STR0-1, SH-2, SH-3, SH-4, SB10 and HOP-26.[1] Therefore, the use of antibodies cannot be used to confirm the presence of committed osteoprogenitors but can be used to identify MSCs which may potentially differentiate into OBs. **Cellular and Molecular markers**[13]: The expression of the bone/kidney/liver isoform of alkaline phosphatase (ALP) is seen in the more mature cells of the OB lineage i.e. mature osteoprogenitors, preosteoblasts and mature OBs. Its expression increases with differentiation. It is not expressed by MSCs or by more immature osteoprogenitors. Type I collagen is expressed in all the cells in the OB differentiation lineage succeeding the immature osteoprogenitor. Although it is not expressed in MSCs, other cells differentiated from MSCs outside the OB lineage express it. Hence, it is not a definitive test for the presence of OBs. Osteocalcin is expressed (in varying amounts) by mature OBs but not all mature OBs express it. Therefore, osteocalcin is perhaps useful qualitatively but its use quantitatively is questionable. Cells express bone sialoprotein at various stages in the OB lineage including preosteoblasts and OBs, but in varying proportions also. The mineralized matrix secreted by OBs contains calcium phosphate. It can be detected by Alizarin Red stain.[14]

2.2. Culture condition to maximise OB lineage differentiation

Different investigators use slightly different culture methods to favor differentiation of MSCs into the OB cell lineage. Figure 2, used by Wickham et al[14] and Zuk et al[15] is a typical osteogenic culture condition.

Dulbecco's modified eagle medium (DMEM)
10% fetal bovine serum
0.01 μM 1,25-dihydroxyvitamin D3
or 0.1μM dexamethasone
50 μM ascorbate-2-phosphate
10 mM β-glycerophosphate
1% antibiotic (e.g. penicillin, streptomycin)

Figure 2. The most used osteogenic culture conditions.

3. Established and Potential Sources of MSCs

3.1. Bone marrow

The bone marrow (BM) was the first source of MSCs identified. Friedenstein was the first to describe MSCs in the BM, although he did not call them MSCs.[16] Since then the BM has become the most established source of MSCs. Although majority of research into MSCs has been performed on MSCs from BM, one cannot assume that MSCs collected from other tissues will behave in exactly the same way as the BM-MSCs. Therefore, using BM-MSCs, as a standard to study and compare with MSCs of other tissues is necessary. The number of MSCs found in the BM was relatively high in comparison to other sources, with one in 3.4×10^4 cells in BM aspirate is a MSC.[17] BM contains more committed osteoprogenitors in addition to uncommitted MSCs. Assessment of the committed progenitors showed that 30% had osteo/chondro/adipo potentials and the remainder had osteo/chondro or pure osteogenic potential.[18] The disadvantages of using BM as a source of MSCs are largely practical. The procedure of BM aspiration requires a highly skilled professional and lasts 20-30 minutes. A local anesthetic (e.g. Lidocaine) must be used and the patient must be supine for one hour after the procedure.[19] The majority of patients experience pain during aspiration and over a third patients experience moderate to severe pain for a prolonged period afterwards.[20] The amount of BM that can be aspirated at one site is usually less than 2 ml.[14] Therefore; culture expansion of BM-MSCs is usually required before they are used for therapeutic purposes.

The use of BM-MSCs to regenerate bone *in vivo* has been well documented in animals. In combination with a 3D scaffold, cultured BM-MSCs have been shown to form highly vascularised primary bone tissue in mice.[21] Clinical trials are at phase II in using BM-MSCs for osteogenesis in humans. Phase I has reportedly produced successful results where MSCs were implanted into the alveolar region of the jaw in preparation for dental implants.[22] Overall, it can be seen that BM is an excellent source of MSCs and it is relatively well understood. But practical problems of harvesting BM are a significant disadvantage to its use.

3.2. Adipose tissue

A considerable amount of research in recent years has supported adipose tissue as a source of MSCs. Adipose tissue derived MSCs (A-MSCs) have been collected from animals[23] and numerous human sources, including the infrapatellar fat pad of the knee,[14] and lipoaspirate.[15] However, only a fraction of the cultured A-MSCs under osteogenic conditions will differentiate into osteogenic cells, and the rest remain as adipogenic cells. This fact is not detrimental to the use of A-MSCs for osteogenesis, as the differentiated OBs can be selected from the culture for therapeutic use. Adipose tissue has a number of advantages over BM as a source of MSCs. From a practical perspective, human adipose tissue is plentiful and can be removed more easily than BM (although anesthetic is usually required), larger amounts can be collected than from BM, and less pain is experienced by the patents.[24] Under some culture conditions, A-MSCs were found to produce more ALP than BM-MSCs; however, the difference was only significant in the early stages (four days).[25] Although bone formation from A-MSCs has been observed in rats,[24] it has been noted that the MSCs found in adipose tissue have several distinctions. A-MSCs do not undergo chondrogenic or myogenic differentiation under the same conditions as BM-MSCs, no osteocalcin is expressed by A-MSCs without 1,25-dihydroxyvitamin D3 unlike BM-MSCs and there are discretions in a small number of surface markers.[15] Therefore, it may not be assumed that the OBs derived from adipose tissue will act identically to BM derived OBs.

3.3. Periosteum

It has been shown as early as 1962 that the cells at the outer layer of the periosteum differentiate into osteoblasts.[26] The periosteum has been used as a source of osteogenic cells in two distinct ways – grafting of periosteal tissue, and use of periosteum as a source of MSCs (P-MSCs). Grafts of periosteal tissue have been observed to regenerate the damaged mandibular head of rabbits,[27] and the zygomatic arch of rats.[28] However, this is not dissimilar to standard bone grafting because the entire tissue is used, not just the osteoprogenitors. As a source of MSCs, the periosteum had excellent osteogenic potential.[29] 100% of equine tibial P-MSCs differentiate into osteogenic cells when cultured in osteogenic conditions. There has also been some success in using cultured human P-MSCs for bone regeneration. Osteogenic cells from biopsies of human calvarial periosteum seeded into nude mice stained positive for osteocalcin after six weeks.[30] In another study, P-MSCs were collected from rabbits and cultured to multiply and differentiate, and were subsequently seeded into a calvarial defect. This resulted in newly formed bone and repair.[31] In comparison to BM-MSCs and A-MSCs, there is relative little research performed on P-MSCs. It is unclear if the cells should 'earn' the name MSCs because they have only been observed differentiating into OBs and chondroblasts so far. Removal of periosteum also presents similar practical difficulties to BM aspiration. Local anesthetic is required and the amount can be removed is small. However, the high osteogenic potential of the P-MSCs means that they may have a possible future use in orthopaedics and tissue engineering.

3.4. Skeletal muscle

Cases of ectopic bone formation are seen clinically in conditions such as heterotrophic ossification and during fracture healing.[32] This suggests that there may be osteoprogenitors present in the muscle. Indeed, it has been found that muscle satellite cells have the multipotential properties similar to BM-MSCs. These muscle satellite cells (which are only found in skeletal muscle) have been observed to express myogenic, adipogenic and osteogenic potential.[33,34] Levey et al[32] collected healthy adult skeletal muscle and cultured it to enrich satellite cell numbers in osteogenic conditions. Greater than 70% of the cultured cells expressed ALP and osteocalcin. The majority of recent research into osteogenesis using muscle satellite cells concerns gene therapy. One approach is gene therapy using BMPs to stimulate osteogenesis.[35] Muscle, as a source of osteoprogenitors, has an advantage because its removal is a little more convenient than BM aspiration although anesthetic is required and there may be some pain experienced for up to a few weeks. However, the osteogenic potential of the muscle-MSCs cells appear not to be as osteogenic as those derived from bone marrow, periosteum, or adipose tissue at this stage of investigation and more research into this area is required.

3.5. Adult peripheral blood

A small number of investigations have examined the possibility of adult peripheral blood as a source of MSCs with somewhat mixed results. There have been several investigations of note recently, one of which detected MSCs in patients with breast cancer.[36] However, if MSCs were found in the peripheral blood of healthy individuals, it would be of greater significance. Zvaifler et al[37] centrifuged peripheral blood from normal individuals in order to obtain MSC rich elutriation fractions. In the appropriate fraction, MSCs were found in over 100 individuals. It was reported that 0.3-0.7% of this blood cell fraction consisted of MSCs, i.e. 1 in 2 x 10^9 blood cells are MSCs. After culture for 20 days in osteogenic conditions, about a third of the selected cells expressed ALP, osteocalcin and other markers for OBs. However, no *in-vivo* data on bone formation potentials of these blood MSCs was presented. Kuznetsov et al[38] furthered the investigation by transplanting human osteogenic cells derived from blood on ceramic particles into the subcutis of immunocompromised mice. Bone formation was found at the transplant sites after eight weeks, and with the use of a human DNA probe the osteocytes in the newly formed bony tissues were identified as human origin. But the number of MSCs found in human blood is very rare.[38] More recently, Li et al[39] have reported that the number of MSCs in the peripheral blood of patients with long bone fractures and non-unions increased significantly, and the BMP-2 expression was also significantly unregulated in the peripheral blood mononuclear cells in the fracture patients compared with the normal controls. Shirley et al[40] have further confirmed that in a rabbit ulna fracture model, BM-MSCs were recruited to the fracture sites from remote bone marrow sites via peripheral circulation. These findings point to a possible direction of research into the use of blood MSCs for tissue repair and engineering.

However, there were studies, which have not succeeded in obtaining MSCs from adult peripheral blood. These include Lazarus et al,[41] and Wexler et al.[17] Neither of these studies could identify MSCs in the peripheral blood. These may be due to the low prevalence of the MSCs in blood described by Zvaifler et al[37] and Kuznetsov et al[38], and the inappropriate culturing techniques used. The advantages of using peripheral blood as a potential source of MSCs are obvious, as the procedure for collecting blood is one of the most common procedures conducted in clinical practice. This would make therapy utilizing MSCs very accessible to almost the entire population. However, the numbers of MSCs in the adult peripheral blood are very low, and more research on the enrichment and recruitment of blood MSCs are needed

3.6. Umbilical cord blood

Umbilical cord blood (UCB) is currently used as a source of heamopoeitic stem cells.[42] It has been suggested that it may also contain MSCs. There have been a number of investigations that support this. Rosada et al[43] successfully cultured MSCs from full-term human UCB and under osteogenic conditions, these cells differentiated and stained for ALP, osteocalcin and mineralised matrix. When transplanting the UCB-MSCs into the subcutaneous tissue of mice, it resulted in a greater amount of stroma-like tissue formation and a lesser amount of bone formation compared to BM-MSCs. It was noted that UCB-MSCs were slower to establish in culture, had a lower precursor frequency and a lower level of bone antigen expression than that of BM-MSCs.[44, 45] However, some studies[17, 46] failed to identify MSCs in UCB, which suggested that this source is difficult to work with to obtain MSCs reliably. Use of UCB-MSCs as a source would involve allogeneic transplantation, as it is not possible to obtain autologous UCB cells. It has been reported that using UCB transplantation for haemopoietic stem cells does not require a close human leukocyte antigen match,[47] but it is not known if this is the case for MSCs transplantation. In effect, the cross matching that may be required along with the long-term storage considerations of UCB, are the two great disadvantages to use UCB as a possible therapeutic source of MSCs.

3.7. *Other sources*

A number of other tissues have been investigated as potential sources of MSCs, with varying success and variable potential for therapeutic use. Vascular pericytes have been investigated intermittently over the years. In one of the more recent studies, transplanted bovine pericytes into athymic mice were found to form various tissues including bone.[48] In culture conditions to encourage OB differentiation, the pericyte derived MSCs stained for a wide host of OB markers including osteocalcin and bone sialoprotein. A number of earlier investigations also support the *in vitro* differentiation of pericytes into OB-like cells.[49, 50] The source of the pericytes in the studies were the vessels of the retina and therefore would not be a sensible option for human collection. Bone tissue has also been identified as a source of committed osteoprogenitors.[51, 52] But to remove bone from the body may be the most difficult procedure to gain osteoprogenitors. Recently, a study reported second trimester amniotic fluid to be a source of MSCs that had a greater expansion potential than BM-MSCs.[53] A more thorough investigation into this would be warranted but the therapeutic use is limited due to the possible need for cross matching, long-term storage and possible (although small) risks to the fetus. The fetal blood and liver have also been reported to contain MSCs,[54] although this is unlikely to be used for therapeutic purposes. The spleen has recently been identified as a source of MSCs in rats.[55] Although experiments failed to produce any bone matrix *in vivo*, the spleen cells cultured under osteogenic conditions stained positive for osteocalcin, ALP and bone sialoprotein. Once again, the use of spleen as a potential source of MSCs would be rather limited. Dermis has also been shown to contain MSCs[56] and this area may also be worthwhile for further investigation as the skins are relatively easy to remove and plentiful.

4. Conclusion

A wide variety of sources, both established and novel, were discussed and their advantages and disadvantages examined (Table 1). The most established, best-understood and most reliable source of skeletal stem

Table 1. Summary of sources of skeletal stem cells

Source	**Advantages**	**Disadvantages**	**Comment**
Bone marrow	Reliable, relatively high numbers of MSCs,[17] well established, high OB potential.[18]	Collection – time consuming, painful, small amount removable.	At present the best option for clinical use.
Adipose tissue	Reliable, plentiful supply, less difficult collection than BM, though not ideal.	Lower OB potential than BM,[14] subtle differences in A-MSCs and BM-MSCs.[15]	Fairly good therapeutic potential. More research needed.
Periosteum	Excellent OB potential.[30]	Collection – time consuming small amount removable.	Fair therapeutic potential. More research needed.
Skeletal muscle	Plentiful supply.	Low OB potential.[33, 34] Collection difficulties.	Fair therapeutic potential.
Adult peripheral blood	Ease of collection.	Low numbers of MSCs, low OB potential.[37,38]	Worthy of more research. If reliable techniques developed, it would become source of choice.
Umbilical cord blood	Convenient collection. No pain experienced.	Low OB potential,[43] slower to culture than BM.[44] Difficult to isolate, allogeneic transplantation needed.	Low therapeutic potential.
Vascular pericytes	No obvious advantages.	Difficult collection, small amounts.	Fairly low therapeutic potential. More research needed.
Bone tissue	Source of committed osteoprogenitors.[51,52]	Removal of bone is a difficult procedure.	Fair therapeutic potential.
Amniotic fluid	Greater expansion potential than BM-MSCs.[53]	Possible risks to fetus. Allogeneic transplantation.	Fairly low therapeutic potential.
Spleen	No obvious Advantages.	Little research[55] Difficulty of collection.	Low therapeutic potential.
Dermis	Easy collection.	Difficult to isolate MSCs, little research.[56]	Low therapeutic potential.

cells is the bone marrow (BM). However, the collection of BM is quite possible the most significant drawback to its future therapeutic use. Both adipose tissue, due to its high availability and relatively easy collection, and periosteum, due to its high osteogenic properties, have possible future potentials. Skeletal muscle, umbilical cord blood, vascular pericytes, bone tissue, amniotic fluid, spleen and dermis are all less significant sources of MSCs. With development and further investigation, there may be some possible merits to these sources.

The identification of MSCs in adult peripheral blood is a significant finding. Although not all studies have confirmed MSCs in the blood and the studies, which were successful, identified only a very small numbers of MSCs in peripheral blood, the ease of collection of peripheral blood could transform the clinical use of MSCs in the future. It is essential to further study techniques of enriching, isolating and differentiating skeletal stem cells from adult peripheral blood. Were this to be successful, clinical use of MSCs could become widespread in orthopaedics and many other lines of medicine. Until then, bone marrow will probably remain the primary source of choice for skeletal stem cells.

References

1. J. E. Aubin and J. T. Triffitt, in *Principles of Bone Biology 3rd Ed*n, Ed. J. P. Bilezikian, L. G. Raisz and G. A. Rodan (Academic Press, San Deigo, USA,2002), p. 59-83.
2. B. E. Petersen, W. C. Bowen, K. D. Patrene, *et al*, *Science,*284 (1999)..
3. E. Lagasse,H. Connors, M. Al-Dhalimy, *et al*, *Nat. Med.,* 6 (2000).
4. C. R. Bjornson, R. L. Rietze, B. A. Reynolds, M. C. Magli and A. L. Vescovi, *Science,* 283 (1999)..
5. F. Parhami, S. M. Jackson, Y. Tintut Y, *et al*, *J. Bone Miner. Res.,* 14 (1999).
6. L. M. Hoffman, A. D. Weston and T. M. Underhill, *J. Bone Joint Surg.* 85A suppl 2 (2003).
7. C. Rauch, A. C. Brunet, J. Deleule and E. Farge, *Am. J. Physiol. Cell Physiol.,* 283 (2002).
8. Sanchez-Ramos J, Song S, Cardozo-Pelaez F, *et al*, *Exp. Neurol*., 164 (2000).
9. P. Bianco, M. Riminucci, S. Kuznetsov and P. G. Robey, *Crit. Rev. Eukaryot. Gene Expr.,* 9 (1999).
10. A. Yamaguchi, K. Toshihisa and S. Tatsuo, *Endocr. Rev.,* 21 (2000).
11. T. Komori, H. Yagi, S. Nomura S, *et al. Cell,* 89 (1997).
12. B. St-Jacques, M. Hammerschmidt and A. P. McMahon, *Genes Dev.,* 13 (1999).
13. G. S. Steinand J. B. Lian JB, *Endocr. Rev.,* 14 (1993).

14. M. Q. Wickham, G. R. Erickson, J. M. Gimble, T. P. Vail and F. Guilak, *Clin. Orthop.*, 412 (2003).
15. P. A. Zuk, M. Zhu, P. Ashjian P, *et al*, *Mol. Biol. Cell*, 13 (2002).
16. A. J. Friedenstein, N. W. Latzinik, A. G. Grosheva and U. F. Gorskaya, *Exp. Hematol.*, 10 (1982).
17. S. A. Wexler, C. Donaldson, P. Denning-Kendall, C. Rice, B. Bradley and J. M. Hows, *Br. J. Heamatol.* 121 (2003).
18. J. J. Minguell, A. Erices A and P. Conget, *Exp. Biol. Med.*, 226 (2001).
19. G. R. Lee, J. Foerster, J. Lukens, F. Paraskevas, J. P. Greer and G. M. Rodgers, in *Wintrobe's Clinical Hematology. 10th Edn.* (Williams and Wilkins, Baltimore, USA, 1999), p 102.
20. P. Vanhellputte, K. Nijs, M. Delforge, G. Evers and S. Vanderschueren, *J. Pain Symptom Manage.*, 26 (2003).
21. R. Cancedda, M. Mastrogiacomo, G. Bianchi, A. Derubeis, A. Muraglia and R. Quarto, *Novartis Found Symp.*, 249 (2003).
22. Osiris Inc., *Nature,* 414 (2001).
23. J. I. Huang, S. R. Beanes, M. Zhu, P. Lorenz, M. H. Hedrick and P. Benhaim, *Plast. Reconstr. Surg.*,109 (2002).
24. H. Mizuno and H. Hyakusoku, *J. Nippon Med. Sch.*, 70 (2003).
25. J. L. Dragoo, J. Y. Choi, J. R. Lieberman JR, *et al*, *J. Orthop. Res.*, 21 (2003).
26. E. A. Tonna and E. P. Cronkite, *Lab Invest.*, 11 (1962).
27. T. Ueno, T. Kagawa and J. Fukunaga J, *Ann. Plast. Surg.*,51 (2003).
28. D. Ozcelik, T. Turan, F. Kabukcuoglu, *et al*, *Arch. Facial. Plast. Surg.*, 5 (2003).
29. T. Fukumoto, J. W. Sperling, A. Sanyal, *et al.*, *Osteoarthr. Cartil.*, 11 (2003).
30. J. T. Schantz, D. W. Hutmacher, H. Chim, K. W. Ng, T. C. Lim and S. H. Teoh, *Cell Transplant*, 11 (2002).
31. A. S. Breitbart, D. A. Grande, R. Kessler, J. T. Ryaby and R. J. Fitzsimmons, *Plast. Reconstr. Surg.*, 101 (1998).
32. M. M. Levey, C. J. Joyner, A. S. Virdi AS, *et al*, *Bone* 29 (2001).
33. A. Asakura, M. Komaki and M. Rudnick, *Differentiation,* 68 (2001).
34. M. R. Wada, M. Inagawa-Ogashiwa, S. Shimizu, S. Yasumoto and H. Hashimoto, *Development*, 129 (2002).
35. D. S. Musgrave, R. Pruchnic, V. Wright, *et al*, *Bone*, 28 (2001).
36. M. Fernandez, V. Simon, G. Herrera, C. Cao, H. Del-favero and J. J. Minguell, *Bone Marrow Transplant.*, 20 (1997).
37. N. J. Zvaifler, L. Marinova-Mutafchieva, G. Adams, *et al*, *Arthritis Res.*,2 (2000).
38. S. A. Kuznetsov, M. H. Mankani, S. Gronthos, K. Satomura, P. Bianco and P. G. Robey, *J. Cell Biol.*, 153 (2001).
39. G. Li, D. Shirley, G. Burke and D Marsh, *J. Bone Miner. Res.*, Suppl 1 (2002), p 579.
40. D. S. L. Shirley, D. Marsh, G. Jordan and G. Li, *J. Bone Miner. Res.*, Suppl 1 (2003), p 235.
41. H. M. Lazarus, S. E. Haynesworth, S. L. Gerson and A. I. Caplan, *J. Hematother.*, 6 (1997).
42. F. Lazurier, M. Doedens, O. I. Gan and J. E. Dick, *Ann. N. Y. Acad. Sci.*, 996 (2003).

43. C. Rosada, J. Justesen, D. Melsvik, P. Ebbesen and M. Kassem, *Calcif. Tissue Int.*, 72 (2003).
44. H. S. Goodwin, A. R. Bicknese, S. N. Chien, B. D. Bogucki, C. O. Quinn and D. A. Wall, *Biol. Blood Marrow Transplant*, 11 (2001).
45. O. K. Lee, T. K. Kuo, W. M. Chen, K. D. Lee, S. L. Hsieh and T. H. Chen, *Blood,* 102 (2003).
46. K. Mareschi, E. Biasin, W. Piacibello, M. Aglietta, E. Madon and F. Fagioli, *Haematologica*, 86 (2001).
47. E. Gluckman, *Curr. Opin. Hematol.*, 2 (1995).
48. M. J. Doherty, B. A. Ashton, S. Walsh, J. N. Beresford, M. E. Grant and A. E. Canfield, *J. Bone Miner. Res.,*13 (1998).
49. M. J. Doherty and A. E. Canfield, *Crti. Rev. Eukaryot. Gene Expr.,* 9 (1999).
50. L. Diaz-Flores, R. Getierrez, A. Lopez-Alonso, R. Gonalez and H. Varela, *Clin. Orthop. Rel. Res.,* 275 (1992).
51. M. E. Nuttall, A. J. Patton, D. L. Olivera, D. P. Nadeau and M. Gowen, *J. Bone Miner. Res.,*13 (1998).
52. R. Tuli, M. R. Seghatoleslami, S. Tuli, *et al*, *Mol. Biotechnol.,* 23 (2003).
53. P. S. Anker, S. A. Scherjon, C. K. van der Keur, *et al. Blood*, 102 (2003).
54. C. Campagnoli, I. A. Roberts, S. Kumar, P. R. Bennett, I. Bellantuono and N. M. Fisk, *Blood*, 98 (2001).
55. A. R. Derubeis, M. Mastrogiacomo, R. Cancedda and R. Quarto, *Eur. J. Cell Biol.,* 84 (2003).
56. L. Lecoeur and J. P. Ouhayoun, *Biomaterials,* 18 (1997).

CHAPTER 4

OSTEOCLAST BIOLOGY

Xu Feng[1] and Hong Zhou[2]

[1]*Department of Pathology, University of Alabama at Birmingham, Alabama, USA*
Email: xfeng@path.uab edu

[2]*Bone Biology Lab, ANZAC Research Institute, University of Sydney, Sydney, Australia*
Email: hzhou@med.usyd.edu.au

Osteoclasts, the principal bone-resorbing cells, not only play an essential role in skeletal development and maintenance but are also implicated in the pathogenesis of various bone disorders. Osteoclasts are giant cells that possess distinct morphological features, including mutlinucleation, polarization, ruffled border membrane, a unique cell-bone matrix attachment and special cytoskeleton organization. Osteoclasts differentiate from mononuclear precursors of monocyte/macrophage lineage upon the stimulation of two key factors: M-CSF and RANKL. In addition, osteoclast differentiation is also modulated by a variety of other osteotropic factors such as 1α,25(OH)$_2$ vitamin D_3, dexamethasone, IL-1, TNF-α, prostaglandin E_2, IL-11, PTH and estrogen. Osteoclastic bone resorption involves several major steps. It starts with the establishment of a functional resorption compartment and then the formation of ruffled border membrane facing bone. The ruffled border membrane is highly rich in proton pumps which transport protons into the resorption compartment to dissolve inorganic components of bone matrix. Onorganic component of bone is then degraded by various proteolytic enzymes also released through the ruffled border membrane. Finally, degraded products are removed outside of the resorption compartment by transcytosis.

1. Introduction

Bone is a crucial connective tissue that plays an essential role in providing mechanical support for locomotion, protecting vital organs such as brain and bone marrow, and maintaining mineral homeostasis. Bone is also a dynamic tissue that undergoes constant remodeling (bone remodeling) after its development. Bone remodeling is a physiological process in which old bone is degraded and subsequently replaced by newly formed bone. Both bone development and adult bone remodeling requires two types of cells: osteoblasts and osteoclasts. Osteoblasts are the bone-forming cells while osteoclasts are the bone-resorbing cells.

Since a separate chapter in this book is already devoted to osteoblast biology (Chapter 5), we will primarily focus on the osteoclast biology in this chapter. Moreover, Chapter 19 describes in detail how these two types of bone cells coordinate to regulate bone remodeling. Given that this chapter is intended to provide a basic reading for readers with very different backgrounds ranging from researchers to clinicians, we have tried our best to make it comprehensive and concise. In addition, to keep up with the recent advances in osteoclast biology, we have also made every effort to integrate new findings into this chapter. Finally, like in every other field in biomedical sciences, numerous discrepancies and controversies have emerged in the course of discovery in osteoclast field. We feel that it is important to highlight these discrepancies and controversies here not only to emphasize their existence but, more importantly, to stimulate the readers' thinking.

2. Physiological and Pathological Roles of Osteoclasts

2.1. Physiological role of osteoclasts in bone development and maintenance

Physiologically, osteoclasts not only play a pivotal role in bone development but they also continue to be critically involved in bone maintenance even after skeleton maturation. During bone development, osteoclasts are required to degrade the cartilage matrix for vascular invasion in endochondral ossification (1) (also see chapters on bone formation in this book). Furthermore, osteoclasts are also involved in bone growth and shape modification in endochondral ossification. For instance, as shown in Fig 1, in order for this long bone to grow from A to B, it has to expand in length, to change in diameter and shape. In this process, both bone formation and bone resorption are involved.

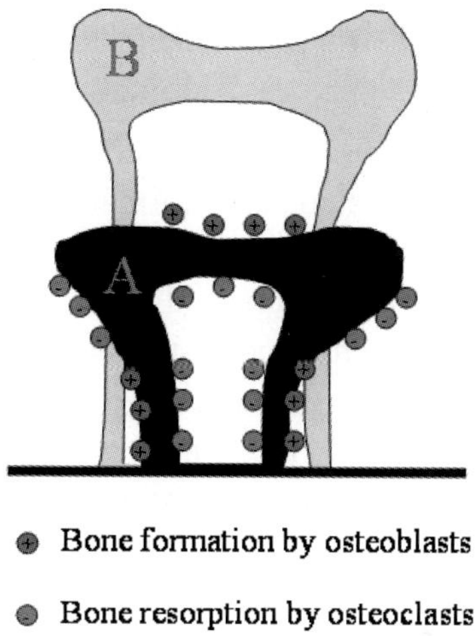

Fig. 1. Role of Osteoclasts in Bone Development

After skeletal maturation, osteoclasts continue to play a critical role in bone maintenance by participating in bone remodeling. Bone remodeling is a lifelong process of bone renewal in which old bone is resorbed by osteoclasts and then replaced by new bone formed by osteoblasts (2) (also see Chapter 19 in this book). The constant remodeling is important in three ways. (1) Bone remodeling is needed to repair fatigue damage. Bone, like other structural materials that undergo repetitive cyclical loadings, are subject to fatigue. After a number of loading cycles, tiny cracks may form. If these tiny cracks are not fixed in a timely fashion, they will accumulate and eventually lead to structural failures. Bone remodeling will replace bone containing cracks and prevent structural failure. (2) Bone remodeling is needed to adapt bone material properties to the mechanical demands that are placed on bones. For instance, bone remodeling will help you obtain stronger bone with exercise. (3) Finally, bone remodeling play a critical role in regulating calcium homeostasis which is critical for life. Bone is a major reservoir for calcium. When serum calcium becomes low and there is no intake calcium, calcium will be released from bone by osteoclasts to meet the demand.

2.2. Pathological role of osteoclasts in various disorders

Bone remodeling requires balanced coupling of bone formation and bone resorption. Disruption of this balance results in bone disease. In many pathological conditions, osteoclast formation and/or function are elevated, leading to excessive destruction of bone. The most prevalent metabolic bone disease, postmenopausal osteoporosis, is caused in part by increased osteoclast activity due to the decline in estrogen resulting from the cessation of ovary function in postmenopausal women (3). Furthermore, osteoclasts are also implicated in bone erosion in rheumatoid arthritis (4). In rheumatoid arthritis, osteoclast formation and function are primarily enhanced by activated T-cells through the inflammatory cytokines they secrete (5). Finally, osteoclasts have been shown to play an essential role in osteolysis associated with the bone metastases of several tumors including breast and prostate cancers (6). We would also like to refer readers to Chapters 7 and 8, which will provide greater details on the role of osteoclasts in the pathogenesis of various bone disorders.

3. Osteoclast Morphology

The osteoclast is not only functionally a unique cell that is specialized to degrade bone matrix but it also possesses several distinct morphological features that support its unique function (7;8). The most obvious features of the osteoclast are its large size and multinucleation. Osteoclasts are giant cells in comparison to other cell types in the body and specifically they are usually more than 100 um in diameter (7). Moreover, osteoclasts are multinucleated cells and normal osteoclasts contain up to 10 nuclei (7).

Secondly, in addition to the large size and multinucleation, osteoclasts are also highly polarized cells (8). Osteoclast polarization is characterized by the localization of nuclei toward the apical membrane and also by the formation of the ruffled border membrane facing bone (7). The presence of the ruffled border membrane represents another one of the distinguishing features of the osteoclasts. It comprises a series of deeply interfolded finger-like projections of the plasma and cytoplasmic membranes adjacent to the bone surface. Functionally, this represents the resorbing apparatus of the cell (9). It represents an extensive area of cell surface where secretion of enzymes and uptake of matrix components can take place.

Thirdly, the cell surface next to bone is characterized by close apposition of the plasma membrane to bone and an adjacent, organelle-free area, rich in actin filaments, called the clear zone (10). The clear zone is also known as the "sealing zone", as the plasma membrane in this region comes into tight apposition with the bone surface during resorption encircles the ruffled border completely so that the site of resorption is isolated and localized (11) (12;13).

Finally, the osteoclast maintains distinct cytoskeleton organization. The specialized ruffled border and sealing zone appear in osteoclasts only during resorption and disappear when the cells are motile. When osteoclasts are plated on bone surfaces, a characteristic resorption pit is formed below the cell at the site of attachment of the ruffled border (14). This attachment involves the specific interaction between adhesion molecules in the cell membrane (integrins) and some bone matrix proteins. The integrins are a family of transmembrane proteins whose cytoplasmic domains interact with the cytoskeleton while their extracellular domains bind to bone matrix proteins, enabling them to mediate cell-substratum and cell-cell interactions (15). The space contained inside this ring of attachment and between the osteoclast and the bone matrix constitutes the bone-resorbing compartment. The cell membrane of the apical pole is invaginated to form a ruffled border.

In summary, the osteoclast possesses several important morphological features such as large size, mutlinucleation, polarization, presence of ruffled border membrane, a unique cell-bone matrix attachment and special cytoskeleton organization. Importantly, as we will discuss in detail below, these morphological features are required to support the function, degradation of bone matrix.

4. Osteoclast Differentiation

4.1. Hematopoietic origin of osteoclasts

In the 1960's, osteoclasts were thought to arise by fusion of osteoblasts (16). Many studies since then have shown decisively that the osteoclast is hematopoietic in origin and unrelated to the stromal lineage. The conclusive evidence came from three different types of *in vivo* experiments in 1970's. Gothlin *et al* used a parabiotic method to join the circulation of two rats together and found that osteoclasts migrating to a fracture in an irradiated rat were derived from the bloods of its non-irradiated partner (17;18). In addition, evidences from chick/quail chimera experiments (19), and the restoration of bone resorption by transplanting normal marrow cells or spleen cells (20;21), indicated that osteoclast precursors were present in hematopoietic tissues such as bone marrow, spleen and peripheral blood. This notion was further supported by in vitro experiments. Burger *et al* (1982) showed that osteoclasts formed when mouse marrow cells were co-cultured with stripped fetal bone rudiments (22). Using co-cultures of hematopoietic stem cells purified from mouse bone marrow and fetal bone rudiments, Scheven and co-workers (1986) reported that some of the stem cell populations differentiated into osteoclasts (23). Co-cultures of mouse osteoblasts and spleen cells provided additional evidence that osteoclast progenitors were present in spleen cell but not in osteoblastic cells (24). Treatment of mouse spleen cells or human peripheral blood monocytes with soluble RANKL further confirmed that osteoclast derived from haemopoietic cell origin, in monocyte/macrophage lineage (25;26).

4.2. Stages of osteoclast differentiation

The elucidation of the origin of osteoclasts profoundly facilitated our investigation of molecular mechanism underlying osteoclast differentiation. Particularly, in the 1990's we have witnessed many great advances in the understanding of osteoclast differentiation (8;27;28). Now it has been established that the osteoclast differentiation involves several critical stages outlined in Fig. 2. First, the hematopoietic stem cells (HSC) give rise to circulating mononuclear cells called colony forming unit-granulocyte/macrophages (CFU-GM). M-CSF supports the proliferation of CFU-GM and keeps the mononuclear cells in monocyte/macrophage lineage. The mononuclear precursors are attracted to prospective resorption sites by an unknown mechanism (presumably by chemotaxis) and they will then attach onto bone matrix to further differentiate into prefusion osteoclasts with the stimulation of two critical factors: M-CSF and RANKL. With continuous presence of M-CSF and RANKL, these prefusion osteoclasts will continue to differentiate by fusion to become multinucleated cells. The multinucleated osteoclasts are not functional osteoclasts since they lack the ruffled border membrane that is critical for bone resorption. It is now clear that RANKL plays an important role in activating osteoclasts by stimulating formation of the ruffled border membrane within plasma membrane facing bone.

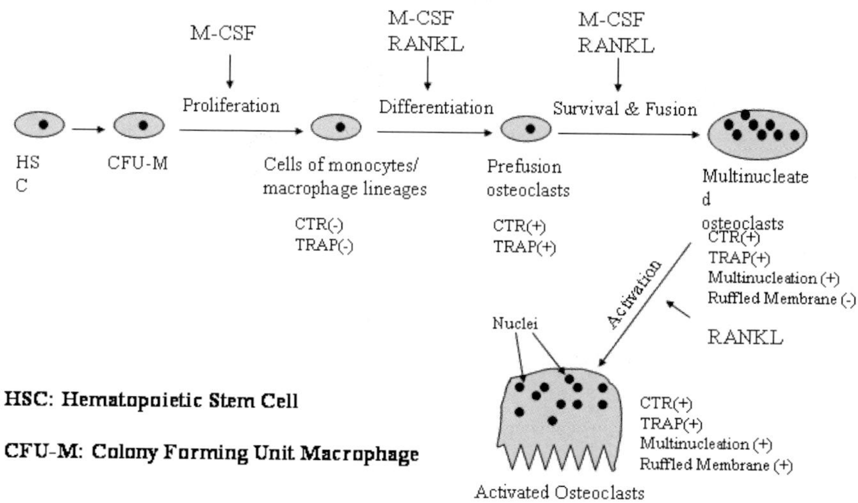

Fig. 2. Osteoclast Differentiation Pathway

As also shown in Fig. 2, many markers are used to distinguish cells at different stages of the osteoclast differentiation pathway (28). These markers are either the cellular morphological characteristics or proteins expressed by the cells at distinct stages of the osteoclast differentiation. The widely used markers include calcitonin receptor (CTR), tartrate-resistant acid phosphotase (TRAP), the multinucleation and the presence of the ruffled border membrane. For instance, while the mononuclear precursor cells of monocyte/macrophage lineage lack both CTR and TRAP, the prefusion osteoclasts become positive for CTR and TRAP. Among these markers, TRAP is the most widely used marker for osteoclasts, partially because they are easy to be detected and relatively more specific. To further understand the mechanism of the osteoclast differentiation, we will provide more details on the major steps of the osteoclast differentiation below.

4.3. Attachment of osteocalst precursors and osteoclasts on bone matrix

After the circulating osteoclast precursor cells are attracted to prospective resorption sites, these precursor cells need to attach on bone matrix to differentiate. In addition, it is also essential for fully differentiated mature osteoclasts remain attached on bone to resorb bone. The identification and characterization of adhesion molecules involved in mediating these attachments has been a major focus of bone biology research. As a result, it has now been established that integrins play a central role in this process (29;30). Mature osteoclasts express a variety of integrins, including integrin $\alpha_v\beta_3$, $\alpha_2\beta_1$, $\alpha_v\beta_1$, $\alpha_M\beta_2$ (31-34). Among these integrins, integrin $\alpha_v\beta_3$ was shown to play a role in osteoclast attachment and bone resorption. Initial evidence supporting this notion came from an in vitro study showing that a monoclonal antibody against an antigen on osteoclasts inhibits bone resorption (35) and the antigen was later identified as integrin $\alpha_v\beta_3$ (36). Consistently, an independent study showed that LM609, a blocking antibody recognizing avian integrin $\alpha_v\beta_3$, not only blocks the avian osteoclast attachment onto bone but also bone resorption (37). Subsequently, integrin $\alpha_v\beta_3$ was shown to mediate osteoclast attachment by recognizing the RGD sequence present in various bone matrix proteins such as osteopontin, vitronectin, and bone sailoprotein (38-41). Consistent with the in vitro data, integrin β_3 knockout mice exhibited an osteoscloretic phenotype due to a functional defect in osteoclasts, confirming that integein $\alpha_v\beta_3$ is important for osteoclast function *in vivo* (42). Nonetheless, the $\beta3^{-/-}$ osteoclasts did not completely lose capacity to attach on bone and exhibited residual bone resorption, suggesting other adhesion molecules may also be involved in mediating the interaction between mature osteoclasts and bone matrix.

In contrast, the adhesion proteins involved in regulating the attachment of osteoclast precursors on bone matrix largely remain unknown. It was previously thought that integrin αvβ5 may play a functional role in this process because this integrin β5 subunit is highly homologous to integrin β3 and more importantly it is abundantly expressed by osteoclast precursors (30;43). More interestingly, integrin β3 and β5 are reciprocally expressed during osteoclast differentiation (44-46). However, integrin β5 knockout failed to show bone phenotype with normal osteoclast differentiation (47), suggesting that other unidentified adhesion molecules may play a role in attachment of osteoclast precursor on bone matrix.

4.4. Osteoclast fusion and multinucleation

An essential facet of osteoclast differentiation is the fusion of committed mononuclear precursors to form mature multinucleated cells. Normal osteoclasts usually possess up to 10 nuclei. It is believed that the number of nuclei may reflect the osteoclast activity. For example, osteoclasts in Paget's disease, which is characterized as elevated osteoclastic bone resorption, contain as many as 100 nuclei (48). Osteoclast nuclei are distinct from one another and this characteristic distinguishes the osteoclast from the megakaryocyte. The osteoclast multinucleation occurs through cellular fusion rather than nuclear division (49). Furthermore, the nuclei in osteoclasts have special cellular localization and they are preferentially polarized away from the plasma membrane facing the bone, residing close to the antiresorptive surface of the cell. Functional significance of the osteoclast nuclei polarization is not clear. Although the osteoclast fusion represents an important part of osteoclast differentiation, the molecular mechanism controlling this process has not been elucidated. Future studies aimed at delineating this event is needed since the effective inhibition of the process represents an attractive therapeutic strategy for treating bone disease involving elevated osteoclast activity.

4.5. Formation of Ruffled border membrane

The ruffled border membrane is not only a morphological characteristic of the osteoclast but it is also an important functional organelle of the cell. Formation of the ruffled border membrane probably results from the insertion of proton pump (H-ATPase)-bearing vesicles into the plasma membrane facing the bone (50). In unattached cells, acidifying vesicles containing the proton pumps distribute diffusely throughout the cytoplasm. Upon attachment of cells to bone, matrix-derived signals prompt the acidifying vesicles to migrate and insert into the plasma membrane facing the bone. As a result, the ruffled border membrane is rich in proton pumps, which play a critical role in bone resorption by transporting proton to the resorption compartment to dissolve the inorganic components of bone.

4.6. Regulation of osteoclast formation

Osteoblasts/stromal cells are derived from bone marrow mesenchymal stem cells (51;52) while osteoclasts differentiate from cells of hematopoietic origin (7;8;27). Although osteoblasts/stromal cells and osteoclasts differentiate from different precursors, evidence was already accumulated in early 1980's to suggest that osteoblasts/stromal cells play a central role in mediating osteoclastogenesis (53). Specifically, experimental evidence suggests that osteoblasts mediate osteoclast formation and bone resorption by producing soluble factors and by signaling to osteoclasts via cell-cell contact (24;54). Thus, in vitro, osteoclasts can be generated by co-culturing mononuclear precursors with osteoblasts or stromal cells in the presence of osteotropic factors such as $1\alpha,25(OH)_2$ vitamin D_3 and dexamethasone (54-56). Notably, this co-culture system was the only method available to prepare murine osteoclasts in vitro prior to the discovery of the RANKL/RANK system.

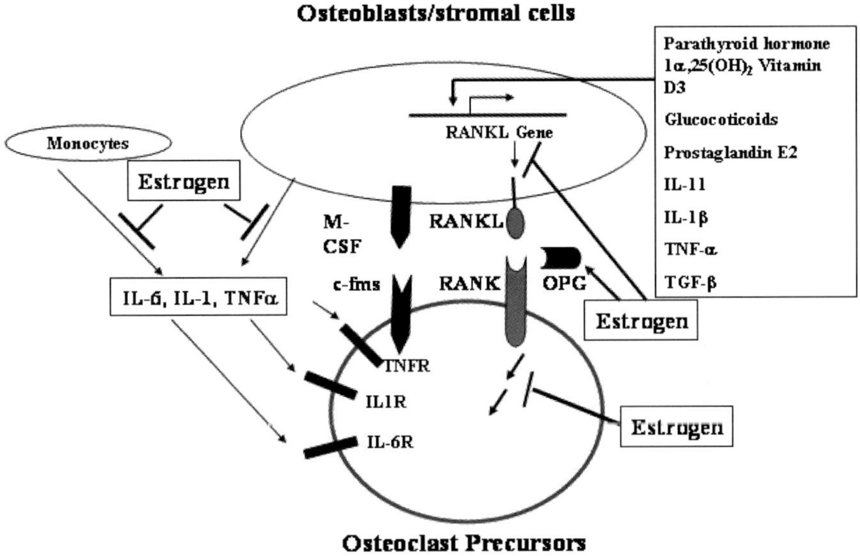

Fig. 3. Regulation of RANKL Gene Expression

Since the discovery the RANKL/RANK system, it has been established not only that osteoclast formation and function require two key osteoclastogenic cytokines: RANKL and M-CSF (25;57), but also that osteoblasts/stromal cells support osteoclast differentiation primarily by serving as a source of M-CSF and RANKL (27). Fig. 3 summarizes the current understanding of osteoclast formation and function involving osteoblasts/stromal cells. Osteoblasts/stromal cells express both M-CSF and RANKL (membrane-bound RANKL and soluble RANKL). M-CSF and RANKL will bind to their respective receptor c-fms and RANK expressed on osteoclast precursors to stimulate osteoclast formation. In mature osteoclasts, RANKL, but not M-CSF, is required to mediate osteoclast function and survival. In addition, osteoblasts/stromal cells also produce a factor called OPG, which is decoy receptor for RANKL. OPG inhibits RANKL function by competing with RANK for RANKL (8;27).

Moreover, the unraveling of the RANKL/RANK system has also helped reveal that many osteotropic hormones and cytokines regulate osteoclast formation and function through modulating RANKL expression by osteoblasts/stromal cells [see recent reviews in (58-60)]. For instance, it has been known for a quite long time that *in vitro* generation of osteoclasts by co-culturing osteoblasts/stromal cells and osteoclast precursors requires 1α,25-$(OH)_2$ vitamin D_3 and dexamethonsone. However, it was not clear until the discovery of the RANKL/RANK system that 1α,25$(OH)_2$ vitamin D_3 and dexamethasone stimulate osteoclast formation in the co-culture system by up-regulating RANKL production by osteoblasts/stromal cells (25;61). In addition, other osteotropic hormones and cytokines such as IL-1, TNF-α, prostaglandin E_2, IL-11 and PTH have also been shown to stimulate RANKL gene expression in osteoblasts/stromal cells (25;62;63). In contrast, TGF-β suppresses RANKL gene expression (64). To summarize, the factors that have been shown to be involved in the regulation of RANKL gene expression are listed in Fig. 3.

Finally, estrogen is critically implicated in regulation of bone remodeling and the decline in estrogen level resulting from the cessation of ovary function in postmenopausal women underlies the pathogenesis of postmenopausal osteoporosis. It has been shown that estrogen regulates bone remodeling in part by modulating osteoclast formation. As shown in Fig. 3, osteoblasts/stromal cells and monocytes are major sources of IL-1, IL-6 and TNF-α, which all exert positive effects on osteoclast formation (65-69). Estrogen is a potent factor that inhibits the production of these three cytokines by osteoblasts/stromal cells and monocytes (70;71). In addition, since the unraveling of RANKL and its decoy receptor OPG, it has been shown that estrogen is also implicated in osteoclast differentiation by inhibiting RANKL expression and stimulating OPG expression (72;73) Finally, a recent study indicated that estrogen can also negatively affect the RANK-mediated intracellular signaling in osteoclasts (74).

4.7. *In vitro generation of osteoclasts*

Early studies of osteoclasts were greatly facilitated by the establishment of methods for isolation and enrichment of pure populations of mature osteoclasts. Importantly, much of the understanding of the mechanism by which osteoclasts degrade bone resulted from the ability to obtain highly enriched osteoclasts from chickens (75), rabbits (76), rats (77), and mice (78). However, study of the mechanism by which osteoclast differentiate requires generation of osteoclasts from its precursors *in vitro*.

Prior to the discovery of the RANKL/RANK system, osteoclasts were primarily prepared *in vitro* by co-culture system, which was initially established by Udagawa and co-worker in late 1980's (55). To use the coculture system to prepare osteoclasts *in vitro*, mononuclear precursors (either spleen macrophages or bone marrow macrophages) are cultured together with osteoblasts or stromal cells in the presence of osteotropic factors such as $1\alpha,25(OH)_2$ vitamin D_3 and dexamethasone and osteoclasts start to form around day 6 or 7 (54-56). The method contributed considerably to our understanding of the mechanism underlying osteoclast differentiation. However, this coculture system has a shortcoming. The osteoclasts formed in the system are contaminated with osteoblasts or stromal cells, making it unsuitable for certain studies. Although efforts were made to remove the osteoblasts/stromal cells, the outcome has never been satisfactory. Upon the unravelling of the RANKL, it has been established that osteoclasts can be easily generated in vitro by treating either spleen macrophages or bone macrophages with M-CSF and RANKL. Significantly, this improved method for preparing osteoclast generation *in vitro* is able to produce highly pure populations of osteoclasts, which will greatly facilitate our investigation of the osteoclast differentiation in the future.

5. Mechanism of Osteoclast Bone Resorption

5.1. Overview

Osteoclastic bone resorption is a complicated process involving several major steps (8;79;80) (summarized in Fig. 4). The initial step of the resorption process is the establishment of a functional resorption compartment. A functional resorption compartment has two important features. First, osteoclasts attach on bone through a special structure called sealing zone, which seals the resorption site from its surroundings. The second feature is the formation of ruffled border membrane facing bone. The ruffled border membrane plays an important role in the

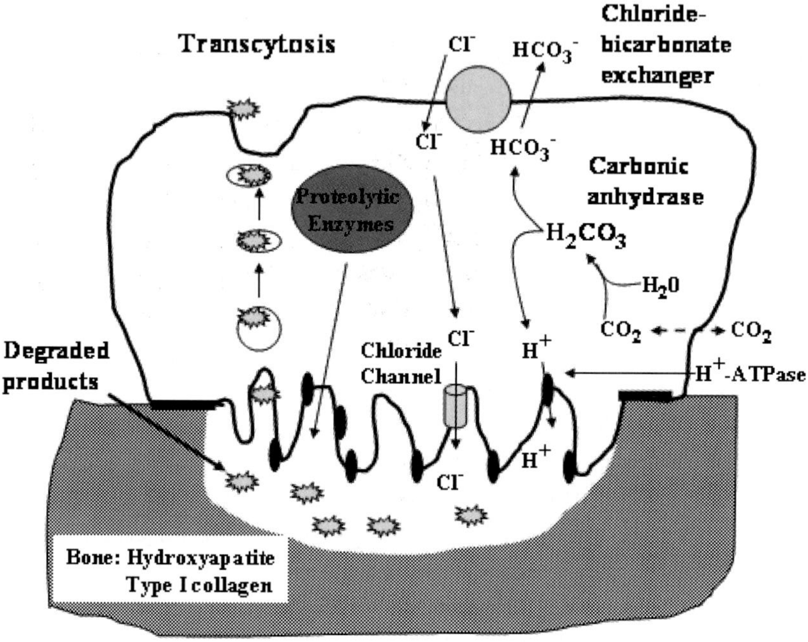

Fig. 4. Mechanism of Bone Resorption

degradation of bone matrix. The ruffled border membrane is highly rich in proton pumps which transport protons into the resorption compartment to maintain a low-pH environment that is critical for the dissolution of inorganic components of bone matrix. Dissolution of inorganic components of bone is then followed by the degradation of organic components of bone, which depends on the action of various proteolytic enzymes also released through the ruffled border membrane. Finally, several lines of recent evidence suggest that degraded products are removed outside of the resorption compartment by transcytosis. Thus, the bone resorption process involves four major events.

5.2. Formation of a functional resorption compartment

A functional resorption compartment is a critical structural requirement for the bone resorption. As described in Section 4 (osteoclast differentiation) above, one of the important end results of osteoclast differentiation and activation is the formation of a functional resorption compartment, which includes the establishment of the sealing zone and formation of the ruffled border membrane. Upon the activation of mature osteoclasts, they will start to degrade bone matrix in the location where they form. Often osteoclasts will migrate to new sites to start more cycles of bone resorption after they complete the bone resorption in the primary site. Every time osteoclasts migrate to a new site, they will need to reassemble the resorption compartment.

The sealing zone is very important to osteoclast function. However, the precise adhesion molecules involved in formation of sealing zone have not been definitely identified. A variety of integrins including integrin $\alpha_v\beta_3$, $\alpha_2\beta_1$, $\alpha_v\beta_1$, $\alpha_M\beta_2$ are expressed in mature osteoclasts (31-34). Many previous studies supported that integrin $\alpha v\beta 3$ may play a central role in forming the sealing zone, because antibody against this integrin as well as RGD-containing peptides blocked both attachment of osteoclasts to bone and bone resorption (35;37). However, numerous other research groups failed to localize this integrin in sealing zone by immunostaining (81;82). Furthermore, two $\beta 1$ integrins ($\alpha_2\beta_1$ and $\alpha_v\beta_1$) are able to recognize collagens (15), suggesting that these integrins are possible candidate for adhesion molecules mediating the formation of sealing zone. However, since early blocking experiments indicate $\alpha v\beta 3$ is a major molecule mediating osteoclast attachment to bone. The role of these two integrins in osteoclast attachment to bone has not been confirmed *in vivo*, primarily due to the fact that integrin $\beta 1$ knockout mice are embryonic lethal (83). Interestingly, many studies demonstrated that Pan-cadherin antibodies recognize sealing zones, suggesting that some members of the cadherins family might mediate the tight attachment of osteoclasts to bone (80).

As discussed above, the last step of osteoclast differentiation (osteoclast activation) is primarily characterized by the formation of the ruffled border membrane. This is achieved by migration and insertion of acidifying vesicles into the plasma membrane facing the bone, driven by matrix-derived signals. It is worthwhile to emphasize here that every time when osteoclasts start to migrate, they will detach from the bone matrix, resulting in losing the ruffled border membrane. Once the migrating osteoclasts are settled at new site, they will need to form the ruffled border membrane to start the new cycle of bone resorption by the similar mechanism.

5.2. Degradation of inorganic components of bone matrix

Bone matrix consists of both inorganic and organic components (84;85). The inorganic component is primarily crystalline hydroxyapatite: $[Ca_3(PO_4)_2]_3Ca(OH)_2$. The organic component of bone contains about 20 proteins with type I collagen as the most abundant one (>90%) (86). Thus, degradation of bone matrix involves two events: 1) dissolution of crystalline hydroxyapatite and 2) proteolytice cleavage of the organic component of bone matrix. Dissolution of crystalline hydroxyapatite precedes proteolytic cleavage of the organic component since the collagen and other bone matrix proteins will not be efficiently accessible to proteolytic degradation until these proteins embedded in crystalline hydroxyapatite are released upon dissolution of hydroxyapatite (87).

The dissolution of the inorganic content of bone involves acidification of the extracellular bone-resorbing compartment, which represents one of the most important features of osteoclast action. The acidification of the resorption compartment is mediated by a vacuolar H+-adenosine triphosphatase (H+-ATPase) which is abundantly present in the ruffled border membrane (88-90). H+-ATPase transports protons (H+) into the resorption compartment to create and maintain a very low-pH environment (~4.5). The low-pH condition helps deposit high concentrations of acid onto a strongly basic mineral to liberate calcium: $[Ca_3(PO_4)_2]_3Ca(OH)_2 + 8\ H+ \leftrightarrow 10\ Ca_3\ + 6\ HPO_4^{2-} + 2\ H_2O$. The source of the cytoplasmic protons for secretion into the extracellular bone-resorbing compartment is carbonic acid. The protons are produced by carbonic anhydrase, which is enriched in the cytoplasm of the osteoclast (91)(Fig. 4). Carbonic anhydrase generates protons and bicarbonate from carbon dioxide and water ($CO_2 + H_2O \leftrightarrow H_2CO_3 \leftrightarrow H^+ + HCO_3^-$). Importantly, in order to maintain elcetroneutrality, Cl^- are also transported into the resorption compartment via Cl^- channels present in the ruffled border membrane (92;93), which is charge-coupled to the H+-ATPase. It is then worthy to note that the secretion of protons across the ruffled border membrane into the extracellular resorbing compartment leaves behind the conjugate base (HCO_3^-) inside the osteoclast, which needs to timely be removed out of the cell. Furthermore, the osteoclast also needs to continuously supply Cl^- ions for secretion into the resorption compartment. These two jobs are accomplished by a passive chloride-bicarbonate exchanger in the basolateral membrane (94)(Fig. 4).

5.4. Degradation of organic components of bone matrix

The degradation of the organic component of bone matrix is accomplished by two major classes of proteolytic enzymes: lysosomal family of acidic proteinases and matrix metalloproteinases (MMPs) (79) (80) (8). It has been shown that the degradation of the bone matrix proteins involves several lysosomal proteinases, including cathepsin K, cathepsin D, cathepsin B and cathepsin L (95-100). Among them, cathepsin K has been shown to play a predominant role in degrading bone matrix proteins (98;99). In keeping with this notion, patients with a mutation in the cathepsin K gene develop pycnodystosis (101). Moreover, cathepsin K knockout mice exhibit osteopetrosis due to a defect in matrix degradation (102), further supporting a functional role for cathepsin K in bone resorption. Notably, given that many acidic proteinases are involved in the degradation of bone matrix proteins, the acidification of the resorption compartment also contributes to the degradation of the organic component in addition to its predominant role in the dissolution of the inorganic component of bone matrix. The second class of proteolytic enzymes that are involved in the degradation of the organic matrix are MMPs. In particular, MMP-9 is highly expressed in osteoclasts and is functionally implicated in bone resorption (103). Finally, these proteolytic enzymes are formed in the Golgi region and then vectorially transported toward the ruffled border membrane through their association with mannose-6 phosphate receptors. At their destination, these enzymes bound to mannose-6-phosphate receptors fuse with the ruffled border membrane and their contents discharged into the bone-resorbing compartment (104).

5.5. *Removal of degraded products*

The last remaining question concerning the osteoclast bone resorption is how degraded products are removed out of the osteoclast. Many proposed models existed describing how the removal of degraded products might be achieved. While one possibility that the degraded products are removed through a detachment and reattachment process that can allow degraded products to be leaked out of the cell, another proposed model is that the removal of the products is accomplished by transcytosis via a vesicular process. Significantly, available evidence indicate that transcytosis is the key, if not exclusive, mechanism by which fragmented bone matric products are transported outside of an actively resorbing osteoclast (105;106). The transcytosis process includes several steps. First, degraded products are endocytosed, then transported along a transcytotic vesicular pathway toward anti-resorptive side of the cell, and finally released out of the cell at anti-resorptive side (**Fig. 4**).

6. Conclusions and Perspectives

As the principal bone-resorbing cell, the osteoclast not only plays a pivotal role in skeletal development and maintenance but it is also implicated in the pathogenesis of various bone diseases. In the past three decades, there have been many great advances in our understanding of the molecular and cellular mechanism underlying osteoclast differentiation and function. In particular, the establishment of the hematopietic origin of the osteoclast represents the first hallmark in our investigation and understanding of the mechanism controlling osteoclast differentiation. Significantly, this historical finding was then followed by many important discoveries, including the unraveling of a functional role for osteoblasts/stromal cells in osteoclastogenesis, the subsequent establishment of the co-culture system for preparing osteoclasts in vitro), and the most recent cloning of RANKL as an essential and sufficient factor involved in osteoclast differentiation. Moreover, in the last several decades, we have also made significant progress in the elucidation of the mechanism of osteoclast-mediated bone resorption.

Nonetheless, as discussed throughout the chapter, many unsolved issues and controversies concerning osteoclast differentiation and function still remain to be addressed. With the development of our knowledge in molecular and cell biology in general and future improvements in various techniques, we expect to see further exponential expansion in our understanding of the osteoclast biology. More importantly, the new and expected advances in our knowledge in osteoclast biology will provide us a greater opportunity to reveal and define new therapeutic strategies for preventing and treating bone disorders involving osteoclasts.

References

1. Lee, C. A. and Einhorn, T. A. (2001) The bone organ system: Form and Function. In Marcus, R., Feldman, D., and Kelsey, J., editors. *Osteoporosis*, Academic Press, San Diego
2. Martin, T. J. and Rodan, G. A. (2001) Coupling of bone resorption an formation during bone remodeling. In Marcus, R., Feldman, D., and Kelsey, J., editors. *Osteoporosis*, Academic Press, San Diego
3. Pacifici, R. (2001) Postmenopausal Osteoporosis: How the Hormonal Changes of Menopausal Cause Bone Loss. In Marcus, R., Feldman., and Kelsey, J., editors. *Osteoporosis*, Academic Press, San Diego
4. Goldring, S. R. (2003) *Rheumatology* **42 Suppl 2,** ii11-ii16
5. Goldring, S. R. (2003) *Calcif.Tissue Int.* **73,** 97-100
6. Mundy, G. R. (2002) *Nature Reviews* **Cancer. 2,** 584-593
7. Ross, F. P. and Teitelbaum, S. L. (2001) Osteoclast Biology. In Marcus, R., Feldman, D., and Kelsey, J., editors. *Osteoporosis* , Academic Press, San Diego
8. Teitelbaum, S. L. (2000) *Science* **289,** 1504-1508
9. Blair, H. C. (1998) *BioEssays* **20,** 837-846
10. Vaananen, H. K. and Horton, M. (1995) *J.Cell Sci.* **108,** 2729-2732
11. Holtrop, M. E. and King, G. J. (1977) *Clinical Orthopaedics & Related Research* 177-196
12. Baron, R., Neff, L., Louvard, D., and Courtoy, P. J. (1985) *J.Cell Biol.* **101,** 2210-2222
13. Yoshida, S., Domon, T., and Wakita, M. (1989) *Archives of Histology & Cytology* **52,** 513-520
14. Boyde, A. and Jones, S. J. (1979) *Scanning Electron Microscopy* 393-402
15. Hynes, R. O. (1992) *Cell* **69,** 11-25
16. Tonna, E. A. and Cronkite, E. P. (1961) *Naturwissenschaften* **190,** 459
17. Gothlin, G. and Ericsson, J. L. E. (1976) *Clin.Orthop.Rel.Res.* **120,** 201-231
18. Walker, D. G. (1973) *Science* **180,** 275
19. Kahn, A. J. and Simmons, D. J. (1975) *Nature* **258,** 325-327

20. Walker, D. G. (1975) *Science* **190,** 785-787
21. Walker, D. G. (1975) *Science* **190,** 784-785
22. Burger, E. H., Van Der Meer, J. W. M., Van De Gevel, J. K., Gribnau, J. C., Thesingh, C. W., and Van Furth, R. (1982) *J.Exp.Med.* **156,** 1604-1614
23. Scheven, B. A. A., Visser, J. W. M., and Nijweide, P. J. (1986) *Nature* **321,** 79-81
24. Takahashi, N., Akatsu, T., Udagawa, N., Sasaki, T., Yamaguchi, A., Moseley, J. M., Martin, T. J., and Suda, T. (1988) *Endocrinol.* **123,** 2600-2602
25. Yasuda, H., Shima, N., Nakagawa, N., Yamaguchi, K., Kinosaki, M., Mochizuki, S., Tomoyasu, A., Yano, K., Goto, M., Murakami, A., Tsuda, E., Morinaga, T., Higashio, K., Udagawa, N., Takahashi, N., and Suda, T. (1998) *Proc.Natl.Acad.Sci.U.S.A.* **95,** 3597-3602
26. Matsuzaki, K., Udagawa, N., Takahashi, N., Yamaguchi, K., Yasuda, H., Shima, N., Morinaga, T., Toyama, Y., Yabe, Y., Higashio, K., and Suda, T. (1998) *Biochem.Biophys.Res.Commun.* **246,** 199-204
27. Suda, T., Takahashi, N., Udagawa, N., Jimi, E., Gillespie, M. T., and Martin, T. J. (1999) *Endocr.Revs.* **20,** 345-357
28. Boyle, W. J., Simonet, W. S., and Lacey, D. L. (2003) *Nature* **423,** 337-342
29. Duong, L., Lakkakorpi, P., Nakamura, I., and Rodan, G. A. (2000) *Matrix Biology* **19,** 97-105
30. Teitelbaum, S. L. (2000) *Journal of Bone & Mineral Metabolism* **18,** 344-349
31. Clover, J., Dodds, R. A., and Gowen, M. (1992) *J Cell Sci* **92,** 267-271
32. Hughes, D. E., Salter, D. M., Dedhar, S., and Simpson, R. (1993) *J.Bone Miner.Res.* **93,** 527-533
33. Nesbitt, S., Nesbit, A., Helfrich, M., and Horton, M. (1993) *J.Cell Biol.* **268,** 16737-16745
34. Duong, L. T., Tanaka, H., Wesolowski, G., and Rodan, G. A. Role of $\alpha_M\beta_2$ integrin in osteoclast differentiation. Journal of Bone & Mineral Research 8, S378. 1993. Ref Type: Abstract
35. Chambers, T. J., Fuller, K., Darby, J. A., Pringle, J. A. S., and Horton, M. A. (1986) *Bone Miner.* **1,** 127-135
36. Davies, J., Warwick, J., Totty, N., Philp, R., Helfrich, M., and Horton, M. (1989) *J.Cell Biol.* **109,** 1817-1826
37. Miyauchi, A., Alvarez, J., Greenfield, E. M., Teti, A., Grano, M., Colucci, S., Zambonin-Zallone, A., Ross, F. P., Teitelbaum, S. L., and Cheresh, D. (1991) *J.Biol.Chem.* **266,** 20369-20374
38. Ross, F. P., Alvarez, J. I., Chappel, J., Sander, D., Butler, W. T., Farach-Carson, M. C., Mintz, K. A., Robey, P. G., Teitelbaum, S. L., and Cheresh, D. A. (1993) *J.Biol.Chem.* **268,** 9901-9907
39. Flores, M., Norgard, M., Heinegard, D., Reinholt, F. P., and Andersson, G. (1992) *Exp.Cell Res.* **201,** 526-530
40. Helfrich, M. H., Nesbitt, S. A., Dorey, E. L., and Horton, M. A. (1992) *J.Bone Miner.Res.* **7,** 335-343
41. Helfrich, M. H., Nesbitt, S. A., Lakkakorpi, P. T., Barnes, M. J., Bodary, S. C., Shankar, G., Mason, W. T., Mendrick, D. L., Vaananen, H. K., and Horton, M. A. (1996) *Bone* **19,** 317-328

42. McHugh, K. P., Hodivala-Dilke, K., Zheng, M. H., Namba, N., Lam, J., Novack, D., Feng, X., Ross, F. P. , Hynes, R. O., and Teitelbaum, S. L. (2000) *J.Clin.Invest.* **105,** 433-440
43. Sago, K., Teitelbaum, S. L., Venstrom, K., Reichardt, L. F., and Ross, F. P. (1999) *J.Bone Miner.Res.* **14,** 32-38
44. Inoue, M., Ross, F. P., Erdmann, J., Abu-Amer, Y., and Teitelbaum, S. L. (2000) *Endocrinol.* **141,** 284-290
45. Inoue, M., Namba, N., Chappel, J., Teitelbaum, S. L., and Ross, F. P. (1998) *Mol.Endocrinol.* **12,** 1955-1962
46. Feng, X., Teitelbaum, S. L., Quiroz, M., Towler, D. A., and Ross, F. P. (1999) *J.Biol.Chem.* **274,** 1366-1374
47. Huang, X., Griffiths, M., Wu, J., Farese, R. V., Jr., and Sheppard, D. (2000) *Mol.Cell.Biol.* **20,** 755-759
48. Fallon, M. D. and Schwamm, H. A. (1989) *Path.Annual* **24,** 115-159
49. Teti, A., Volleth, G., Carano, A., and Zambonin-Zallone, A. (1988) *Calcif.Tissue Int.* **42,** 302-308
50. Abu-Amer, Y., Teitelbaum, S. L., Schlesinger, P., and Ross, F. P. (1999) *J.Bone Miner.Res.* **14,** 1855-1860
51. Karsenty, G. (1999) *Genes Dev.* **13,** 3037-3051
52. Lian, J. B. and Stein, G. S. (2001) Osteoblast Biology. In Marcus, R., Feldman, D., and Kelsey, J., editors. *Osteoporosis* , Academic Press, San Diego
53. Rodan, G. A. and Martin, T. J. (1981) *Calcif.Tissue Int.* **33,** 349-351
54. Suda, T., Takahashi, N., and Martin, T. J. (1992) *Endocr.Revs.* **13,** 66-80
55. Udagawa, N., Takahashi, N., Akatsu, T., Sasaki, T., Yamaguchi, A., Kodama, H., Martin, T. J., and Suda, T. (1989) *Endocrinol.* **125,** 1805-1813
56. Udagawa, N., Takahashi, N., Akatsu, T., Tanaka, H., Sasaki, T., Nishihara, T., and Suda, T. (1990) *Proc.Natl.Acad.Sci.U.S.A.* **87,** 7260-7264
57. Lacey, D. L., Timms, E., Tan, H. L., Kelley, M. J., Dunstan, C. R., Burgess, T., Elliott, R., Colombero, A., Elliott, G., Scully, S., Hsu, H., Sullivan, J, Hawkins, N., Davy, E., Capparelli, C., Eli, A., Qian, Y. X., Kaufman, S., Sarosi, I, Shalhoub, V., Senaldi, G., Guo, J., Delaney, J., and Boyle, W. J. (1998) *Cell* **93,** 165-176
58. Hofbauer, L. C., Khosla, S., Dunstan, C. R., Lacey, D. L., Boyle, W. J., and Riggs, B. L. (2000) *J.Bone Miner.Res.* **15,** 2-12
59. Ross, F. P. (2000) *J.Clin.Invest.* **105,** 555-558
60. Hofbauer, L. C. (1999) *European Journal of Endocrinology* **141,** 195-210
61. Kitazawa, R., Kitazawa, S., and Maeda, S. (1999) *Biochim.Biophys.Acta* **1445,** 134-141
62. Hofbauer, L. C., Lacey, D. L., Dunstan, C. R., Spelsberg, T. C., Riggs, B. L., and Khosla, S. (1999) *Bone* **25,** 255-259
63. Lee, S. K. and Lorenzo, J. A. (1999) *Endocrinol.* **140,** 3552-3561
64. Takai, H., Kanematsu, M., Yano, K., Tsuda, E., Higashio, K., Ikeda, K., Watanabe, K., and Yamada, Y. (1998) *J.Biol.Chem.* **273,** 27091-27096
65. Gowen, M., Wood, D. D., Ihrie, E. J., McGuire, M. K., and Russell, R. G. (1983) *Nature* **306,** 378-380
66. Gowen, M., Wood, D. D., and Russell, R. G. (1985) *J.Clin.Invest.* **75,** 1223-1229

67. Jilka, R. L., Hangoc, G., Girasole, G., Passeri, G., Williams, D. C., Abrams, J. S., Boyce, B., Broxmeyer, H., and Manolagas, S. C. (1992) *Science* **257,** 88-91
68. Bertolini, D. R., Nedwin, G. E., Bringman, T. S., Smith, D. D., and Mundy, G. R. (1986) *Nature* **319,** 516-518
69. Thomson, B. M., Mundy, G. R., and Chambers, T. J. (1987) *J.Immunol.* **138,** 775-779
70. Manolagas, S. C. and Jilka, R. L. (1995) *N.Engl.J.Med.* **332,** 305-311
71. Pacifici, R. (1996) *J.Bone Miner.Res.* **11,** 1043-1051
72. Hofbauer, L. C., Khosla, S., Dunstan, C. R., Lacey, D. L., Spelsberg, TC , and Riggs, B. L. (1999) *Endocrinol.* **140,** 4367-4370
73. Eghbali-Fatourechi, G., Khosla, S., Sanyal, A., Boyle, W. J., Lacey, D. L., and Riggs, B. L. (2003) *J.Clin.Invest.* **111,** 1221-1230
74. Shevde, N. K., Bendixen, A. C., Dienger, K. M., and Pike, J. W. (2000) *Proc.Natl.Acad.Sci.U.S.A.* **97,** 7829-7834
75. Blair, H. C., Teitelbaum, S. L., Grosso, L. E., Lacey, D. L., Tan, H.-L., McCort, D. W., and Jeffrey, J. J. (1993) *Biochem.J.* **290,** 873-884
76. Tezuka, K., Sato, T., Kamioka, H., Nijweide, P. J., Tanaka, K., Matsuo, T., Ohta, M., Kurihara, N., Hakeda, Y., and Kumegawa, M. (1992) *Biochem.Biophys.Res.Commun.* **186,** 911-917
77. Hefley, T. and Stern, P. (1982) *Calcif.Tissue Int.* **34,** 480-487
78. Zimolo, Z., Wesolowski, G., Tanaka, H., Hyman, J. L., Hoyer, J. R., and Rodan, G. A. (1994) *Am.J.Physiol.* **266,** C376-C381
79. Darnay, B. G., Ni, J., Moore, P. A., and Aggarwal, B. B. (1999) *J.Biol.Chem.* **274,** 7724-7731
80. Vaananen, H. K., Zhao, H., Mulari, M., and Halleen, J. M. (2000) *J.Cell Sci.* **113,** 377-381
81. Lakkakorpi, P. T., Horton, M. A., Helfrich, M. H., Karhukorpi, E.-K., and Vaananen, H. K. (1991) *J.Cell Biol.* **115,** 1179-1186
82. Masarachia, P., Yamamoto, M., Leu, C. T., Rodan, G., and Duong, L. (1998) *Endocrinol.* **139,** 1401-1410
83. Fassler, R. and Meyer, M. (1995) *Genes Dev.* **9,** 1896-1908
84. Baron, R. (2003) General principle of bone biology. In Favus, M. J., editor. *Primer on the metabolic bone diseases and disorders of mineral metabolism*, ASBMR, Washington DC
85. Gokhale, J. A., Robey, P. G., and Boskey, A. L. (2001) The biochemistro fo bone. In Marcus, R. , Feldman, D., and Kelsey, J., editors. *Osteoporosis*, Academic Press, San Diego
86. Boskey, A. L. (1989) *Bone Miner.* **6,** 111-123
87. Blair, H. C., Kahn, A. J., Crouch, E. C., Jeffrey, J. J., and Teitelbaum, S. L. (1986) *J.Cell Biol.* **102,** 1164-1172
88. Silver, I. A., Murrills, R. J., and Etherington, D. J. (1988) *Exp.Cell Res.* **175,** 266-276
89. Blair, H. C., Teitelbaum, S. L., Ghiselli, R., and Gluck, S. (1989) *Science* **245,** 855-857

90. Chatterjee, D., Chakraborty, M., Leit, M., Neff, L., Jamsa-Kellokumpu, S., Fuchs, R., and Baron, R. (1992) *Proc.Natl.Acad.Sci.U.S.A.* **89,** 6257-6261
91. Gay, C. V. and Mueller, W. J. (1974) *Science* **183,** 432-434
92. Blair, H. C., Teitelbaum, S. L., Tan, H.-L., Koziol, C. M., and Schlesinger, P. H. (1991) *Am.J.Physiol.* **260,** C1315-C1324
93. Schlesinger, P. H., Blair, H. C., Teitelbaum, S. L., and Edwards, J. C. (1997) *J.Biol.Chem.* **272,** 18636-18643
94. Teti, A., Blair, H. C., Schlesinger, P., Grano, M., Zambonin-Zallone, A., Kahn, A. J., Teitelbaum, S. L., and Hruska, K. A. (1989) *J.Clin.Invest.* **84,** 773-780
95. Tezuka, K., Tezuka, Y., Maejima, A., Sato, T., Nemoto, K., Kamioka, H., Hakeda, Y., and Kumegawa, M. (1994) *J.Biol.Chem.* **269,** 1106-1109
96. Inaoka, T., Bilbe, G., Ishibashi, O., Tezuka, K., Kumegawa, M., and Kokubo, T. (1995) *Biochem.Biophys.Res.Commun.* **206,** 89-96
97. Li, Y. P., Alexander, M., Wucherpfennig, A. L., Yelick, P., Chen, W., and Stashenko, P. (1995) *J.Bone Miner.Res.* **10,** 1197-1202
98. Bossard, M. J., Tomaszek, T. A., Thompson, S. K., Amegadzie, B. Y., Hanning, C. R., Jones, C., Kurdyla, J. T., McNulty, D. E., Drake, F. H., Gowen, M., and Levy, M. A. (1996) *J.Biol.Chem.* **271,** 12517-12524
99. Drake, F. H., Dodds, R. A., James, I. E., Connor, J. R., Debouck, C., Richardson, S., Leerykaczewski, E., Coleman, L., Rieman, D., Barthlow, R., Hastings, G., and Gowen, M. (1996) *J.Biol.Chem.* **271,** 12511-12516
100. Delaisse, J. M., Ledent, P., and Vaes, G. (1991) *Biochem.J.* **279,** 167-174
101. Gelb, B. D., Shi, G. P., Chapman, H. A., and Desnick, R. J. (1996) *Science* **273,** 1236-1238
102. Green, D. M. (1990) *J Acoust Soc Am* **87,** 2662-2674
103. Vu, T. H., Shipley, J. M., Bergers, G., Berger, J. E., Helms, J. A., Hanahan, D., Shapiro, SD, Senior, R. M., and Werb, Z. (1998) *Cell* **93,** 411-422
104. Baron, R., Neff, L., Brown, W., Courtoy, P. J., Louvard, D., and Farquhar, M. (1988) *J.Cell Biol.* **106,** 1863-1872
105. Nesbitt, S. A. and Horton, M. A. (1997) *Science* **276,** 266-269
106. Salo, J., Lehenkari, P., Mulari, M., Metsikko, K., and Vaananen, H. K. (1997) *Science* **276,** 270-273

CHAPTER 5

INTERCELLULAR COMMUNICATION OF OSTEOBLAST AND OSTEOCLAST IN BONE DISEASES

Jiake Xu, Tony C.A. Phan and Ming H. Zheng

Molecular Orthopaedic Laboratory
School of Surgery and Pathology
The University of Western Australia
QE II Medical Centre, M Block
Nedlands, Western Australia, 6009 Australia
Tel: 618 9346 4051
Fax: 618 9346 3210
Email: jiakexu@cyllene.uwa.edu.au

Bone is a living tissue and is maintained by the coordinate action of osteoblasts and osteoclasts. The intercellular communication between these two cells is the quintessential mechanism in bone remodelling. The importance of this interaction is increasingly evident as disruption of this "cross-talk" results in debilitating bone diseases. It has been well established that osteoblasts can regulate osteoclast formation and activation via the production of osteoclastogenic factors. There are now several emerging investigations that have identified novel osteoclast-derived factors that have the potential to control osteoblastic growth and function. This chapter highlights the intercellular communication between osteoblasts and osteoclasts and the impact of this "cross-talk" in bone diseases, including giant cell tumour of bone and other osteolytic bone tumours, osteoporosis, osteopetrosis, osteogenesis imperfecta, Paget's disease, periodontitis, osteoarthritis and aseptic loosening.

1. Introduction

Bone is a crucial tissue that provides internal skeletal support for every organ as well as forming and structuring the entire human frame. In addition, bone is the home for the formation of haematopoietic cells and the regulation of blood calcium. Due to its importance in the human body, bone needs to be continuously replenished in order to maintain its strength and structural integrity. This replenishment, also known as bone remodelling, is controlled by two equal, but opposing, forces: bone formation by osteoblasts and bone destruction or resorption by osteoclasts. Intimate communication between these cells is an integral element in maintaining bone homeostasis. Despite the vigorous regulation and control of bone equilibrium, changes in remodelling can occur, either by defects in the cell or obstruction of the intercellular communication between the cells, which leads to debilitating bone diseases such as osteoporosis, osteopetrosis, osteogenesis imperfecta and even Paget's Disease. The treatment of these diseases can cost billions of dollars a year, and diseases such as osteoporosis have a relatively high incidence rate.

This chapter focuses on the biology of osteoblasts and osteoclasts, the intercellular communication between these two cells and the role of this "cross-talk" in bone diseases.Osteoblasts and Bone Formation

2. Intercellular Communication Between Osteoblasts and Osteoclasts

2.1. Osteoblasts regulate the differentiation and function of osteoclasts

It has been well established that osteoblasts plays a pivotal role in the growth and differentiation of osteoclasts by producing several factors that directly bind to osteoclastic precursors (1). Until recently, it was unclear how osteoclast differentiation was controlled. In fact, it was not even certain whether osteoblasts play an important role in osteoclastogenesis. However, Yoshida et al (2) partially solved this enigma by examining osteoclastogenesis through their osteopetrotic mice model (3). Osteoblastic cells, extracted from these osteopetrotic mice, could not induce osteoclastogenesis when cultured together with normal osteoclast precursor cells *in vitro*. However osteoclast precursor cells extracted from the same mice could generate into osteoclasts when exposed to normal osteoblastic cells. This reinforced the importance of osteoblasts in osteoclastogenesis. Furthermore, the osteopetrotic mice model allowed the characterisation of a paracrine factor secreted by osteoblasts, known as macrophage colony stimulating factor (M-CSF), that would bind to its receptor, *c-fms*, on osteoclastic precursors (4). Their experiments demonstrated osteopetrotic mice that lacked M-CSF had very few osteoclasts. Interestingly, the osteopetrotic phenotype was reduced when Bcl2, an anti-apoptotic gene, was over-expressed in these mice, indicating that M-CSF may regulate the survival of osteoclastic precursors (5). Although the importance of M-CSF in bone remodelling can not be denied, this factor, alone, can not induce osteoclastogenesis. Therefore, the search was on for a potent "hypothetical" molecule, expressed by osteoblasts, which could directly bind to osteoclastic precursors and directly differentiate them into mature bone resorbing osteoclasts.

Unfortunately, a different molecule was discovered, one that inhibited osteoclastogenesis. Osteoprotegerin, OPG, also known as osteoclast inhibitory factor (OCIF), is a secreted protein that is part of the TNF superfamily (6,7). It is expressed by a wide range of cells including those in the heart, kidney, lung and especially by osteoblastic cells (6). Studies revealed that overexpression of OPG induces severe osteopetrosis in transgenic mice (6) with completely absent osteoclasts, while mice deficient in OPG had osteoporosis (8,9). Furthermore, OPG strongly inhibited osteoclast formation induced by vitamin D3 and parathyroid hormone in vitro (6,10). Interestingly, when OPG was added to osteoblast-osteoclast co-culture, osteoclast formation was further inhibited (11). These reports demonstrated that OPG is a potent negative regulator of osteoclastogenesis. To further characterise OPG, researchers looked towards its receptor. This search for the receptor came to an end when Lacey et al (12) discovered a molecule called receptor activator of NF-kB ligand (RANKL) that could bind to OPG (13). Further characterisation of this molecule revealed very surprising results. RANKL, also known as osteoprotegerin ligand (OPGL) and TNF related activation induced cytokine (TRANCE), is a part of the TNF superfamily of ligands. RANKL is expressed in a wide range of tissues, including the skin, kidney and heart, although the function of RANKL in these tissues is unknown (14). Structurally, RANKL is smaller then OPG and naturally exists as a transmembrane protein on the surface of osteoblastic cells, although it can be cleaved to a secreted form by MMPs (15). RANKL, unlike OPG, is a potent, positive regulator of osteoclastogenesis, by binding to its own receptor, RANK (16). For instance, soluble recombinant RANKL, together with M-CSF, can directly induce the formation of osteoclasts in vitro, even without the presence of osteoblasts and 1,25$[OH]_2$ vitamin D3 (17,18). This function was further reinforced by injection of soluble recombinant RANKL into mice, which caused osteoporosis due to the increased activation of osteoclastogenesis (19). On the other hand, mice deficient in RANKL revealed an osteopetrotic phenotype and no osteoclasts (20). Thus, RANKL was the so-called potent "hypothetical" molecule that could bind directly to osteoclastic precursors and induce their differentiation. These finding revitalised bone biology. The discovery of the OPG-RANK-RANKL axis defined the hallmark mechanism of osteoclastogenesis. RANKL, expressed on osteoblasts, forces the cell to physically interact with osteoclastic precursors in order for RANKL to bind to its receptor RANK. To negatively regulate this mechanism, OPG is expressed by osteoblasts and acts as a decoy receptor to compete with RANK for RANKL. It is important to note that the common theme for OPG- and RANKL-mediated osteoclastogenesis is that osteoblasts directly regulate, both positively and negatively.

The interaction of RANKL with RANK activates a deluge of signal transduction pathways that allow the commitment of osteoclastic precursors to mature bone-resorbing osteoclasts. The interaction of RANKL and RANK induces the recruitment of the TNF receptor activating factor (TRAF) family. Investigations have demonstrated that several TRAF molecules, especially TRAF 6, are recruited and interact with the C-terminal of the RANK receptor, as well as its cytoplasmic region (21,22). Kadono et al (23) discovered that TRAF6, compared with TRAF 1-3, can induce osteoclastogenesis independently of other TRAFs and therefore is essential for osteoclastogenesis (24). Furthermore, mice deficient in TRAF6 have severe osteopetrosis (25,26). The recruitment of TRAF to RANK induced the activation of the NF-kB and mitogen-activated protein kinase (MAPK) pathway. Both pathways are essential for cellular survival, growth and apoptosis, and especially the induction of osteoclast differentiation (27). Darnay et al (12) have revealed that over-expression of RANK stimulates the NF-kB and JNK (regulator of MAPK) pathway (21). Interestingly, other studies have further demonstrated that RANK can induce activation of the Erk and Akt pathway (28).

Several other molecules, expressed by osteoblasts, can also facilitate the regulation of osteoclastogenesis via the RANKL-RANK pathway. TNFα and TNFβ play an important role in osteoclastogenesis. Early investigations demonstrated that TNF can induce DNA synthesis of osteoclasts (29). Further studies reported that TNFα, and to a lesser extent TNFβ, increased bone resorption when bone explants were treated with TNF (29-31). *In vivo* studies also reported that the TNF family was the major cytokine constituent and the blockage of this family resulted in lower bone resorption (32,33). Johnson et al (34) discovered that cells over-expressed with TNFα when injected into nude mice increased osteoclast number, as well as bone resorption (35). TNFα has also been shown to induce the expression of the RANK receptor on osteoclasts, leading to higher activation of RANKL (36). Interestingly, the binding of TNFα to its receptor, TNFR1, produces a signal transduction cascade that mirrors that of the RANK-RANKL interaction. For instance, TNF instigates the recruitment of TRAF molecules to TNFR1, which leads to the activation of the NF-kB pathway and to a lesser extent, the MAPK pathway (37). TNFα, together with 1,25$[OH]_2$ vitamin D3 and stromal osteoblasts, has also been demonstrated to directly mediate osteoclastogenesis in bone marrow culture (38). Interestingly, recent studies revealed that mice deficient in both TNFα and TNFβ genes, does not produce an osteopetrotic phenotype (37). This indicates that TNFα and TNFβ are involved in modulating osteoclastogenesis rather than directly controlling it.

Interleukin-1 (IL-1) is another possible regulator of osteoclastogenesis produced by osteoblasts. The effect of IL-1 was recognised when estrogen deficient mice models were used. It was demonstrated that removal of estrogen from mice elevated the number of the number of osteoclasts, which coincided with an increase in IL-1 activity (27,39). Conversely, when estrogen was augmented it lowered osteoclast number by inhibiting IL-1 (27,39). It has been postulated that estrogen affects the IL-1 receptor, which indirectly leads to an affect on IL1 (40). Investigations have revealed that IL-1 can induce osteoclast differentiation in vitro (41), through a special domain present on the cytokine called gp130. Interestingly, despite gp130-deficient mice experiencing mild symptoms of osteopetrosis, the average number of osteoclastic cells was still high (42), indicating that like TNF, interleukins are not obligatory for osteoclastogenesis but are simply involved in modulation. Reinforcing this study is the finding that specific deletion of IL-1 does not produce severe osteopetrotic phenotype (43).

The control of osteoclastogenesis by osteoblastic cells mentioned above, indirectly leads to the modulation of bone resorption. For instance studies have discovered osteoclastic bone resorption pits increase in the presence of osteoblasts due to the maintenance of osteoclast survival and induction of osteoclastogenesis by the RANKL-RANK-OPG axis (44,45). Furthermore, administration of RANKL to mice leads to heightened bone resorption due to increased activation of osteoclastogenesis (19). However, there are several molecules expressed by the osteoblast, which bind directly to mature osteoclasts and either inhibit or enhance bone resorption. M-CSF, described above as a modulator of osteoclastogenesis, has been found to have an effect on bone resorption. For instance, osteopetrotic mice that lack M-CSF also have impaired bone resorption (4). Although there is increasing evidence that osteoblasts modulate bone resorption, a recent study has shown that bone resorption continues even after complete abolishment of bone formation (46). However, it must be noted that in the presence of osteoblasts, bone resorption does increase, indicating that although osteoblasts are not obligatory for bone resorption, they do modulate the process.

2.2. Can osteoclasts control the differentiation and function of osteoblasts?

It is well established that bone resorption and formation is a coupled process: If one increases, so does the other. Logically, this leads to the conclusion that if both cells are coupled it must mean their mechanism of control is intertwined. The preceding section has revealed a snippet of the large plethora of studies into the control of osteoclastic growth, differentiation and function by osteoblast-derived molecules (Figure 1). However, when the reverse case is examined, only a small handful of reports can be found in the current literature. Despite this apparent paradox, recent research has uncovered some osteoclast-derived molecules that may play an important role in the control of osteoblastic growth, differentiation and function.

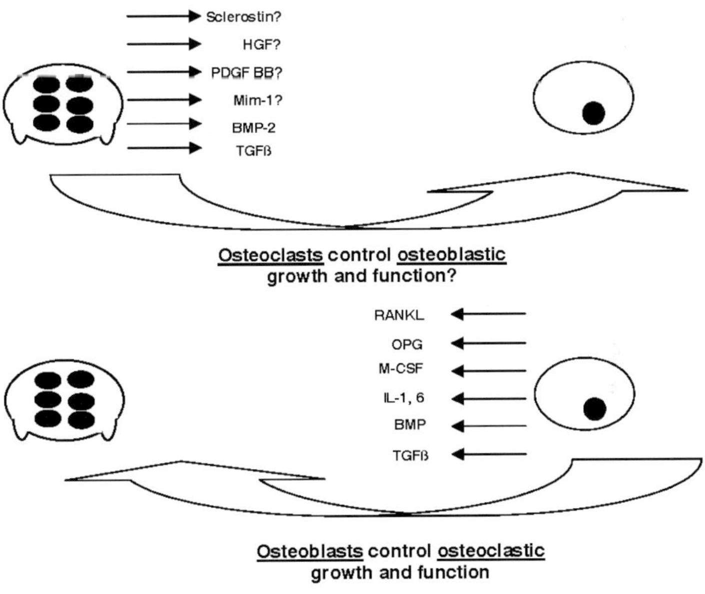

Figure 1. Schematic representation of the proteins involved in the intercellular communication between osteoblasts and osteoclasts. A plethora of studies have been reported that the osteoblast can secrete molecules that can bind to osteoclasts and control their function and growth. The reverse case is still in question. Proteins with a question mark indicate that further study will need to be done in order to elucidate their function.

Hepatocyte growth factor (HGF) has recently been described as a potential paracrine regulator of osteoblastic growth, differentiation and function. HGF is a heterodimeric protein that was originally discovered in nonparachymal hepatocytic cells (47,48) and has been well characterised as an inducer of cell division, motility and as a potent morphogen. Recently, HGF was identified in cells of the bone lineage, namely osteoblasts and osteoclasts (49). In this study, active HGF proteins, expressed by osteoclasts, were found to have both a paracrine and an autocrine mode of action, due to the presence of the HGF receptor for both cell types. Functionally, HGF was able to influence DNA synthesis and cellular proliferation in osteoblasts and osteoclasts (49). The paracrine function of HGF was further reinforced when D'lppolito et al (50) demonstrated that HGF, together with 1,25$[OH]_2$ vitamin D3, could induce the proliferation and differentiation of bone marrow stromal cells (51). Furthermore, the treatment of both HGF and 1,25$[OH]_2$ vitamin D3 in bone marrow stromal cells elevated bone mineralisation but not with 1,25$[OH]_2$ vitamin D3 or HGF alone. Together these observations demonstrate that HGF, expressed by osteoclasts, could specifically be used to control osteoblastic growth and differentiation. Although the results for HGF are promising there are several conflicting studies showing that HGF may not be as important in osteoclast-derived control of osteoblastic growth as first thought. Taichman et al (23) recently discovered that specific stromal osteoblast-like cells could secrete HGF (52). This indicates that the original affect of HGF on osteoblasts (49) may be due to HGF secreted by surrounding stromal osteoblastic precursor cells, rather than the osteoclast alone. If this is true, HGF may act mainly through an autocrine regulation. Furthermore, HGF seems to have a more potent affect on osteoclastic cells, especially since it can direct changes in morphology and motility in osteoclastic cells, as well as the ability to induce intracellular calcium elevation (49). Its dual affect on osteoclasts and osteoblasts may render its effectiveness as a therapeutic treatment for bone disease useless, since administration of HGF to osteoporotic patients may further increase bone resorption. To unequivocally confirm the function of HGF, *in vivo* work will need to be examined.

Sclerostin is another, recently characterised osteoclast-derived molecule that may have a specific paracrine mode of action on osteoblastic cells. A mutation within the *sclerostin* (SOST) gene, which codes for the Sclerostin protein, has been attributed to a type of bone disease Sclerosteosis (23,53). This disease is an autosomal recessive, inherited disease characterised by abnormal hardening and thickening of developmental bone tissue (sclerosis dysplasia). Recently, Kusu et al (54) identified a novel protein that is expressed by osteoclastic cells that acts as a negative regulator of osteoblast differentiation (55). In this study, sclerostin was expressed within intramembranous and endochondral bones in the mouse embryo skeleton, and colocalises within cells expressing MMP-9, a marker for osteoclasts. Due to the presence of a secretory signal on sclerostin, and the secretion of the protein from insect cells, it was characterised as an osteoclast-derived molecule (55). Furthermore, sclerostin was demonstrated to inhibit alkaline phosphatase activity (marker for osteoblastogenesis) by binding to BMP-6 and BMP-7. Unfortunately, the results of this study were questioned when Winkler et al (54), discovered that sclerostin is expressed by cells in the osteoblast lineage, especially osteoblasts and osteocytes (56). In fact, they further showed that sclerostin is not expressed in the human osteoclast. Therefore, in this study sclerostin acts as an autocrine factor that is secreted by osteoblasts, and negatively regulates osteoblastogenesis by binding to BMPs. The conflicting results produced by these two laboratories could be due to different methodology. Kusu et al (54) used in situ hybridisation and co-localisation with MMP-9 to characterise the expression of sclerostin in osteoclasts (55). On the other hand Winkler et al (54) used specific reverse transcriptase polymerase chain reaction (RT-PCR) and examined a range of osteoclast-like and osteoblast-like cells, including specific human osteoclasts and osteoblasts. These conflicting observations, similar to the ones associated with hepatocyte growth factor, suggest that more *in vivo* work will need to be done in order to unequivocally confirm the function of this protein.

Platelet-Derived Growth Factor BB (PDGF BB) has recently been demonstrated as a potential paracrine factor that may control osteoblastogenesis. The PDGF molecule exists as a dimer composed of a variation of two subunits, A and B. PDGF AA, AB and BB have been well characterised in bone remodelling, and have been shown to have a mitogenic affect, mainly on osteoblasts and osteoblast-like cells (57-59). PDGF is expressed by platelets of the blood and osteoblastic cells. Recently, Kubota et al 2002 recently demonstrated that the PDGF BB isoform is expressed by osteoclastic cells and has an inhibitory affect on osteoblast growth and differentiation (60). In this study, PDGF BB was expressed and secreted by $RAW_{264.7}$-derived osteoclastic cells, a macrophage cell line differentiated into osteoclasts by soluble RANKL. In addition, PDGF BB was demonstrated to inhibit alkaline phosphatase activity in MC3T3 cells, a cell line that is committed to the osteoblastic lineage. To confirm the paracrine function of PDGF BB, antibodies to the growth factor were used in order to neutralise the activity of PDGF BB. The antibodies neutralised over 83% of the PDGF BB activity (60), suggesting that PDGF BB did have an inhibitory affect on the MC3T3 cells. Although these results are promising, further investigations will need to be ascertained taking into account cells that are closer to the osteoclast and osteoblast lineage, as well as *in vivo* work. Recent reports have demonstrated that PDGF BB may influence the motility and attachment of osteoblastic cells to the bone surface (61). Although PDGF has already been implicated in cell motility (59,62), this would be the first protein derived from osteoclasts that affects osteoblast motility. It would be interesting to see whether other studies can support this observation. It is important to note that in vivo work is essential for this type of study. Since PDGF BB has been ascertained to affect both osteoblasts and osteoclasts (57), it is possible that therapeutic administration of this growth factor in osteoporotic patients may further increase bone resorption.

Apart from the above studies mentioned, several other reports have provided evidence for the intimate communication between osteoclasts and osteoblasts, and further demonstrated that osteoclasts can control osteoblastic growth and function. Transgenic overexpression of cathepsin K resulted in elevation of osteoblast cell number and mineralising surfaces as well as an increase in bone resorption (63). The increase in bone resorption was expected, as cathepsin K is crucial for dissolution of the organic matrix. However, what was not expected was an increase in the number of osteoblastic cells and mineralising surface. It has been suggested that the increase in bone resorption from the overexpression of cathepsin K lead to the induction of several pro-osteoblastic growth factors, such as BMPs and RANKL (63). Recently, Falany et al (23) discovered that osteoclasts secrete a chemokine called myeloid protein-1 precursor (Mim-1) that can induce the proliferation of osteoblastic precursor cells (64). Although it is not known whether Mim-1 can directly bind to osteoblasts or is simply antagonising osteoblast-specific molecules, it will be interesting to see follow up studies of this cytokine.

Despite these recent discoveries into the control of osteoblasts by osteoclasts, the specific mechanism remains to be elucidated. It is noteworthy to mention that most of the proteins mentioned above have an autocrine function in addition to their proposed paracrine effect. In other words, the protein has an affect on osteoclastic cells, as well as osteoblasts. Care must be taken when using these proteins as possible anabolic treatments for bone disease, such as osteoporosis. Administration of these drugs may increase bone resorption in osteoporotic patients, further increasing the severity of the disease. For instance, parathyroid hormone (PTH) can induce bone resorption when continuously administered to the patient, while PTH increases bone formation if an intermittent dose is given (65). Presently, there is no protein that is expressed by the osteoclast that has a specific paracrine function for osteoblastic cells.

3. The Role of the Osteoclast-Osteoblast Interaction in Bone Diseases

3.1. Giant cell tumour of bone and other osteolytic bone tumors

Many tumors including osseous and cartilage tumors display pathological bone destruction. The osteolytic condition was caused by excessive osteoclast formation induced by stromal cells or osteoblastic cells. In the case of giant cell tumor of bone (GCT), the interaction of osteoblasts and osteoclasts is likely to play an important role in bone destruction. GCT is characterized by abundant multinuclear osteoclast-like giant cells scattered among mononuclear cells. It is generally believed that the osteoblastic stromal-like cells have the ability to recruit circulating monocytes to become multinuclear osteoclast-like giant cells in GCT (66,67). We have found that stromal-like cells of GCT express RANKL whereas macrophage-like cells and osteoclast-like multinuclear giant cells express RANK. OPG, the decoy receptor for RANKL, was also ubiquitously expressed in the stromal-like cells, indicating that a negative feedback loop may exist in which the stromal cells of GCT themselves may modulate the presentation of RANKL molecules on their surface, which in turn can be inhibited by OPG (67). Thus the ratio of RANKL and OPG gene expression in stromal-like cells may determine local osteoclastogenesis and osteoclastic bone resorption (67). In addition, GCT stromal-like cells also produce a number of cytokines such as VEGF, TGF-β and MCP-1 to regulate osteoclastogenesis (66). Recent studies have established that GCT stromal cells may have an osteoblastic lineage expressing Cbfa-1 and Osterix genes and retain the ability to differentiate into osteoblasts (68). These studies indicates that interaction of osteoblastic cells and osteoclastic like cells might contribute to the osteolytic lesions in GCT. This process is illustrated in Fig. 2.

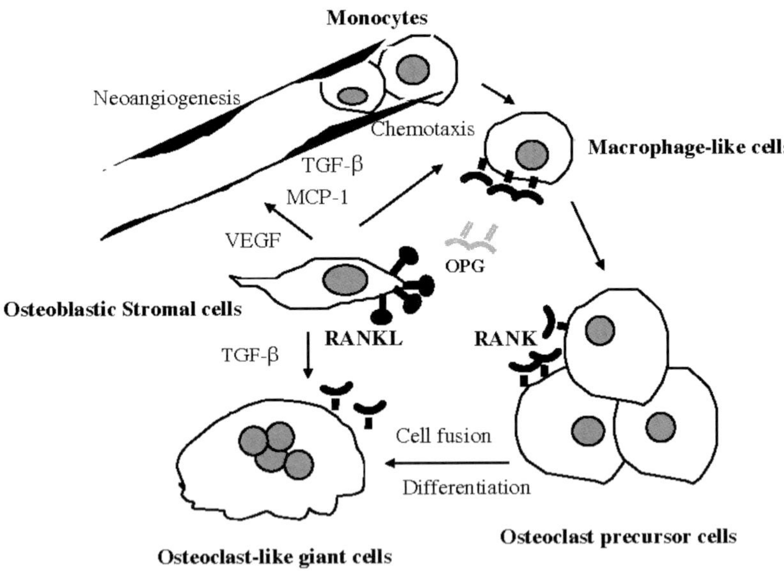

Figure 2. A model of the interaction of osteoblastic stromal cells and osteoclasts in Giant cell tumour of bone. Osteoblastic stromal cells of GCT express RANKL, whereas macrophage-like cells and multinuclear osteoclast-like giant cells express the receptor for RANKL, the RANK, and OPG, the decoy receptor for RANKL. It appears that the ratio of RANKL/OPG produced by osteoblastic stromal cells may contribute to the degree of osteoclast-like giant cell formation and bone destruction in GCT.

3.2. Osteoporosis

Osteoporosis is regarded as a systemic, skeletal disorder, mostly characterised by low bone mass and high susceptibility to fractures. Despite intense research, the molecular mechanism and etiology of osteoporosis is still an enigma. Fortunately, there are several universal themes that seem to be constant in osteoporosis. Firstly, the disease is principally defined by an imbalance of bone remodelling favouring osteoclastic bone resorption. Secondly, osteoporosis is commonly associated with postmenopausal women, and to a lesser extent elderly men. For instance, a third of all men and half of all women over the age of 60 suffer from fractures due to osteoporosis. The main reason for the high incidence in postmenopausal women is the change in hormonal balance associated with menopause. Estrogen, a potent endocrine hormone linked to osteoporosis, has been demonstrated to bind to osteoblastic cells and suppress the expression of several paracrine pro-osteoclast factors, such as IL-1, IL-6 and TNFα (27,69). Moreover, estrogen can further increase OPG (70,71) and inhibit RANKL expression (72,73). Therefore, since menopause is associated with a decreased level of estrogen, the above hormonal action is reversed, leading to an increase in osteoclast formation, and thus bone resorption. For example, a study demonstrated that estrogen could inhibit the transcription factor Egr-1, which is involved in increasing production of M-CSF (69,74). In this case, the menopausal state increases Egr-1 activation leading to higher M-CSF production, eventually leading to elevated bone resorption. Studies have now shown that osteoporotic (ovariectomized) mice that are deficient in IL-1, IL-6, M-CSF, TNFα or combinations of these cytokines recover from osteoporosis (43,69). Thus, estrogen-mediated bone resorption is accomplished via these osteoblastic-derived cytokines. Interestingly, all of these molecules are expressed and secreted by osteoblasts, which bind to osteoclasts. This indicates that estrogen indirectly affects the osteoclast by affecting the intercellular communication between the two cell types. This theory has one fundamental flaw: absence of osteoblast-osteoclast coupling. If osteoclastic bone resorption is increased, osteoblastic bone formation should be further activated to counteract the heightened level of resorption. Obviously this is not seen in osteoporotic patients. It is possible that estrogen withdrawal may further have an effect on osteoblast function via the inhibition of pro-osteoblastic factors expressed by osteoclasts. Clearly, researchers still do not know the true effect of estrogen. Furthermore, it is unlikely that our understanding of this disease will ever become clear if research focuses on osteoclastic bone resorption and production of pro-osteoclastic factors by osteoblasts, rather than bone formation and production of pro-osteoblastic factors from osteoclasts.

3.3. Osteopetrosis

Osteopetrosis is the general name given to a group of diseases where the rate of bone formation is higher than the rate of bone resorption. Due to the varying pathology, etiology and clinical features of these patients, osteopetrosis is categorised into specific disease groups. Carbonic Anhydrase II-deficient osteopetrosis has been largely established as a specific disease associated with osteoclast defects (75). Patients with this type of disease exhibit short stature, frequent fractures, developmental delays, and in severe cases, stillbirths (75). The abnormal bone formation in these patients can often block the respiratory path in the body, leading to severe lung complications. Infantile malignant osteopetrosis is one of the most severe forms of osteopetrosis, with most infants that inherit this disease dying before birth unless a bone marrow transplant can be performed (76). Like carbonic anhydrase II-deficient osteopetrosis, the infantile form can have severe complications with breathing, due to abnormally large bone formation. Infantile malignant osteopetrosis has been associated with a defect in osteoclastic bone resorption (76). For instance, osteoclastic precursor cells, extracted from patients with malignant osteopetrosis, can be differentiated into mature osteoclasts in the presence of osteoblastic cells (77). However, these osteoclasts lack the ability to resorb bone, indicating that lower bone resorption may be a common factor in osteopetrosis. Unfortunately, the disease gene(s) that causes malignant osteopetrosis has yet to be elucidated. Recent studies have demonstrated the possibility of osteopetrosis being caused exclusively by a defect in the intercellular communication between osteoblasts and osteoclasts, rather than attributing it solely to osteoclasts. Lajeunesse et al (78) extracted and cultured osteoblast-like cells from patients with osteopetrosis (79). These osteoblast cells had normal expression of alkaline phosphatase, but no expression of osteocalcin when compared to controls. Since osteocalcin is a marker for mature osteoblasts, the results indicate that osteopetrosis occurs due to a defect in the differentiation of early osteoblasts to mature osteoblasts, rather than differentiation of mesenchymal cells to early osteoblasts. The study further demonstrated that osteopetrotic-derived osteoblasts did not significantly induce M-CSF secretion. Inhibition of M-CSF causes the down-regulation of osteoclastogenesis, which eventually leads to lower bone resorption. Lajeunesse et al (78) then offered a unique twist in their study. They gave the osteopetrotic patients bone marrow transplants to reverse the osteopetrotic phenotype. Osteoblastic cells extracted from these patients revealed normal osteoblast activity as well as normal expression of alkaline phosphatase and osteocalcin. Furthermore, the production and secretion of M-CSF was also normal. The proposed hypothesis is that the original defect in osteopetrosis was the osteoclast, possibly a mutation in a gene that regulates the production of pro-osteoblastic factors from osteoclastic cells. This lead to the creation of defective osteoblastic precursors that were unable to mature into proper osteocalcin-producing osteoblasts. Replacement of the defective osteoclast precursors and osteoclast population with normal ones, by bone marrow transplant, corrected the defective intercellular communication between the two cells (79). The osteoblasts formed normally again and normal bone formation followed. Although more studies are required to confirm this report, it shows promise that osteoclasts do control the growth and function of osteoblastic cells.

3.4. Osteogenesis imperfecta

Osteogenesis Imperfecta is an inherited, usually autosomal dominant, bone disease caused by defects within the type 1 collagen matrix the most abundant protein in bone tissue. Due to these defects, patients with this disease are often characterised by lower bone mass, skeletal deformities and re-occurring fractures due to the fragility of the bone tissue (80). The clinical severity of osteogenesis imperfecta ranges from being asymptomatic to mild localised skeletal deformities to morbidity at birth. Due to the variability of the symptoms, patients with osteogenesis imperfecta are clinically categorised into four types. Type 1 is the least severe with infrequent fractures and low probability of having congenital skeletal deformities. Small proportions of children with this type of disease exhibit deafness during early adulthood, although they are physically normal. Type II increases in severity with a small proportion of infants being born prematurely or dying before birth. Those that do survive have shorter limbs, fragile skulls and mandible (dentinogenesis) and frequently have fractures. Type III is similar to type II, however, patients exhibit more signs of skeletal deformities and, due to defects in the growth plate, are usually shorter. Type IV patients frequently have hearing loss and fractures in the mandible and skull area. All four types further exhibit blue sclera or blue discolouration of the sclera within the eyeball. Despite all the varying signs and symptoms, all of these four groups are caused by mutations within the Type 1 collagen gene, which can result in either complete abolishment of collagen synthesis or structural deficiencies in the collagen protein (80). Recently, three new types of osteogenesis imperfecta have been described, and all of them are believed to be caused by defects in osteoblastic cells, and possibly by defects in the intercellular mechanism between osteoblasts and osteoclasts (81-83). Type V, VI and VII do not have mutations within type 1 collagen, although they share similar symptoms to other types of osteogenesis imperfecta (80). Histomorphometric studies in patients with Type V have been shown to have normal rates of bone formation and resorption in the general skeleton, compared to healthy control individuals (81). However, localised sites within the skeleton had abnormal bone formation activity, with lower rate of bone matrix deposition. Additionally, biochemical results show higher alkaline phosphatase activity but normal osteocalcin levels in these patients. This suggests that the defect may be in osteoblast differentiation, rather than bone formation directly (81). The same biochemistry and histomorphometric results are seen for Type VI and, to a lesser extent, Type VII (82-84). It is not unreasonable to hypothesise that there could also be defects within the intercellular communication between osteoblasts and osteoclasts, since disruption of osteoblast differentiation alone cannot produce the varying symptoms seen in osteogenesis imperfecta. Obviously more work will need to be done in order to characterise the pathology of Type V, VI and VII osteogenesis imperfecta. Furthermore, work is under way to scan families of cytokines and growth factors that may be involved in regulating osteoblastogenesis to determine whether these proteins may be involved in the pathogenesis of this disease (81). It will be interesting to see whether any novel osteoclast-derived osteoblastic proteins are discovered as a result of this work.

Paget's Disease

Paget's Disease is usually characterised by localised areas of abnormally large and expanded bone growth, usually weaker and more sensitive to fractures than other bones in the skeleton. The severity of Paget's disease also called osteitis deformans, ranges from asymptomatic to more severe bone deformity, and pain. The more severe patients often have secondary diseases, such as arthritis. The cause and etiology of Paget's disease is not known, although Paget's disease has been described as a specific disease of the osteoclast (85,86). The osteoclasts are larger than normal and contain many more nuclei. This, however, does not rule out the possibility of a defect in osteoblastic cells. For instance, one of the clinical features of Paget's disease is very high alkaline phosphatase activity, indicating unusually high osteoblast activity (85,86). However, it is the disruption of the osteoclast-osteoblast interaction that may be the defining cause of Paget's disease. The disease process occurs when signals are given to osteoclasts to increase bone resorption. Due to the coupling nature of osteoclasts and osteoblasts, the increase in bone degradation induces the activation and subsequent elevation of bone formation. The dual activation of these two forces, and the abnormality of the osteoclast, give rise to the large but weak bone growth seen by patients with this disease. The molecular mechanism of Paget's disease is not known. In fact, it is not even known whether this disease is inherited or occurs due to environmental factors, such as viral infections. Recent studies, however, are just beginning to unravel the molecular mechanism, and as expected it involves the osteoclast-osteoblast interaction (87). Unusually high RANKL mRNA transcripts have been discovered in osteoblastic-like cell lines in Pagetic lesions and marrow (88). In addition, osteoclastic precursors extracted from patients with Paget's disease are much more sensitive to RANKL stimulation compared to cells from normal patients (89). Interestingly, most patients with Paget's disease have significantly lower amounts of OPG, which inhibits osteoclastogenesis (87). Therefore, it is possible that a defect in the RANKL-OPG system results in abnormal osteoclast activity. The abnormal osteoclast may further secrete its own factors to increase osteoblastic activity, in order to compensate for the elevation in bone resorption. Interestingly, only a few patients have been reported to have a specific mutation in the RANKL gene (90). This may suggest that the specific defect or mutation may be in other genes related to the intercellular communication between osteoblasts and osteoclasts. For instance, studies have implicated p62 or sequestasome-1 in Paget's disease, an important signal transducer of NF-kB that is activated via the RANK-RANKL-TRAF interaction (91). Other investigations have implicated the role of IL-6, a potent cytokine expressed and secreted by the osteoblasts, in Paget's Disease. The abnormal activities of the osteoclast result in increased activity of IL-6 from the osteoblasts that bind to osteoclasts, which increases their activities further (92). In addition, reports have ascertained high levels of IL-6 in Pagetic lesions (92). The pathogenesis of Paget's disease still remains an enigma, despite all the recent research. However, it is possible that our understanding of this disease will become clearer once the focus of bone biology shifts towards the production of pro-osteoclastic factors by osteoblasts.

3.5. Periodontitis

Periodontitis, and to a lesser extent gingivitis, is the most common cause of tooth loss in human adults. Both diseases have a common etiology, with gingivitis often viewed as a pre-cursor stage during the evolution of periodontitis (93). Infection and accumulation of bacteria on the gingival tissues (gums) initiate the pathology of periodontitis. The progressive nature of this disease results in expansion of the bacterial infection from gums to ligaments and bone that support the teeth. The inflammatory response accompanies the spread of infection. Bacterial invasion of the supporting bone structure elevate osteoclastic bone resorption which reduces the rigidity and strength of bone, ultimately leading to tooth loss (93,94). Several investigations have identified the molecular players in periodontitis and as expected it involves the intercellular communication between osteoblasts and osteoclasts. Bacterial infection of bone induces the expression and secretion of several autocrine factors by osteoclasts, such as IL-1 and TNFα (95). Moreover, IL-1 has been demonstrated to bind osteoblasts and increase production of pro-osteoclastic molecules, such as monocyte chemoattractant protein-1 (MCP-1). Interestingly, MCP-1 does not directly activate osteoclasts, rather it stimulate and attract monocytes and pro-inflammatory leuokocytes to the site of infection, which results in further osteoclastic bone resorption (95,96). In this model the over-expression of pro-osteoclastic factors outweighs the growth and function of osteoblasts, leading to heightened levels of bone resorption. Interestingly, the inflammations of gingival tissue and bone have resulted in the identification of a potential link between periodontitis and rheumatoid arthritis (93). Furthermore, since RANKL expression is high in both diseases (87,97), several investigations have postulated the use of recombinant OPG as a therapeutic approach to treating these disorders. Currently the common preventative treatment for periodontitis is proper oral hygiene, while palliative treatments are performed by the specific removal of dental plaque from inflamed gums, which is often uncomfortable. The use of OPG to possibly treat periodontitis will open a new avenue of therapeutic approaches for this disease. However, more clinical research will be needed to ascertain whether OPG can be used in this manner.

3.6. Osteoarthritis

Osteoarthritis is a chronic joint disease that can lead to severe pain and sclerosis of the subchondral bone. Currently, there are no direct treatments for sufferers of this affliction, however, new insights into the pathology of osteoarthritis as well as rheumatoid arthritis, are revealing promising results. Osteoarthritis is initiated by inflammation of the joints and possibly by over activation of T-cells, although the latter mechanism is more commonly attributed to rheumatoid arthritis. Several key players have been discovered, most of which play an important role in osteoclast function. Recently, Clements et al (54) ascertained that MMP-3 knockout mice demonstrated accelerated articular cartilage breakdown compared to wild type (98). Additionally, it has been shown that MMP-3 plays an important role in the breakdown of aggregan, an important constituent of articular cartilage and bone (99,100). Several investigations have implicated an indirect role of RANKL mediated arthropathy and have suggested the utilisation of recombinant OPG as a therapeutic agent to inhibit osteoclast bone resorption in the joint area (87). More clinical investigations will be needed in order to ascertain whether OPG can be used in this manner.

3.7. Aseptic loosening

Aseptic loosening is a common condition that occurs due to the failure of the bone and implant interface, used in joint replacement therapy such as total hip arthoplasty. The loss of prosthetic implants cost several millions dollars a year due to the need for replacement. The current view is that aseptic loosening is due to the presence of wear particles and surrounding resorptive macrophages. Wear particles, consisting mainly of ultra-high polyethylene, lead to the activation of specific pro-osteoclastic cytokines from osteoblasts such as TNFα and IL-1. This results in an elevation of osteoclastic resorption. Several investigations have utilised bisphosphonates and BMP-2 to augment bone formation and prevent osteolysis-induced aseptic loosening (101,102). This is particularly novel since a considerable number of reports have implicated bisphosphonates in inhibiting osteoclastic bone resorption. It would be interesting to see whether more clinical trials can ascertain if increasing bone formation directly can alleviate aseptic loosening of prosthetic joint implants.

4. Conclusion

The intercellular communication between osteoblasts and osteoclasts is crucial to bone homeostasis. Unfortunately, the importance of this interaction is often overlooked when studying bone biology. In fact, the significance is only realised when there are changes or disruption to the intimate conversation between the two cells that results in severe, often life threatening, debilitating bone diseases. Due to the severity of these disorders and the high cost associated with treatment, medical research into the control of bone remodelling and disease is becoming increasingly important. The mechanism in which osteoblasts and osteoclasts communicate and coordinate bone remodelling has not been fully elucidated. However, it is well established that osteoblastic cells directly control the growth and function of osteoclastic cells by expressing specialised molecules that bind directly to osteoclasts (103-105). Moreover, most investigations and treatments in bone disease focus entirely on osteoclasts and bone destruction, even though the remodelling of bone is also attributed to osteoblasts and bone formation. In fact, it is surprising how little is known about how osteoclasts regulate osteoblastic growth and function. This discrepancy leads to the question: Can osteoclastic cells directly control the growth and function of osteoblastic cells by expressing specific proteins that bind directly to osteoblasts? If so, is it possible to use these proteins to control and, possibly, treat bone disease? An emphatic yes can be answered for these questions. The only road left to travel is to elucidate these novel osteoclast-derived factors.

References

1. Rodan GA, Martin TJ 1981 Role of osteoblasts in hormonal control of bone resorption--a hypothesis. Calcif Tissue Int 33(4):349-51.
2. Anakwe OO, Gerton GL 1990 Acrosome biogenesis begins during meiosis: evidence from the synthesis and distribution of an acrosomal glycoprotein, acrogranin, during guinea pig spermatogenesis. Biol Reprod 42(2):317-28.
3. Yoshida H, Hayashi S, Kunisada T, Ogawa M, Nishikawa S, Okamura H, Sudo T, Shultz LD 1990 The murine mutation osteopetrosis is in the coding region of the macrophage colony stimulating factor gene. Nature 345(6274):442-4.
4. Wiktor-Jedrzejczak W, Bartocci A, Ferrante AW, Jr., Ahmed-Ansari A, Sell KW, Pollard JW, Stanley ER 1990 Total absence of colony-stimulating factor 1 in the

macrophage-deficient osteopetrotic (op/op) mouse. Proc Natl Acad Sci U S A 87(12):4828-32.
5. Lagasse E, Weissman IL 1997 Enforced expression of Bcl-2 in monocytes rescues macrophages and partially reverses osteopetrosis in op/op mice. Cell 89(7):1021-31.
6. Simonet WS, Lacey DL, Dunstan CR, Kelley M, Chang MS, Luthy R, Nguyen HQ, Wooden S, Bennett L, Boone T, Shimamoto G, DeRose M, Elliott R, Colombero A, Tan HL, Trail G, Sullivan J, Davy E, Bucay N, Renshaw-Gegg L, Hughes TM, Hill D, Pattison W, Campbell P, Boyle WJ, et al. 1997 Osteoprotegerin: a novel secreted protein involved in the regulation of bone density. Cell 89(2):309-19.
7. Yasuda H, Shima N, Nakagawa N, Mochizuki SI, Yano K, Fujise N, Sato Y, Goto M, Yamaguchi K, Kuriyama M, Kanno T, Murakami A, Tsuda E, Morinaga T, Higashio K 1998 Identity of osteoclastogenesis inhibitory factor (OCIF) and osteoprotegerin (OPG): a mechanism by which OPG/OCIF inhibits osteoclastogenesis in vitro. Endocrinology 139(3):1329-37.
8. Bucay N, Sarosi I, Dunstan CR, Morony S, Tarpley J, Capparelli C, Scully S, Tan HL, Xu W, Lacey DL, Boyle WJ, Simonet WS 1998 osteoprotegerin-deficient mice develop early onset osteoporosis and arterial calcification. Genes Dev 12(9):1260-8.
9. Mizuno A, Amizuka N, Irie K, Murakami A, Fujise N, Kanno T, Sato Y, Nakagawa N, Yasuda H, Mochizuki S, Gomibuchi T, Yano K, Shima N, Washida N, Tsuda E, Morinaga T, Higashio K, Ozawa H 1998 Severe osteoporosis in mice lacking osteoclastogenesis inhibitory factor/osteoprotegerin. Biochem Biophys Res Commun 247(3):610-5.
10. Tsuda E, Goto M, Mochizuki S, Yano K, Kobayashi F, Morinaga T, Higashio K 1997 Isolation of a novel cytokine from human fibroblasts that specifically inhibits osteoclastogenesis. Biochem Biophys Res Commun 234(1):137-42.
11. Fuller K, Wong B, Fox S, Choi Y, Chambers TJ 1998 TRANCE is necessary and sufficient for osteoblast-mediated activation of bone resorption in osteoclasts. J Exp Med 188(5):997-1001.
12. Bateman A, Bennett HP 1998 Granulins: the structure and function of an emerging family of growth factors. J Endocrinol 158(2):145-51.
13. Lacey DL, Timms E, Tan HL, Kelley MJ, Dunstan CR, Burgess T, Elliott R, Colombero A, Elliott G, Scully S, Hsu H, Sullivan J, Hawkins N, Davy E, Capparelli C, Eli A, Qian YX, Kaufman S, Sarosi I, Shalhoub V, Senaldi G, Guo J, Delaney J, Boyle WJ 1998 Osteoprotegerin ligand is a cytokine that regulates osteoclast differentiation and activation. Cell 93(2):165-76.
14. Kartsogiannis V, Zhou H, Horwood NJ, Thomas RJ, Hards DK, Quinn JM, Niforas P, Ng KW, Martin TJ, Gillespie MT 1999 Localization of RANKL (receptor activator of NF kappa B ligand) mRNA and protein in skeletal and extraskeletal tissues. Bone 25(5):525-34.
15. Lum L, Wong BR, Josien R, Becherer JD, Erdjument-Bromage H, Schlondorff J, Tempst P, Choi Y, Blobel CP 1999 Evidence for a role of a tumor necrosis factor-alpha (TNF-alpha)-converting enzyme-like protease in shedding of TRANCE, a

TNF family member involved in osteoclastogenesis and dendritic cell survival. J Biol Chem 274(19):13613-8.

16. Hsu H, Lacey DL, Dunstan CR, Solovyev I, Colombero A, Timms E, Tan HL, Elliott G, Kelley MJ, Sarosi I, Wang L, Xia XZ, Elliott R, Chiu L, Black T, Scully S, Capparelli C, Morony S, Shimamoto G, Bass MB, Boyle WJ 1999 Tumor necrosis factor receptor family member RANK mediates osteoclast differentiation and activation induced by osteoprotegerin ligand. Proc Natl Acad Sci U S A 96(7):3540-5.
17. Burgess TL, Qian Y, Kaufman S, Ring BD, Van G, Capparelli C, Kelley M, Hsu H, Boyle WJ, Dunstan CR, Hu S, Lacey DL 1999 The ligand for osteoprotegerin (OPGL) directly activates mature osteoclasts. J Cell Biol 145(3):527-38.
18. Quinn JM, Elliott J, Gillespie MT, Martin TJ 1998 A combination of osteoclast differentiation factor and macrophage-colony stimulating factor is sufficient for both human and mouse osteoclast formation in vitro. Endocrinology 139(10):4424-7.
19. Karsenty G 1999 The genetic transformation of bone biology. Genes Dev 13(23):3037-51.
20. Kong YY, Yoshida H, Sarosi I, Tan HL, Timms E, Capparelli C, Morony S, Oliveira-dos-Santos AJ, Van G, Itie A, Khoo W, Wakeham A, Dunstan CR, Lacey DL, Mak TW, Boyle WJ, Penninger JM 1999 OPGL is a key regulator of osteoclastogenesis, lymphocyte development and lymph-node organogenesis. Nature 397(6717):315-23.
21. Darnay BG, Haridas V, Ni J, Moore PA, Aggarwal BB 1998 Characterization of the intracellular domain of receptor activator of NF-kappaB (RANK). Interaction with tumor necrosis factor receptor-associated factors and activation of NF-kappab and c-Jun N-terminal kinase. J Biol Chem 273(32):20551-5.
22. Wong BR, Josien R, Lee SY, Vologodskaia M, Steinman RM, Choi Y 1998 The TRAF family of signal transducers mediates NF-kappaB activation by the TRANCE receptor. J Biol Chem 273(43):28355-9.
23. Balemans W, Ebeling M, Patel N, Van Hul E, Olson P, Dioszegi M, Lacza C, Wuyts W, Van Den Ende J, Willems P, Paes-Alves AF, Hill S, Bueno M, Ramos FJ, Tacconi P, Dikkers FG, Stratakis C, Lindpaintner K, Vickery B, Foernzler D, Van Hul W 2001 Increased bone density in sclerosteosis is due to the deficiency of a novel secreted protein (SOST). Hum Mol Genet 10(5):537-43.
24. Kadono Y, Akiyama T, Ogata T, Nakamura I, Oda H, Nakamura K, Tanaka S 2001 TRAF6 expression pormoted by RANKL/RANK signaling plays an important role in osteoclast differentiation. JBMR 16 (Suppl. 1)(S176).
25. Lomaga MA, Yeh WC, Sarosi I, Duncan GS, Furlonger C, Ho A, Morony S, Capparelli C, Van G, Kaufman S, van der Heiden A, Itie A, Wakeham A, Khoo W, Sasaki T, Cao Z, Penninger JM, Paige CJ, Lacey DL, Dunstan CR, Boyle WJ, Goeddel DV, Mak TW 1999 TRAF6 deficiency results in osteopetrosis and defective interleukin-1, CD40, and LPS signaling. Genes Dev 13(8):1015-24.
26. Naito A, Azuma S, Tanaka S, Miyazaki T, Takaki S, Takatsu K, Nakao K, Nakamura K, Katsuki M, Yamamoto T, Inoue J 1999 Severe osteopetrosis,

defective interleukin-1 signalling and lymph node organogenesis in TRAF6-deficient mice. Genes Cells 4(6):353-62.

27. Troen BR 2003 Molecular mechanisms underlying osteoclast formation and activation. Exp Gerontol 38(6):605-14.
28. Wong BR, Besser D, Kim N, Arron JR, Vologodskaia M, Hanafusa H, Choi Y 1999 TRANCE, a TNF family member, activates Akt/PKB through a signaling complex involving TRAF6 and c-Src. Mol Cell 4(6):1041-9.
29. Canalis E 1987 Effects of tumor necrosis factor on bone formation in vitro. Endocrinology 121(5):1596-604.
30. Bertolini DR, Nedwin GE, Bringman TS, Smith DD, Mundy GR 1986 Stimulation of bone resorption and inhibition of bone formation in vitro by human tumour necrosis factors. Nature 319(6053):516-8.
31. Stashenko P, Dewhirst FE, Peros WJ, Kent RL, Ago JM 1987 Synergistic interactions between interleukin 1, tumor necrosis factor, and lymphotoxin in bone resorption. J Immunol 138(5):1464-8.
32. Chabaud M, Miossec P 2001 The combination of tumor necrosis factor alpha blockade with interleukin-1 and interleukin-17 blockade is more effective for controlling synovial inflammation and bone resorption in an ex vivo model. Arthritis Rheum 44(6):1293-303.
33. Kimble RB, Bain S, Pacifici R 1997 The functional block of TNF but not of IL-6 prevents bone loss in ovariectomized mice. J Bone Miner Res 12(6):935-41.
34. Blair HC, Teitelbaum SL, Ghiselli R, Gluck S 1989 Osteoclastic bone resorption by a polarized vacuolar proton pump. Science 245(4920):855-7.
35. Johnson RA, Boyce BF, Mundy GR, Roodman GD 1989 Tumors producing human tumor necrosis factor induced hypercalcemia and osteoclastic bone resorption in nude mice. Endocrinology 124(3):1424-7.
36. Kobayashi K, Takahashi N, Jimi E, Udagawa N, Takami M, Kotake S, Nakagawa N, Kinosaki M, Yamaguchi K, Shima N, Yasuda H, Morinaga T, Higashio K, Martin TJ, Suda T 2000 Tumor necrosis factor alpha stimulates osteoclast differentiation by a mechanism independent of the ODF/RANKL-RANK interaction. J Exp Med 191(2):275-86.
37. Nanes MS 2003 Tumor necrosis factor-alpha: molecular and cellular mechanisms in skeletal pathology. Gene 321:1-15.
38. Pfeilschifter J, Chenu C, Bird A, Mundy GR, Roodman GD 1989 Interleukin-1 and tumor necrosis factor stimulate the formation of human osteoclastlike cells in vitro. J Bone Miner Res 4(1):113-8.
39. Spelsberg TC, Subramaniam M, Riggs BL, Khosla S 1999 The actions and interactions of sex steroids and growth factors/cytokines on the skeleton. Mol Endocrinol 13(6):819-28.
40. Sunyer T, Lewis J, Collin-Osdoby P, Osdoby P 1999 Estrogen's bone-protective effects may involve differential IL-1 receptor regulation in human osteoclast-like cells. J Clin Invest 103(10):1409-18.
41. Vaananen HK 1996 Osteoclast function: Biology and mechanisms. In: Bilezikian JP, Raisz LG, Rodan GA (eds.) Principles of Bone Biology. Academic Press, San Diego, pp 103-113.

42. Kawasaki K, Gao YH, Yokose S, Kaji Y, Nakamura T, Suda T, Yoshida K, Taga T, Kishimoto T, Kataoka H, Yuasa T, Norimatsu H, Yamaguchi A 1997 Osteoclasts are present in gp130-deficient mice. Endocrinology 138(11):4959-65.
43. Pacifici R 1996 Estrogen, cytokines, and pathogenesis of postmenopausal osteoporosis. J Bone Miner Res 11(8):1043-51.
44. Grano M, Colucci S, Cantatore FP, Teti A, Zambonin Zallone A 1990 Osteoclast bone resorption is enhanced in the presence of osteoblasts. Boll Soc Ital Biol Sper 66(11):1051-7.
45. Teti A, Grano M, Colucci S, Argentino L, Zambonin Zallone A 1990 Osteoblast-osteoclast interaction in bone resorption. Preliminary results. Boll Soc Ital Biol Sper 66(5):427-31.
46. Corral DA, Amling M, Priemel M, Loyer E, Fuchs S, Ducy P, Baron R, Karsenty G 1998 Dissociation between bone resorption and bone formation in osteopenic transgenic mice. Proc Natl Acad Sci U S A 95(23):13835-40.
47. Nakamura T, Nawa K, Ichihara A 1984 Partial purification and characterization of hepatocyte growth factor from serum of hepatectomized rats. Biochem Biophys Res Commun 122(3):1450-9.
48. Michalopoulos G, Houck KA, Dolan ML, Leutteke NC 1984 Control of hepatocyte replication by two serum factors. Cancer Res 44(10):4414-9.
49. Grano M, Galimi F, Zambonin G, Colucci S, Cottone E, Zallone AZ, Comoglio PM 1996 Hepatocyte growth factor is a coupling factor for osteoclasts and osteoblasts in vitro. Proceedings of the National Academy of Sciences of the United States of America 93(15):7644-8.
50. Armstrong AP, Tometsko ME, Glaccum M, Sutherland CL, Cosman D, Dougall WC 2002 A RANK/TRAF6-dependent signal transduction pathway is essential for osteoclast cytoskeletal organization and resorptive function. J Biol Chem 277(46):44347-56.
51. D'Ippolito G, Schiller PC, Perez-stable C, Balkan W, Roos BA, Howard GA 2002 Cooperative actions of hepatocyte growth factor and 1,25-dihydroxyvitamin D3 in osteoblastic differentiation of human vertebral bone marrow stromal cells. Bone 31(2):269-75.
52. Taichman R, Reilly M, Verma R, Ehrenman K, Emerson S 2001 Hepatocyte growth factor is secreted by osteoblasts and cooperatively permits the survival of haematopoietic progenitors. Br J Haematol 112(2):438-48.
53. Brunkow ME, Gardner JC, Van Ness J, Paeper BW, Kovacevich BR, Proll S, Skonier JE, Zhao L, Sabo PJ, Fu Y, Alisch RS, Gillett L, Colbert T, Tacconi P, Galas D, Hamersma H, Beighton P, Mulligan J 2001 Bone dysplasia sclerosteosis results from loss of the SOST gene product, a novel cystine knot-containing protein. Am J Hum Genet 68(3):577-89.
54. Abe E, Marians RC, Yu W, Wu XB, Ando T, Li Y, Iqbal J, Eldeiry L, Rajendren G, Blair HC, Davies TF, Zaidi M 2003 TSH is a negative regulator of skeletal remodeling. Cell 115(2):151-62.
55. Kusu N, Laurikkala J, Imanishi M, Usui H, Konishi M, Miyake A, Thesleff I, Itoh N 2003 Sclerostin is a novel secreted osteoclast-derived bone morphogenetic protein antagonist with unique ligand specificity. J Biol Chem 278(26):24113-7.

56. Winkler DG, Sutherland MK, Geoghegan JC, Yu C, Hayes T, Skonier JE, Shpektor D, Jonas M, Kovacevich BR, Staehling-Hampton K, Appleby M, Brunkow ME, Latham JA 2003 Osteocyte control of bone formation via sclerostin, a novel BMP antagonist. Embo J 22(23):6267-76.
57. Zhang L, Leeman E, Carnes DC, Graves DT 1991 Human osteoblasts synthesize and respond to platelet-derived growth factor. Am J Physiol 261(2 Pt 1):C348-54.
58. Canalis E, McCarthy TL, Centrella M 1989 Effects of platelet-derived growth factor on bone formation in vitro. J Cell Physiol 140(3):530-7.
59. Lind M, Deleuran B, Thestrup-Pedersen K, Soballe K, Eriksen EF, Bunger C 1995 Chemotaxis of human osteoblasts. Effects of osteotropic growth factors. Apmis 103(2):140-6.
60. Kubota K, Sakikawa C, Katsumata M, Nakamura T, Wakabayashi K 2002 Platelet-derived growth factor BB secreted from osteoclasts acts as an osteoblastogenesis inhibitory factor. J Bone Miner Res 17(2):257-65.
61. Ahlen K, Ring P, Tomasini-Johansson B, Holmqvist K, Magnusson KE, Rubin K 2004 Platelet-derived growth factor-BB modulates membrane mobility of beta(1) integrins. Biochem Biophys Res Commun 314(1):89-96.
62. Westermark B, Siegbahn A, Heldin CH, Claesson-Welsh L 1990 B-type receptor for platelet-derived growth factor mediates a chemotactic response by means of ligand-induced activation of the receptor protein-tyrosine kinase. Proc Natl Acad Sci U S A 87(1):128-32.
63. Kiviranta R, Morko J, Uusitalo H, Aro HT, Vuorio E, Rantakokko J 2001 Accelerated turnover of metaphyseal trabecular bone in mice overexpressing cathepsin K. J Bone Miner Res 16(8):1444-52.
64. Falany ML, Thames AM, 3rd, McDonald JM, Blair HC, McKenna MA, Moore RE, Young MK, Williams JP 2001 Osteoclasts secrete the chemotactic cytokine mim-1. Biochem Biophys Res Commun 281(1):180-5.
65. Dobnig H, Turner RT 1995 Evidence that intermittent treatment with parathyroid hormone increases bone formation in adult rats by activation of bone lining cells. Endocrinology 136(8):3632-8.
66. Zheng MH, Robbins P, Xu J, Huang L, Wood DJ, Papadimitriou JM 2001 The histogenesis of giant cell tumour of bone: a model of interaction between neoplastic cells and osteoclasts. Histol Histopathol 16(1):297-307.
67. Huang L, Xu J, Wood DJ, Zheng MH 2000 Gene expression of osteoprotegerin ligand, osteoprotegerin, and receptor activator of NF-kappaB in giant cell tumor of bone: possible involvement in tumor cell-induced osteoclast-like cell formation. Am J Pathol 156(3):761-7.
68. Huang L, Teng XY, Cheng YY, Lee KM, Kumta SM 2004 Expression of preosteoblast markers and Cbfa-1 and Osterix gene transcripts in stromal tumour cells of giant cell tumour of bone. Bone 34(3):393-401.
69. Zaidi M, Blair HC, Moonga BS, Abe E, Huang CL 2003 Osteoclastogenesis, bone resorption, and osteoclast-based therapeutics. J Bone Miner Res 18(4):599-609.
70. Hofbauer LC, Khosla S, Dunstan CR, Lacey DL, Spelsberg TC, Riggs BL 1999 Estrogen stimulates gene expression and protein production of osteoprotegerin in human osteoblastic cells. Endocrinology 140(9):4367-70.

71. Khosla S, Atkinson EJ, Dunstan CR, O'Fallon WM 2002 Effect of estrogen versus testosterone on circulating osteoprotegerin and other cytokine levels in normal elderly men. J Clin Endocrinol Metab 87(4):1550-4.
72. Shevde NK, Bendixen AC, Dienger KM, Pike JW 2000 Estrogens suppress RANK ligand-induced osteoclast differentiation via a stromal cell independent mechanism involving c-Jun repression. Proc Natl Acad Sci U S A 97(14):7829-34.
73. Srivastava S, Toraldo G, Weitzmann MN, Cenci S, Ross FP, Pacifici R 2001 Estrogen decreases osteoclast formation by down-regulating receptor activator of NF-kappa B ligand (RANKL)-induced JNK activation. J Biol Chem 276(12):8836-40.
74. Srivastava S, Weitzmann MN, Cenci S, Ross FP, Adler S, Pacifici R 1999 Estrogen decreases TNF gene expression by blocking JNK activity and the resulting production of c-Jun and JunD. J Clin Invest 104(4):503-13.
75. Key LL, Ries WL 1996 Osteopetrosis. In: Bilezikian JP, Raisz LG, Rodan GA (eds.) Principles of Bone Biology. Academic Press, San Diego, pp 941 950.
76. de Vernejoul MC, Benichou O 2001 Human osteopetrosis and other sclerosing disorders: recent genetic developments. Calcif Tissue Int 69(1):1-6.
77. Flanagan AM, Sarma U, Steward CG, Vellodi A, Horton MA 2000 Study of the nonresorptive phenotype of osteoclast-like cells from patients with malignant osteopetrosis: a new approach to investigating pathogenesis. J Bone Miner Res 15(2):352-60.
78. Asahina I, Sampath TK, Hauschka PV 1996 Human osteogenic protein-1 induces chondroblastic, osteoblastic, and/or adipocytic differentiation of clonal murine target cells. Exp Cell Res 222(1):38-47.
79. Lajeunesse D, Busque L, Menard P, Brunette MG, Bonny Y 1996 Demonstration of an osteoblast defect in two cases of human malignant osteopetrosis. Correction of the phenotype after bone marrow transplant. J Clin Invest 98(8):1835-42.
80. Roughley PJ, Rauch F, Glorieux FH 2003 Osteogenesis imperfecta--clinical and molecular diversity. Eur Cell Mater 5:41-7; discussion 47.
81. Glorieux FH, Rauch F, Plotkin H, Ward L, Travers R, Roughley P, Lalic L, Glorieux DF, Fassier F, Bishop NJ 2000 Type V osteogenesis imperfecta: a new form of brittle bone disease. J Bone Miner Res 15(9):1650-8.
82. Glorieux FH, Ward LM, Rauch F, Lalic L, Roughley PJ, Travers R 2002 Osteogenesis imperfecta type VI: a form of brittle bone disease with a mineralization defect. J Bone Miner Res 17(1):30-8.
83. Ward LM, Rauch F, Travers R, Chabot G, Azouz EM, Lalic L, Roughley PJ, Glorieux FH 2002 Osteogenesis imperfecta type VII: an autosomal recessive form of brittle bone disease. Bone 31(1):12-8.
84. Labuda M, Morissette J, Ward LM, Rauch F, Lalic L, Roughley PJ, Glorieux FH 2002 Osteogenesis imperfecta type VII maps to the short arm of chromosome 3. Bone 31(1):19-25.
85. Siris ES, Roodman GD 2003 Paget's Disease. In: Favus MJ (ed.) Primer on the Metabolic Bone Diseases and Disorders of Mineral Metabolism. American Society for Bone and Mineral Research, Chicago, pp 495-508.
86. Bender IB 2003 Paget's disease. J Endod 29(11):720-3.

87. Buckley KA, Fraser WD 2002 Receptor activator for nuclear factor kappaB ligand and osteoprotegerin: regulators of bone physiology and immune responses/potential therapeutic agents and biochemical markers. Ann Clin Biochem 39(Pt 6):551-6.
88. Menaa C, Reddy SV, Kurihara N, Maeda H, Anderson D, Cundy T, Cornish J, Singer FR, Bruder JM, Roodman GD 2000 Enhanced RANK ligand expression and responsivity of bone marrow cells in Paget's disease of bone. J Clin Invest 105(12):1833-8.
89. Neale SD, Smith R, Wass JA, Athanasou NA 2000 Osteoclast differentiation from circulating mononuclear precursors in Paget's disease is hypersensitive to 1,25-dihydroxyvitamin D(3) and RANKL. Bone 27(3):409-16.
90. Sparks AB, Peterson SN, Bell C, Loftus BJ, Hocking L, Cahill DP, Frassica FJ, Streeten EA, Levine MA, Fraser CM, Adams MD, Broder S, Venter JC, Kinzler KW, Vogelstein B, Ralston SH 2001 Mutation screening of the TNFRSF11A gene encoding receptor activator of NF kappa B (RANK) in familial and sporadic Paget's disease of bone and osteosarcoma. Calcif Tissue Int 68(3):151-5.
91. Laurin N, Brown JP, Morissette J, Raymond V 2002 Recurrent mutation of the gene encoding sequestosome 1 (SQSTM1/p62) in Paget disease of bone. Am J Hum Genet 70(6):1582-8.
92. Manolagas SC, Jilka RL 1995 Bone marrow, cytokines, and bone remodeling. Emerging insights into the pathophysiology of osteoporosis. N Engl J Med 332(5):305-11.
93. Mercado FB, Marshall RI, Bartold PM 2003 Inter-relationships between rheumatoid arthritis and periodontal disease. A review. J Clin Periodontol 30(9):761-72.
94. Wiebe SH, Hafezi M, Sandhu HS, Sims SM, Dixon SJ 1996 Osteoclast activation in inflammatory periodontal diseases. Oral Dis 2(2):167-80.
95. Graves DT 1999 The potential role of chemokines and inflammatory cytokines in periodontal disease progression. Clin Infect Dis 28(3):482-90.
96. Wysocki SJ, Zheng MH, Smith A, Lamawansa MD, Iacopetta BJ, Robertson TA, Papadimitriou JM, House AK, Norman PE 1996 Monocyte chemoattractant protein-1 gene expression in injured pig artery coincides with early appearance of infiltrating monocyte/macrophages. J Cell Biochem 62(3):303-13.
97. Liu D, Xu JK, Figliomeni L, Huang L, Pavlos NJ, Rogers M, Tan A, Price P, Zheng MH 2003 Expression of RANKL and OPG mRNA in periodontal disease: possible involvement in bone destruction. Int J Mol Med 11(1):17-21.
98. Clements KM, Price JS, Chambers MG, Visco DM, Poole AR, Mason RM 2003 Gene deletion of either interleukin-1beta, interleukin-1beta-converting enzyme, inducible nitric oxide synthase, or stromelysin 1 accelerates the development of knee osteoarthritis in mice after surgical transection of the medial collateral ligament and partial medial meniscectomy. Arthritis Rheum 48(12):3452-63.
99. Tortorella MD, Malfait AM 2003 The usual suspects: verdict not guilty? Arthritis Rheum 48(12):3304-7.
100. Tortorella MD, Pratta M, Liu RQ, Austin J, Ross OH, Abbaszade I, Burn T, Arner E 2000 Sites of aggrecan cleavage by recombinant human aggrecanase-1 (ADAMTS-4). J Biol Chem 275(24):18566-73.

101. Kenney J, Goldring S, Manning C, Rubash H, Shanbhag A 2000 Anabolic Effects of Alendronate on osteoblasts. Genzyme Tissue Repair.
102. Shanbhag A, Rubash H 1998 Bisphosphanate therapy for the prevention of osteolysis in total joint replacemnts. Current Opinion in Orthopaedics 9:81-87.
103. Teitelbaum SL 2000 Bone resorption by osteoclasts. Science 289(5484):1504-8.
104. Teitelbaum SL, Ross FP 2003 Genetic regulation of osteoclast development and function. Nat Rev Genet 4(8):638-49.
105. Teitelbaum SL 2000 Osteoclasts, integrins, and osteoporosis. J Bone Miner Metab 18(6):344-9.

CHAPTER 6

OSTEOCLASTS AND INFLAMMATORY OSTEOLYSIS

Lianping Xing, M.D., Ph.D; Qian Zhang, M.D. and Zhenqiang Yao, M.D.
Department of Pathology
University of Rochester Medical Center, Rochester, New York
E-mail: Lianping_Xing@URMC.Rochester.Edu

1. Introduction

Bone remodeling is a dynamic process that allows the skeleton to adapt to local biomechanical changes and to repair micro-damaged region. This process requires the coordinated actions of osteoclasts and osteoblasts. When new bone formation matches bone resorption, bone mass is maintained. However, when bone remodeling is not regulated appropriately, such as elevated osteoclast function, bone loss occurs. Osteoclasts are bone resorbing cells that are derived from the multi-potent progenitor cells in the myeloid lineage. The progenitors proliferate and differentiate to form mature osteoclasts under the influence of a number of proteins, including PU.1, c-Fos, monocyte colony stimulating factor (M-CSF), nuclear factor kappa B (NF-κB) p50 and p52, receptor activator of NF-κB ligand (RANKL) [1]. PU.1 commits stem cells to the myeloid lineage and is thought to regulate expression of c-Fms, the receptor for M-CSF. Expression of M-CSF by osteoblasts/stromal cells is critical for the differentiation of the myeloid progenitor cells to osteoclast precursors. M-CSF interaction with c-Fms on the surface of RANK negative osteoclast precursors simulates their proliferation and survival to expand osteoclast precursor pool. Under influence of M-CSF, these early stage precursors progress to RANK positive cells [2]. RANKL/RANK interaction leads to complete differentiation of the cells and to their subsequent fusion to form multinucleated tartrate-resistant acid phosphatase (TRAP)+ cells. Thus, M-CSF and RANKL are two essential factors for osteoclast formation. Genetic studies have proved that disruption any element within the M-CSF and RANKL signal pathways will lead to defect in osteoclastogenesis [3] [4] [5]. Under physiological condition, osteoclastogenesis is tightly controlled by systemic and local factors to maintain a healthy skeleton.

Disturbing normal osteoclast function will lead to abnormal bones. A defect in osteoclast generation or function in vivo leads to osteopetrosis, a feature with increase in bone density. In contrast, accelerated osteoclast generation and activity cause bone loss, which represents the majority forms of adult skeletal diseases [6]. Such diseases would include, but not exclude, osteoporosis, Paget's disease, bone metastatic cancers, and inflammatory osteolysis. In this chapter, we will focus on the role of osteoclasts in pathogenesis of inflammatory bone diseases including rheumatoid arthritis (RA), periodontal disease, and aseptic loosening of orthopedic implants.

2. Inflammatory Osteolysis

Inflammatory osteolysis represents a group of diseases with a common feature of massive local bone destruction mediated by accelerated osteoclastic bone resorption. The patients may suffer from mild general osteoporosis, but the persistent inflammation and local bone destruction is the predominant phenotype. Unlike osteoporosis which results from a nature happened estrogen deficiency during aging, the production of inflammatory cytokines is perhaps central to the pathogenesis of inflammatory osteolysis. Among all the cytokines studied, IL-1 and TNFα are two of the critical factors to mediate osteoclast-induced local bone loss. Both IL-1 and TNFα regulate osteoclasts generation, activity and survival using RANKL-dependent and –independent pathways.

2.1. TNFα

TNFα, named for its ability to cause rapid necrotic tumor regression, is the founding member of the TNF ligand super-family, which is known to have pleiotropic functions including cell proliferation, differentiation, activation, and apoptosis. TNFα is mainly produced by activated macrophages. Other cell types known to synthesize TNFα at much lower levels include activated T cells, nature killer cells, and mast cells. During acute inflammation, synovial fibroblasts produce a large amount of TNFα. TNF exerts its biological effects via two receptors, namely TNF receptor 1(p55r) and 2 (p75r), which each is capable of mediating distinct intra-cellar signals, to mediate various, often opposing biological effects in a cell-specific manner. The importance of TNFα in the pathogenesis of various forms of local bone loss is supported by several lines of experimental and clinical evidence: 1) TNFα produced by monocytes is elevated in patients with osteoporosis and in joints of RA patients [7, 8]; 2) addition of anti-TNFα antibody inhibits the production of other pro-inflammatory cytokines including IL-1, IL-6, IL-8, and GM-CSF [9, 10]; 3) TNFα can induce joint inflammation and proliferation of synovial cells that can secrete pro-inflammatory cytokines [11]; 4) TNFα stimulates bone resorption by inducing osteoclastogenesis [12]; and most importantly, 5) several human clinical trials of anti-TNFα therapy for RA have demonstrated their efficacy in halting the progression of joint damage [13, 14].

TNFα is a major mediator of inflammation and play a critical role in inflammation-induced bone loss. In past 10 years, following the generation of null mutant and transgenic mice of TNFα, coupled with the successful introduction of anti-TNFα treatment in clinical practice, understanding of the molecular and cellular pathways through which TNF orchestrates disease process has increased dramatically. Now, it is clear that TNFα not only is major mediator in inflammation and immune system, it also actively participates in regulating osteoclast formation, activation and survival [15] [16]. Reorganization of the influence of TNFα in osteoclasts biology has explored the underscored role of osteoclasts in inflammation-induced bone loss and leads to develop osteoclast-targeting therapy.

TNFα and osteoclastogenesis

TNFα stimulates the production of M-CSF by T cells and of RANKL by osteoblasts and other cell types [6], thereby indirectly affects osteoclast formation. Recently, accumulated data support that TNFα also directly stimulates osteoclast formation in vitro, which is independent of RANKL/RANK signaling pathway because TNFα induces osteoclasts formation in the present of RANKL blockers, osteoprotegerin (OPG) or RANK: Fc [17, 18] and it induces osteoclast formation in RANK-/- cells [19]. However, since administration of TNFα into RANK-/- mice cannot induce osteoclast formation in vivo, the clinical significance of findings on TNFα-induced osteoclasts formation is not clear [20]. To directly investigate this question, we have used two in vivo animal models in which TNFα is over-expressed under RANK blockade condition: 1) to apply the RANK antagonist, RANK: Fc, into TNFα transgenic mice (TNF-Tg) mice and 2) to generate the TNF-Tg/RANK-/- mice. In both models, we found that there is no TRAP positive osteoclast formed, confirming that in vivo in the absence of RANK signaling, TNFα cannot mediate osteoclast formation [21].

TNFα and osteoclasts activation

TNFα. on mature osteoclasts activation has not been well studied. A general accepted concept is that TNFα stimulates the formation of osteoclasts, but these osteoclasts need other cytokines, such as IL-1, to be activated [17], implying that TNFα does not affect the activation of existing osteoclasts, at least in vitro. Recently, Dr. Fuller et al reported that TNFα directly stimulates actin formation of mature osteoclasts, which is not blocked by OPG, suggesting a direct stimulatory effect of TNFα on osteoclasts activation [22]. We have found that TNFα-induced osteoclasts are capable to resorb bones, but the bone resorption is greatly enhanced if cells are treated with IL-1 or infected with c-Fos (our unpublished observation). We feel that TNFα may increase the resorptive ability of mature osteoclasts. However, the major effect of TNFα on osteoclasts function is through controlling the final number of osteoclasts present in the resorption sites, rather existing osteoclasts.

TNFα and osteoclasts survival

Osteoclasts are the cells derived from terminally differentiation, which are short-lived. In *in vitro* cell cultures, mature osteoclasts die within 24 hours in the absence of survival factors. Prolong osteoclast survival is an important determinant of how much bone will be removed by a given osteoclast [23, 24]. Other and we have demonstrated that TNFα supports osteoclasts survival in vitro. This effect cannot be blocked by RANKL blockade but it is abolished by addition of TNFα antagonist. Furthermore, in osteoclasts derived from RANK-/- mice, TNFα mediated osteoclasts survival persists, indicating that TNFα directly supports osteoclast survival independent of the RANK signaling pathway. TNFα increases the phosphorylation of Akt in osteoclasts, which is suppressed by a phosphatidylinositol 3-kinase inhibitor LY294002 and an Src family kinase selective inhibitor PP1. In support to this, TNF-mediated osteoclast survival is abolished in Src-/- osteoclasts. TNF also stimulates the phosphorylation of extra-cellular signal-regulated kinase. Thus, TNF promotes the survival of osteoclasts by engaging the Src/PI-3K/Akt and MEK/ERK pathways [25, 26].

TNFα blockade in clinical practice

Therapeutic administration of TNFα antagonists in patients with arthritis results in deactivation of the pro-inflammatory cytokine cascade, diminished recruitment of inflammatory cells from blood to the rheumatoid joint, and alterations in chemokines, acute phase proteins, vascular permeability, and angiogenesis. However, in vivo TNFα blockade have some unwanted clinical complications including infections, autoimmunity, and probably oncogenesis [27]. Three TNFα antagonists used in the clinical practice are etanercept (Enbrel; Amgen/Wyeth), infliximab (Remicade; Centocor/Johson& Johnson/Schering-Plough), and adalimumab (Humira; Abbott). Infliximab and adalimumab are both monoclonal anti-TNFα antibodies and work by binding to TNFα and blocking its interaction with the TNFα receptor. However, infliximab is chimerical, whereas adalimumab is fully human origin. Thus, its immunogenicity should be very low. In contrast, etanercept is the combination of recombinant soluble form of human p75 TNF receptor to the Fc portion of immuno globin IgG. It acts as a competitive inhibitor of TNFα and prevents the binding of TNFα to the cell surface TNFα receptor to reduce the effect of circulating TNFα̃

An interesting aspect of the clinical studies of TNFα antagonists in arthritis is related to the different outcome obtained when particular agents are used. Etanercept, infliximab and adalimumab, all potent inhibitors of TNFα, have overall comparable efficacy in the treatment of rheumatoid arthritis [28]. Because their structures and mechanisms of action are different, however, Etanercept neutralizes lymphotoxin in addition to TNFα, whereas both anti-TNFα monoclonal antibodies do not and may impact other in vivo functions and binding characteristics. Infliximab may have a more pronounced effect on the immune system because it may lead to apoptosis of TNFα-expressing T cells and macrophages [29, 30]. Interestingly, etanercept and infliximab have a differential efficacy in inhibiting the inflammation in Crohn disease [31], underscoring the dissimilarities of TNFα-mediated pathology between these diseases and rheumatoid arthritis. Such information may offer new insights into the heterogeneous pathogenic processes operating in different inflammatory immune disorders. Finally, pharmacodynamic studies of TNFα antagonists have further clarified how important is the fine-tuning of TNFα neutralization, in relation to the differential drug bioavailability, in each patient.

2.2. IL-1

Interleukin-1 (IL-1) is a multifunctional cytokine that regulates various cellular and tissue functions [32]. Bone is one of the most sensitive tissues to IL-1, which exhibits potent bone-resorbing activity in vivo and in vitro [33]. Several lines of evidence suggest that IL-1 is involved in the bone destruction associated with inflammatory osteolysis: 1) the IL-1 concentration in plasma of RA patients is significantly higher than that of healthy individuals and IL-1 levels in the blood and in the synovial fluids correlate with RA disease activity; 2) chronically elevated intra-articular levels of IL-1 beta by gene transfer alone are sufficient to produce virtually all the pathologies found in rheumatoid arthritis [34];

3) Deletion of the IL-1 receptor antagonist (ra) gene, an endogenous inhibitor of IL-1 that is supposed to regulate IL-1 activity, leads to the spontaneous development of a chronic polyarthropathy in BALB/cA mice [35]; and anti-IL-1 therapy attenuates the severity of joint disease of TNF transgenic mice.

IL-1 and osteoclasts

Osteoclasts, the bone-resorbing cells, are macrophage-related multinucleated cells that play a critical role in bone remodeling [16, 36]. Osteoclastic bone resorption consists of multiple steps: proliferation of osteoclast progenitors, differentiation of progenitors into mononuclear prefusion osteoclasts, fusion of prefusion osteoclasts into osteoclast-like multinucleated cells, sealing zone and ruffled border formation, the active resorption, and eventually apoptosis. IL-1 is directly or indirectly involved in all the steps listed above.

In bone organ culture, bone marrow culture and co-culture system using primary osteoblasts and spleen cells, IL-1 stimulates osteoclast formation that is thought to be indirect by stimulating PGE_2 and RANKL synthesis in osteoblasts/stromal cells [37] [38]. IL-1 generated in the bone micro-environment plays a stimulatory role in PBMC mobilization from the peripheral circulation and their subsequent differentiation into osteoclast-like cells in the bone tissue [39]. Osteoclasts are terminally differentiated cells with a limited lifespan. When osteoclasts is purified by removing osteoblasts in the co-culture system, the osteoclasts died due to apoptosis within 24 hours. The addition of IL-1 to purified osteoclasts cultures prolonged the survival of osteoclasts without the help of osteoblasts/stromal cells [40, 41]. Therefore, IL-1 is involved in prolonging the lifespan of osteoclasts as well as in recruiting osteoclasts. In normal pre-osteoclasts, IL-1 induces actin ring formation and tyrosine phosphorylation of $p130^{Cas}$, a known substrate of c-Src ([41, 42]. c-Src is required for IL-1-induced tyrosine phosphorylation of $p130^{Cas}$, which has been shown to take part in actin ring formation in osteoclasts in vitro [42]. After IL-1 binding, the IL-1R-associated kinase, a serine/threonine kinase, becomes autophosphorylated and is recruited to the receptor complex by binding to MyD88. Another adapter, TNFR-associated factor

6 that is required for osteoclastogenesis [43], then interacts with IL-1R-associated kinase [44]. IL-1 cross-regulates the tyrosine kinase pathway via the association of TNFR-associated factor 6 and c-Src, leading to osteoclast cyto-skeletal rearrangement and activation [45].

In addition its indirectly role in osteoclast function, IL-1 directly stimulates osteoclast function through the IL-1 type-1 receptor expressed by osteoclasts. Unlike TNF that can directly mediate osteoclastogenesis from wild type or RANK-/- spleen cells, IL-1 alone does not induce osteoclasts formation from osteoclast progenitors in vitro while some TRAP+ mononuclear cells were formed (our unpublished data). However, IL-1 plus c-Fos over-expression induces osteoclast formation from wild type and RANK-/- mice in the absence of RANKL, suggesting that IL-1 and c-Fos over-expression promotes the differentiation of osteoclast precursors to mature osteoclasts. Furthermore, IL-1 promotes osteoclasts formation and resorption mediated by TNF.

IL-1 and TNF are implicated in postmenopausal and inflammation-mediated bone loss, and increased concentrations of these cytokines are present in inflamed joints of patients with rheumatoid arthritis. Their expression is regulated by NF-κB and vice versa. Thus, once increased cytokine production is initiated, a self-sustained up-regulatory cycle of increasing cytokine release can ensue leading to progressive bone loss and cartilage destruction. To examine the role of NF-κB p50 and p52 in IL-1-induced bone resorption, we used various NF-κB knockout (KO) mice, including p50-/- and p52-/- single KO, 3/4KO, and dKO mice [46]. IL-1 increased blood calcium and bone resorption in wt, p50 and p52 single KO mice, but not in 3/4KO or dKO mice. Osteoclast formation was impaired in bone marrow cultures from 3/4KO compared with single KO and wt mice treated with IL-1, indicating that full expression of p50 and p52 is required for IL-1 induced osteoclast formation. IL-1 receptor expression was similar in CFU-GM colony cells from wt and dKO mice. However, IL-1 promoted CFU-GM colony formation and survival as well as the formation, activity and survival of osteoclasts generated from these colonies from wt mouse splenocytes, but not from dKO splenocytes. No difference in expression of the osteoclast regulatory cytokines, RANKL and OPG, was observed in osteoblasts from wt and dKO mice. Thus, expression of either NF-κB p50 or p52 is required in osteoclasts and their precursors, rather than osteoblasts, for IL-1-mediated bone resorption [46].

IL-1 blockade and therapeutic effect

As for the role of osteoclasts proliferation of osteoclast progenitors, differentiation of progenitors into mononuclear prefusion osteoclasts, fusion of prefusion osteoclasts into osteoclast-like multinucleated cells, the active resorption, and prolonging mature osteoclasts survival, targeted blocking the effect of IL-1 have already been tried both in experiment and in clinic. In animal models of RA, blocking the effects of IL-1 with either IL-1 receptor antagonist (IL-1Ra; endogenous), anti-Il-1 monoclonal antibodies or soluble type II IL-1 receptors significantly reduce joint destruction and bone erosion. Direct in vivo transfer of the IL-1Ra gene into osteoarthritis knee cells using intra-articular injections of a plasmid vector and lipids can significantly reduce the progression of experimental osteoarthritis [47]. Double-blind, randomized, placebo-controlled, multi-center clinical study have confirmed both the efficacy and the safety of recombinant human IL-1Ra to reduce radiological progression and to provide significantly greater clinical improvement than placebo of RA in large cohort of patients with active and severe RA [48, 49]. Anakinra, a human recombinant IL-1 receptor antagonist significantly reduced clinical symptoms when used alone or in addition to existing methotrexate therapy. These results demonstrate that blocking IL-1 protects bone and cartilage from progressive destruction in RA.

3. Rheumatoid Arthritis (RA)

RA is the most severe chronic joint disease by virtue of persistent inflammation and destruction of cartilage and bone. The latter is a characteristic feature of RA usually not observed in other forms of inflammatory arthritis and constitutes a major cause of progressive disability and crippling of RA patients [50]. As a typical histopathological feature, a hyperplastic and hypercellular synovial membrane is built up in which lymphocytes, macrophage-like cells, and fibroblast-like cells accumulate. Recently the role of osteoclasts in bone destruction of RA has attracted growing interest [51]. Osteoclast precursors and mature osteoclasts are abundant at sites of arthritic bone erosions [51, 52]. The critical role of osteoclasts in local bone destruction of RA was elucidated by manipulating osteoclast in the animal models of arthritis. The central question is that if osteoclast generation or bone resorbing function is blocked, whether local bone erosion still occurs in RA. To date, three lines of experiment evidence have been reported: 1) blockade of osteoclast formation by OPG in collage II-induced arthritis prevents local bone destruction [53]; 2) generating arthritis by serum transfer approach in RANKL-/- mice and no bone erosion was observed; 3) crossing TNF transgenic mice with c-Fos-/- and RANK-/- mice. Both c-Fos and RANK knockout mice are severe osteopetrosis due to a defect in osteoclastogenesis during development. In vivo, no single TRAP+ve osteoclast can be detected in these mice [20]. The unique feature of the mutant mice is an elevated TNF levels in a background of no osteoclasts formation. As expected that both c-Fos-/-/TNF-Tg or RANK-/-/ TNF-Tg mice have no bone erosion in their inflamed joints [21, 54, 55], supporting the essential role of osteoclasts in the bone destruction in RA patients. Interestingly, in all three models, inflammation still exists, suggesting that osteoclasts are solely cell type which is responsible for local bone erosion.

We used TNF-Tg mice to study the effects of chronic TNF exposure on osteoclast progenitor numbers and the mechanisms involved in this process. TNF transgenic mice was generated by George Kollias' laboratory which contain a human TNF transgene (hTNF-Tg) in which the AU rich 3' UT region (which shortens mRNA half life) was replaced by the stable □globin 3' UT [56]. This mutation results in the over-expression of human TNF transgene. By itself, this transgene engenders multiple lines of mice that develop severe polyarthritis with histological lesions resembling human rheumatoid arthritis as well as with remarkable osteoclastic bone resorption. Like RA patients, TNF-Tg mice also develop mild osteoporosis [56]. We found that transgenic TNFα primed splenocytes and peripheral blood mononuclear cells for osteoclastogenesis, and increased the frequency of $CD11b^{hi}$ osteoclast progenitors in the spleen by more than 4-fold. This increase in peripheral osteoclast progenitors was also observed in wild type mice injected with

TNFα. Cell sorting experiments revealed that all of the osteoclast progenitors in spleen are in the CD11b^{hi} population, which contains both c-Fms^{-} and c-Fms^{+} pre-osteoclasts. Interestingly, this increase in peripheral osteoclast progenitors was directly associated with the TNF levels in the blood and completely reversible by in vivo administration of a TNFα antagonist, etanercept [57]. Towards elucidating the mechanism for this increase in osteoclast progenitors, we showed that TNFα did not affect proliferation, differentiation or survival of CD11b^{hi} splenocytes in vivo or in vitro. However, administration of TNFα in vivo caused a rapid increase in CD11b^{+} cells in the peripheral blood and spleen. These data support a model in which TNFα stimulates focal bone loss by inducing the mobilization of osteoclast progenitors from the bone marrow into the systemic circulation. This leads to a marked increase in circulating osteoclast progenitors, which are then recruited and differentiate into active osteoclasts at sites of chronic inflammation within the bone microenvironment.

Therapeutic interventions have demonstrated to slow or arrest local bone erosions by inhibiting bone erosions target molecules involved in the differentiation and activation of osteoclasts. Inhibition of TNF and IL-1 in human RA by antibodies, soluble receptors, or receptor antagonists have already used in clinical examples [58] and experimental therapy with OPG, a decoy receptor of RANKL, slows or even completely blocks bone erosion in animal models of arthritis [53]. Other osteoclast inhibitors such as pamidronate and RANK:Fc have been used in animal models and humans ([21, 54]. However, blockade of osteoclasts may not be sufficient in repair of erosions. Redlich [59] reported that OPG and anti-TNF alone led to arrest of bone erosions from TNF-Tg mice with established erosive arthritis and systemic bone loss but did not achieve repair. However, local bone erosions almost fully regressed, on combined treatment with anti-TNF and PTH and/or OPG. Therefore, local joint destruction and systemic inflammatory bone loss because of TNF can regress and repair requires a combined approach by reducing inflammation, blocking bone resorption, or stimulating bone formation.

4. Loosening of Orthopedic Implant

Aseptic loosening is the most common complication of orthopedic implants worldwide. Up to 20% of patients with total joint arthroplasty will develop radiographic evidence of loosening within 10 years [60]. Prosthesis failure results from multiple factors including materials, biomechanics and host responses. The quest for more durable and wear resistant materials, as well as better implant designs, and the study of the forces involved in implant integration and prosthesis failure, continue to be areas of active investigation. Several groups however, have focused on the host response to wear debris postulating that wear debris-induced osteolysis is the major cause of prosthetic implant failure [61]. In this model, wear debris generated from the prosthesis is phagocytosed by macrophages and initiates an inflammatory response that leads to the recruitment of activated osteoclasts and osteolysis at the bone-implant interface. Several lines of evidence support this model including: i) as great as 10^9 particles per gram of tissue can be recovered from the inflammatory membrane attached to the failed prosthesis following revision surgery [62]. ii) Ingestion of wear-debris particles induces cytokine production by mononuclear phagocytes in vitro [63]. iii) High levels of cytokines including TNFα which are produced by macrophages, are found in the fluid and tissue surrounding loose implants [64]. iv) Conditioned medium, from wear debris stimulated monocytes can stimulate increased bone resorption in vitro [65]. And, v) animal models of wear debris-induced osteolysis have documented the importance of cytokines in this process [66] [8].

We used TNF-Tg mice and mice genetically deficient for the TNF receptor 1 (p55r-/-) [67] in this study. As expected we found that the titanium induced osteolysis in the TNF-Tg mice was much greater (approximately 4-fold) than that observed in normal mice. Consistently, there was virtually no osteolysis observed in the TNF p55r-/- mice. The surprise of the experiment came from our results with the sham surgery controls, in which mice received an incision without titanium. The normal mice failed to exhibit any osteolysis from this treatment, however the sham TNF-Tg mice had just as much inflammation and osteolysis as their littermates that received the titanium. These experiments formally demonstrated in vivo that wear debris particles do not directly induce osteolysis. Rather, it is the pro-inflammatory cytokines produced in response to the wear debris that directly signal for osteoclastic bone resorption; as it is possible to get osteolysis without particles (sham TNF-Tg mice) but it is impossible to get osteolysis without TNF signaling (titanium treated TNF p55r-/-mice). Similar studies with polymethylmethacrylate particles that corroborate our findings have also been published by Dr. Stephen Teitelbaum's lab [68].

IL-1 and TNF are two key cytokines that play a critical role in the development of inflammatory bone resorption. Macrophages exposed to wear debris rapidly express mRNA followed by protein of TNF and IL-1 in vitro. Wear debris in mouse inflammation model provokes a significant tissue inflammation associated with significant increase of IL-1 and TNF synthesis and release [69]. NF-κB signaling is activated in response to particles in macrophage lineage [70, 71]. Brief exposure of peripheral blood monocytes obtained from healthy human donors to titanium-alloy particles activated the transcription factors NF-κB and followed by the release of TNF and IL-6 [70]. It has been reported that the production of TNF by macrophages in response to titanium particles was preceded by the translocation of NF-κB to the nucleus, accompanied by the increase of inhibitory factor-kappa B. Their continuing study [71]. They further confirmed that titanium particles-induced NF-κB activation is dependent upon proteasome activity and associated with the degradation of p105, a precursor of p50 that binds to p65. Glant and coworkers [72] found that exposure of osteoblast-like osteosarcoma cells or bone marrow-derived primary osteoblasts to titanium particles rapidly increased NF-κB DNA binding activity before the phagocytosis of these particles. The NF-κB activation was associated with lower mRNA levels of procollagen alpha1 [I] and procollagen alpha1 [III]. Interestingly, pre-treatment of these cells with NF-κB inhibitor pyrrolidine dithiocarbamate can significantly reduce the suppressive effect of titanium on collagen gene expression, suggesting particles suppress collagen gene expression through the NF-κB signaling pathway. Ren et al [73] found that Erythromycin, a macrolide antibiotic, suppresses wear debris-induced osteoclastic bone resorption by down-regulation of NF-κB signaling pathway, suggesting that Erythromycin represents a potential therapeutic candidate for the treatment and prevention of aseptic loosening. Therefore, modification of the NF-κB signaling may be important to retard wear debris-induced inflammatory osteolysis and improve the performance of orthopaedic implants.

5. Periodontal Disease

The active remodeling of alveolar bone in periodontal tissue is well known to occur in various kinds of conditions such as patients with a homeostatic status, inflammation, aging and orthodontic treatment. Osteoblasts and osteoclasts are the main cells that are responsible for the remodeling of alveolar bone. The inflammation and tissue destruction seen in periodontitis is associated with granulomatous tissue containing inflammatory cells, such as T and B lymphocytes, plasma cells and many cells of the monocytes/macrophage lineage. These cells are thought to produce a variety of inflammatory mediators. High levels of several inflammatory cytokines, such as IL-1, IL-6, prostaglandin E_2 and TNF, have been found in the tissue and gingival cervical fluid of patients suffering advanced periodontitis [74]. Similar findings have been reported in animal models of periodontitis [75]. Gingival cervical fluid is reported to stimulate bone resorption in vitro indicating that the mediators required to promote osteoclast formation are present in this disease [76].

The function of osteoclasts that is known to be cells that cause resorption of the alveolar bone in periodontitis are also regulated by RANKL/RANK signaling. Inflammatory cytokines presented in cervical fluids of patients with periodontitis are reported to stimulate production of RANKL [77]. An increased concentration of RANKL and a decreased concentration of OPG were detected in Gingival cervical fluid from patients with periodontitis and the ratio of the concentration of RANKL to that of OPG in the Gingival cervical fluid was significantly higher for periodontal disease patients than for healthy subjects [78]. Periodontal ligament cells (PDLs) are thought to play some role in alveolar bone remodeling as well as osteoblasts and osteoclasts. The function of osteoclasts is regulated by interaction with these PDLs. Fibroblasts derived from the PDLs are able to induce osteoclastogenesis in vitro [79] by the production of RANKL and OPG [80]. In a rat pathological bone resorption model, osteoblasts, PDLs express RANKL that activate osteoclast resorption and it was observed an autocrine mechanism of RANKL-RANK IL-1β and TNFα exist in osteoclast, which is heightened in the pathological conditions [81].

In addition to mechanical scaling and root planning, surgical and adjunctive antimicrobial treatments, bisphosphonates, the drug to inhibit osteoclast resorption, are tried to use in the treatment of periodontal disease both in animal [82] and in clinic [83]. It is believed that IL-1/TNFα antagonist and modification of NF-B signaling might be the potential applications in the therapeutic management of periodontal diseases as did in rheumatoid arthritis.

6. Conclusion

Recent studies in experimental of arthritis have much improved our understanding of the role of TNFα and IL-1 in an inflammation, cartilage and bone metabolism. It is now clear that TNFα-derived joint inflammation is not sufficient to induce cartilage destruction and bone loss in animal models, while osteoclast-targeted therapies may be prevent joint destruction in the presence of inflammation. The concept that bone loss is mediated by osteoclasts in the presence of excessive TNFα and IL-1 production is further supported by clinical observations indicating that TNFα and IL-1 promotes osteoclast survival and differentiation via key factors such as RANK and RANKL. Further study is needed on the apparent differential bioactivities of the soluble and trans-membrane form of TNFα, as well as the possible heterogeneity in the function of its two trans-membrane receptors. Such information is necessary in order to develop safer and more effective therapies aiming to specifically target the deleterious pro-inflammatory properties of TNFα, without compromising its protective role in host defense and immunity. For therapy targeting IL-1, it is clear that epidemiological studies will be needed to document the lone-term benefits and risks associated with blocking IL-1.

References

1. Karsenty, G. and E. Wagner, Reaching a genetic and molecular understanding of skeletal development. Dev Cell, 2002. **2**(4): p. 389-406.
2. Arai, F., et al., Commitment and differentiation of osteoclast precursor cells by the sequential expression of c-Fms and receptor activator of nuclear factor kappaB (RANK) receptors. Journal of Experimental Medicine, 1999. **190**(12): p. 1741-54.
3. Wang, Z.-Q., et al., Bone and haematopoietic defects in mice lacking c-Fos. Nature, 1992. **360**: p. 741-5.
4. Kong, Y.Y., W.J. Boyle, and J.M. Penninger, Osteoprotegerin ligand: a common link between osteoclastogenesis, lymph node formation and lymphocyte development. Immunology & Cell Biology, 1999. **77**(2): p. 188-93.
5. Dougall, W.C., et al., RANK is essential for osteoclast and lymph node development. Genes & Development, 1999. **13**: p. 2412-2424.
6. Boyle, W.J., W.S. Simonet, and D.L. Lacey, Osteoclast differentiation and activation. Nature, 2003. **423**(6937): p. 337-42.
7. Sabokbar, A., et al., Human arthroplasty derived macrophages differentiate into osteoclastic bone resorbing cells. Ann Rheum Dis, 1997. **56**(7): p. 414-20.
8. Schwarz EM, L.R., O'Keefe RJ, Anti-TNFa therapy as a clinical intervention for periprosthetic osteolysis. Arthritis Research, 2000. **2**(3): p. 165-168.
9. Hernigou, P., et al., Interleukin-6 in the blood of patients with total hip arthroplasty without loosening. Clin Orthop, 1999(366): p. 147-54.
10. Itonaga, I., et al., Rheumatoid arthritis synovial macrophage-osteoclast differentiation is osteoprotegerin ligand-dependent. J Pathol, 2000. **192**(1): p. 97-104.
11. Hsu, H., et al., Tumor necrosis factor receptor family member RANK mediates osteoclast differentiation and activation induced by osteoprotegerin ligand. Proceedings of the National Academy of Sciences of the United States of America, 1999. **96**(7): p. 3540-5.
12. Abu-Amer, Y., et al., Tumor necrosis factor receptors types 1 and 2 differentially regulate osteoclastogenesis. J Biol Chem, 2000. **275**(35): p. 27307-10.
13. Fender, D., W.M. Harper, and P.J. Gregg, Outcome of Charnley total hip replacement across a single health region in England: the results at five years from a regional hip register. J Bone Joint Surg Br, 1999. **81**(4): p. 577-81.
14. Callaghan, J.J., et al., Charnley total hip arthroplasty in patients less than fifty years old. A twenty to twenty-five-year follow-up note. J Bone Joint Surg Am, 1998. **80**(5): p. 704-14.
15. Goldring, S.R., The final pathogenetic steps in focal bone erosions in rheumatoid arthritis. Ann Rheum Dis, 2000. **59 Suppl 1**: p. i72-4.
16. Teitelbaum, S.L., Bone resorption by osteoclasts. Science, 2000. **289**(5484): p. 1504-8.
17. Kobayashi, K., et al., Tumor necrosis factor alpha stimulates osteoclast differentiation by a mechanism independent of the ODF/RANKL-RANK interaction. Journal of Experimental Medicine, 2000. **191**(2): p. 275-86.

18. Azuma, Y., et al., Tumor necrosis factor-alpha induces differentiation of and bone resorption by osteoclasts. Journal of Biological Chemistry, 2000. **275**(7): p. 4858-64.
19. Teruhito, Y., et al., c-Fos over-expression induces osteoclastogenesis independent of RANK signaling. J Bone Miner Res, 2002. **17**: p. S131.
20. Li, J., et al., RANK is the intrinsic hematopoietic cell surface receptor that controls osteoclastogenesis and regulation of bone mass and calcium metabolism. Proc Natl Acad Sci U S A, 2000. **97**(4): p. 1566-71.
21. Li, P., et al., RANK signaling is not required for TNFa-mediated increase in CD11bhi osteoclast precursors, but is essential for mature osteoclast formation in TNFa-mediated inflammatory arthritis. J Bone Miner Res, 2004. **19**: p. in press.
22. Fuller, K., et al., TNFalpha potently activates osteoclasts, through a direct action independent of and strongly synergistic with RANKL. Endocrinology, 2002. **143**(3): p. 1108-18.
23. Hughes, D.E., et al., Bisphosphonates induce osteoclast apoptosis in vivo and in vitro, but calcitonin does not. J Bone Miner Res, 1994b. **9**: p. S347.
24. Hughes, D.E., et al., Estrogen promotes apoptosis of murine osteoclasts mediated by TGF-β. Nature Med, 1996. **2**: p. 1132-1136.
25. Lee, S.E., et al., Tumor necrosis factor-alpha supports the survival of osteoclasts through the activation of Akt and ERK. J Biol Chem, 2001. **276**(52): p. 49343-9.
26. Xing, L., et al., Genetic evidence for a role for Src family kinases in TNF family receptor signaling and cell survival. Genes Dev, 2001. **15**(2): p. 241-53.
27. Sfikakis, P.P. and G. Kollias, Tumor necrosis factor biology in experimental and clinical arthritis. Curr Opin Rheumatol, 2003. **15**(4): p. 380-6.
28. Furst, D., B. FC, and K. JR, Updated consensus statement on biological agents for the treatment of rheumatoid arthritis and other rheumatic diseases. Ann Rheum Dis, 2002. **61**: p. ii2-ii7.
29. Scallon, B.J., et al., Chimeric anti-TNF-alpha monoclonal antibody cA2 binds recombinant transmembrane TNF-alpha and activates immune effector functions. Cytokine, 1995. **7**(3): p. 251-9.
30. Scott, K.A., et al., An anti-tumor necrosis factor-alpha antibody inhibits the development of experimental skin tumors. Mol Cancer Ther, 2003. **2**(5): p. 445-51.
31. van Deventer, S.J., Transmembrane TNF-alpha, induction of apoptosis, and the efficacy of TNF-targeting therapies in Crohn's disease. Gastroenterology, 2001. **121**(5): p. 1242-6.
32. Dinarello, C.A., The interleukin-1 family: 10 years of discovery. Faseb J, 1994. **8**(15): p. 1314-25.
33. Boyce, B.F., et al., Effects of interleukin-1 on bone turnover in normal mice. Endocrinology, 1989. **125**: p. 1142-1150.
34. Ghivizzani, S.C., et al., Constitutive intra-articular expression of human IL-1 beta following gene transfer to rabbit synovium produces all major pathologies of human rheumatoid arthritis. J Immunol, 1997. **159**(7): p. 3604-12.
35. Horai, R., et al., Development of chronic inflammatory arthropathy resembling rheumatoid arthritis in interleukin 1 receptor antagonist-deficient mice. J Exp Med, 2000. **191**(2): p. 313-20.

36. Suda, T., et al., Modulation of osteoclast differentiation and function by the new members of the tumor necrosis factor receptor and ligand families. Endocrine Rev., 1999. **20**: p. 345-57.
37. Akatsu, T., et al., Role of prostaglandins in interleukin-1-induced bone resorption in mice in vitro. J Bone Miner Res, 1991. **6**(2): p. 183-9.
38. Yasuda, H., et al., Osteoclast differentiation factor is a ligand for osteoprotegerin/osteoclastogenesis-inhibitory factor and is identical to TRANCE/RANKL. Proc Natl Acad Sci U S A, 1998. **95**: p. 3597-602.
39. Tokukoda, Y., et al., Interleukin-1beta stimulates transendothelial mobilization of human peripheral blood mononuclear cells with a potential to differentiate into osteoclasts in the presence of osteoblasts. Endocr J, 2001. **48**(4): p. 443-52.
40. Jimi, E., T. Shuto, and T. Koga, Macrophage colony-stimulating factor and interleukin-1α maintain the survival of osteoclast-like cells. Endocrinology, 1995. **136**: p. 808-811.
41. Jimi, E., et al., Interleukin 1 induces multinucleation and bone-resorbing activity of osteoclasts in the absence of osteoblasts/stromal cells. Exp. Cell. Res, 1999. **247**: p. 84-93.
42. Nakamura, I., et al., Tyrosine phosphorylation of p130Cas is involved in actin organization in osteoclasts. J Biol Chem, 1998. **273**(18): p. 11144-9.
43. Lomaga, M.A., et al., TRAF6 deficiency results in osteopetrosis and defective interleukin-1, CD40, and LPS signaling. Genes & Development, 1999. **13**(8): p. 1015-24.
44. Kopp, E.B. and R. Medzhitov, The Toll-receptor family and control of innate immunity. Curr Opin Immunol, 1999. **11**(1): p. 13-8.
45. Nakamura, I., et al., IL-1 regulates cytoskeletal organization in osteoclasts via TNF receptor-associated factor 6/c-Src complex. J Immunol, 2002. **168**(10): p. 5103-9.
46. Xing, L., et al., Expression of either NF-kappaB p50 or p52 in osteoclast precursors is required for IL-1-induced bone resorption. J Bone Miner Res, 2003. **18**(2): p. 260-9.
47. Fernandes, J., et al., In vivo transfer of interleukin-1 receptor antagonist gene in osteoarthritic rabbit knee joints: prevention of osteoarthritis progression. Am J Pathol, 1999. **154**(4): p. 1159-69.
48. Bresnihan, B., et al., Treatment of rheumatoid arthritis with recombinant human interleukin-1 receptor antagonist. Arthritis Rheum, 1998. **41**(12): p. 2196-204.
49. Jiang, Y., et al., A multicenter, double-blind, dose-ranging, randomized, placebo-controlled study of recombinant human interleukin-1 receptor antagonist in patients with rheumatoid arthritis: radiologic progression and correlation of Genant and Larsen scores. Arthritis Rheum, 2000. **43**(5): p. 1001-9.
50. Scott, D.L., et al., The links between joint damage and disability in rheumatoid arthritis. Rheumatology (Oxford), 2000. **39**(2): p. 122-32.
51. Romas, E., M.T. Gillespie, and T.J. Martin, Involvement of receptor activator of NFkappaB ligand and tumor necrosis factor-alpha in bone destruction in rheumatoid arthritis. Bone, 2002. **30**(2): p. 340-6.

52. Gravallese, E.M., et al., Identification of cell types responsible for bone resorption in rheumatoid arthritis and juvenile rheumatoid arthritis. Am J Pathol, 1998. **152**(4): p. 943-51.
53. Kong, Y.Y., et al., Activated T cells regulate bone loss and joint destruction in adjuvant arthritis through osteoprotegerin ligand. Nature, 1999. **402**(6759): p. 304-9.
54. Redlich, K., et al., Osteoclasts are essential for TNF-alpha-mediated joint destruction. J Clin Invest, 2002. **110**(10): p. 1419-27.
55. Li, P., et al., Systemic tumor necrosis factor alpha mediates an increase in peripheral CD11bhigh osteoclast precursors in tumor necrosis factor alpha-transgenic mice. Arthritis Rheum, 2004. **50**(1): p. 265-76.
56. Keffer, J., et al., Transgenic mice expressing human tumour necrosis factor: a predictive genetic model of arthritis. Embo J, 1991. **10**(13): p. 4025-31.
57. Ping, L., et al., Systemic TNFa mediates an increase in peripheral CD11bhi osteoclast precursors in TNFa transgenic mice. Arthritis Research, 2004: p. in press.
58. Lipsky, P.E., et al., Infliximab and methotrexate in the treatment of rheumatoid arthritis. Anti-Tumor Necrosis Factor Trial in Rheumatoid Arthritis with Concomitant Therapy Study Group. N Engl J Med, 2000. **343**(22): p. 1594-602.
59. Redlich, K., et al., Repair of local bone erosions and reversal of systemic bone loss upon therapy with anti-tumor necrosis factor in combination with osteoprotegerin or parathyroid hormone in tumor necrosis factor-mediated arthritis. Am J Pathol, 2004. **164**(2): p. 543-55.
60. Kim, Y.H., J.S. Kim, and S.H. Cho, Primary total hip arthroplasty with a cementless porous-coated anatomic total hip prosthesis: 10- to 12-year results of prospective and consecutive series. J Arthroplasty, 1999. **14**(5): p. 538-48.
61. Goldring, S.R., et al., Formation of a synovial-like membrane at the bone-cement interface. Its role in bone resorption and implant loosening after total hip replacement. Arthritis Rheum, 1986. **29**(7): p. 836-42.
62. Margevicius, K.J., et al., Isolation and characterization of debris in membranes around total joint prostheses. J Bone Joint Surg Am, 1994. **76**(11): p. 1664-75.
63. Blaine, T.A., et al., Increased levels of tumor necrosis factor-alpha and interleukin-6 protein and messenger RNA in human peripheral blood monocytes due to titanium particles. J Bone Joint Surg Am, 1996. **78**(8): p. 1181-92.
64. Horowitz, S.M., et al., Studies of the mechanism by which the mechanical failure of polymethylmethacrylate leads to bone resorption. J Bone Joint Surg Am, 1993. **75**(6): p. 802-13.
65. Glant, T.T. and J.J. Jacobs, Response of three murine macrophage populations to particulate debris: bone resorption in organ cultures. J Orthop Res, 1994. **12**(5): p. 720-31.
66. Merkel, K., et al., Tumor necrosis factor-alpha mediates orthopedic implant osteolysis. American Journal of Pathology, 1999. **154**: p. 203-210.
67. Pfeffer, K., et al., Mice deficient for the 55 kd tumor necrosis factor receptor are resistant to endotoxic shock, yet succumb to L. monocytogenes infection. Cell, 1993. **73**(3): p. 457-67.
68. Teitelbaum, S.L., Y. Abu-Amer, and F.P. Ross, Molecular mechanisms of bone resorption. J Cell Biochem, 1995. **59**(1): p. 1-10.

69. Wooley, P.H., et al., Inflammatory responses to orthopaedic biomaterials in the murine air pouch. Biomaterials, 2002. **23**(2): p. 517-26.
70. Schwarz, E.M., et al., Tumor necrosis factor-alpha/nuclear transcription factor-kappaB signaling in periprosthetic osteolysis. J Orthop Res, 2000. **18**(3): p. 472-80.
71. Soloviev, A., et al., The role of p105 protein in NFkappaB activation in ANA-1 murine macrophages following stimulation with titanium particles. J Orthop Res, 2002. **20**(4): p. 714-22.
72. Vermes, C., et al., Particulate wear debris activates protein tyrosine kinases and nuclear factor kappaB, which down-regulates type I collagen synthesis in human osteoblasts. J Bone Miner Res, 2000. **15**(9): p. 1756-65.
73. Ren, W., et al., Erythromycin inhibits wear debris-induced osteoclastogenesis by modulation of murine macrophage NF-kappaB activity. J Orthop Res, 2004. **22**(1): p. 21-9.
74. Rasmussen, L., L. Hanstrom, and U.H. Lerner, Characterization of bone resorbing activity in gingival crevicular fluid from patients with periodontitis. J Clin Periodontol, 2000. **27**(1): p. 41-52.
75. Assuma, R., et al., IL-1 and TNF antagonists inhibit the inflammatory response and bone loss in experimental periodontitis. J Immunol, 1998. **160**(1): p. 403-9.
76. Lerner, U.H., et al., Gingival crevicular fluid from patients with periodontitis contains bone resorbing activity. Eur J Oral Sci, 1998. **106**(3): p. 778-87.
77. Nakashima, T., Y. , et al., Protein expression and functional difference of membrane-bound and soluble receptor activator of NF-kappaB ligand: modulation of the expression by osteotropic factors and cytokines. Biochem. Biophys. Res. Commun., 2000. 275: p. 768-75.
78. Mogi, M., et al., Differential Expression of RANKL and Osteoprotegerin in Gingival Crevicular Fluid of Patients with Periodontitis. J Dent Res, 2004. **83**(2): p. 166-9.
79. Kanzaki, H., et al., Periodontal ligament cells under mechanical stress induce osteoclastogenesis by receptor activator of nuclear factor kappaB ligand up-regulation via prostaglandin E2 synthesis. J Bone Miner Res, 2002. **17**(2): p. 210-20.
80. Hasegawa, T., et al., Expression of receptor activator of NF-kappa B ligand and osteoprotegerin in culture of human periodontal ligament cells. J Periodontal Res, 2002. **37**(6): p. 405-11.
81. Ogasawara, T., et al., In situ expression of RANKL, RANK, osteoprotegerin and cytokines in osteoclasts of rat periodontal tissue. J Periodontal Res, 2004. **39**(1): p. 42-9.
82. Kaynak, D., et al., A histopathological investigation on the effect of systemic administration of the bisphosphonate alendronate on resorptive phase following mucoperiosteal flap surgery in the rat mandible. J Periodontol, 2003. **74**(9): p. 1348-54.
83. El-Shinnawi, U.M. and S.I. El-Tantawy, The effect of alendronate sodium on alveolar bone loss in periodontitis (clinical trial). J Int Acad Periodontol, 2003. **5**(1): p. 5-10.

CHAPTER 7

ENDOCHONDRAL BONE FORMATION AND EXTRACELLULAR MATRIX

Qian Chen, Lei Wei, Zhengke Wang, Xiaojuan Sun, Junming Luo, and Xu Yang

Department of Orthopaedics, Brown Medical School/Rhode Island Hospital
Suite 402, 1 Hoppin Street, Providence, RI 02903
E-mail: Qian_Chen@Brown.edu

This chapter summarizes the recent advances of research in endochondral bone formation, and in cartilage extracellular matrix. Specific emphasis is placed on molecular pathways that regulate endochondral bone formation and a novel extracellular matrix protein family called matrilin. The different aspects of matrilin research are described in detail to serve as an example of the current state of extracellular matrix research with multidisciplinary approaches involving genetics, biochemistry, cell biology, and pathology.

1. Introduction

All the long bones in vertebrates are developed through a process called endochondral bone formation, in which a bone is formed from cartilage anlagen. Thus, the growth of the cartilage anlagen, which is called a growth plate, determines the size and shape of a long bone. Furthermore, tissues adjacent to the growth plate cartilage exquisitely regulate this morphogenic process. This regulation is essential to coordinate growth of different tissues comprising the limb. These neighboring tissues include peri-articular cartilage that caps the growth plate, subchondral bone marrow at the base of the growth plate, and skeletal muscles circumscribing the growth plate. How do these neighboring tissues regulate endochondral bone formation? This external regulation may be achieved by specific mediators produced by the adjacent tissues that interact with the growth plate (Fig. 1). In particular, the peri-articular cartilage layer adjacent to the proliferation zone of the growth plate produces Parathyroid Hormone-related Peptide (PTHrP), which stimulates chondrocyte proliferation and inhibits hypertrophy. Conversely, the subchondral bone marrow adjacent to the hypertrophic zone produces cytokines and chemokines including Stromal Cell-Derived Factor-1 (SDF-1), which maintain chondrocyte hypertrophy and stimulate matrix resorption. The growth of the cartilage anlagen is affected by the circumscribing skeletal muscle, whose growth exerts contracting forces to skeleton. These regulations are achieved, at least in part, by modulating mitogen-activated protein kinase (MAPK), including p38 MAPK activity within growth plate chondrocytes. The experimental evidences leading to the formulation of these concepts are presented in the following.

2. Endochondral Bone Formation

The process of endochondral bone formation is a complex one that consists of multiple stages, and the progression from one stage to the next requires precise regulation and coordination. In the first stage, a resting chondrocyte is activated and enters into a dividing cycle. The immature chondrocytes at this stage exhibit elongated and flattened cell morphology both in vivo and in vitro [1], and synthesize collagen type II and aggrecan [2]. In the second stage, chondrocyte cease proliferation, start the maturation process, and increase their matrix production. The mRNA of matrilin-1 is synthesized abundantly by mature chondrocytes, but very little by either proliferating chondrocytes or the up-coming hypertrophic chondrocytes [1]. Cells at this stage have a medium cuboidal shape and gradually enlarge their sizes. In the third stage, chondrocytes become hypertrophic and synthesize type X collagen [3], before the matrix is degraded, removed, and replaced by bone and bone marrow.

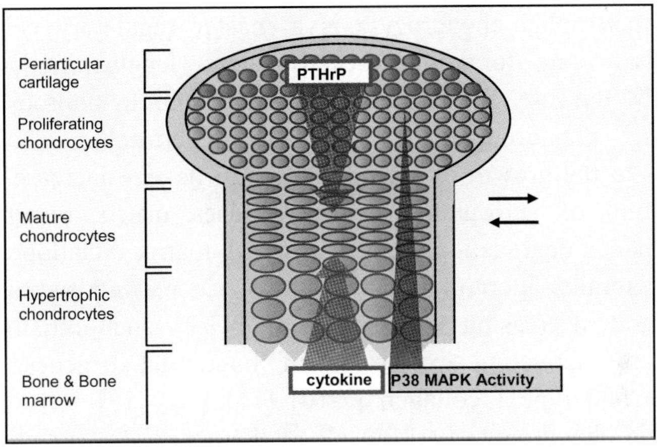

Fig. 1. Schematic diagram depicting our theoretic model: the increasing gradient of p38 MAPK activity during chondrocyte differentiation is essential to endochondral bone formation. This gradient is established by the inhibitory effect of PTHrP from periarticular cartilage and the stimulatory effect of cytokines and chemokines such as SDF-1 from subchondral bone marrow, and modulated by the growth of skeletal muscle in the limb. Understanding the important role of p38 MAPK may suggest new pharmacological means to control bone formation.

The direction of sequential progression of the stages during chondrocyte differentiation is essential to endochondral bone formation. On the one hand, to achieve longitudinal growth, proliferating chondrocytes adjacent to peri-articular cartilage layer have to keep proliferating. On the other hand, to achieve ossification, hypertrophic chondrocytes at the other end of the growth plate have to be converted into bone. Disrupting this sequential progression of chondrocyte differentiation leads to pathological consequences such as chondrodysplasia. For example, acceleration of chondrocyte hypertrophy by ablating the gene encoding parathyroid hormone-related peptide (PTHrP), which potently inhibits chondrocyte hypertrophy, causes chondrodysplasia in mice [4]. Interestingly, delaying chondrocyte hypertrophy by overexpression of (PTH-rP) or its constitutively active receptor also causes chondrodysplasia [5]. Therefore, controlling the pace of chondrocyte hypertrophy is essential for endochondral ossification.

The hypertrophic chondrocyte is a master regulatory cell during endochondral bone formation, because bone lengthening is driven primarily by the rate of production of hypertrophic chondrocytes from proliferating chondrocytes [6]. The hypertrophic chondrocyte contributes to the growth of cartilage through its size increases, directs mineralization of surrounding matrix, causes matrix resorption by releasing matrix degradation enzymes such as matrix metalloproteinases (MMP), commits apoptotic cell death by increasing expression of caspases and decreasing expression of bcl-2, and initiating bone formation by activating expression of bone-related genes such as osteocalcin and type I collagen [7-10] [11] [12] [13]. How does a hypertrophic chondrocyte perform all of these tasks? Our recent data indicate that one of the mitogen-activated protein kinase (MAPK) pathways--p38 MAPK pathway—may be a key regulator of these multiple processes during chondrocyte hypertrophy [14]. Thus, our study will focus on the role of chondrocyte p38 MAPK activity as an intrinsic regulator of endochondral bone formation.

3. Intrinsic Regulators

At least two major classes of intracellular molecules that are essential to bone formation have been identified in recent years. The first contains transcriptional factors including Cbfa1 (also called Runx2) and osteorix (Osx).

3.1. Transcriptional factors

Cbfa1 is a master transcription factor that regulates expression of osteoblast differentiation genes such as osteocalcin, osteopontin, and type I collagen [15], and chondrocyte hypertrophy genes such as Col X [16]. Mice missing Cbfa1 have no osteoblasts and also exhibit abnormalities of chondrocyte maturation [17]. Most bones either lack or have decreased numbers of hypertrophic chondrocytes in the growth plate. Expression of hypertrophic chondrocyte markers such as type X collagen and MMP13 is diminished [18]. These abnormalities are rescued by transgenic expression of Cbfa1 in chondrocytes under Col II promoter. Transgenic expression of Cbfa1 accelerates hypertrophy of chondrocytes and induces ectopic bone formation in permanent cartilage [18].

While Cbfa1 regulates differentiation of both chondrocytes and osteoblasts, Osx regulates osteoblast differentiation only [19]. Osx-null mice lack osteoblasts, but have normal hypertrophic chondrocytes. Furthermore, Cbfa1 regulates Osx expression because Osx is not expressed in Cbfa1 KO mice. However, it is not known what regulates Cbfa1 expression. This regulator is of paramount importance because it regulates downstream transcription factors Cbfa1 and Osx thereby controlling bone formation process. Some experimental data indicate tha MAPK activity may be such an upstream regulator in chondrocytes.

3.2. MAP kinases

There are at least three families of MAP kinases: ERK, JNK, and p38. While ERK is primarily activated by mitotic signals, JNK and p38 MAPK pathways are activated in response to cellular stress and proinflammatory cytokines [20]. Recent studies indicate that JNK and p38 MAPK may mediate apoptosis and other differentiation events by transmitting extracellular signals to the nucleus and activating transcriptional factors. It has been shown that p38 and ERK1/2 may be involved in regulating chondrogenesis in a micromass culture system using chick limb bud cells [21].

Many questions arise from the involvement of MAP kinase in regulating bone formation. First, how does MAPK, a ubiquitously expressed molecule regulate a tissue specific function such as endochondral bone formation? The answer may come from observations that MAPK regulates the expression and/or activation of transcriptional factors that are important in regulating bone formation, such as Cbfa1[22]. Thus MAPK may regulate endochondral bone formation through its downstream transcriptional factors. Second, how does MAPKs regulate different aspects of endochondral bone formation including cartilage growth, matrix resortion, and bone formation at the same time? MAPK may accomplish this by regulating different downstream molecules in a chondrocyte. For example, it has been shown that elevated p38 activity may induce matrix resorption through release of MMP, and also induce apoptosis. Third, do tissues adjacent to the growth plate such as skeletal muscle regulate bone formation? Development of adjacent tissues in the limb, such as muscle and bone, needs to be tightly coordinated, and therefore, is dependent on each other. The possible roles for the adjacent tissues to regulate endochondral bone formation are discussed in the following.

4. Extrinsic Regulators

4.1. Peri-articular layer

Capping the growth plate, peri-articular cartilage layer is situated next to proliferating chondrocytes. The perichondral cells in the layer are the major source for production of PTHrP, which forms a concentration gradient decreasing from proliferating to hypertrophic chondrocytes. PTHrP is a potent stimulator of chondrocyte proliferation, and a potent inhibitor of chondrocyte hypertrophy. PTHrP and PTH both bind to PTH receptors, which belong to G-protein-coupled receptors [23]. In a growth plate, PTH receptors are expressed at low levels in proliferating chondrocytes and at the highest level at the prehypertrophic to hypertrophic transition [24], thereby regulating chondrocyte differentiation from the prehypertrophic state to the hypertrophic state. During identification of intrinsic molecules that transduce PTHrP signals, we discovered that PTHrP inhibits chondrocyte hypertrophy by diminishing p38 MAPK activity in chondrocytes [14]. Since PTHrP decreases p38 activity in chondrocytes in a concentration dependent manner [14], p38 may link PTHrP signaling to its regulation of chondrocyte differentiation by controlling Cbfa1 expression.

4.2. Subchondral bone marrow

If the peri-articular cartilage induces proliferation and inhibits hypertrophy of the adjacent proliferating chondrocytes by secreting PTHrP, is there any induction of hypertrophy and matrix resorption at the base of the growth plate where those events take place? Previous studies have shown that the degradation of hypertrophic cartilage requires the presence of bone and marrow [25]. However, the molecular nature of this induction remains elusive.

There are some evidence to suggest that this induction is achieved, at in part, by a chemokine stromal cell-derived factor 1 (SDF-1) that is secreted by bone marrow adjacent to hypertrophic cartilage.

It has been shown that SDF-1 regulates chondrocyte catabolic activities [26]. Chemokines are a soluble peptide family that regulates cell movement, morphology, proliferation, and differentiation [27]. Chemokines achieve their regulation by signaling through a family of seven transmembrane G-protein coupled receptors. SDF-1 is an 8 KDa peptide originally isolated from a bone marrow stromal cell linc [28]. It activates a wide variety of primary cells through binding to its receptor CXCR4 [29]. It has been shown previously that SDF-1 and its receptor CXCR4 play an important role in cell migration, embryonic development, and human immunodeficiency virus infection. CXCR4 is the major HIV type 1 (HIV-1) coreceptor required for viral entry into the host cells [30]. SDF-1, as a ligand of the coreceptor, can block virus from entering the cells [30].

SDF-1 signaling is also important during development and morphogenesis, because both SDF-1 and CXCR4 knockout mice exhibit significant developmental abnormalities that lead to embryonic lethality [31]. Interestingly, SDF-1 and its receptor CXCR4 are expressed in a complementary pattern in a wide variety of apposed tissue pairs during development including gastrular mesoderm/ectoderm, vascular endothelium/mesoderm, thyroid endodermal epithelium/mesenchyme, and nasal ectodermal epithelium/mesenchyme [32]. Such complementary gene expression pattern results in a paracrine regulatory mechanism in which a SDF-1 producing tissue induce development of the opposing tissue that expresses CXCR4. Intriguingly, such complementary expression pattern exists for hypertrophic cartilage, which is positive for CXCR4, and its adjacent subchondral bone marrow, which secret SDF-1. Interaction of SDF-1 and CXCR4 in chondrocytes results in up-regulation of MMP-13, a major collagenase that cleaves collagens in hypertrophic cartilage.

4.3. Skeletal muscle

If both ends of the growth plate are regulated by adjacent tissues, how about the sides of a growth plate, which are circumscribed by skeletal muscle? In adult, muscle weakness is closely associated with osteoporosis (loss of bone), and osteoarthritis (degeneration of cartilage). The health of muscle and bone is so inter-related that a "muscle-bone unit" has been proposed [33]. However, interdependence of muscle and bone during limb development has not been analyzed in a systematic way. Normal growth of the skeleton requires the presence of viable, actively contracting skeletal muscle throughout limb development. Studies from a chick embryo model indicate that chemical paralysis and secondary muscle atrophy result in a decline of bone growth rate of 20-30%, with significant inhibition of chondrocyte proliferation in the growth plate [34]. One explanation is that mechanical load from contraction of muscle is important for cartilage development. Indeed, cartilage formation can be induced by intermittent articulation of embryonic chick membranous bone [35]. Our study using an in vitro 3D cell culture system demonstrated that mechanical stress stimulates proliferation and differentiation of growth plate chondrocytes [36]. However, such studies have limitations. Chemicals inducing paralysis may affect other tissues besides skeletal muscle. In addition, the developmental interdependence of muscle and skeleton may be more than mechanical. Muscle atrophy may affect formation of the blood vessels that supply nutrients to the growth plate, thereby affecting bone formation. Future systematic studies in vivo are needed to determine the relationship between muscle growth and bone formation during limb development.

5. Extracellular Matrix in Cartilage

Normal cartilage is one of the most efficient weight-bearing tissues in a vertebrate body. It provides flexibility, shock absorbance, and other articular functions in different locations of a body including joints and the vertebral column. The strength and resilience of cartilage is determined and provided by its uniquely organized matrix. The matrix of cartilage contains thin fibrils that run in various directions and interconnect with each other to form an integrated network. Any defects in such a matrix network may cause osteoarthritis (OA). For instance, mutations in collagens type II and type IX, extracellular matrix molecules that are components of this network, result in OA [37] [38]. Recently, it has been shown that mutations of a non-collagenous molecule in a novel extracellular matrix protein family called matrilins (MATN) are associated with Multiple Epiphyseal Dysplasia (MED) manifesting with joint pain and early onset OA [39].

In cartilage matrix, there are two major structural units, which have been studied extensively. One is the heterotypic collagen fibrils [40] consisting of collagen types II, XI, IX, and type X, the latter being restricted to hypertrophic cartilage [41]. The other major structural unit is the ternary complex of hyaluronic acid (HA), aggrecan, and link protein [42]. However, very little is known about how these units are connected to form an integrated matrix network. For example, how is one collagen fibril connected to another fibril that is oriented in a different direction? How does a collagen fibril connect with the HA-proteoglycan complex? Recent progress in the field including those made in our laboratory indicates that matrilins may play an important role in connecting different matrix molecules and connecting matrix molecules to chondrocytes.

6. Matrilins: A Novel Family of Extracellular Matrix Proteins

Matrilin-1, a prototype of matrilins formerly called cartilage matrix protein or CMP, forms a filamentous network to bridge different matrix components [43]. Since then, multiple laboratories have shown that formation of such filamentous network is a general property of all members of matrilin family [44] [45] [46]. Matrilin-1 interacts with both collagen fibrils [47] and proteoglycans [48], thereby integrating collagenous components with non-collagenous components in cartilage matrix. Furthermore, matrilin-1 also interacts with chondrocytes via integrin $\alpha 1\beta 1$ [49], thereby forming pericellular filaments to link chondrocytes with interstitial matrix. Defects of the interactions of matrilins in cartilage, therefore, may lead to a defective cell-matrix network.

6.1. Expression

Matrilins consist of four members (Fig. 2), all of which are expressed in skeletal tissues. Matrilin-1 and –3 are more restricted to cartilage tissue, while matrilin-2 and –4 are widely expressed in many connective tissues, including bone, lung, and muscle [50]. So far, studies of the matrilins in cartilage have focused on MATN1 and 3, the cartilage-specific matrilins. For example, we have also shown that MATN1 mRNA is a marker of the maturation zone of a growth plate [1], while MATN3 mRNA is mainly expressed in the proliferation zone [51]. MATN3 is also expressed in the maturation zone, overlapping with that of the mature chondrocyte-abundant MATN1 mRNA. This suggests that, in addition to self-assembly of MATN1 in the maturation zone and self-assembly of MATN3 in the proliferation zone, MATN1 and 3 may co-assemble in the maturation zone [51]. Both MATN1 and 3 levels are increased in OA articular cartilage [52] [53]. Furthermore, MATN1 protein amount in cartilage increases dramatically with age, which may have considerable influence on the physical properties of mature cartilage [54]. Recent evidence suggest that MATN2, in addition to MATN1 and 3, may also play a role in cartilage matrix assembly. MATN2 together with MATN1

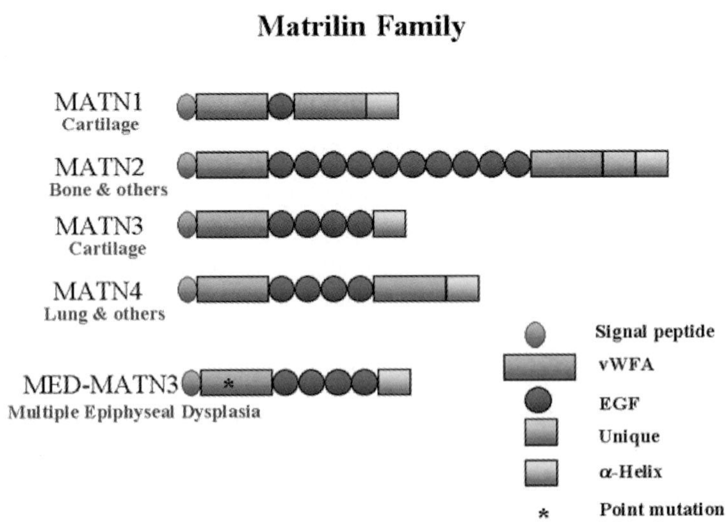

and 3 are among the most up-regulated genes during chondrocyte maturation, as indicated by recent gene microarray analysis [55] [56].

6.2. Assembly

The assembly process of matrilin network determines how different matrix components are connected in matrix. Analysis of the matrilin assembly process so far indicates that assembly of matrilin filaments requires all the domains composing a matrilin molecule [43] [57], including vWF A domains, EGF-like domains, and a heptad repeat coiled-coil domain at the carboxyl-terminal end (Fig. 2). Assembly of matrilins consists of at least two steps. First, a matrilin monomer assembles into an oligomer. Second, matrilin oligomers interact with ligands in the matrix through all subunits to build a network. The carboxyl coiled-coil domain is essential to the first step, because it acts as a nucleation site for oligomerization [58]. It is also required for the second step, because network formation cannot occur without oligomerization of matrilins. EGF-like domain is not required for oligomerization of matrilins. Nevertheless it is required for filament assembly, since deletion of EGF-like domain alters the configuration of a matrilin. The vWF A domain is essential to the second step, because it contains metal ion-dependent adhesion sites (MIDAS) responsible for matrilin interacting with matrix molecules [57]. Intergrins and matrix molecules such as some collagens contain the same adhesion site, which is responsible for matrix adhesions of these molecules [59]. In addition, the A2 domain abutting the coiled-coil modulates oligomerization of matrilins. For example, deletion of A2 from MATN1 converts the trimeric MATN1 into a mixture of monomers, dimers, and trimers. This suggests that the domain abutting the coiled-coil may modulate oligomerization.

6.3. Degradation

Excessive proteolysis of matrilins, if occurs, may have catastrophic effect on cartilage matrix structure. The damage could be two fold. First, cleaving matrilin oligomers may destroy a structural link between collagen fibrils and aggrecan complex, and weaken a connection between chondrocytes and interstitial matrix. Second, cleavage fragments of matrilins that contain the adhesive vWFA domains may bind intergrins and matrix ligands including collagens, thereby preventing adhesion of full-length matrilins to these molecules. Degradation events with such important implication clearly warrant investigation. It has been reported recently that MATN4 can also be cleaved at a site between the vWF A2 domain and the coiled-coil domain [45]. The mechanism for extracellular proteolysis of matrilins may be conserved among different members of the matrilin family.

Currently it is not known which extracellular matrix protease cleaves matrilins. The major enzymes cleaving extracellular matrix include matrix metalloproteinase (MMP) and aggrecanases. Aggrecanases are members of the ADAMTS (a disintegrin and metalloprotease with thrombospondin motifs) family. The ADAMTS family members are extracellular matrix proteins with ADAM-like protease domain and matrix-binding thrombospondin type 1-repeat. Among them, two cartilage aggrecanases, aggrecanase-1 (ADAMTS4) and aggrecanase-2 (ADAMTS5) are present in OA cartilage, and responsible for aggrecan degradation without the participation of other matrix metalloproteinases [60]. Thus, they are potential target for the treatment of OA. It is believed that aggrecanase only cleaves the core protein of a proteoglycan such as aggrecan, versican, and brevican [61, 62]. Both MMP and aggrecanases are candidates for extracellular degradation of matrilins in cartilage.

6.4. Disease

There are two major types of matrilin diseases discovered so far. One is multiple epiphyseal dysplasia (MED), which may associate with abnormal matrilin synthesis and/or assembly. The other is relapsing polychondritis (RP), which may associate with excessive matrilin proteolysis. RP is a rare inflammatory disease that mainly affects cartilage tissue in the auricle, nose, and trachea. The disease is occasionally fatal when tracheolaryngeal cartilage is involved. MATN1 has been identified as an antigen that causes this immune response from both RP patients and animal models [63]. Excessive MATN1 in sera produces autoantibodies to induce respiratory distress and nasal destruction. Proteolysis of MATN1 from matrix releases MATN1 peptides from cartilage into circulation. Thus, study of proteolysis of MATN1 may shed light on the mechanism of RP. In contrast, MED is a relatively common and heterogeneous disease of cartilage manifesting with joint pain and early onset OA. MED shares many features with bilateral Perthes disease and OA. Genetic linkage studies indicated that mutations of COMP, type IX collagen ($\alpha1$, $\alpha2$, and $\alpha3$), and MATN3 cause MED and related conditions. However, it is not known how mutations in MATN3 lead to MED. The MED mutations of MATN3 occur in the vWF A domain that contains MIDAS. Thus, mutations in the A domain may affect MATN adhesion to matrix ligand, thereby impairing filamentous network formation. Alternatively, MATN3 mutants may have secretion defects similar to COMP mutants, which are sequestered within a chondrocyte, leading to absence of the molecule in matrix [64].

6.5. Function

Matrilin assembly defect or excessive proteolysis may disrupt the formation of an integrated cell-matrix network, thereby affecting tissue integrity of cartilage. However, in MATN1 knockout mice, the cartilage phenotype is very mild according to two published reports [65] [66]. No obvious effects from the lack of MATN1 were found on either organization of chondrocytes, or cartilage formation during skeletal development. Type II collagen fibrillogenesis and fibril organization is abnormal in MATN1 KO mice [65], which confirms the interactions between matrilins and collagens. Neither study has examined articular cartilage in detail during aging; therefore, it is not known whether the lack of MATN1 would induce early onset OA. The lack of MATN1 KO phenotype could also be due to functional redundancy among different matrilin forms in cartilage. Indeed, MATN3 is expressed predominantly by the proliferating chondrocytes, prior to the synthesis of MATN1 by mature chondrocytes. In addition, MATN2 is widely distributed in developing cartilage. Other types of matrilins may compensate for at least part of the MATN1 functions. Thus, a key to determining the roles of matrilins in structural organization of cartilage is to disrupt function of different matrilins in the tissue, and determine the consequences of disrupting matrix interactions of different matrilins on cartilage matrix organization.

7. Conclusions and Outlook

In summary, significant progress has been made in understanding the regulation of endochondral bone formation and the properties of extracellular matrix in skeletons in the last several years. The rapid progress of our knowledge of matrilins is a good example. In the future, in vivo studies using transgenic or knockout mouse models are needed to verify the conclusions drawn from biochemical studies in vitro.

Acknowledgments

The original studies in this article were supported by funding from NIH (AG14399, AG00811), and from Arthritis Foundation.

References

1. Chen, Q., et al., Progression and recapitulation of the chondrocyte differentiation program: cartilage matrix protein is a marker for cartilage maturation. Developmental Biology, 1995. **172**(1): p. 293-306.
2. Sandell, L.J., J.V. Sugai, and S.B. Trippel, Expression of collagens I, II, X, and XI and aggrecan mRNAs by bovine growth plate chondrocytes in situ. Journal of Orthopaedic Research, 1994. **12**(1): p. 1-14.
3. Schmid, T.M. and T.F. Linsenmayer, *Developmental acquisition of type X collagen in the embryonic chick tibiotarsus.* Developmental Biology, 1985. **107**(2): p. 373-81.
4. Karaplis, A.C., et al., Lethal skeletal dysplasia from targeted disruption of the parathyroid hormone-related peptide gene. Genes Dev., 1984. **8**: p. 277-289.
5. Schipani, E., et al., Targeted expression of constitutively active receptors for parathyroid hormone and parathyroid hormone-related peptide delays endochondral bone formation and rescues mice that lack parathyroid hormone-related peptide. Proceedings of the National Academy of Sciences of the United States of America, 1997. **94**(25): p. 13689-94.

6. Hunziker, E.B., *Mechanism of longitudinal bone growth and its regulation by growth plate chondrocytes.* Microscopy Research & Technique, 1994. **28**(6): p. 505-19.
7. Kirsch, T., et al., Annexin V-mediated calcium flux across membranes is dependent on the lipid composition: implications for cartilage mineralization. Biochemistry, 1997. **36**(11): p. 3359-67.
8. Aubin, J.E., et al., *Osteoblast and chondroblast differentiation.* Bone, 1995. **17**(2 Suppl): p. 77S-83S.
9. Bianco, P., et al., Bone formation via cartilage models: The "borderline" chondrocyte [Review]. Matrix Biology, 1998. **17**(3): p. 185-192.
10. Gerstenfeld, L.C. and F.D. Shapiro, Expression of bone-specific genes by hypertrophic chondrocytes: Implications of the complex functions of the hypertrophic chondrocyte during endochondral bone development. Journal of Cellular Biochemistry. 62(1):1-9, 1996 Jul.
11. Gibson, G., D.L. Lin, and M. Roque, *Apoptosis of terminally differentiated chondrocytes in culture.* Experimental Cell Research, 1997. **233**(2): p. 372-82.
12. Pacifici, M., et al., Hypertrophic chondrocytes. The terminal stage of differentiation in the chondrogenic cell lineage? Annals of the New York Academy of Sciences, 1990. **599**: p. 45-57.
13. Wang, Y., et al., Expression of Bcl-2 protein in the epiphyseal plate cartilage and trabecular bone of growing rats. Histochemistry & Cell Biology, 1997. **108**(1): p. 45-55.
14. Zhen, X., Wei, L., Wu, Q., Zhang, Y., Chen, Q., *Mitogen-activated protein kinase p38 mediates regulation of chondrocyte differentiation by parathyroid hormone.* The Journal of Biological Chemistry, 2001. **276**(7): p. 4879-4885.
15. Ducy, P. and G. Karsenty, *Genetic control of cell differentiation in the skeleton.* Current Opinion in Cell Biology, 1998. **10**(5): p. 614-9.
16. Leboy, P., et al., *Smad-Runx interactions during chondrocyte maturation.* Journal of Bone & Joint Surgery - American Volume., 2001. **83-A Suppl 1**(Pt 1): p. S15-22.
17. Komori, T., et al., Targeted disruption of Cbfa1 results in a complete lack of bone formation owing to maturational arrest of osteoblasts.[see comment]. Cell., 1997. **89**(5): p. 755-64.
18. Takeda, S., et al., Continuous expression of Cbfa1 in nonhypertrophic chondrocytes uncovers its ability to induce hypertrophic chondrocyte differentiation and partially rescues Cbfa1-deficient mice. Genes & Development., 2001. **15**(4): p. 467-81.
19. Nakashima, K., et al., The novel zinc finger-containing transcription factor osterix is required for osteoblast differentiation and bone formation. Cell., 2002. **108**(1): p. 17-29.
20. Raingeaud, J., et al., MKK3- and MKK6-regulated gene expression is mediated by the p38 mitogen-activated protein kinase signal transduction pathway. Molecular & Cellular Biology, 1996. **16**(3): p. 1247-55.
21. Oh, C.D. and J.S. Chun, Signaling mechanisms leading to the regulation of differentiation and apoptosis of articular chondrocytes by insulin-like growth factor-1. Journal of Biological Chemistry., 2003. **278**(38): p. 36563-71.

22. Xiao, G., et al., MAPK pathways activate and phosphorylate the osteoblast-specific transcription factor, Cbfa1. Journal of Biological Chemistry., 2000. **275**(6): p. 4453-9.
23. Abou-Samra, A.B., et al., Expression cloning of a common receptor for parathyroid hormone and parathyroid hormone-related peptide from rat osteoblast-like cells: a single receptor stimulates intracellular accumulation of both cAMP and inositol trisphosphates and increases intracellular free calcium. Proceedings of the National Academy of Sciences of the United States of America, 1992. **89**(7): p. 2732-6.
24. Lee, K., et al., In situ localization of PTH/PTHrP receptor mRNA in the bone of fetal and young rats. Bone, 1993. **14**(3): p. 341-5.
25. Cole, A.A., et al., The influence of bone and marrow on cartilage hypertrophy and degradation during 30-day serum-free culture of the embryonic chick tibia. Developmental Dynamics, 1992. **193**(3): p. 277-85.
26. Kanbe, K., K. Takagishi, and Q. Chen, Stimulation of matrix metalloprotease 3 release from human chondrocytes by the interaction of stromal cell-derived factor 1 and CXC chemokine receptor 4.[see comment]. Arthritis & Rheumatism., 2002. **46**(1): p. 130-7.
27. Pulsatelli, L., et al., *Chemokine production by human chondrocytes.* Journal of Rheumatology, 1999. **26**(9): p. 1992-2001.
28. Jo, D.Y., et al., Chemotaxis of primitive hematopoietic cells in response to stromal cell-derived factor-1. Journal of Clinical Investigation, 2000. **105**(1): p. 101-11.
29. Mohle, R., et al., The chemokine receptor CXCR-4 is expressed on CD34+ hematopoietic progenitors and leukemic cells and mediates transendothelial migration induced by stromal cell-derived factor-1. Blood, 1998. **91**(12): p. 4523-30.
30. Mbemba, E., et al., Glycans and proteoglycans are involved in the interactions of human immunodeficiency virus type 1 envelope glycoprotein and of SDF-1alpha with membrane ligands of CD4(+) CXCR4(+) cells. Virology, 1999. **265**(2): p. 354-64.
31. Ma, Q., et al., *Impaired B-lymphopoiesis, myelopoiesis, and derailed cerebellar neuron migration in CXCR4- and SDF-1-deficient mice.* Proceedings of the National Academy of Sciences of the United States of America, 1998. **95**(16): p. 9448-53.
32. McGrath, K.E., et al., *Embryonic expression and function of the chemokine SDF-1 and its receptor, CXCR4.* Developmental Biology (Orlando), 1999. **213**(2): p. 442-56.
33. Frost, H.M. and E. Schonau, *The "muscle-bone unit" in children and adolescents: a 2000 overview.* Journal of Pediatric Endocrinology & Metabolism., 2000. **13**(6): p. 571-90.
34. Germiller, J.A. and S.A. Goldstein, *Structure and function of embryonic growth plate in the absence of functioning skeletal muscle.* Journal of Orthopaedic Research, 1997. **15**(3): p. 362-70.
35. Hall, B.K., In vitro studies on the mechanical evocation of advenitious cartilage in the chick. Journal of Experimental Zoology, 1968. **168**(3): p. 283-305.
36. Wu, Q., Zhang, Y., Chen, Q., Indian hedgehog is an essential component of mechanotransduction complex to stimulate chondrocyte proliferation. The Journal of Biological Chemistry, 2001. **276**(38): p. 35290-35296.

37. Metsaranta, M., et al., Chondrodysplasia in transgenic mice harboring a 15-amino acid deletion in the triple helical domain of pro a1(ii) collagen chain. J. Cell Biol., 1992. **118**: p. 203-212.
38. Nakata, K., et al., Osteoarthritis associated with mild chondrodysplasia in transgenic mice expressing a1(IX) collagen chains with a central deletion. Proc. Natl. Acad. Sci. U.S.A., 1993. **90**: p. 2870-2874.
39. Chapman, K.L., et al., Mutations in the region encoding the von Willebrand factor A domain of matrilin-3 are associated with multiple epiphyseal dysplasia. Nature Genetics, 2001. **28**(4): p. 393-6.
40. Mendler, M., et al., Cartilage contains mixed fibrils of collagen types II, IX, and XI. J. Cell Biol., 1989. **108**: p. 191-198.
41. Gibson, G.J. and M.H. Flint, Type X collagen synthesis by chick sternal cartilage and its relationship to endochondral development. Journal of Cell Biology, 1985. **101**(1): p. 277-84.
42. Goetinck, P.F., *Proteoglycans in development.* Current Topics in Developmental Biology 1991;25:111-31.
43. Chen, Q., et al., Cartilage matrix protein forms a type II collagen-independent filamentous network: analysis in primary cell cultures with a retrovirus expression system. Molecular Biology of the Cell, 1995. **6**: p. 1743-1753.
44. Piecha, D., et al., *Expression of matrilin-2 in human skin.* Journal of Investigative Dermatology., 2002. **119**(1): p. 38-43.
45. Klatt, A.R., et al., *Molecular structure, processing, and tissue distribution of matrilin-4.* Journal of Biological Chemistry, 2001. **276**(20): p. 17267-75.
46. Klatt, A.R., et al., Molecular structure and tissue distribution of matrilin-3, a filament-forming extracellular matrix protein expressed during skeletal development. Journal of Biological Chemistry, 2000. **275**(6): p. 3999-4006.
47. Winterbottom, N., et al., *Cartilage matrix protein is a component of the collagen fibril of cartilage.* Developmental Dynamics, 1992. **193**(3): p. 266-76.
48. Paulsson, M. and D. Heinegard, Matrix proteins bound to associatively prepared proteoglycans from bovine cartilage. Biochemical Journal, 1979. **183**(3): p. 539-45.
49. Makihira, S., et al., Enhancement of cell adhesion and spreading by a cartilage-specific noncollagenous protein, cartilage matrix protein (CMP/Matrilin-1), via integrin alpha1beta1. Journal of Biological Chemistry, 1999. **274**(16): p. 11417-23.
50. Deak, F., et al., The matrilins: a novel family of oligomeric extracellular matrix proteins. Matrix Biology., 1999. **18**(1): p. 55-64.
51. Zhang, Y. and Q. Chen, Changes of matrilin forms during endochondral ossification - Molecular basis of oligomeric assembly. Journal of Biological Chemistry, 2000. **275**(42): p. 32628-32634.
52. Okimura, A., et al., *Enhancement of cartilage matrix protein synthesis in arthritic cartilage.* Arthritis & Rheumatism, 1997. **40**(6): p. 1029-1036.
53. Pullig, O., et al., Matrilin-3 in human articular cartilage: increased expression in osteoarthritis. Osteoarthritis & Cartilage, 2002. **10**(4): p. 253-63.
54. Paulsson, M., S. Inerot, and D. Heinegard, Variation in quantity and extractability of the 148-kilodalton cartilage protein with age. Biochem. J., 1984. **221**: p. 623-630.

55. Sekiya, I., et al., In vitro cartilage formation by human adult stem cells from bone marrow stroma defines the sequence of cellular and molecular events during chondrogenesis. Proceedings of the National Academy of Sciences of the United States of America, 2002. **99**(7): p. 4397-402.
56. Stokes, D.G., et al., Assessment of the gene expression profile of differentiated and dedifferentiated human fetal chondrocytes by microarray analysis. Arthritis & Rheumatism, 2002. **46**(2): p. 404-19.
57. Chen, Q., et al., Assembly of a novel cartilage matrix protein filamentous network: Molecular basis of differential requirement of von Willebrand factor A domains. Molecular Biology of the Cell., 1999. **10**(7): p. 2149-2162.
58. Haudenschild, D.R., et al., The role of coiled-coil alpha-helices and disulfide bonds in the assembly and stabilization of cartilage matrix protein subunits. A mutational analysis. Journal of Biological Chemistry, 1995. **270**(39): p. 23150-4.
59. Lee, J.O., et al., Crystal structure of the A domain from the alpha subunit of integrin CR3 (CD11b/CD18). Cell, 1995. **80**(4): p. 631-8.
60. Malfait, A.M., et al., Inhibition of ADAM-TS4 and ADAM-TS5 prevents aggrecan degradation in osteoarthritic cartilage. Journal of Biological Chemistry., 2002. **277**(25): p. 22201-8.
61. Caterson, B., et al., *Mechanisms involved in cartilage proteoglycan catabolism.* Matrix Biology., 2000. **19**(4): p. 333-44.
62. Flannery, C.R., et al., *Autocatalytic cleavage of ADAMTS-4 (Aggrecanase-1) reveals multiple glycosaminoglycan-binding sites.* Journal of Biological Chemistry., 2002. **277**(45): p. 42775-80.
63. Hansson, A.S., D. Heinegard, and R. Holmdahl, *A new animal model for relapsing polychondritis, induced by cartilage matrix protein (matrilin-1).* Journal of Clinical Investigation, 1999. **104**(5): p. 589-98.
64. Delot, E., et al., Physiological and pathological secretion of cartilage oligomeric matrix protein by cells in culture. Journal of Biological Chemistry., 1998. **273**(41): p. 26692-7.
65. Huang, X., D.E. Birk, and P.F. Goetinck, Mice lacking matrilin-1 (cartilage matrix protein) have alterations in type II collagen fibrillogenesis and fibril organization. Developmental Dynamics, 1999. **216**(4-5): p. 434-41.
66. Aszodi, A., et al., Normal skeletal development of mice lacking matrilin 1: redundant function of matrilins in cartilage? Molecular & Cellular Biology, 1999. **19**(11): p. 7841-5.

CHAPTER 8

BONE MORPHOGENETIC PROTEINS IN BONE FORMATION AND DEVELOPMENT

Xiu-Jie Qi[1], Philip R. Bell[2] and Gang Li[2]

[1]*Department of Anesthetics, The Central Hospital of Chang Chun, Chang Chun City, Ji-Lin Province, 130051, PR China*

[2]*Department of Trauma and Orthopaedic Surgery, Queen's University Belfast, Musgrave Park Hospital, Belfast, BT9 7JB, UK.*

The discovery, purification, and recombinant synthesis of bone morphogenetic proteins (BMPs) constitute a major milestone in the understanding of bone physiology, and this chapter discusses the history of BMPs from Senn's accidental discovery in the 1880s to Urist's monumental discoveries in the 1960s through to the present day and the FDAs decision on the use of BMPs in mainstream medicine, their classification and functions. The role of BMPs in the development and formation of bone in the embryo and in the adult, their clinical applications in orthopaedics, dentistry, gene therapy and in other medical fields, the dosage, carriers for BMPs, and the potential risks that accompany the use of BMPs, are all reviewed and discussed.

1. The History and Classification of BMPs

During 1889, Senn[1] noted while he was treating osteomyelitis defects in bone using a decalcified residue of ox bone and iodoform that the decalcified bone induced healing in the bone defect. In the 1930's Levander[2, 3] noted that crude alcohol extracts of bone induced new bone formation when injected into muscle tissue. In 1961 Sharrard and Collins[4] reported the use of ethylenediaminetetraacetic acid (EDTA) decalcified allograft of bone for spinal fusion in children. The idea was supported by laboratory studies carried out by Ray and Holloway.[5]

Probably one of the most significant discoveries in this field was made by Marshall Urist. In 1965 Urist[6] showed the ability of bone matrix to induce bone formation. Urist did this by implanting HCL-decalcified homogenous diaphyseal bone from animal donors into ectopic sites, e.g. a pouch in the belly of the rectus abdominus, quadriceps, or erector spinae muscles. Urist found that the implanted bone extracts induced new bone formation and he named the active ingredient "bone morphogenetic protein" or "osteogenic protein". However, this research was hampered by the fact that there was no reproducible assay for the protein and that it was not conclusively determined that the putative protein was responsible for new bone induction at the ectopic sites. In 1983 Reddi and Sampath created a crude but highly reproducible assay for ectopic bone formation, and they showed that when the protein component was isolated from the rest of the matrix the remaining matrix did not induce new bone formation. However when the protein was reconstituted with the matrix it was as effective at inducing new bone formation as the original matrix responsible for ectopic bone formation.[7]

Table 1: Bone morphogenetic protein family

BMP	Main Functions
BMP-1	Release of BMPs from bone matrices
BMP-2	Osteoinductive, osteoblast differentiation
BMP-3 (Osteogenin)	Most abundant BMP in bone, inhibits osteogenesis
BMP-4	Osteoinductive, lung and eye development
BMP-5	Chondrogenesis
BMP-6	Osteoblast differentiation, Chondrogenesis
BMP-7 (OP-1)	Osteoinductive, development of kidney and eye
BMP-8 (OP-1)	Osteoinductive
BMP-9	Nervous system, hepatogenesis
BMP-10	Cardiac system development
BMP-11 (GDF-8)	Mesodermal and neuronal tissues patterning
BMP-12 (GDF-7)	Tendon and ligament formation
BMP-13 (GDF-6)	Tendon and ligament formation
BMP-14 (GDF-5)	Chondrogenesis, enhancing tendon and bone healing
BMP-15	Modifies follicle-stimulating hormone activity

GDF: Growth Differentiation Factor.

The first clinical study was by Johnson et al in 1992, who studied purified human BMP[8]. Intense competition followed in gene sequencing for BMP during the 1990s. The final landmark in this saga is the FDA approval in 2002 for OP-1 (BMP-7) for long bone defects treatment and rhBMP-2 in a collagen carrier within a cage for anterior lumbar interbody fusions.

BMPs are members of the TGF-beta superfamily, as classified on the basis of similarities in the amino acid sequence, which includes TGF-beta activins and inhibins, and Müllerian inhibiting substance (MIS).[9, 10]

2. Roles of BMPs in Bone Development and Formation

2.1. Ectopic bone formation and BMPs

Cells located in periosteum, bone marrow, and other extraskeletal sites, have the capacity for bone formation.[11–13] The differentiation of an unspecialized mesenchymal cell population into bone tissue is initiated by a process known as bone induction. Histologically, formation of bone from a transplanted bone chip (which contain BMPs) resembles the classic picture of endochondral ossification. The initial phase is characterized by attraction of mesenchymal stem cells to the site of implantation. These stem cells surround the chip and within 1–3 days there is a powerful wave of mitogenic activity followed by differentiation into cartilage around the bone fragment. The cartilage becomes calcified and new bone forms. It has been accepted that this process demonstrates the cartilage model system for bone formation, but closer inspection of the temporal events has revealed otherwise. Caplan[14] reports that there is a layer of osteogenic cells that form a sheet covering the bone chip and that this layer of cells, in intimate contact with invading capillaries, forms the first osteoid, which is mineralized onto the surface of the bone fragment. The hypertrophic cartilage is, however, replaced by marrow, and there are accounts of marrow formation associated with these bone chips.[15] Ectopic bone formation is usually used a functional assay of the true bone induction capacity of BMPs.

2.2. BMPs and the embryonic skeleton

Over the last number of years BMPs have been localized in developing skeletal structures. This has provided evidence that the role of BMPs can be linked to the patterning and differentiation of skeletal cells. *In situ* hybridization has confirmed that BMP-2 to BMP-7 and GDF-5 to GDF-7 transcripts are present in the developing embryo. This is of particular relevance as the various transcripts are present at times and sites within the embryo that are consistent with their participation in mesenchymal condensation and cartilage differentiation.[16] Although many of the upstream signals of BMP expression at specific sites are unknown, studies suggest that BMPs, fibroblast growth factors (FGFs) and sonic hedgehog (SHH) interact in a hierarchical way to pattern skeletal elements.[17–19]

BMP-2 and 4 were found in the apical ectodermal ridge and the zone of polarizing activity which are two important signalling centres involved in defining limb patterning,[20] however mice carrying the null mutations for BMP-2/4 die at a stage before limb patterning occurs in embryogenesis. This means that there is little know about the specific roles played by BMP-2/4 in early limb development except that mice deficient for BMP-2 are non-viable and have defects in amnion /chorion and cardiac development.[21, 22] It is likely that unlike other BMPs the role of BMP-2 cannot be compensated for, as BMP-2/4 null mice die before birth but BMP-7 null mice only had mild skeletal deformities. This maybe related in part to the role of BMP-4 in the development of lung tissue; however as the full spectrum of BMP-2/4s functions is unknown, other possibilities may exist. BMP-2 has been proven to stimulate bone and cartilage growth in numerous clinical trials and therefore we can hypothesize that BMP-2 should have a major role to play in bone induction and mesenchymal cell differentiation during embryogenesis if indeed other BMPs cannot compensate for the role it plays. It is important to note that without a cartilage precursor for bone induction life could not occur, as the majority of our vital organs are protected either by bone or by cartilaginous tissues at birth. Without these structural protectors our vital organs would not be able to withstand any form of trauma and numerous birth defects could occur, but this is not sufficient to explain why BMP-2/4 null mice die early during embryogenesis.

BMP-3 is the most abundant BMP in bone matrix and is thought to be an antagonist to BMP-2 activity.[23] The mechanism by which this occurs may relate to the activins receptor pathway. BMP-3 null mice showed increased bone density and increased trabecular volume, therefore BMP-3 maybe responsible for halting bone growth at appropriate sites and times. BMP-5 plays a central role in the formation of cartilaginous structures in the outer ear, sternum and ribs, mice with non-functioning BMP-5 genes showed defects in these structures.[23] BMP-5 was found in other sites but there was no apparent abnormality when BMP-5 was not present, presumably due to other BMPs compensating effects for its absence.[24] BMP-6 is involved in chondrocyte hypertrophy and replacement by bone,[25] however in BMP-6 null mice there was no apparent abnormalities at birth.[26] BMP-7 is present in the early limb bud, however mice lacking BMP-7 still survive until birth but there were mild skeletal abnormalities present. BMP-7 is also an inducer of nephrogenesis, and is required for eye development and skeletal patterning, [27, 28] and it may be compensated by other BMPs e.g. BMP-2/4,[29] suggesting that the roles and functions of BMPs may overlap in skeletal development.[30] In the GDF-5 (BMP-14) null mouse there was a clear shortening of the long bones, a reduction in the number of digits in the paws and misshapen bones in the front and hind feet.[31, 32] GDF-5 expression normally occurs at the sites where these malformations and abnormalities where present. Therefore it is most probable that GDF-5 is responsible for joint morphogenesis between individual bones and for maintaining regularity in the size and shape of mesenchymal condensations. These can be observed in humans with GDF-5 gene mutations where joint dysmorphogenesis occurs.[33–35] GDF-11 is thought to be linked to the development of the axial skeleton and in palate development, mice lacking functional GDF-11 had developed additional thoracic and lumbar vertebrae and the complete absence of a tail.[36] GDF-11 may be a negative regulator in skeletal planning and inhibiting chondrongenesis and myogenesis, as ectopic application of GDF-11 in the developing limbs of chicks resulted in shortening of the limbs.[37]

2.3. BMPs in bone remodelling

The size, shape and location of various skeletal elements are determined during embryogenesis; however the adult skeleton undergoes a continuous turnover, bone remodelling, which occurs in response to various systemic and local signals and to mechanical stimulus. Bone remodelling requires the differentiation of osteoblasts and osteoclasts from bone marrow and other precursors.[38] BMPs are local signals which are thought to induce the differentiation of mesenchymal stem cells into osteoprogenitors and osteoblast.[39, 40] Osteoblasts synthesise and secrete BMP both *in vivo* and *in vitro* suggesting that BMPs initiate mesenchymal cell differentiation and create a positive feedback loop allowing the production of additional BMP signals. When recombinant human bone morphogenetic protein 2 (rhBMP-2) was used in segmental bone defect models in rats and in rabbits, it induced endochondral bone formation. Antagonists to BMPs exist, noggin and gremlin are also present in bone and are made by osteoblasts, suggesting that there maybe local control over mesenchymal cell activation and differentiation.[41, 42] BMP-1 is thought to relate to the release of BMPs from collagenous matrix providing an addition source of exogenous BMPs for site specific remodelling.[43] Bone formation and resorption in remodelling cycle maybe linked through BMPs as BMPs regulate the transcription of several osteoblast specific transcription factors.[44]

3. Therapeutic Applications of BMPs

3.1. Matrices and carriers for BMPs

Matrices and carriers are usually needed for BMPs delivery to a particular site. The purpose of using a carrier is not only to control the distribution of BMPs to a specific site but also to retain BMPs long enough for a cellular response to occur. Matrices and carriers are also helpful as they can be used to define the shape and volume of the bone being induced and their physical form can be manipulated and optimized for particular therapeutic applications. Matrices or carriers can be made out of numerous materials such as collagen, hyaluronic acid, allograft of bone, calcium phosphates, hydroxyapatite/tricalcium-phosphate or synthetic materials like polylactide.[45–48]

3.2. *Orthopaedic applications*

The main application of BMPs in relation to orthopaedics is in fracture repair, where BMPs could be used when insufficient repair has occurred or simply to accelerate the rate of fracture repair. They may also be used to treat large segmental bone defects and in spinal fusions where large amounts of bone are needed. Many animal studies have demonstrated the ability of BMP-2 to accelerate fracture repair in rabbit, goat, and dog models.[45, 46, 47, 48] These studies showed that BMP-2 was able to reduce the fracture healing time by 30-50% and that addition of BMP-2 increased the amount of fracture callus formed and accelerated the maturation of the callus. Several clinical studies have also shown that BMP-2 was useful in treating tibial non-unions and acute tibial fractures.[49] BMP-2 has been shown to heal large segmental bone defects in rats, dogs, rabbits and monkeys.[50] rhBMP-2 has been tested successfully in preclinical animal models of spinal fusion (interbody and intertransverse process) by radiographic, mechanic and histologic criteria.[51–55] These studies have demonstrated that rhBMP performance in spinal fusion is equal to or better than the current autograft procedure.[56]

3.3. *Dental applications*

In dentistry there is a need for bone regeneration to fill tooth extraction sockets, for bone lost to periodontal disease, and to augment the alveolar bone that has decreased with age for dental implants and restoration. Preclinical studies have demonstrated BMPs ability to induce bone formation in segmental defects in the jaw in animal models.[57, 58] Augmentation of maxillary bone and mandibular bone by rhBMP-2 has been shown in goat, dog, monkey and human models respectively,[59–62] and in all scenarios the newly formed bone behaves like the native bone, are capable of supporting dental implants.

3.4. Gene therapy

This involves vectoring of a gene into cells, which will then synthesize the BMPs of interest. Major problems arise from gene therapy relating to BMPs as they are only needed for a short period of time to heal the fracture/instigate bone induction and finding a mechanism by which the expression of BMPs can be "switched off" will be the largest challenge. However adenoviral vectors carrying a BMP-2 gene have been shown in animal models to enhance fracture healing and enhance spinal fusions.[63–67] BMP carrying adenoviral vectors inserted into bone marrow mesenchymal cells have been shown to induce new bone formation and repair bone and cartilage defects.[68–71] Direct application of DNA containing a BMP gene construct has also been shown to enhance long bone repair in a rat model.[72]

3.5. Other uses of BMPs

rhBMP-2 has been shown to inhibit proliferation of vascular smooth muscle cells without stimulating extra cellular matrix synthesis, and this suggested the possibility of therapeutic application of rhBMP-2 for the treatment and prevention of vascular proliferative disorders.[73] BMP-6 is believed to be a brain and muscle protective agent, in an ischemic rat model, BMP-6 has been shown to reduce the size of the infarct in hear and brain.[74] It is an intriguing possibility that in the future BMP may be useful as a protective agent in severe head trauma and stroke. In patients with chronic renal disease levels of BMPs are lower because kidneys are their primary source in the human adult. Renal osteodystrophy syndrome occurs in patients undergoing long-term dialysis in cases of end stage kidney disease. It is possible that systemic administration of BMPs may restore some of the renal functions in patients with chronic renal failure.[74]

4. Potential Risks of using BMPs

As BMPs occur naturally in the body we already know that their presence is tolerated, however they are only present in small quantities and therefore the dosage of BMPs must be considered. Toxicity studies in rat and rabbit models using 1,000 times the human rhBMP-2 dose have not found any systemic effects. However, there was evidence that BMP-4 over-expression was associated with heterotopic ossification in fibrodysplasia ossificans progressiva.[75] Recent studies have indicated that the increased levels of BMP-4 mRNA in fibrodysplasia ossificans progressiva cells are attributable to an increased rate of transcription of the BMP-4 gene.[75] The increased activation of BMP-4 in fibrodysplasia ossificans progressiva cells may be attributable to a mutation within the BMP-4 gene itself or to a mutation in another genetic locus that causes over-expression of BMP-4 in the cells of fibrodysplasia ossificans progressiva patients. Therefore, over expression of BMP-4 is related to a disabling disease and that it is entirely possible that large doses of BMPs may elicit a similar response even if it is on a smaller scale. At present a dose of between 6 mg and 12 mg of rhBMP-2 is recommended for treatment of open tibial fractures.[49] It has been estimated that normal bone contains approximately 0.002 mg of BMPs per kilogram of pulverized bone, although at a fracture site, the BMPs may be at a higher concentration as a result of release from the injured bone and inflammatory cells, but the exact concentration of the BMPs at the fracture site as opposed to physiological concentration in the normal bone is unknown. However, this means the recommended dosage is at least a magnitude of over 1000 times greater than the amount of BMP present during normal fracture healing. At a cost of approximately £1000/mg of rhBMP-2 (at 2003), the dosage being recommended raises the question of whether or not the use of rhBMPs is economically viable.

Although BMPs are human proteins, there is still a risk of human body developing an immune reaction to the recombinant proteins. This risk increases if rhBMPs are administered repeatedly. Although the magnitude of this risk remains unknown, the FDA has concerns on the potential immune response, which may cause adverse effects on embryogenesis, and maternal immune response. One study has shown that the level of anti-BMP antibody in the serum significantly increased in the animals implanted with BMPs, but returned to normal after 6 weeks.[76] Previous research work have found that there are several other molecules which can inhibit osteoinduction by BMPs in addition to noggin and gremlin. TNP-470, a synthetic analog of fumagillin, is an antiangiogenic agent that strongly inhibits neovascular formation *in vivo*. TNP-470 reversibly inhibited the biological activity of rhBMP-2 in the early stage of bone induction, suggesting that angiogenesis may play an essential role in the recruitment of BMP-receptor-positive cells that can respond to rhBMP-2 and differentiate into chondrocytes and/or osteoblasts.[77] The major implication is that TNP-470 like molecule may have a transient inhibitory effect on BMPs and thereby lessen the effectiveness of BMPs, increases the uncertainty of the delayed effects of rhBMP-2 on osteogenesis.[77]

The other major concern is that BMPs may initiate tumors as they were found at higher concentrations in osteosarcomas.[74] Although some tumours express BMP-2 and have BMP-2 receptors, tumour biology studies have found no evidence that rhBMP-2 would initiate tumour. No cytotoxic or mutagenic activity has been found *in vitro*, and no evidence of abnormal cell biology has been found in implant toxicity studies of rhBMP-2. *In vitro* testing of 51 tumour cell lines resulted in growth promotion only in 3 lines (2 pancreas, 1 prostate) and no effect on the remaining cell lines. The preclinical evaluations on carcinogenicity of BMPs, however, are not sufficient to reveal the proteins' effect on tumorigenesis.[74] Another concern is that without appropriate containment of BMPs within carriers, the BMPs may "leak" into inappropriate areas and stimulate bone growth. This may lead to malignancy or a loss of function due to ectopic bone formation which may impede joint mobility if bone is laid down within a joint capsule or more severely disturb metabolic and renal processes if it is laid down in the liver or kidneys. However the systemic availability of rhBMP-2 is low and minimal exposure to the protein occurs outside the implantation site as rhBMP-2 is rapidly cleared from the body through a renal pathway.[74]

5. Conclusion

BMPs are still relatively new to us despite their accidental discovery over 100 years ago. Many of their functions are still unknown, but at present we have the capabilities to hypothesize and assess the positive and negative effects of BMP use in transgenic animal models and animal/human trails. More research still needed before we can safely recommend rhBMPs use in clinical situations. Not only are there the potentials for some severe negative effects may occur, but the use of rhBMPs may be severely restricted by the fact that they are too costly as the current recommended dosage for most procedures is extremely high. However, many research works continue to evaluate the signalling pathways, which BMPs are involved in, in the hope that a smaller but more potent molecule involved in the pathway that is cheaper to produce, and easier to use maybe found. As with all clinical procedures it will come down to a battle between the pros and cons, whether or not the benefits outweigh the risks. Although BMPs have promising therapeutic potentials in many clinical aspects, the long-term effects of rhBMPs usage on human body and health are yet to be defined.

References

1. N. Senn, *Am. J. Med. Sci.,* 98 (1889).
2. G. Levander, *Acta. Chir. Scand.*, 74 (1934).
3. G. Levander G, *Surg. Gynecol. Obstet.*, 67 (1938).
4. W. J. Sharrard and D. H. Collins, *Proc. Roy. Soc. Med.*, 54 (1961).
5. R. D. Ray and J. A. Holloway, *J. Bone Joint Surg. Am.,* 39 (1957).
6. M. R. Urist, *Science*, 150 (1965).
7. T. K. Sampath and A. H. Reddi, *Proc. Natl. Acad. Sci. USA.*, 78 (1981).
8. E. E. Johnson, M. R. Urist and G. A. Finerman, *Clin. Orthop.*, 277 (1992).
9. A. T. Dudley, K. M. Lyons and E. J. Robertson, *Genes Dev.*, 9 (1995).
10. T. Sakou, *Bone,* 22 (1998).
11. J. Connors, (Springer-Verlag, Berlin, 1983), p.20-39.
12. R. Smith and J. T. Triffitt, *Q. J. Med.*, 61 (1986).
13. M. R. Urist, R. J. DeLange and G. A. Finerman, *Science*; 220 (1983).
14. A. I. Caplan in *Cell and Molecular Biology of Vertebrate Hard Tissues*, Ed. A. I. Caplan and D. G. Pechak (Ciba Foundation Symposium, 1988), p.3–21.
15. A. H. Reddi and K. E. Kuettner, *Dev. Biol.*, 82 (1981).
16. L. W. Gamer and V. Rosen, in *Osteoporosis: Genetics, Prevention and Treatment*, Ed. J. Adams and B. Lukert (Kluwer Academic, New York, 2000), p.7-23.
17. M. J. Bitgood and A. P. McMahon, *Dev. Biol.*, 172 (1995).
18. E. Laufer, C. E. Nelson, R. L. Johnson, B. A. Morgan and C. Tabin, *Cell,* 79 (1994).
19. L. Niswander and G. R. Martin, *Nature*, 361(1993).
20. K. M. Lyons, R. W. Pelton and B. L. M. Hogan, *Development*, 109 (1990).
21. G. Winnier, M. Blessing, P. A. Labosky and B. L. M. Hogan, *Genes Dev.*, 9 (1995).
22. H. Zhang and A. Bradley, *Development*, 122 (1996).

23. J. A. King, P. C. Marker, K. J. Seung and D. M. Kingsley, *Dev. Biol.*, 166 (1994).
24. D. M.Kingsley, A. E. Bland, J. M. Grubber, P. C. Marker, L. B. Russel, N. G. Copeland and N. A. Jenkins, *Cell*, 71 (1992).
25. A. Vortkamp, K. Lee, B. Lanske, G. V. Segre, H. M. Kronenberg and C. J. Tabin, *Science*, 273 (1996).
26. M. J. Solloway, A. T. Dudley, E. K. Bikoff, K. M. Lyons, B. L. Hogan and E. J. Robertson, *Dev. Genet.*, 22 (1998).
27. A. J. Dudley, K. M. Lyons and E. J. Robertson, *Genes Dev.*, 9 (1995).
28. G. Luo, C. Hofmann, A. L. J. J. Bronkers, M. Sohocki, A. Bradley and G. Karsenty, *Genes Dev.*, 9 (1995).
29. A. J. Dudley and E. J. Robertson, *Dev. Dyn.*, 208 (1997).
30. S. E. Yi, A. Daluski, R. Pederson, V. Rosen and K. M. Lyons, *Development*, 127 (2000).
31. H. Grunberg and A. J. Lee, *J. Embryol. Exp. Morphol.,*30 (1973).
32. E. E. Storm, T. V. Huynh, N. A. Copeland, D. M. Kingsley and S. J. Lee, *Nature,* 386 (1994).
33. A. Polinkovsky, N. H. Robin, J. T. Thomas, M. Irons, A. Lynn, F. R. Goodman, W. Reardon, S. G. Kant, H. G. Brunner and I. van der Burgt, et al., *Nature Genet.*, 17 (1997).
34. J. T. Thomas, M. W. Kilpatrick, K. Lin, L. Erlacher, P. Lemessis, T. Costa, P. Tsipouras, and F. P. Lutyen, *Nature Genet.*, 17 (1997).
35. M. Weinstein, X. Yang and C. X. Deng, *Cytokine Growth Fac. Rev.*, 11 (2000).
36. A. C. McPherson, M. Lawler and S. J. Lee, *Nature Genet.*, 22 (1999).
37. L. W. Gamer and V. Rosen in *Osteoporosis: Genetics, Prevention and Treatment*, Ed. J. Adams and B. Lukert (Kluwer Academic, New York, 2000), p.7-23.
38. S. C. Manolagas and R. L. Jilka, *N. Engl. J. Med.*, 332 (1995).
39. E. Abe, M. Yamamoto, Y. Taguchi, B. Lecka-Czernick, C. A. O'Brien, A. N. Economides, N. Stahl, R. L. Jilka and S. C. Manolagas, *J. Bone miner. Res.,* 15 (2000).
40. S. C. Manolagas and R. S. Weinstein, *J. Bone Miner. Res.*, 14 (1999).
41. E. Gazzarro, V. Ganji and E. Canalis, *J. Clin. Invest.*, 102 (1998).
42. R. C. Pereira, A. N. Economides and E. Canalis, *Endocrinology*, 141 (2000).
43. F. C. Wardle, L. M. Angerer, R. C. Angerer and L. Dale, *Dev. Biol.*, 206 (1999).
44. S. C. Manolagas and R. S. Weinstein, *J. Bone Miner. Res.*, 14 (1999).
45. M. L. Bouxsein, T. J. Turek, C. Blake, D. D'Augusta, H. Seeherman, and J. Wozney, *Trans. Orthop. Res. Soc.*, 24 (1999).
46. L. S. Popich, S. L. Salkeld, D. C. Rueger, M. Tucker and S. D. Cook, *Trans. Orthop. Res. Soc.*, 22 (1997).
47. T. J. Turek, M. P. G. Bostrum, N. Camacho, C. A. Blake, R. Palmer, H. Seeherman and J. Wozney, *Trans. Orthop. Res. Soc.*, 22 (1997).
48. R. D. Welch, A. L. Jones, R. W. Bucholz, C. M. Reinert, J. S. Tija, W. A. Pierce, J. M. Wozney and X. J. Li, *J. Bone Miner. Res.*, 13 (1998).
49. S Govender, C. Csimma, H. K. Genant and A. Valentin-Opran, *J. Bone Joint Surg.,* 84 (2002).

50. V. Rosen and J. M. Wozney in *Skeletal Growth Factors*, Ed. E. Canalis (Lippincott and Williams and Wilkens, Philadelphia, 2000), p. 299-310.
51. S.D. Boden, P. A. Moskovitz, M. A. Morone and Y. Toribitake, *Spine,* 21 (1996).
52. S. M. David, H. E. Gruber, R. A. J. Meyer, T. Murakami, O. B. Tabor, B. A. Howard, J. M. Wozney and E. N. J. Hanley, *Spine*, 24 (1999).
53. H. S. Sandhu, L. E. Kanim, J. M. Kabo, E. N. Zeegan, D. Lui, L. L. Seeger and E. G. Dawson, *Spine,* 20 (1995).
54. H. S. Sandhu, L. E. Kanim, J. M. Kabo, E. N. Zeegan, D. Lui, R. B. Delamarter and E. G. Dawson, *Spine*, 21 (1996).
55. J. H. Schimandle, S. D. Boden and W. C. Hutton, *Spine*, 20 (1995).
56. S. D. Boden, T. A. Zdeblick, H. S. Sandhu and S. E. Heim, *Spine*, 25 (2000).
57. P. J. Boyne, *Bone,* 19 (1996).
58. D. M. Toruimi, H. S. Kotler, D. P. Luxenberg, M. E. Holtrop and E. A. Wang, *Arch. Otolaryngol. Head Neck Surg.,*117 (1991).
59. O. Hanisch, D. N. Tatakis, M. M. Boslovic, M. D. Rohrer and U. M. E. Wilkesjo, *Int. J. Oral Maxillofac. Implants,* 12 (1997).
60. M. D. Margolin, A. G. Cogan, M. Taylor, D. Buck, T. N. McAllister, C. Toth and B. S. McAllister, *J. Peridontol.*, 69 (1998).
61. B. S. McAllister, M. D. Margolin, A. G. Cogan, M. Taylor and J. Wollins, *Int. J. Peridontol. Restorat. Dent.*, 18 (1998).
62. M. Nevins, C. Kirker-Head, J. M. Wozney, R. Palmer and D. Graham, *Int. J. Peridontol. Restorat. Dent.*, 16 (1996).
63. T. D. Alden, D. D. Pittman, E. J. Beres, G. R. Hankins, D. F. Kallmes, B. M. Wisotsky, K. M. Kerns and G. A. Helm, *J. Neurosurg.*, 90 (1999).
64. A. W. Baltzer, C. Lattermann, J. D. Whalen, P. Wooley, K. Weiss, M. Grimm, S. C. Ghivizzani, P. D. Robbins and C. H. Evans, *Gene Ther.*, 7 (2000).
65. R. T. Franceschi, D. Wang, P. H. Krebsback and R. B. Rutherford, *J. Cell Biochem.*, 78 (2000).
66. G. A. Helm, T. D. Alden, E. J. Beres, S. B. Hudson, S. Das, J. A. Engh, D. D. Pittmann, K. M. Kerns and D. F. Kallmes, *J. Neurosurg.*, 92 (2000).
67. D. S. Musgrave, P. Bosch, S. Ghivizzani, P. D. Robbins, C. H. Evans and J. Huard, *J. Bone Miner. Res.*, 9 (1999).
68. J. R. Lieberman, L. Q. Le, L. Wu, G. A. Finerman, A. Berk, O. N. Wite and S. Stevenson, *J. Orthop. Res.*, 16 (1998).
69. J. R. Lieberman, A. Daluiski, S. Stevenson, L. Wu, P. McAllister, Y. P. Lee, J. M. Kabo, G. A. Finerman, A. J. Berk and O. N. Witte, *J. Bone Joint Surg. Am.*, 81 (1999).
70. J. M. Mason, A. S. Breitbart, M. Barcia, D. Porti, R. G. Pergolizzi and D. A. Grandle, *Clin. Orthop.*, 414 (2000).
71. K. D. Riew, N. M. Wright, S. Cheng, L. V. Avioli and J. Lou, *Calcif. Tissue Int.*, 63 (1998).
72. J. M. Fang, Y. Y. Zhu, E. Smiley, J. Bonadio, J. P. Rouleau, S. A. Goldstein, L. K. McCauley, B. L. Davidson and B. J. Roessler, *Proc. Natl. Acad. Sci. USA,* 93 (1996).

73. T. Nakaoka, K. Gonda, T. Ogita, Y. Otawara-Hamamoto, F. Okabe, Y. Kira, K. Harii, K. Miyazono, Y. Takuwa and T. Fujita, *J. Clin. Invest.*, 100 (1997).
74. K. Azari, B. A. Doll, C. Sfeir, et al., *Expert Opin. Investig. Drugs*, 10 (2001).
75. T. K. Sampath, K. E. Rashka, J. S. Doctor, et al., *Proc Natl Acad Sci USA*, 90 (1993)
76. V. V. Viljanen, T. J. Gao, T. C. Lindholm, T. S. Lindholm, B. Kommonen, *Int. J. Oral Maxillofac. Surg.*, 25 (1996).
77. S. Mori, H. Yoshikawa, J. Hashimoto, T. Ueda, H. Funai, M. Kato and K. Takaoka K, *Bone,* 22 (1998).

CHAPTER 9

MECHANICAL TESTING FOR BONE SPECIMENS

Ling[1] Qin and Ming[2] Zhang

[1]*Musculoskeletal Research Laboratory, Dept. of Orthopaedics & Traumatology, The Chinese University of Hong Kong*
Email: lingqin@cuhk.edu.hk

[2]*Jockey Club Rehabilitation Engineering Centre, The Hong Kong Polytechnic University*
Email: rcmzhang@polyu.edu.hk

Biomechanical tests are designed as an end-point measure in many experimental studies to confirm both pharmaceutical and non-pharmaceutical treatment effects on musculoskeletal tissues, especially on skeletons for the prevention and treatment of osteoporosis and osteoporotic fractures. This chapter introduces basic knowledge and concepts of mechanical testing methodology in studying mechanical properties of bone specimens at organ, tissue and matrix levels. The mechanical testing methods are categorized into both destructive and non-destructive or non-contact ones using both conventional and state-of-the-art testing facilities. Computational modeling method is also briefly introduced. In order to obtain the most update information of testing methods, homepages of some relevant or known research societies and devices suppliers are listed as reference sources to facilitate direct network communication. Practical tips are suggested for preparation and preservation of bone specimens. Factors affecting testing results or data interpretations are summarized to avoid inconsistent and incomparable testing results.

1. Introduction

Osteoporosis is one of the most deteriorating musculoskeletal problems associated with increased risk of osteoporotic fractures. Such problems get severe with aging population.[1-3] Research and development (R&D) of both pharmaceutical and non-pharmaceutical approaches is one of the most effective ways in prevention of osteoporosis and osteoporotic fractures.[2,3] Many established evaluation methods and procedures are useful to prove its efficacy before clinical trials or application. These include biochemistry, hiotomorphometry, bone densitometry and not the last bone biomechanical testing. Although bone densitometry is often used as a surrogate to evaluate bone fragility, direct biomechanical testing of bone undoubtedly provides more information about mechanical integrity. Especially in preclinical evaluations, biomechanical tests are often used as an end-point measurement.[3-9]

Standard mechanical testing methods are typically conducted in line with ASTM (American Society for Testing and Materials) or BS (British Standards) guidelines.[4,5] However bone specimens are generally inhomogenous, anisotropic, and porous materials with widely varying mechanical properties.[4-9] This implies that special considerations should be taken during bone testing when adopting ASTM and BS standard testing methods. Depending on the nature of the specimen and the type of testing to be conducted, the equipment and methodology used in mechanical testing of bone tissues vary widely. Several approaches may be equally valid, and the strategy adopted generally depend on the particular requirements of individual studies. This chapter serves the purpose to provide basic knowledge on bone biomechanics, testing techniques for bone specimens at organ, tissue and matrix levels using both destructive and non-destructive or non-contact methods. Computational modeling method is also briefly introduced. In order to obtain the most update information of testing methods, homepages of some relevant or known research societies and devices suppliers are listed as reference sources to facilitate direct network communications. Practical tips are mentioned for preparation and preservation of bone specimens. Factors affecting testing results or data interpretations are summarized to avoid inconsistent and incomparable testing results.

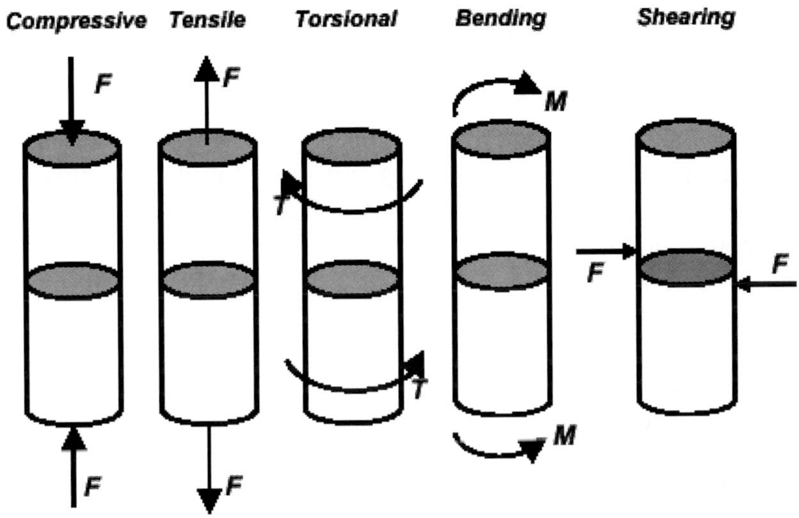

Fig. 1. A schematic illustration of loading modes for mechanical testing. F: force, T: torque, M: bending moment.

2. Basic Biomechanical Concepts[4-9]

2.1. *Force and deformation*

Force is the mechanical interactions between bodies. The physiological forces applied on bone are complicated. However, they can be divided into simple modes or combinations. Those simple modes are centrically axial loading (tension or compression), flexural loading (bending), torsional loading, and shearing loading (Fig. 1). To understand the mechanical properties, a single mode of loading is often applied on the specimen, the relationship between load applied to structure and deformation response to the load called force-deformation curve can be obtained.

2.2. *Stress and strain*

Because the force-deformation relationship depends not only on the material properties, but also on the size of the specimen and the load mode applied, it is necessary to transfer it into stress-strain relationship for the understanding of material properties.

Stress is defined as the force per unit area (N/m^2 or Pa) and strain is defined as deformation per unit length. Both stress and strain have normal (perpendicular to the surface) and shear (in plane) components, referring to the surface specified. As shown in Fig. 2, the normal component is defined as

$$\text{Stress } \sigma = F/A$$
$$\text{Strain } \varepsilon = \Delta L/L_0$$

where L_0 is original length, ΔL is elongation, A is original area and F is force applied on the surface. The shear component is defined as

$$\text{Shear stress } \tau = F/A$$
$$\text{Shear strain is } \gamma = \Delta L/L_0$$

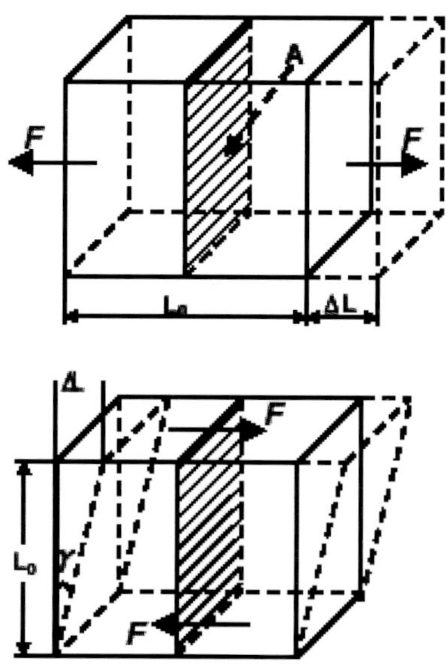

$$\sigma = \frac{F}{A}$$

$$\varepsilon = \frac{\Delta L}{L_0}$$

$$\tau = \frac{F}{A}$$

$$\gamma = \frac{\Delta L}{L_0}$$

Fig. 2. A schematic illustration for definitions of normal stress and strain, and shear stress and strain. F: force applied, A: cross-section area, L_0: initial length, ΔL: deformation, σ: normal stress, τ: shear stress, ε: normal strain, and γ: shear strain.

2.3. Definitions of material properties

A typical stress-strain curve is shown in Fig. 3. The curve can be divided into elastic region and plastic region. The transition point between two regions is called yield point. Sometimes it is not easy to determine the transition point. In engineering, normally a straight line parallel to the elastic region with 0.2% offset strain is drawn, and the cross point is the yield point (Fig. 3). The slope of the line is called elastic modulus or Young's modulus ($E = \sigma/\varepsilon$). The maximum stress σ_u is called ultimate strength and the stress at the yield point is called yield strength σ_y. If the curve is shear stress and shear strain, the slope of the elastic region is called as shear modulus ($G = \tau/\gamma$), or modulus of rigidity.

The area beneath stress-strain curve is the energy per unit volume. The energy required to break the material is referred as toughness. The energy under the linear elastic portion is called modulus of resilience, indicating the energy absorbed without plastic deformation.

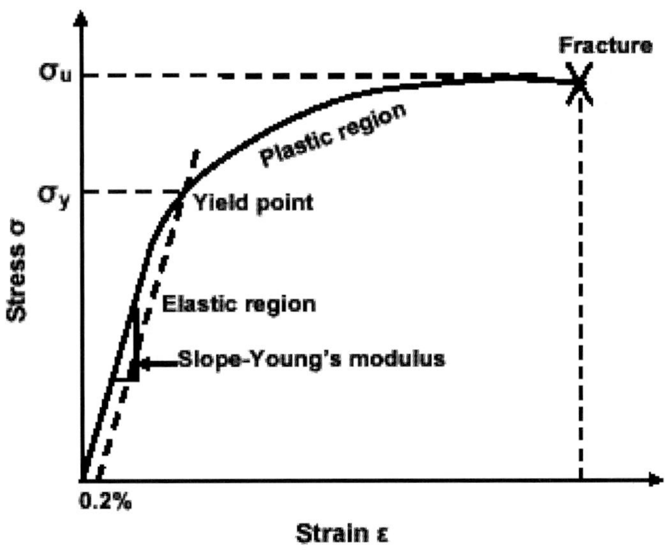

Fig. 3. A schematic illustration of a typical stress-strain curve. σ_u is ultimate strength and σ_y is yield strength

3. Bone Sample Storage and Preparation

3.1. Storage

Sample documentation

A documented history of the intact bone or donor should be kept, including data like age, gender, level of activity, and any diseases affecting bone metabolism that could confound the results of testing. For cadaveric specimens, body weight, body height, gender and the cause of death should be recorded.[4-10]

Sample inspection

A careful visual inspection of specimens should be carried out to check for abnormalities, structural damage, etc. Radiographic inspection is also recommended to check for any bony defects, such as fracture or other morphological abnormalities that may not be found under visual inspection.[4-10]

Storage

Tissue autolysis begins within hours following removal of bone from the body.[10] If no immediate testing is possible, normally the intact bone specimens are wrapped in gauze with 0.9% saline (such as Ringers solution) or in ethanol (50% ethanol / saline), sealed in freezer bags to avoid drying or dehydration, and stored at -20^{0}C freezer until tests are performed, not longer than 2 years. Refrigeration (4°C) is not appropriate for storing bone specimens for extended periods of longer than approximately 12-24 hours.[4-10]

3.2. Bone specimen preparation for testing

Sampling

Specimens shall be thawed out at room temperature. Packing material should only be removed once the specimen has thawed. If a piece of bone specimen is required for testing, such as cortical beam or trabecular cube, careful sectioning using low speed diamond saw or other cutting machines is needed under cooling conditions as soon as taken out from the freezer.[4-10]

Anatomical measurements

Calculation of material parameters such as stress and strain requires conversion by the dimensions of the material. This is not always straightforward, as the cross-sectional areas (CSA) of many tissues vary widely along their lengths. Specimens may be prepared to uniform areas or mean values for CSA can be used. There are several methods to measure CSA, including both contact methods and non-contact methods. Example of contact method includes the use of fine caliber. On the other hand, laser-meter, computer tomography (CT), high-resolution radiography and histomorphometry on cross-sectioned bone specimens after mechanical testing are samples of non-contact methods.[11-14]

The recent use of quantitative CT (QCT) provides data on bone geometry including not only CSA but also cross-sectional moment of inertia (CSMI).[11-13] Taking bone mineral density (BMD) into account, QCT can provide bone strength index (BSI), which is derived from both BMD and CSMI. The BSI has found to have better correlation with bone mechanical strength than either alone.[11,12] This suggests that apart from currently used BMD data, the bone geometry or structure plays an important role in noninvasive prediction of bone mechanical strength.

Mounting

Mounting of the specimen may require more time and effort than the testing itself. Careful mounting is essential to assure an accurate testing, as insecure, irregular or misaligned specimens will skew the results. In order to avoid dehydration or over heating of specimens, preparation and mounting is often performed during thawing, so that once the specimen has thoroughly thawed, it is ready for testing. Overnight storage in a refrigerator of 4°C to thaw partially and then warm up to testing temperature (normally room temperature) during preparation and mounting is often a convenient option.[4-9]

Custom made mounting jigs and frames, such as clamps, supporting or fixation devices are commonly used to attach specimens to the test equipment. A well-designed jig, commonly made from stainless steel, aluminum, non-porous plastics etc. can save time while providing a consistent physiological alignment and loading pattern.

Larger bone specimens often need to be mounted or potted. Bone cement, dental cement, or fast drying epoxy resin adhesives are all suitable, but curing times and temperatures may need to be considered, and augmentation with bone screws, k-wires, etc. is recommended. Potting material such as epoxy, bone cement, or dental cement are often difficult to be removed, and should not be used to fix specimens directly to testing machines (Fig. 4).

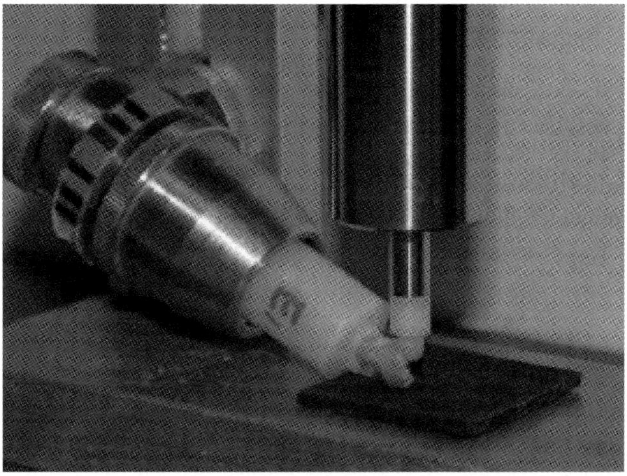

Fig. 4. The jig for fixation of rat femoral shaft for testing compression strength of the femoral neck. Epoxy was used as bone cement (white cylinder) to fix the femoral shaft directly and mounted to a custom-made fixation device.

4. Bone Biomechanical Testing Methods

Bone fracture may be caused by tensile or compressive, or shearing stress. Different testing methods have been developed to understand various behaviors under different load modes. The mechanical behaviors can be described as stiffness, strength, hardness and toughness, etc. Conventional mechanical testing methods include compressive, tensile, torsional, bending, and shear tests. Some non-contact or non-destructive techniques such as ultrasound, scanning acoustic microscopy (SAM) and other imaging methods, provide data correlated with mechanical properties of the bone specimen. The following part introduces methods which are often used in bone mechanical testing and data analyses.

4.1. Axial loading test

If the external force is applied along the axis of the specimen through the center of the transverse cross section without inducing a moment on the specimen, it is called centrically axial loading, including tension and compression (Fig. 5). Such a load produces normal stress without shearing component on the transverse cross section. Under the assumption of uniform distribution of the force, the stress and strain can be expressed as $\sigma = F/A$ and $\varepsilon = \Delta L/L_0$. The stress-strain curve (σ-ε) can be easily obtained from the force-deformation curve (F-ΔL).

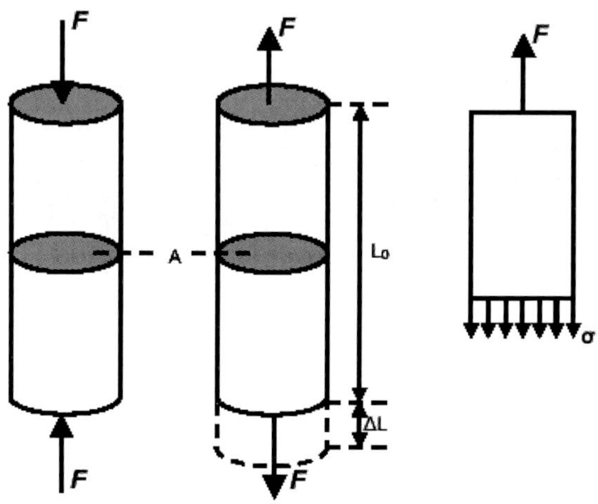

Fig. 5. A schematic illustration of axial loading test. F: Force applied, A: cross-section area, L_0: initial length, ΔL: deformation, σ: normal stress.

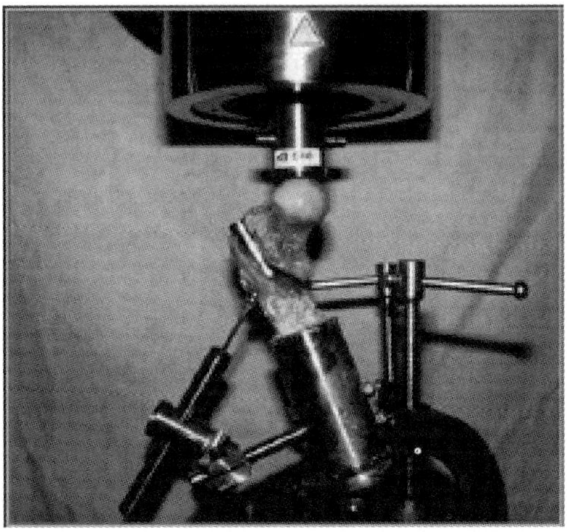

Fig. 6. Compressive test on proximal femur after fixation using intramedullar nail in human cadaver.

Some special compressive tests on both intact bone and fractured bone have been conducted,[4-9] such as compressive tests of femoral head to understand the bone strength, hip fracture mechanism experimentally (Fig 4) and testing quality of fixation after hip fracture in cadaver model (Fig. 6).

4.2. Bending test (flexural loading)

As many bone fractures are related to bending load, it is necessary to understand the bone responses to bending. Bending tests are usually for the testing of intact bone shaft, especially long bones, or fabricated cortical beam. The modes most often used for bone are three-point bending and four-point bending tests.[4-8, 11, 12, 14-16]

Three-point bending test

A test of a long bone supported at the two ends and loaded at the middle point is called three-point bending (Fig. 7a, 8). The force-deflection curve for specified specimen can be obtained experimentally. The conversion of the force-deflection curve to a stress-strain curve is not straightforward. It is better to understand how such loading produces stress and strain within the specimen.

This load configuration produces resultant shear force and moment for each transverse cross section. The magnitudes of resultant shear force and moment for cross section along the bone long axis are called load diagrams (Fig. 7b,c). The load diagram can indicate how much load is applied over each cross section and where the maximum load and fracture may occur.

The resultant shear force Vr produces shear stresses over the cross section distributed as shown in Fig. 7d. The maximum shear stress is located at the middle layer of the cross section. The maximum value of the shear stress, depending on the cross-section geometry is 3Vr/2A, 4Vr/3A and 2Vr/A for the rectangular, cylindrical and hollow cylindrical cross sections, respectively (A is the CSA). The shear stress is normally relatively small in contribution to fracture and is not considered in data analysis of bending test.

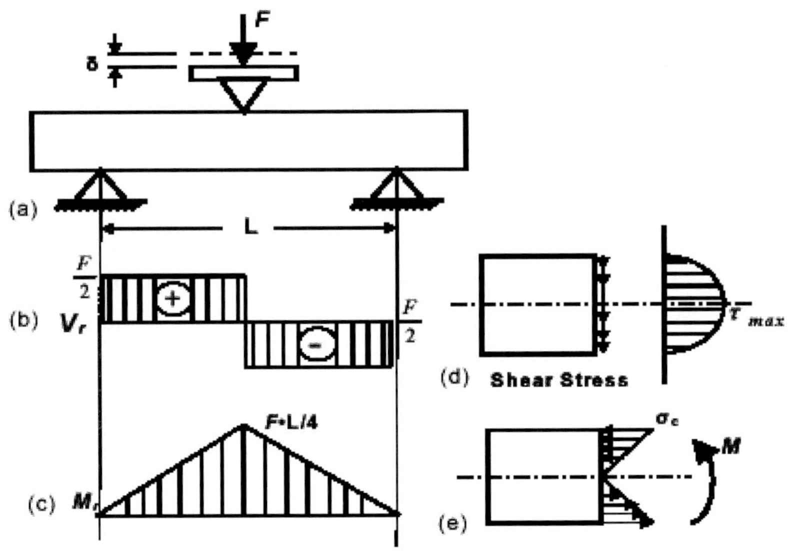

Fig. 7. A schematic illustration of mechanical testing. a) Three-point bending test, b) load diagram for resultant shear Vr, c) load diagram for bending moment Mr, d) shear stress (τ) distribution over cross section and e) normal stress (σ) distribution over cross section. F: force applied, L: span of the beam, δ: deflection.

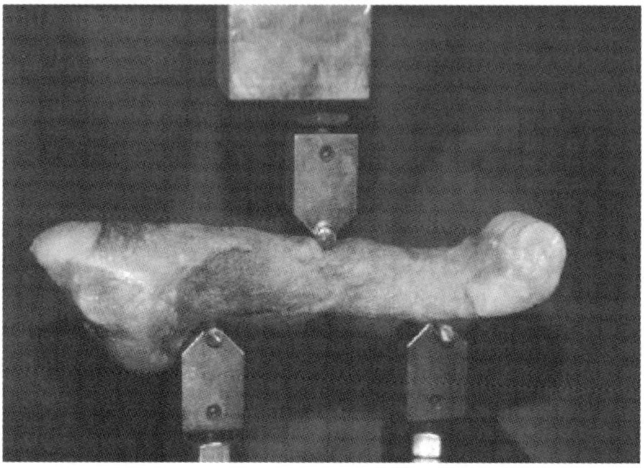

Fig. 8. Three-point-bending test on an intact long bone. The self-rotating metal bar placed on the two lower supporting jigs reduces share stress during bending.

As shown in Fig. 7e, the moment Mr produces normal stresses over the cross section and their distribution can be expressed as

$$\sigma = \frac{My}{I}$$

where M is the moment in the specified cross section, y is the distance to the neutral surface, and I is cross sectional moment of inertia.

Under bending moment, the bone is compressed over one side and extended over the other side, and the peak stress occurs where the point are distant from the neutral surface. For three-point bending, the maximum moment occurs over the cross section under the load application point, so the maximum normal stress occurs over this cross section.

$$\sigma_c = M_{\max}\frac{c}{I} = \frac{FLc}{4I}$$

For a cylinder with a radius R, $I=\pi R^4/4$, $c=R$, maximum normal stress is FL/R^3.

The deflection of the bone at the loading point is $\delta = \dfrac{FL^3}{48EI}$, so the Young's modulus can be obtained using the following formula,

$$E = \frac{L^3}{48I}\frac{F}{\delta}$$

For three-point bending, the most often fracture point is over the cross section where the force is applied and where the largest moment occurs. The contact stress caused by the pressing head may affect the accuracy of testing, because the contact stress produces the compressive stress there.

Four-point bending

The loading configuration shown in Fig 9a, 10 is called four-point bending and it produces a constant uniform moment between two loading points (Fig 9b). The deflection of the bone is,

$$\delta_{max} = \frac{F}{EI}\frac{a}{48}(3L^2 - 4a^2)\text{, at the middle;}$$

$$\delta = \frac{F}{EI}\frac{a}{12}(3aL - 4a^2)\text{, at the loading point.}$$

The calculation of the Young's modulus depends on which deflection is measured,

either $E = \dfrac{F}{\delta_{max}}\dfrac{a}{48I}(3L^2 - 4a^2)$, if the deflection at the middle is measured,

or $E = \dfrac{F}{\delta}\dfrac{a}{12I}(3aL - 4a^2)$, if the deflection at the loading point is measured.

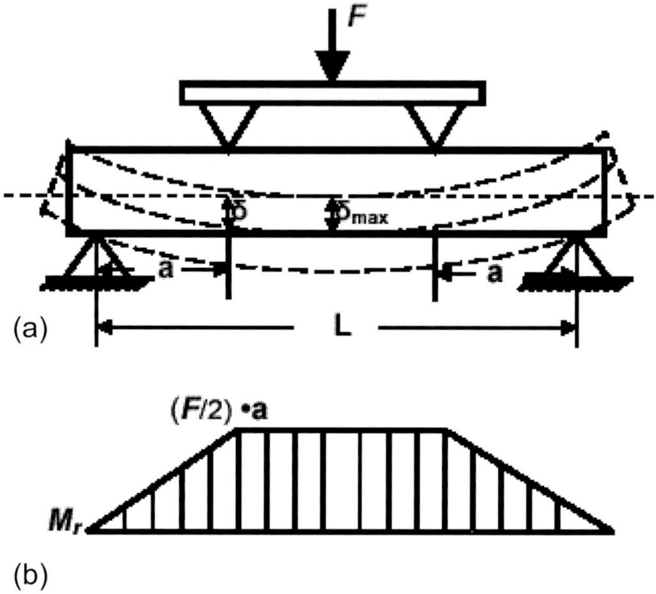

Fig. 9. (a) A schematic illustration of four-point bending test, (b) load diagram for bending moment (M_r). F: force applied, L: span of the beam, a: distance loading point to support, δ: deflection.

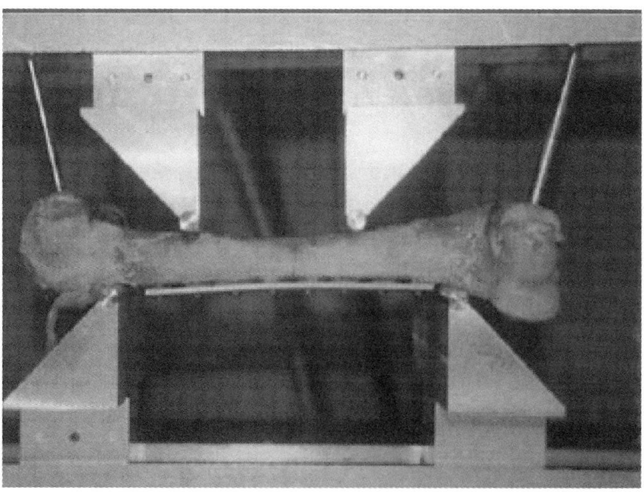

Fig. 10. Four-point bending test of a long bone with fracture fixation.

During bending text, the concave site sustains for tension while convex site endures more compression (Fig. 9a). As bone sustains mainly compressive load in normal physiological conditions, normally it breaks first at the tension site. As compared with three-point bending test, the stress distribution in four-point bending test is rather homogenous along the tested region and failure occurs mostly around the weakest region of the bone of fracture-fixation device complex. In addition, single osteon has also been prepared for bending test at micro level, however, it is not popular to evaluate entire bone structure because of small sampling and variations in degrees of its mineralization, which affect the testing results.[17]

4.3. *Torsional testing*

Axial loading and bending tests produce normal stresses over the transverse cross section, and the mechanical behaviors of normal stress and normal strain can be obtained. In order to measure the shear properties, torsional or pure shearing loading test can be conducted.

If the bone is subject to twisting moment called torque applied at the two ends (Fig. 11a, 12), shear stresses are produced over the transverse cross section, distributed as (Fig. 11b,c),

$$\tau = \frac{T\rho}{J}$$

where T is the torque applied, ρ is a specified radius, and J is the polar moment of inertia. For a cylindrical cross section, J is equal to $\pi R^4/2$, and for a hollow cylinder, J is $\pi(R^4 - r^4)/2$. The maximum shear stress $\tau_{max} = TR/J$ occurs over the outer surface.
The angular deformation is given by

$$\phi = \frac{TL}{JG}$$

, so the shear modulus can be expressed as:

$$G = \frac{T}{\phi}\frac{L}{J}$$

The torque-angular deformation (T- Φ) can be measured by experiments.

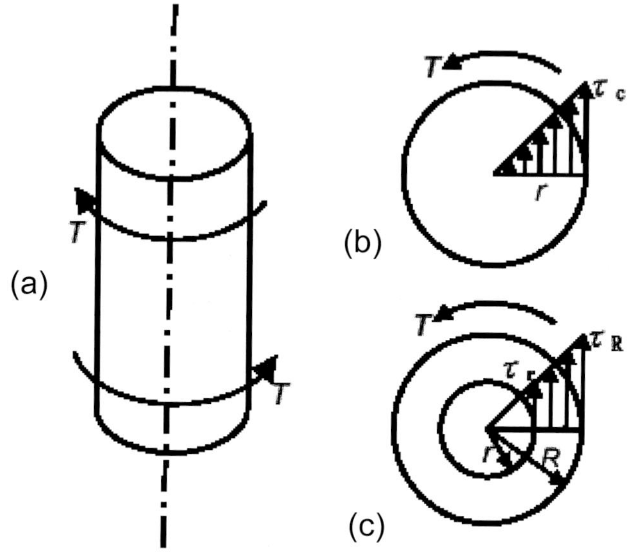

Fig. 11. A schematic illustration of a) Torsional testing, b) shear stress distribution over solid cylindrical cross section and c) shear stress distribution over hollow cylindrical cross section.). T: torque applied, R: outer radius, r: inner radius, τ_c: shear stress on surface.

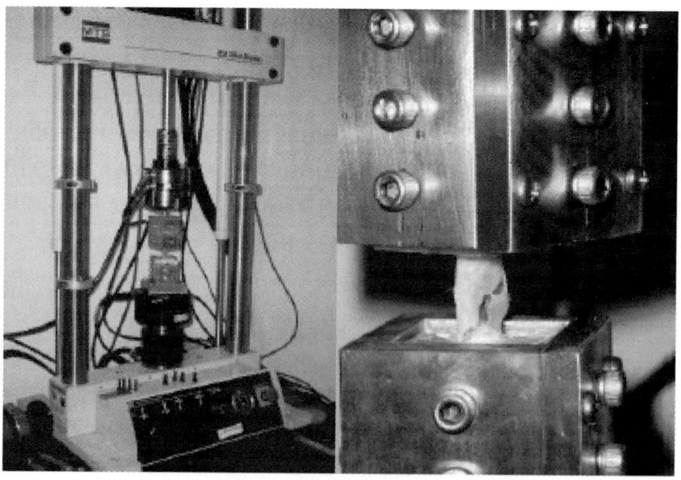

Fig. 12. Tortional test conducted for a long bone. Left. Biaxial testing machine (MTS 858) with testing jig and long bone in place. Right, close-up of the testing jig and long bone after testing to failure with a typical oblique fracture.

4.4. Pure shear test

As mentioned in section 4.2.2, the resultant shear force produces shear stresses over transverse cross section, with maximum value at the middle layer. To conduct an accurate shear testing, normally a special rig is needed. For simplicity, the shear stress is assumed to be distributed uniformly and calculated by $\tau = F/A$, where A is the shear area. The stress is in a complicated situation, caused by compression from the contact load and by shearing in the shearing plane. It is not easy to measure the shear strain. However, as compare with compression, bending or tensile test, pure shear test has rarely been used in mechanical testing of bone specimens.

4.5. Fatigue test

Bone also behaves fatigue. Under repetitive loading lower than the yield point, the bone behaviours change with cycles, such as degradation in strength and stiffness. Fatigue is a slow progressive and accumulated process under cyclic loading. The magnitude of stress applied as a function of the number of cycles when fatigue failure occurs is called S-N curve, relationship of fatigue stress and number of cycles (Fig. 13). Endurance limit is the maximum stress that the material can endure an

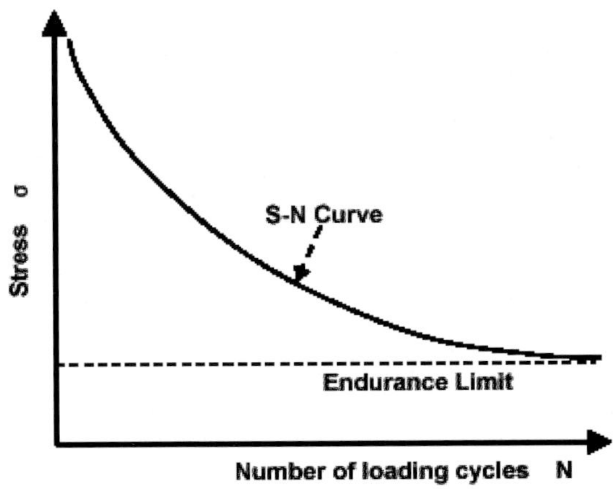

Fig. 13. A schematic illustration of a typical S-N curve under fatigue testing. S-N means stress-number of cycles.

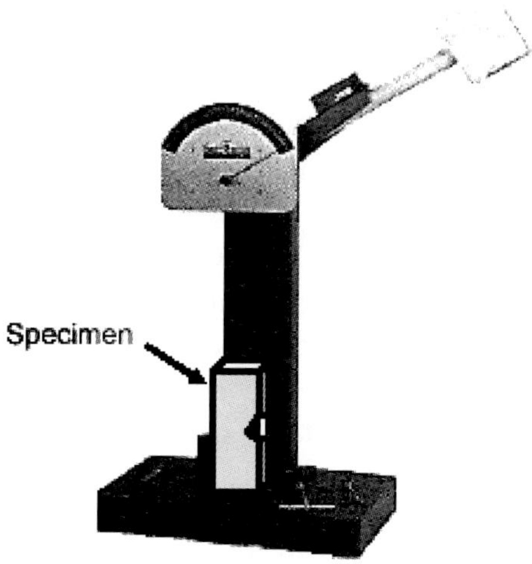

Fig 14. Impact testing machine (Tinus Olsen Testing machine Co.)

infinite number of stress cycles without fatigue failure. Fatigue test can be conducted for tensile, compressive, torsional or bending load. The cyclic loading curve often used is sinusoidal with either zero or non-zero mean.

4.6. Impact test

Impact is defined as a sudden application of a load to a local area of material. Impact test applies impact load to break the material, and gives the indication of the relative toughness of a material. The specimen is normally a standard notched bar. A pendulum swinging strikes and breaks the specimen (Fig. 14). The energy can be calculated by difference of the potential energy at the initial and final heights of the pendulum.

4.7. Indentation test

Indentation tests are used to measure hardness of bone. Hardness of a solid material is a measure of its resistance to penetration by another solid. The measurement of hardness is the indentation force and the permanent deformation remaining on the bone. There are many types of indentations developed, based on the geometry and size of indenters. The indenter shapes often used are sphere, coned, cylindrical and pyramid, etc. Depending on the size of indenter, indentations can be categorized into macro-, micro- and nano-indentation.

The hardness tests often used are named as Brinell, Vickers and Rockwell. Brinell hardness (HB) test uses a sphere indenter and the hardness value can be calculated using HB = P/A, where P is the load applied and A is indentation area after removing the indenter. A modified Brinell hardness MHB is calculated by $2P/\pi dD$, where p is the load applied, d is the diameter of indenter, and D is the depth of the indentation. Vickers hardness (HV) employs a diamond pyramid indenter. Rockwell hardness (HR) testers are direct-reading instrument, using a series of indenters and loads with corresponding scales.[18]

Macro-indentation

Cylindrical indenter is often prepared to test a rather sizable bone region, particularly for testing the hardness of trabecular bone harvested in a uniform diameter and thickness or with intact boundary condition in relation with peripheral cortical shell (Fig. 15).[19]

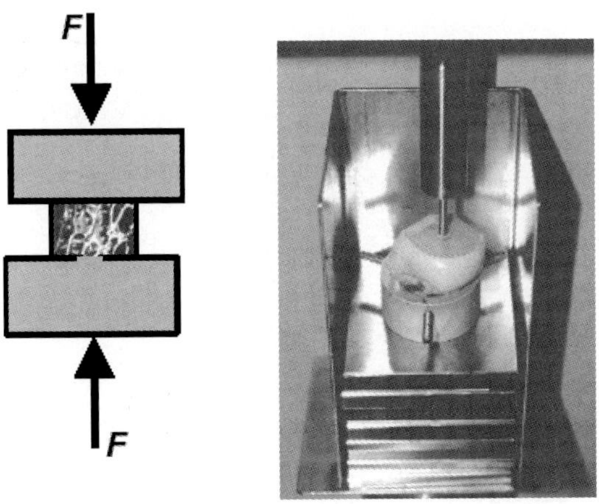

Fig. 15. Micro-indentation tests for isolated trabecular bone sample (left) and trabecular bone with intact boundary condition (Right).

Micro- and nano-indentation

In order to study mechanical properties of bone materials at bone matrix level, micro mechanical testing methods such as nano- or micro-indentations are often used to test hardness and elastic modulus of bone matrix under the guidance of a microscopy.[20-22] Micro-indentations for hardness test range from 20 to 150 μm in length. At this range, the tests evaluate the bone at the microstructure level such as mineral matrix of individual osteons or Harversian systems. Vickers and Knoop are the two main types of microhardness indenters. The Vickers microhardness number (HV) is calculated by using $1.8544P/D^2$, where P is the load applied, and D is the mean length of the two diagonals of the indentation. The knoop microhardness number (HK) is calculated by using P/A, where P is the load applied, and A is the area of indentation. Nano-indentations allow investigation of the mechanical properties in a nano-level, 1μm or smaller in length (Fig. 16). Because many important microstructural components of bone have dimensions of several microns or less, nano-indentation can be used to investigate the mechanical properties of osteonal, interstitial and trabecular lamellar bone. The nano-indentation techniques developed allow us to indent a submicrostructure to record the force-indentation curves. The micro-hardness and elastic modulus for microstructure can be calculated.

Fig. 16. Nano-indentation for testing material properties of region of interests. The arrow points the scratch of the matrix.

4.8. Pullout test

Screw pullout test refers to the measurement of the force required to pull out a screw inserted in a bone (Fig. 17). This measure may indicate the strength of bone, and the bone-implant interface properties. Screws are often used to connect the bone and implant or fixator. The information obtained from pullout tests may help to determine the optimum screw size, insertion technique, angle of insertion and screw hole preparation.[23] The ultimate strength of bone can be calculated using the following formula: $\sigma = P/\pi dh$, where P is the maximum load applied, d is the major diameter of the screw, and h is the length of the effective thread inserted in the bone. Because bone is anisotropic and inhomogeneous, the tests may need to be done in different locations and inserted in different directions.

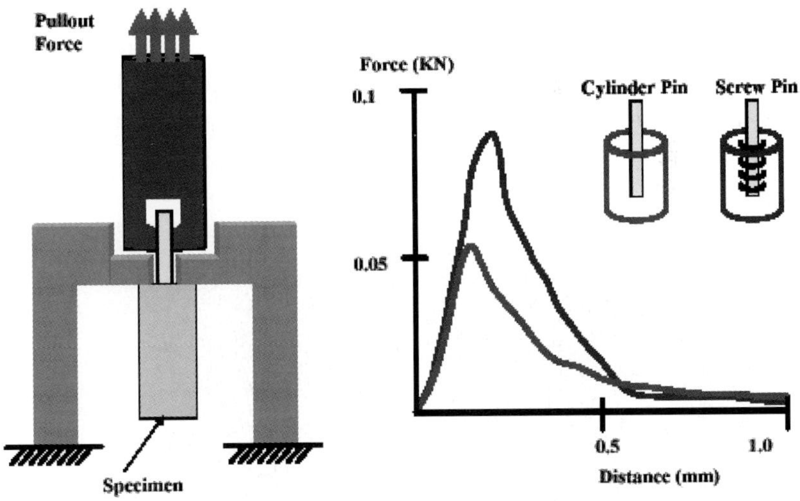

Fig. 17. A schematic illustration of a screw pull-out test (left), force-distance of pullout curves for a cylinder pin and a screw pin (right).

5. Non-destructive Approaches

Besides the conventional mechanical tests, some non-invasive techniques are widely used for investigation of bone mechanical properties.

5.1. Ultrasonic test

Bone specimens with two parallel plan surfaces are needed to be prepared. Ultrasound device with one driving transducer and receiving transducer measures bone structural and material properties (connectivity and stiffness) by detecting either longitudinal or shear wave propagation ultrasound wave propagation. The test could be repeated noninvasively.[8,24]

5.2. High-resolution contact microradiography (CMR)

The degrees of mineralization of a bone section taken by CMR on a high-resolution glass film-plate can be quantified by its brightness intensity using conventional image systems (Fig. 18). It is needed to be pointed out that the limitation is the use of the brightness intensity is only reliable to quantify those osteons before 75% of full-mineralization or maturity and with brightness intensities within so called quasi-linear region of the brightness intensity attenuation (saturated in brightness intensity). Parallel flat surfaces of bone specimens for quantification of CMR bone mineralization are essential.[21,25-27]

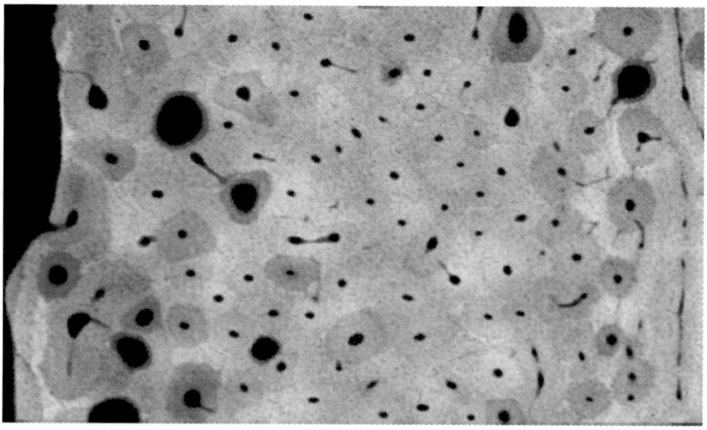

Fig. 18. High-resolution contact microradiography (CMR) of a cortical bone, which shows heterogeneity or anisotrophy of bone matrix in degrees of mineralization.

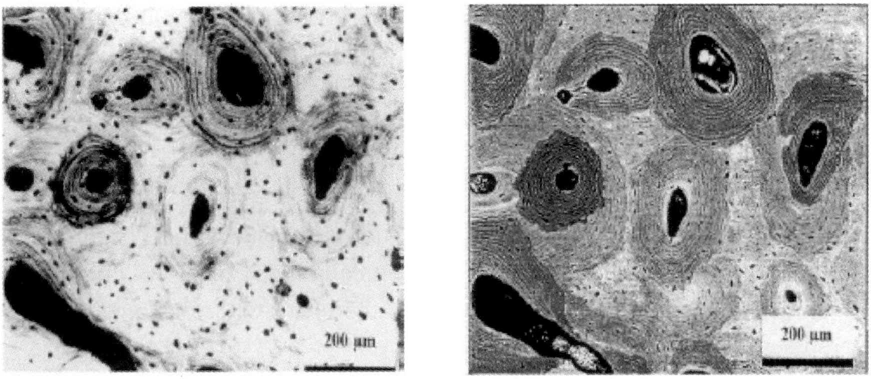

Fig. 19. Micrograph of osteons under Scanning Acoustic Microscopy (SAM) (Left) and Backscatter Scanning Electron Microscopy (bSEM).

5.3. *Scanning Acoustic Microscopy (SAM)*

Bone matrix elastic modulus – Young's modulus and hardness measured by indentation test can be measured indirectly by series of calibration curves using SAM reflection coefficient.[21,25, 28-30] The high resolution image of SAM is even comparable to high resolution optical microscopy or moderate resolution of scanning electron microscopy (Fig. 19). Gray images of SAM micrographs have been reported to have correlation with its reflection of coefficient (r=0.987).[28] The significance of this finding provided evidence for a retrospective measurement of the elastic properties of bone matrix within the areas of interests on SAM micrographs using SAM machine. This will help researchers who are collaborating with one of the few centers equipped with a high resolution SAM, but who have obtained SAM micrographs for imaging quantification.[28]

5.4. *Back scatter Scanning Electron Microscopy (bSEM)*

For bSEM bone specimens shall be prepared and embedded in methyl methacrylate (MMA) without decalcification, and then coated with gold. Degrees and the spatial patterns of bone matrix mineralisation can be revealed by bSEM and quantified by using commercially available imaging analysis systems (Fig. 19). Strong positive correlation between bSEM image gray scale level and mineral content variation has been reported previously.[29,31,32]

5.5. *Bone densitometry and microCT*

The measurement results of bone densitometry or microCT have been widely reported to correlate with bone mechanical strength. This is the focus of other chapters of this book and therefore will not be introduced and discussed in this chapter.

6. Testing and Evaluation Facilities

Much of biomechanical testing involves stress or strain application using some forms of hydraulic material or universal testing machine and related accessories (e.g. testing jigs). The following only provide common market available brands in the internet. Homepages of many other facilities related to mechanical testing and imaging methods have been summarized and published for references.[33]

- ***Instron (Instron Corporation, 100 Royall Street, Canton, Massachusetts 02021-1089; http://www.instron.com/index.asp)***

- ***MTS (MTS Systems Corporation, 14000 Technology Drive, Eden Prairie, MN 55344-2290; http://www.mts.com)***

- ***Hounsfield (Hounsfield Test Equipment, 37 Fullerton Road Croydon CR0 6JD England; http://www.hounsfield.com).***

- ***Nano- and micro-indentors (http://www.enduratec.com)***

More details related to testing and advance in testing devices and jigs may also be referred to from the homepage of the International Society of Biomechanics: http://www.isbweb.org.

7. Computational Modelling

Most diagnostic methods used in clinics to assess the bone mechanical quality are the *in vivo* measurement of bone apparent density, using non-invasive techniques, such as X-ray based techniques or ultrasound. Osteoporosis is often defined purely in terms of bony density and structure, based on the assumption of the mechanical properties being proportional to the density and structure. However, bone mineral density alone cannot be used to determine the bone strength accurately, since it does not account for the role of the bone architecture. For a more accurate and reliable diagnosis, it is necessary to know mechanical parameters of the bone, especially stiffness and strength, because these parameters respect the bone fracture risk. Bone density and morphology can be obtained from in vivo radiological methods.[34,35] However, accurate relationship between morphology and mechanical parameters needs to be studied.

Research has been done to develop computational models based on finite element approach for the understanding of bone mechanical properties in both macro- and micro-levels. Two types of models have been developed to perform micro-finite element analysis of trabecular bones. The first type of model is based on the cellular solid paradigm, to account for some of the complexity of trabecular architecture, while maintaining the computational efficiency that allows for the development of an intuitive understanding of the micro-biomechanics.[36] Models are incorporated with statistical distributions of spacing, angular orientation, and thickness. [37,38] Various lattice-type finite element models have been developed to examine the architectural manifestations of aging and anatomical locations.[38-41] The advantage of such generic models is that it is easy to conduct parametric studies to understand the effects of various parameters, such as trabecular thickness or trabecular number, on mechanical properties of trabecular bone. However, it is still not actual geometric model.

The second type of micro-finite element models uses a high-resolution, three-dimensional image of a specific specimen at up to 10 μm spatial resolution. The digital images are directly converted into a finite element mesh, most using eight-node hexahedral elements or four-node tetrahedral elements.[36,42] The geometry of the model is defined from reconstruction of the micro-CT images by converting all bone voxels to finite element mesh. This method is particularly attractive for trabecular bone. However, the model contains millions of elements. In such models, the mechanical properties of trabecular tissues can be usually assumed to be homogeneous and isotropic to reduce the computation time.

Trabecular bone may be damaged at both apparent and tissue levels. Most recent research extended the high-resolution finite element model to address failure properties.[43] Research showed that at the trabecular tissue level, overloading might cause subtle damage within trabeculae, although no fracture can be seen for the apparent bone, and this damage can cause large reductions in apparent modulus.[44] Using concepts of continuum damage mechanics for brittle materials,[45] the modulus reduction can be interpreted as quantitative measures of effective mechanical damage.

In order to predict bone strength based on computational modeling, three problems have to be solved.[39] First, high-resolution scanners, such as micro-CT or micro-MRI scanners, can be used *in vivo* for clinic. Secondly, the material properties of bone tissue must be quantified based on scanned images, for example by finding relationships between bone mineral content measured from grey-value of image pixels. Micro- or nano-indentation can be used to address the mechanical properties of micro-bone structures. Thirdly, a reliable failure criterion for bone must be established.

8. Considerations for Bone Testing and Result Interpretations

A large variation in the testing results of bone mechanical properties may be due to biological variability, and may also indicate a poor experimental design or set-up.[5-10] Results should always be examined critically. Although many testing documents can be followed, to conduct an accurate and meaningful bone test, it is important to take the special considerations in collection of specimen, specimen preparation and storage, selection and design of test, and data interpretations.

8.1. Subject dependent

Biological variability often dominates the results of mechanical testing. The mechanical properties vary with age, gender, activity levels and health conditions, depending on stages of modeling and remodeling, mechanical or drug interventions, and bone metabolic disorders etc. E.g. the bone of a young, active male donor is likely to be several times stiffer and stronger than that of an aged female donor.

8.2. Site dependent (inhomogeneity)

The mechanical properties of bone are site-dependent or inhomogeneous. Accordingly, the degree of mineralization and hardness or stiffness of bone matrix may vary from region to region and from site to site. When we speak about the mechanical properties of bone, the part or origin of bone must been specified.

8.3. Direction dependent (anisotropy)

Isotropy means the mechanical property is independent of the direction in which the specimen is tested. Bone is an anisotropic or heterogeneous structure because its basic components are assembled in different ways.

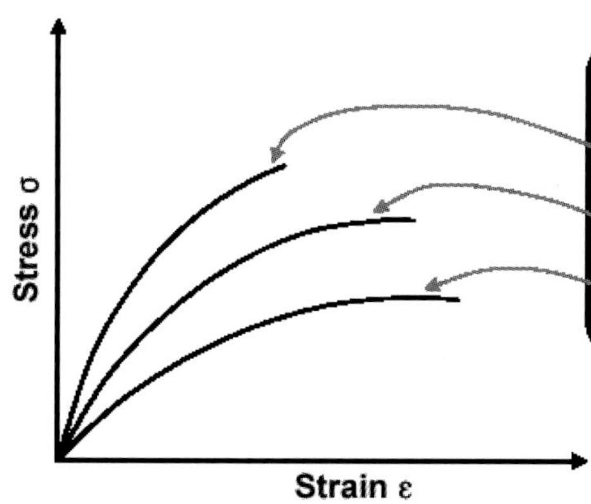

Fig. 20. Anisotropic property of cortical bone specimens if they are obtained and prepared at various angles to the bone long axis.

The mechanical properties are direction-dependent (Fig. 20). Studies showed that the Young's modulus of disphyseal cortical shell is 17 GPa in longitudinal direction and 11.5 GPa in transverse direction.[5-10,46]

8.4. Rate dependent

Viscoelasticity describes the time-dependent mechanical properties. Bone consists of both solid and fluid, and behaves as viscoelastic material. Its mechanical properties are strain-rate dependent (Fig. 21). With increasing strain rate, the material behaves stiffer and stronger. It is necessary to consider and specify the strain rate or load rate in conducting a test and reporting the results.

8.5. Nonlinearity

Calculation of material constants mentioned above is based on linear elasticity model. However, bone properties are much more complicated than linear elasticity. A typical force-deformation curve (Fig. 22) is normally divided into four regions, namely toe (preloading) region, linear elastic region, non-linear elastic (plastic, micro-failure) region and failure region. The toe region with small stiffness may be caused by fiber structure and fixation problem. Some researchers suggested preloading with a small load is useful to tighten the specimen-fixation interface and this is particularly obvious in soft.[47]

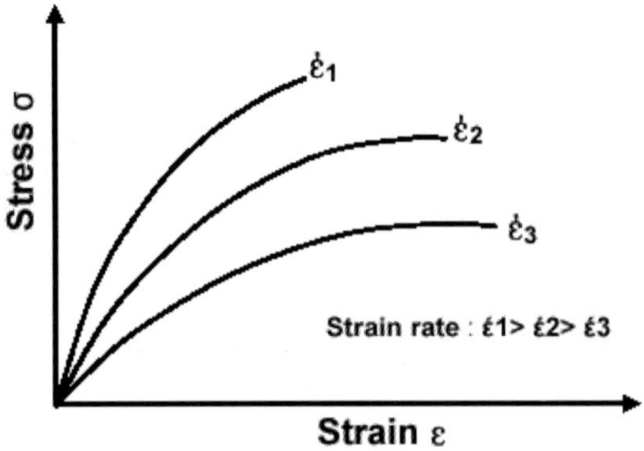

Fig. 21. Strain rate dependent property.

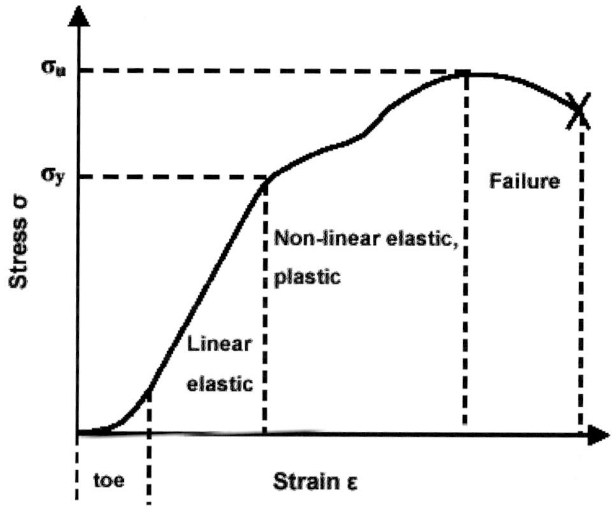

Fig. 22. A typical stress-strain curve showing the nonlinearity, often seen in indentation test of trabecular bone.

8.6. Hysterisis

When a cyclic load is applied on a bone specimen, the loaded and unloaded curves may not be the same and form a loop (Fig. 23). Under larger deformation, this loop gets bigger. This phenomenon is called hysterisis. The area under the loaded curve is the energy absorbed by the bone, and the area under the unloaded curve is the energy returned during unloading. The difference is the energy dissipated by the material during a cyclic loading.

8.7. Preconditioning

If a fresh bone specimen is tested under cyclic loading, the force-deformation curves of cycles are different (Fig. 24). The peak force points move to the right gradually. The stiffness may increase slightly. Normally, the difference between the first and second cycles is bigger and the difference between the sequent successive cycles reduces gradually. After several cycles, the curve reaches a stable situation, then the specimen is said to be preconditioned. This phenomenon is remarkable for fiber dominated tissues, because the collagen fibers are shrunk in a wavy shape before loaded. After loaded and deformed, it takes time to restore to their rest shapes. It is more obvious in soft tissues and least affects testing results of bone.

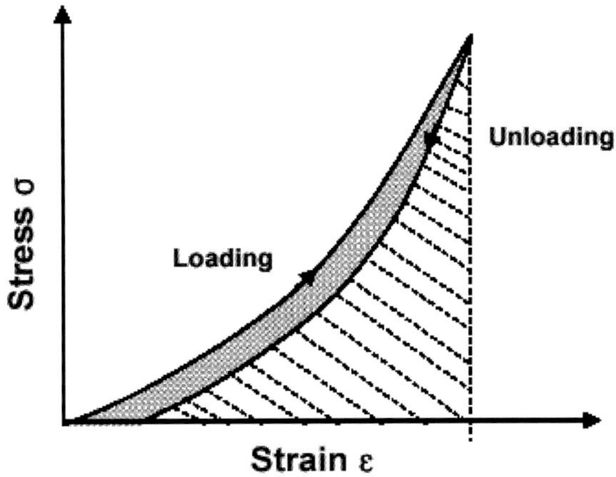

Fig. 23. A hysterisis curve under one cycle of load. The shaded area indicates the energy.

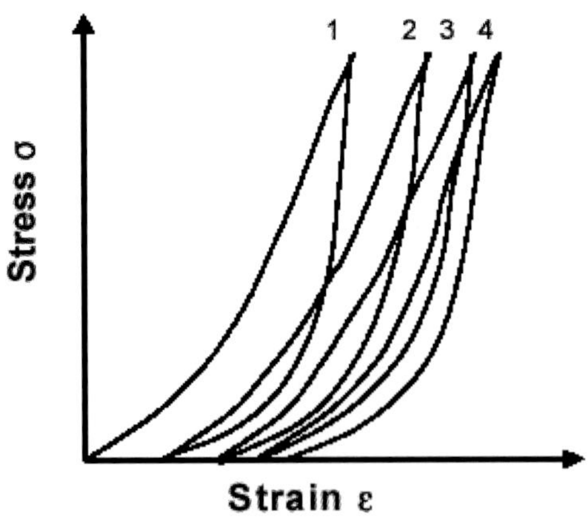

Fig. 24. Stress-strain curves under cyclic loading for a piece of fresh tissues, showing the preconditioning phenomenon. The difference between the first and second cycles is very big, and the difference between the sequent successive cycles reduces gradually. After several cycles, the curve reaches a stable situation, then the specimen is said preconditioned.

8.8. Irregular geometry

All the calculations mentioned above are based on a regular shape of the specimen. Bone may have a complex geometry with irregular cross sections and not uniform along the bone axis. The formula for calculation shall need to be modified slightly by considering a retrospective measurement of the mean CSA of the bone specimen around the region where failure or fracture occurs.

8.9. Gripping and load application effect

It should be noted that any fixation of specimen and contact loading application might produce extra-stresses within the specimen, which was not put into consideration in above calculation and should be considered in fixation design and data interpretations. For a tensile test, the fixture applies forces to hold the specimen. Any loosing or low rigidity may produce errors, while a tight fixation produces extra-stresses and stress concentration on the held part of the specimen. For a uniform specimen, the failure normally occurs at the two ends. Using a dumbbell-shaped specimen or a PMMA (Polymethylmethacrylate) end-coated specimen are common strategies to solve this problem.

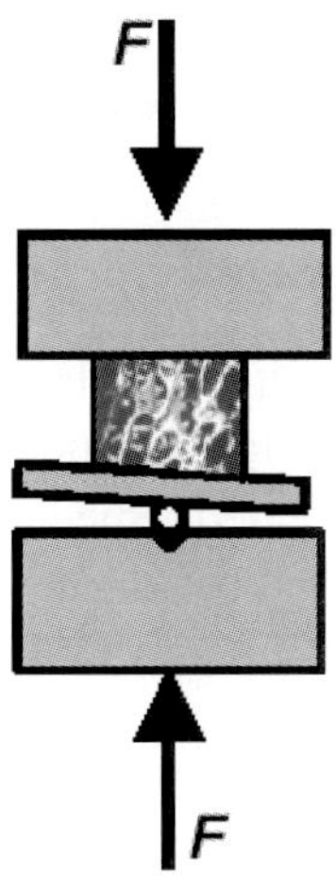

Fig. 25. A schematic illustration of a pivoting platen incorporated into the load train to correct for the misalignment.

For a compression test, uneven or unparallel surfaces of specimen may produce a bending moment, especially for a long specimen, which changes the stress distribution within the specimen. Special designed compression plates may help to solve such problem (Fig. 25). Friction between specimen and platens is another error source. Because compression test is supposed to apply pure compression force in axial direction only, the friction should be reduced to minimal.

8.10. Test environment

Testing environment, such as temperature, humidity, moisture and dryness may influence the mechanical behaviors of bone.

Temperature equilibration of specimens before testing

The frozen specimens shall be thawed overnight in room temperature and at next day the specimens shall be prepared properly for designed mechanical testing. Once thawed, the testing should be completed as soon as possible. 37°C is the physiological or best condition for testing but not always practical. Testing at room temp (23°C) increases the Young's modulus of bone about 2-4% compared to a test at 37C.[48] This becomes more obvious in fatigue test: e.g. at 23°C, bone undergoes twice as many loading cycles prior to failure than to be tested at 37°C.[49]

Humidity or moisture of specimens before testing

Specimens shall always be kept moistened with physiological saline (0.96g NaCl/litre) or Ringers solution, and never allowed to be dried out, unless this is specifically intended. Special care is needed to keep the specimen moist with 0.9% saline during preparation and the whole procedure of mechanical testing to minimize tissue dehydration, particularly in fatigue test which lasts substantial longer than other tests.[4-10,50]

If the bone specimen becomes dry, it will be more brittle than wet bone and its Young's moduli and strength will generally increase, but its toughness will decrease.[50]

If possible, testing shall be conducted in a water bath filled with 0.9% saline and controlled for a constant temperature at 37°C. This approach can avoid both testing variations result from both dehydration and inequilibration of sample temperature.

9. Data Evaluations

Results of biomechanical testing require statistical analysis. All statistical analysis should be already defined in the experimental design, and decided prior to testing. If data sets are to be excluded from analysis (because of damage, technical problems, etc.), the decision to exclude the data should be made before any analysis is carried out. If the data are included in the analysis and gives very different results, then these results should stand unless overwhelming evidence can be found to validate its exclusion.

10. Summary

This chapter introduces basic knowledge and concepts of mechanical testing methodologies in studying mechanical properties of bone specimens at organ, tissue and matrix levels. Computational modeling method has also been briefly introduced. Factors affecting testing results or data interpretations are summarized to avoid inconsistent and incomparable testing. Learning from experiences and perform pilot tests would help to enhance our understanding and quality of related research and testing work.

Acknowledgments

This work was supported by a grant from the Research Grants Council of the Hong Kong Special Administrative Regions, with the reference Number: CUHK4098/01M.

References

1. Melton LJ. Hip fractures: a worldwide problem today and tomorrow. Bone 14:s1-8, 1993
2. NIH: http://www.nlm.nih.gov/pubs/cbm/osteoporosis.html
3. WHO. Guidelines for preclinical evaluation and clinical trials in osteoporosis. WHO, Geneva, 1998
4. Chow DHK, Holmes AD, Qin L. Consideration for in vitro mechanical testing of musculoskeletal tissues. In: Qin L et al. (eds.) Compendium of 2002 International Bone Research Instructional Course & Hands-On Workshop, pp356-77, 2002
5. Bostrom MPG, Boskey A, Kaufman JJ, Einhorn TA. Form and Function of Bone. In: Simon eds. Basic Science in Orthopaedics. American Academy of Orhopaedic Surgeons, pp319-370, 2000
6. An YH, Draughn RA. Mechanical properties and testing methods of bone. In: An YH and Friedman RJ (eds): Animal Models in Orthopaedic Research, pp139-163, CRC Press LLC, Boca Raton, London, New York, 1999
7. Qin L, Leung KS. Application of biomechanical testing in development of drugs for prevention and treatment of osteoporosis. *Chin J Osteoporosis*, 6(1):23-25/68, 2000
8. Turner CH, Burr DB. Basic biomechanical measurements of bone: A tutorial. Bone 14:595-608, 1993
9. Burstein A H, Wright T M. Fundamentals of Orthopaedic Biomechanics. Williams and Wilkins, Baltimore. 1994
10. Sedlin ED, Hirsch C. Factors affecting the determination of the physical properties of femoral cortical bone. Acta Orthop Scand 37:29-48, 1996
11. Siu WS, Qin L, Leung KS. pQCT bone strength index is a better predictor than bone mineral density for long bone breaking strength in goats. J Bone Miner Metabol 21(5):316-22, 2003
12. Ferretti JL. Perspectives of pQCT technology associated to biomechanical studies in skeletal research employing rat models. Bone 17:353-64, 1995
13. Adams DJ. Pedersen DR. Brand RA. Rubin CT. Brown TD. Three-dimensional geometric and structural symmetry of the turkey ulna. J Orthop Res 13(5):690-9, 1995
14. Lieberman DE. Polk JD. Demes B. Predicting long bone loading from cross-sectional geometry. Am J Physical Anthropol 123(2):156-71, 2004
15. Affentranger B, Bauss F, Qin L et al. Mechanical properties of cancellous and cortical bone after long term I bandronatedosingin Beagle dogs. Bone 17:s604, 1995
16. Qin L, Chan KM, Rahn BA, Guo Xia. Bone mineral content measurement using pQCT for predicting the cortical & trabecular bone mechanical properties (in Chinese). *Chin J Orthop* 17(12), 66-70, 1997
17. Choi K, Kuhn JL, Ciarelli MJ et al. The elastic moduli of human sub chondral, trabecular, and cortical bone tissue and the size-dependency of cortical bone modulus. *J Biomech* 23:1103-13, 1990
18. Jacobs JA and Kilduff TF, Engineering materials technology: structures, processing, properties, and selection, 4th edition, Prentice-Hall Inc., pp143-174, 2001
19. Leung KS, Siu WS, EM Cheung, PY Lui, Chow DHK, James A, Qin L. Goats as osteoporotic animal model. *J Bone Miner Res* 16(12):2348-2355, 2001

20. Guo LH, Guo X, Leng Y et al. Nanoindentation study of interfaces between calcium phosphate and bone in an animal spinal fusion model. *J Biomed Mater Res* 54(4):554-9, 2001
21. Qin L, Leng Y, Katz JL. Overview of contact microradiography, back-scattered scanning electron microscopy, acoustic microscopy, and nano-indentation. In: Qin L et al. (eds.) Compendium of 2002 International Bone Research Instructional Course & Hands-On Workshop, pp288-94, 2002
22. Rho JY, Tsui TY, Pharr GM. Elastic properties of human cortical and trabecular lamellar bone measured by nanoindentation. *Biomaterials* 18:1325-30, 1997
23. Chapman JR, Harrington RM, Lee KM et al. Factors affecting the pull out strength of cancellous bone screws. *J Biomech Eng* 118:391-98, 1996
24. Ashman RB, Corin JD, Turner CH. Elastic properties of cancellous bone: Measurement by an ultrasonic technique. *J Biomech* 20:979-86, 1987
25. Qin L, Hung LK, Leung KS, et al. Staining intensity of individual Osteons correlated with elastic properties and degrees of mineralization. *J Bone Miner Metabol* 19(6):359-64, 2001
26. Runkel M. Wenda K. Ritter G. Rahn B. Perren SM. Bone healing after unreamed intramedullary nailing. *Unfallchirurg* 97(1):1-7, 1994
27. Nyssen-Behets C, Arnould V, Dhem A. Hypermineralized lamellae below the bone surface: a quantitative microradiographic study. *Bone* 15(6):685-689, 1994
28. Qin L, Bumrerraj S, Leung KS, Katz JL. Grey levels of osteons correlated with their elastic properties – A scanning acoustic micrography study. *J Bone Miner Metabol* 22(2): 86-89, 2004
29. Katz JL, Meunier A. Scanning acoustic microscopy of human and can in ecortical bone microstructure a thigh frequencies. In Lowet Get al.(eds.) Bone Research in Biomechnanics. IOS Press Amsterdam, pp123-238, 1997
30. Turner CH, Rho J, Takano Y, Tsui TY, Pharr GM. The elastic properties of trabecular and cortical bone tissues are similar: results from two microscopic measurement techniques. *J Biomech* 32:437-441, 1999
31. Bloebaum RD et al. Determining mineral content variations in bone using backscattered electron imaging. *Bone* 20(5):485-90, 1997
32. Skedros JG et al. The meaning of graylevels in backscattered electron images of bone. J Biomed Mater Res 27(1):47-56, 1993
33. Qin L. Useful Internet Homepages in Orthopaedic and Related Research (in Chinese). *J Med Biomech* 18(2):67-71, 2003
34. Genant HK. Current state of bone densitometry for osteoporosis. *Radiograp* 18(4):913-8, 1998
35. Jiang Y, Zhao J, van Holsbeeck MT, Flynn MJ, Ouyang X, Genant HK. Trabecular microstructure and surface changes in the greater tuberosity in rotator cuff tears. *Skeletal Radiol* 31(9):522-8, 2002
36. van Rietbergen B, Weinans H, Huiskes R, Odegaar A, A neew method to determine trabecular bone elastic properties and loading using micromechanical finite element models, *J Biomech* 28: 69-81, 1995
37. Jensen KS, Mosekilde L, A model of vertebral trabecular bone architecture and its mechanical properties, *Bone* 11: 417-423, 1999

38. Yeo JY and Keaveny, Biomechanical effects of intra-specimen variations in trabecular architecture: a three-dimensional finite element study, *Bone* 25: 223-228, 1999
39. van Rietbergen B, Weinans H, Huiskes R, Prospects of computer models for the prediction of osteoporotic bone fracture risk, in Bone Research Biomechanics, eds: G. Lower et al, 1997, page 25-32.
40. Silva MJ, keaveny TM and hayes WC, Load sharing between the shell and centrum in the lumbar vertebral body, *Spine* 22: 140-150, 1997
41. Vajjhala S, kraynik AM, Gibson LJ, A cellular solid model for modulus reduction due to resorption of trabeculae in bone, *J Biomech Eng* 122: 511-515, 2000
42. Hollister SJ, Brennan JM, Kikuchi N, A homogenization sampling procedure for calculating trabecular bone effective stiffness and tissue level tress, *J Biomech* 27: 433-444, 1994
43. Niebur GL, Feldstein MJ, Yuen JC, Chen TJ and Keaveny TM, High-resolution finite element models with tissue strength asymmetry accurately predict failure of trabecular bone. *J Biomech* 30: 1575-1583, 2000
44. Wachtel EF and Keaveny TM, Dependence of trabecular damage on mechanical strain, *J Orthop Res* 15: 781-787, 1997
45. Krajcinovic D, Lemaitre J, Continuum damage mechanics: theory and applications, New York: Springer-Verlag, pp294, 1987.
46. Reilly DT and Burstein AH, The mechanical properties of cortical bone. *J Bone Joint Surg* A56, 1001-22, 1994
47. An YH and Bensen. General considerations of mechanical testing, in An YH, RA Draughn (eds) Mechanical Testing of Bone-implant Interface, 2000
48. Bonfiled W, Li CH. The temperature dependence of the deformation of bone. *J Biomech Eng* 1:323-329, 1968
49. Carter DR, Hayes WC. Fatigue life of compact bone-I. Effects of stress amplitude, temperature and density. J Biomech 9:27-34, 1976
50. Evans FG. Mechanical properties of bone. Springfield, IL: CC Thomas, 1973

CHAPTER 10

ESTROGENS AND ANDROGENS ON BONE METABOLISM

Annie Kung and Jing Gu

Department of Medicine, The University of Hong Kong
Queen Mary Hospital, Hong Kong, PRC
E-mail: awckung@hkucc.hku.hk

1. Introduction

Estrogens (E) and androgens (A) circulate in men and women, and both play an important role in the maintenance of bone homeostasis and bone metabolism. Estrogens have been identified as the major inhibitor of bone resorption in both men and women. Androgen is an important source for estrogen through the action of aromatase, and it has direct effect in stimulating bone formation. These sex steroids play a major role in the sexual dimorphism of the skeleton in mineral homeostasis during reproduction and in bone balance in adults. Estrogens and androgens slow the rate of bone remodeling and protect against bone loss by attenuating the rate of differentiation of osteoblasts and osteoclasts from the precursor cells as well as affecting apoptosis of osteoblasts and osteoclasts.

Estrogens and androgens mediate their effects by binding to their specific nuclear receptors and activate gene transcription by binding to the hormone response elements at the promoter region of target genes. The sex steroids also regulate transcription of genes that do not contain hormone response elements. In this case the ligand-activated receptors form protein-protein complexes with other transcription factors, thus preventing them from interacting with their target gene promoters. Recent data also demonstrated that estrogens and androgens have other nongenotropic actions through their action on a Src/Shc/ extracellular signal-regulated kinase transduction pathway. This chapter will focus on the action of estrogens and androgens on bone cells and the clinical manifestation of sex steroids on the skeleton but will exclude the pharmacology of sex steroids to treat osteoporosis.

2. Effects of Estrogen on Bone

Estrogens are important for the development, maturation and maintenance of both the male and female skeleton as well as functional activity of both osteoclasts and osteoblasts. Estradiol is the biologically active estrogen and the important maintenance of skeletal homeostasis. The metabolism of estrogen is primarily oxidative and occurs predominantly in the liver[1]. Estradiol (E_2) is first oxidized to estrone (E_1) and then hydroxylated at either the A ring (C2 position) or the D ring (C16α position) by the cytochrome P450 enzymes 2-hydroxylase or 16α-hydroxylase[1]. This leads to formation of the two major metabolites of estradiol, 2-hydroxyestrone (2OHE1) and 16α-hydroxyestrone (16αOHE1)[2], which have distinct biological properties. 2OHE1 has little biological activity whereas 16αOHE1 shows estrogen agonistic activity on bone. 16αOHE1 suppressed bone turnover in ovariectomized rats[3] and in postmenopausal women the level of 16αE1 has been shown to be associated with spine and hip bone mineral density[4].

Estrogens have specific functions at the organ, tissue, and cellular levels of the skeleton. At the organ level, estrogens act to conserve bone mass. At the tissue level, estrogens suppress bone turnover and maintain balanced rates of bone formation and bone resorption. At the cellular level, estrogens affect the generation, lifespan, and functional activity of both osteoclasts and osteoblasts. There are four main mechanisms through which estrogens affect bone metabolism: 1. estrogens may stimulate bone formation by direct action on osteoblasts, 2. estrogens affect osteoclastogenesis through its action on cytokines and growth factors, 3. estrogens have direct action on the lifespan of bone cells by anti-apoptotic action on osteoblasts and osteocytes and apoptotic action on osteoclasts, 4. estrogens have direct effect on bone angiogenesis. Sex steroids also have indirect effects through modulation of the secretion and/or the action of calcitropic hormones, as well as other indirect effects such as renal handling of minerals.

3.1. Effects of ovariectomy on cortical bone

Sex steroids are essential for maintenance of normal bone volume. Ovariectomy leads to a deficit in ash weight and bone mineral density in adult rats and monkeys[5-7]. These changes are due to cortical bone modeling and net resorption of cancellous bone[8, 9]. Suppression of endogenous estrogen production by administration of gonadotropin releasing hormone agonist results in similar changes in rats[10] and monkeys[11]. The skeletal changes that follow medical or surgical ovariectomy can be entirely prevented by pharmacological replacement with E_2 [12].

Following ovariectomy the volume of the medullary canal in rat tibiae is enlarged due to a net increase in bone resorption[13, 14]. Osteoclast number is increased and bone formation remains unchanged or increase[9, 15]. In contrast, there is an increase in bone formation at the periosteal surface[9]. As a result of opposing changes in radial growth and endocortical modeling, the cortical bone volume decreases very slowly in ovariectomized rats[9]. In fact, cortical bone volume may increase in rapidly growing rats because the periosteal bone growth may exceed endocortical resorption of bone[15]. The effects of ovariectomy on cortical bone have not been reported for adult rats.

The differential response of the periosteal and endocortical bone surfaces of the midshaft to estrogen depletion appears to be related to the different functions and populations of cells that comprise the two bone envelopes in the rat. Osteoclasts are uncommonly present on the periosteum in the long bone and are not notably increased following ovariectomy, the level of resorption is low within the periosteum. This contrasts with the endocortical surface, which undergoes aggressive bone modeling during growth to increase the volume of the marrow cavity. Estrogen suppresses growth[16] and ovariectomy results in reestablishment of radial bone growth in adult rats and in humans.

3.2. Effect of ovariectomy in cancellous bone

Ovariectomy results in severe cancellous bone loss in long bones and vertebrae of rats[17] and vertebrae of monkeys[18, 19]. The rate of bone loss from the rat vertebrae occurs more slowly than from long bones[12, 20]. There appears to be a regional difference within bones with bone loss being more prominent in the proximal tibial metaphysis than the distal metaphysis or proximal epiphysis. The rate of bone loss may be related to differences in bone marrow and prevailing levels of mechanical loading. Ovariectomy results in increases in osteoblast-lined perimeter, osteoclast-lined perimeter, and osteoclast size in long bones of rats[20, 21]. There are simultaneous increases in the mineral apposition and bone formation rates, suggesting that ovariectomy results in chronic high bone turnover. Increased cancellous bone turnover remains elevated in both rats and monkeys for at least 1 year after ovariectomy[17] which results in osteoclastic perforation and removal of the trabecular plates[22].

4. Effects of Androgens on Bone

Androgens may regulate the male skeleton directly by stimulation of the androgen receptor (AR) or via its metabolites formed locally in bone as a result of local enzyme activities. Similar to estrogens, androgens transfer freely through the plasma membrane into the nucleus to bind to AR on osteoblasts and osteoclasts to mediate their classical action on genomic transcription. Androgens may also regulate osteoblast activity via a more rapid, nongenomic mechanism through receptors on the osteoblast cell surface[23].

Several enzymes play an important role in the metabolism of androgens. These include the aromatase enzyme (converting testosterone to estradiol); 17-β hydroxysteroid dehydrogenase (controlling the androstenedione to T and the estrone to estradiol pathways); 5α-reductase enzyme (metabolized testosterone to dihydrotestosterone (DHT). These enzymes have been identified in bone tissue[24-26], suggesting apart from the primary metabolite sites at the gonads, adrenal cortex and adipose tissue, local conversion and metabolism of androgens also occur in bone tissue[27]. It is estimated that 85% of estradiol in men are produced by peripheral conversion of circulating androgens through the aromatase enzyme.

The importance of the 5α reductase pathway is suggested by the presence of skeletal abnormalities in patients with 5α-reductase deficiency[28]. Aromatase in bone is important in the synthesis of both the potent estrogen estradiol but also the weaker estrogen, estrone, from testosterone and adrenal precursors androstenedione, and dehydroepiandrosterone[29]. The clinical impact of aromatase activity has recently been suggested by the reports of women[30] and a man[31] with enzyme deficiencies who presented with a phenotype that includes an obvious delay in bone age. The presentation of the man with aromatase deficiency was very similar to that of a man with estrogen receptor deficiency, namely lack of epiphyseal closure, tall stature, and osteopenia[32], suggesting that aromatase (and estrogen action) has an important role in male skeletal development.

In premenopausal women androgens play an independent role in the determination of peak bone mass[33-36]. Both trabecular and cortical bone mass in young women are correlated to serum levels of testosterone and androstenedione. Furthermore, in young women with androgen excess but persistent menstruation bone mass is higher than in controls, whereas in amenorrhoeic hyperandrogenic women bone mass is preserved despite low estrogen concentrations[33, 37,38]. The tendency for weight to be increased in hyperandrogenic women was not found to explain the effects on bone mass. These results suggest that androgens are important in regulating bone mass in young women.

Androgens have also been suggested to play a role in bone metabolism in postmenopausal women. The decline in androgen concentrations, especially adrenal androgens, in the postmenopausal period[39] has been suggested to contribute to bone loss in estrogen deficiency postmenopausal women. In fact, testosterone levels correlate with bone mass and rates of fall in bone density in perimenopausal women[40, 41]. There are data to suggest that androgens are important in the control of bone mass in the later postmenopausal period[42, 43]. Finally, the use of small amounts of testosterone with estrogen replacement therapy in postmenopausal women has been reported to enhance the expected positive effects on bone density[44]. In view of the complex interactions between androgens (testosterone and adrenal androgens) and estrogens potentially derived from them via aromatase activity, it is difficult to determine whether the putative associations between bone mass and androgens are the result of direct or indirect effects.

In vitro, androgens have direct effects on osteoblast function. Both testosterone and nonaromatizable androgens (DHT and fluoxymesterone) increase the proliferation of osteoblast-like cells in primary human and mouse osteoblast cell cultures[45-47]. Testosterone and DHT increase creatine kinase activity and [H^3]-thymidine incorporation into DNA in rat diaphyseal bone[48]. Furthermore, androgen treatment increases the proportion of cells expressing alkaline phosphatase activity, suggesting a positive effect of androgens on osteoblastic numbers and differentiation. The effects of androgens on collagen synthesis in osteoblasts are less consistent. Androgens have been shown to increase collagen synthesis in osteoblasts[46, 47] or have no effect[49, 50]. The divergence of these results may to some extent reflect differences in model systems.

5.1. Effects of orchidectomy on cortical bone

In most studies, orchidectomy in young rats results in a reduction in cortical bone mass within 2 to 4 weeks. Calcium content of the femur or tibia[51-55], whole femoral, tibial, or body bone mineral density [52,56,57], and tibial diaphyseal cortical area[58] have been shown to be lower in castrated animalsthan in sham operated controls. Similar trends have been reported in young, castrated male mice[59]. In animals followed for longer periods after castration cortical bone density was slightly reduced, but bone area was clearly smaller in the diaphysis of the femur[55]. The reduction in cortical bone mass appears to result in part from a reduction in periosteal bone formation rate induced by orchidectomy in males[15, 60]. This response is distinctly different than that induced by oophorectomy in females, which results in an increase in periosteal apposition in the period immediately after surgery[15, 60].

While estrogens appear to increase endosteal bone apposition, androgens have little of such action[15, 60, 61]. Gonadectomy in rats at 6 weeks reduced sex difference in bone width and bone strength by halving periosteal bone formations in males and doubling it in females[62].

Gonadectomy had no net effect on the endocortical surface in males but abolished endocortical bone acquisition in females. This divergent trend in the periosteal response to castration in male and female animals abolishes the sexual dimorphism usually present in radial bone growth. Thus cortical thickening occurred almost entirely by acquisition of bone on the outer (periosteal) surface in males and mainly on the inner (endocortical) surface in females. These explain the larger cortical bone and thicker cortex seen in young-adult male than women.

In mature rats androgen withdrawal also results in osteopenia. Castrated mature animals had significant reduction in cortical bone ash weight per unit length, cross-sectional area, thickness, and bone mineral density[63-66]. Periosteal bone accretion is reduced[64] and endocortical bone loss is accelerated in orchiectomized animals[63, 67].

5.2. *Effects of orchidectomy on cancellous bone*

Cancellous bone mass is also reduced in castrated young male rats. Tibial metaphyseal bone volume and vertebral bone mineral density were reduced rapidly after castration[55, 57, 60]. Severe reduction in bone volume up to 40-50% was apparent within 4-10 weeks[68, 69]. Dynamic histomorphometric and biochemical studies of bone remodeling showed rapid[70] increase in osteoclast number within 1 week after castration[64]. However this initial phase of increased bone remodeling subsides with time[70, 71] and by 4 months there is reduction in bone turnover rates in some skeletal areas[71]. Similar to younger animals, there is no evidence that indices of mineral metabolism are altered by these changes in skeletal metabolism in mature animals[66]. Associated with the bone changes was an increase in skeletal blood flow[52, 54], osteoclast numbers and surface[60], serum and urine calcium levels[60], and increased serum tartrate resistant acid phosphatase activity[56]. All these findings strongly suggest an increase in bone remodeling and bone resorption. These animal models therefore suggest an early phase of high turnover bone loss following orchidectomy, followed by a later reduction in remodeling rates. How long bone loss continues, and at what rate, is unclear. Both cortical and trabecular compartments are affected. The remodeling imbalance responsible for loss of bone mass appears complex, as there are changes in rates of both bone formation and resorption, and patterns that vary from one skeletal compartment of another.

6. Estrogen Receptors

There are two pathways in which sex steroids interact with their receptors: classical and nonclassical pathway. The classical pathway is direct interaction of sex steroids with their specific receptors in the nucleus. Once activated, the estrogen-receptor complex can directly mediate gene transcription or interact with transcription factors to influence their activity. The nonclassical pathways depend on the ability of estrogen to interact with either nonsteroid hormone receptors or steroid hormone receptors in the membrane. Nonclassical pathways activate kinases that ultimately regulate transcription of specific genes. (Fig. 1)

The classical action of estrogens and androgens are mediated via their cognate nuclear receptors, the estrogen receptor (ER) and androgen receptor (AR), which are members of the steroid receptor family. All members of this family share similar characteristics including a hormone binding domain, a DNA binding domain, a tendency to form dimers and an enhanced affinity for the cell nucleus in the presence of bound hormone.

The sex steroids mediate their classical action by passive diffusion into the cells and complex with specific intracellular receptor protein. Binding of the ligand induces conformational and posttranslational changes of the receptor itself. This enhances receptor dimerization and the ability of the receptor to bind with high affinity to tissue-specific hormone regulatory DNA elements (the estrogen responsive elements EREs) generally located in the promoter region of target genes.

To date, two ERs, ERα and ERβ have been cloned[72, 73]. The ERα gene was located at chromosome 6q25.1 and consisted of 8 exons and 7 introns[74, 75] while the gene encoding ERβ is mapped to chromosome 14q22-24[76].

ER consists of six structural domains based on their putative functions. The A/B domains contain one of the two transcriptional

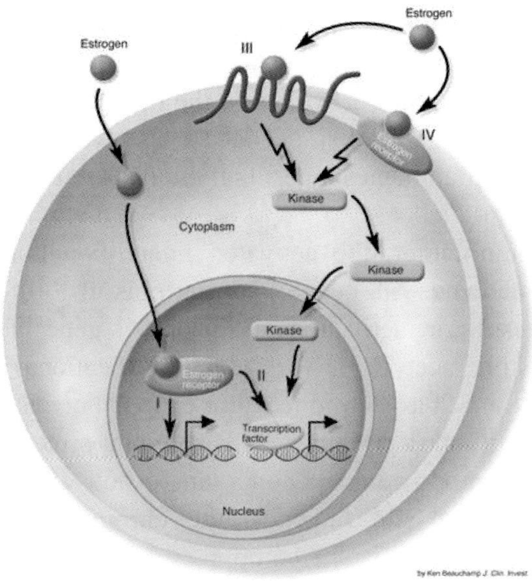

Figure 1
Sex steroid hormones can affect cellular function by a variety of mechanisms. The illustration depicts the mechanisms by which estrogen influences cells. The classical pathways (I and II) depend on direct interaction of estrogen with its receptor in the nucleus. Once activated, the estrogen-receptor complex can directly mediate gene transcription (I) or interact with transcription factors (II) to influence their activity. The nonclassical pathways (III and IV) work more rapidly and depend on the ability of estrogen to interact with either nonsteroid hormone receptors (III) or steroid hormone receptors in the membrane (IV). Both nonclassical pathways activate kinases that ultimately regulate transcription of specific genes. Adapted from ref. 225.

activation factors present in ER. The N-terminus contains the constitutively active activation factor 1(AF-1), while the ligand regulatable AF-2 is located at the C-terminus. The DNA-binding domain, located in the middle of the receptor molecule, contains two zinc fingers that mediate receptor binding to EREs of the target genes. The C region may also bind to heat shock protein and be responsible for nuclear localization of the receptor. In the C-terminal, the E region, is the hormone binding domain (HBD), which contains the AF-2, heat shock protein 90 binding function, a nuclear localization signal (NLS) and a dimerization domain. Once estrogen binds to ER, heat shock proteins dissociate and a change in conformation and homodimerization occurs to form homodimers (α-α, β-β) or heterodimers (α-β). These events trigger an estrogenic response in the cell. The DNA-binding domain of ERβ is 96% conserved compared to ERα, while the ligand-binding domain only shares 58% homology. ERβ also encodes a distinct AF-1 domain which is less active and certain data suggest that the ERβ A/B domain possess a repressive function[77-79].

Once at the promoter, the activated ligand-bound receptor, either ERα or ERβ, interacts with coactivator proteins to form a multiprotein complex that results in activation of the target gene. This involves a sequence of histone acetylation (or other modifications) carried out by histone acetylases such as steroid receptor coactivators as well as cAMP-response element binding protein (CREB)-binding protein (CBP). This is followed by binding of a complex containing coactivator BRG-1/BAF57, which unwinds DNA and remodels the chromatin. This results in formation of stable preinitiation complexes, and enhances the rates of RNA polymerase II reinitiation[80-86]. The ability and the affinity of natural or synthetic compounds to bind to ERs and stimulate gene expression in some but not all cell type and promoter-specific manner leads to the conceptualization of a class of compound called selective estrogen receptor modulators (SERMs).

In the presence of estrogen antagonist, such as tamoxifen, ER interacts with a complex of corepressor rather than coactivator proteins. The corepressor proteins include the silencing mediator of retinoic acid and thyroid hormone receptor (SMRT) and nuclear receptor corepressor (NCoR) that maintains the gene in an inactive state. Thus, the occupation of the ligand binding domain of the receptor, and hence its conformation, determines whether the receptor interacts with coactivators or corepressors and activate or repress transcription[87].

Even in the unliganded state, ER may bind to either corepressor or activator complexes. Intracellular signaling can influence the extent of interaction with these complexes and therefore determine the basal receptor activity: less activity when bound to corepressor complexes and more activity when the equilibrium is shifted to coactivator complex interaction. Selective receptor modulators, which are receptor ligands that exhibit agonistic or antagonistic characteristics in a cell or tissue-dependent manner, induce selective alterations in the confirmation of the ligand-binding domains of the nuclear receptors and influence their abilities to interact with coactivators or corepressors.

Recent data also revealed that sex steroids can be localized to cell membrane and the membrane receptors can mediate the nongenotropic action of sex steroids (see later sections). Both ERα and ERβ can be detected in the cell membrane and the membrane receptors appear to be derived from the same transcripts as the nuclear ones[88]. ER immunoreactivity is detected in caveolae, which are 50-100 nm flask-shaped specialized membrane invaginations enriched in the scaffolding protein caveolin-1 and compartmentalize signal transduction[89]. Caveolae are found in a variety of cell types including osteoblast[90]. The membrane-bound steroid receptors can interact with c-Src and activate MAPK pathways to mediate the nongenotropic action of sex steroids (see later sections).

7. Androgen Receptors

The gene coding for the androgen receptor (AR) is located at chromosome Xq11-12[91,92]. The AR gene is more than 90 kb long and codes for a protein that has 3 major functional domains. Similar to ER, AR has an N-terminal domain which serves a modulatory function, a DNA-binding domain, a nuclear targeting domain and the androgen-binding domain.

Similar to ER, within the amino-terminal domain of AR is the transcription activation region AF-1[93] and the polymorphic polygutamine and polyproline regions, which are important in transcriptional regulation via protein-protein interactions with other transcriptional factors. The amino terminus also contributes to the three-dimensional structure and conformation of the receptor molecule. The ligand-binding domain serves the function of specific, high-affinity binding of androgens. This region is also the binding site for inhibitory proteins such as the 90-kDa heat shock protein and has a role in other receptor function including dimerization and transcriptional regulation via the region referred to as activation function 2 (AF-2). Ligand binding results in conformational changes of AR and creates the surface required for interaction with other transcriptional cofactors that lead to activation of the nuclear receptors [93-96].

A variety of androgens and other steroids are bound by AR with dehydrotestosterone (DHT) having the highest affinity, followed by testosterone. This is mainly because dissociation of the hormone-receptor complex occurs slowly with DHT than with testosterone[97]. AR has low affinity for adrenal androgens such as dehydroepiandrosterone and androstenedione[98] and for non-androgenic steroids such as progesterone and estradiol[97].

8. Expression on ER and AR in Bone Tissue

The 2 isoforms of human ER, ERα and ERβ, occur with district tissue and cell patterns of expression. Additional ER isoforms, generated by alternative mRNA splicing, have been identified in several tissues including bone cells and are postulated to play a role in modulating the estrogen response in both reproductive and non-reproductive tissues[99].

The presence of both ERs and AR has been demonstrated in chondrocytes, bone marrow stromal cells, osteoblasts, and osteoclasts and their progenitors[24, 46,100-107]. A number of studies have definitely demonstrated the presence of ERα and ERβ as well as AR in growth plate tissue at the mRNA and protein level, suggesting that estrogens and androgens can directly regulate processes in the growth plate[108, 109]. However the level of receptor expression in osteoblastic and osteoclastic cells is low, amounting to 10-50 fold less in comparison to reproductive tissues. The binding affinity of ER and AR to their specific ligands in bone cells are the same as other reproductive tissues. Furthermore in bone cells, as similar to other non-reproductive tissues, the level of expression of ER and AR in females is similar to those in males[46, 110]. These characteristics may be important in accounting for the actions of estrogens and androgens in bone cells of both sexes.

Bord et al[107]established the cellular distribution of ERα and ERβ in neonatal human rib bone. ERα and ERβ immunoreactivity was seen in proliferative and prehypertrophic chondrocytes in the growth plate, with lower levels of expression in the late hypertrophic zone. Different patterns of expression of the 2 ERs are seen in bone. In cortical bone, intense staining for ERα is observed in osteoblasts and osteocytes adjacent to the periosteal-forming surface and in osteoclasts on the opposing resorbing surface. In trabecular bone, ERβ is strongly expressed in both osteoblasts and osteocytes, whereas only low expression of ERα is seen in these areas. Nuclear and cytoplasmic staining for ERβ is apparent in osteoclasts. It is believed that distinct patterns of expression for the 2 ER subtypes in developing human bone indicates direct function in both the growth plate and mineralized bone. In the latter, ERα is predominantly expressed in cortical bone, whereas ERβ shows higher levels of expression in trabecular bone.

Recent evidence suggests that estrogens and androgens have different molecular actions on the skeleton accounted partly by the level of ER and AR expression. In normal rat osteoblast cultures, the expression profile for ERα, ERβ and AR was unique during each stage of proliferation. Expression of ERα was increased during matrix maturation and then decreased during mineralization, whereas ERβ levels were relatively constant throughout differentiation which is more suggestive of constitutive expression[111-114]. In contrast, AR levels were lowest during proliferation, and then increased throughout differentiation with highest levels in the more mineralizing cultures. It is believed that this differential expression of steroid receptors determine the action of sex steroids on bone metabolism, that androgens may target the cells during mineralization stage of osteoblast differentiation, while estrogens action through either ERα or ERβ are more likely to affect osteoblast earlier during matrix maturation[114]. There are data to suggest that expression of ERβ may act as a dominant-negative inhibition of ERα and protect the cells from adverse effects of estrogen, with ERβ having the capacity to repress the transcriptional activity of ERα[115]. If both ERα and ERβ reside in the same cell type they can form heterodimers, which would allow ERβ to exert its modulatory actions[116]. The proliferative actions of estrogen seem to require ERα, whereas the differentiative and antiproliferative effect of estrogen may be mediated principally by ERβ[117].

9. Nongenotropic Action of Sex Steroids

Apart from the classical action of sex steroids mediating through their specific receptors, estrogens and androgens can exert effects on mature bone cells outside the nucleus by activating a Src/Shc/excellular signal regulated kinase signal transduction pathway probably within preassembled scaffolds called caveolae. Estrogens and androgen have pro-apopotic effect on osteoclasts but anti-apoptotic on osteoblasts and osteocytes. Importantly, new data suggest that ERα or ERβ or AR can transmit anti-apoptotic signals with similar efficiency, irrespective of whether the ligand is an estrogen or an androgen, i.e. the nongenotropic action is sex non-specific[118].

Studies in mice in vivo as well as in osteoblast and osteocyte cultures in vitro demonstrated that ovariectomy or orchidectomy results in a dramatic increase in the apoptosis of osteoblasts and osteocytes, which can be suppressed by addition of sex steroids[118]. This response is also observed in HeLa cells transfected with ER or AR. Unlike the action mediated via classical receptor pathway which takes a few hours, this anti-apoptotic action of sex steroids is mediated by a rapid action within seconds to minutes on phosphorylation of extracellular signal-regulated kinases (ERKs), a member of the MAPK family. MAPKs are serine/threonine kinases that transduce chemical and physical signals from the cell surface to the nucleus, thereby controling proliferation, differential and survival[119]. The initial event involves phosphorylation and recruitment of assessory proteins such as Ras, Src or Shc. This event leads to a cascade of activation of the intermediate MAPK such as the kinases MAPK or MEK kinase, eventually leading to MAPK activation and PI3 kinase activation. The protective effect of sex steroids on apoptosis can be blocked by specific inhibitors of Src or MEK kinase, and in cells deficient in Src or expressing mutant Src or Shc, the anti-apoptotic effect of sex steroids is abrogated[118].

The anti-apoptotic action of sex steroids on activation of the Src/Shc/ERK pathway is nongenotopic and this effect requires only the ligand-binding domain of the receptor. Using ERα as a paradigm, it was shown that targeting the ligand-binding domain of ERα to the plasma membrane can fully reproduce the ERK-mediated anti-apoptotic function of the full length ERα, whereas targeting the ligand-binding domain exclusively to the cell nucleus results in complete loss of its anti-apoptotic activity, suggesting this nongenotropic activation of signaling pathways of sex steroids is distinct from the classical genotropic actions of the nuclear receptor. The MAPK signaling pathway is localized in the cell membrane invaginations caveolae, to which ERα and ERβ immunoreactivity has been characterized[120].

Furthermore, it is shown that ERα, ERβ or AR mediated this anti-apoptotic action on osteoblast with similar efficiency, irrespective of whether the ligand is an estrogen or an androgen[118]. Activation of the Src/Shc/ERK pathway by sex steroids and modulator of the downstream mediators EIK-1/serum response element (EIK-1/SRE) and activation protein 1 (AP1) can be transmitted by either the ERα or the AR in an interchangeable ligand-receptor interaction manner, in the sense that ERα or AR can mediate them with similar efficiency, irrespective of whether the ligand is 17-β estradiol (E_2) or dihydrotestosterone (DHT). Moreover, using synthetic ligands, the nongenotropic effect can be dissociated from the genotropic actions of the steroid receptors. 4-estren-3α, 17β-diol (estren), a synthetic prototypic ligand, is able to reproduce the activating and repressing actions of E_2 or DHT on transcription factors through ER or AR in vitro. Estren reproduces the nongenotypic effects of E_2 or DHT on the phosphorylation of ERKs, EIK-1 and CCAAT enhancer binding protein-β (C/EBPβ), down regulates C-Jun and upregulates the expression of egr-1, an ERK target gene[118] Administration of estren to ovariectomized mature mice increases serum osteocalcin level, a biochemical marker of osteoblast number and bone formation[121].

This sex nonspecificity of sex steroids on bone cells is similar to other observations and experimental results that the effects of sex steroids on nonreproductive tissues are greatly relaxed. For example, estrogens and androgens are as effective in males as they are in females at protecting against bone loss, lowering cholesterol, or slowing atherosclerosis[122-125]. On the other hand, nonaromatizable androgens prevent osteoblast and promote osteoclast apoptosis and activate endogenous nitric oxide synthase (eNOs) in endothelial cells in vitro irrespective of the gender of donor cell[126]. These observations can be explained by the interchageable profile of ligand-receptor specificity in the regulation of the activity of ubiquitous transcription factors such as C/EBPβ, CREB, EIK-1, and C-Jun/c-Fos, that are involved in the antiapoptic action of sex steroids. The equivalent action of estrogens and androgens on preventing osteoblast apoptosis and stimulating osteoclast apoptosis is further confirmed by the effect of E_2 in stimulating ERK phosphorylation in cells from double estrogen receptor knockout (DERKO) mice lacking both ERα and ERβ[127] and that this effect can be abrogated by silencing the AR. Furthermore, ovariectomy in mature DERKO mice can be prevented by administration of E_2, suggesting an AR-mediated effect of estrogen[128, 129].

Whether these in vitro data can be applicable to the complex interactions that regulate human skeletal in vivo is uncertain. However, as estrogen withdrawal is also associated with an increased number of osteoclast precursor cells in the marrow[130], an effect mediated through the regulation of B-lymphogeniesis[131-133], it is unclear whether the effects of sex steroid withdrawal on the skeleton are medicated predominantly by regulating apoptosis[134] or by regulating the differentiation of osteoblasts and osteoclasts from their precursor cells.

It is noted that the nongenotopic effects of estren, which does not affect classical transcription of ER or AR, reverses bone loss in ovariectomized females or orchidectomized males without affecting the uterus or seminal vesicles, demonstrating that the classical genotropic actions of sex steroid receptors are dispensable for their bone-protective effects, but indispensable for their effects on reproductive organs[118]. The observation of the action of estren has led to the conceptualization of a new group of compounds coined ANGELs "activators of nongenotropic estrogen-like signaling". These ligands lack, completely or partially, the ability to induce the transcriptional activity of the ER and hence avoid the side-effects of estrogens on stimulation of the uterus and breast[135]. These new agents may open up new treatment options for prevention and treatment of bone loss.

10. Effects of Sex Steroids on Osteoblastogenesis

Apart from the action on suppressing bone resorption, estrogens may also be involved in osteoblastogeneis, as estrogen loss may stimulate osteoblastogeneis and increase the number of osteoblast progenitor cells colony-forming unit-osteoblast (SFU-OB) [136]. In a mice model of osteopenia secondary to defective osteoblastogeneis, osteoclastogenesis in ex vivo cultures of the bone marrow of these mice could be restored by addition of osteoblastic cells from normal mice[137], and that ovaricectomy or orchidectomy in these mice did not lead to osteoclastogenesis[138]. It was postulated that increase osteclastogenesis and bone loss following sex steroid loss may be a secondary event to the stimulation of mesenchymal cell differentiation toward osteoblast linkage. This is further supported by observation tthat 17-β-estradiol suppresses differential and self-renewal of early transit amplifying progenitors CFU-OB via ERα[139].

11. Effects of Sex Steroids on Cytokines and Growth Factors

The effect of estrogens on osteoclastogenesis came to light when the critical role of the bone marrow in bone remodeling is appreciated[140, 141]. Osteoclasts are derived from hematopoietic progenitors of the myeloid lineage, colony-forming unit-granulocyte/macrophage (CFU-GM) and CFU-M. Osteoblasts, as well as the hematopoiesis-supporting stromal cells and adipocytes of the bone marrow, are derived from mesenchymal stem cells. The development of osteoclasts depends on a network of autocrine and paracrine factors produced by the stromal and osteoblastic cells. Estrogens play an important role in regulating the production of osteclastogenic cytokines by bone marrow stromal cells and osteoblasts.

The effects of sex steroids on cytokines and growth factors can be illustrated by their action on interleukin-6 (IL-6), one of the important cytokines for osteoclastogenesis. The IL-6 receptor belongs to a family of structurally-related receptors that complex with the gp130 signal transducer. Binding of IL-6 to its specific cell surface receptor gp80 causes recruitment and dimerization of gp130, which is then phosphorylated by members of the JAK tyrosine kinases. This results in activation of the downsteam pathway, including members of the signal transducers and activators of transcription (STAT) family of transcription factors. Phosphorylation of JAK-STAT proteins results in homo- and hetero-dimerization of these factors and translocation to the cell nucleus where they activate target gene transcription. Recent evidence revealed that estrogen and androgen suppress IL-6 production via the classical ER or AR but through ERE-independent mechanism of the IL-6 gene.

The suppressive effect of sex steroids on IL-6 production is strictly dependent on the expression of their classical receptors[105, 142] with estrogen via ER and androgen via AR, but not vice versa. The effect results from an indirect effect of the receptor protein on the transcriptional activity on the proximal 225-bp sequence of the human IL-6 gene promoter[111,149]. Estrogen or raloxifene, one of the selective estrogen receptor modulators (SERMs), suppresses protein-protein interaction between the ER and transcription factors such as nuclear factor kappa beta (NF-κβ) and CCAAT/enhancer binding protein (C/EBP)[143].

Apart from suppressing IL-6 production, sex steroids also suppress the express of the two subunits of the IL-6 receptor, IL-6Rα and gp130, in cells of the bone marrow stromal/osteoblastic lineage[144, 145]. Interestingly, neutralization of IL-6 with antibodies[130] or knockout of the IL-6 gene in mice[146] prevents the upregulation of CFU-GM in the marrow and the expected increase of osteoclast numbers in trabecular bone sections, and also protects the bone loss associated with sex steroids deficiency[105]. Nevertheless, under normal physiological conditions, IL-6 is not required for osteoclast formation even in the presence of sex steroids as osteoclast formation is unaffected by IL-6 neutralizing antibody or in IL-6 knockout mice[130, 146].

The suppressive effect of sex steroids of IL-6 can be extended to other cytokines such as tumour necrosis factor (TNF), IL-1, granulocyte-macrophage-colony-stimulating factor (GM-CSF), M-CSF and prostaglandin-E_2 (PGE_2). In ovariectomized rat, the increase in osteoclastogenesis are attenuated or prevented by measures that impair the synthesis or response to IL-1, IL-6, TNF or PGE_2[147, 148]. Apart from directly suppressing transcription of the TNF gene[149], estrogen also suppresses expansion of TNF-producing T lymphocytes[150, 151]. Evidence suggests that TNF after estrogen deficiency is an early and central contributor to bone loss[147, 152]. TNF may act as a low-grade stimulus after estrogen deficiency, and blockade of TNF action is shown to alleviate bone loss[150, 151]. Moreover, ovariectomy-induced increases in osteoclastogenesis are attenuated or prevented by measures that impair the synthesis or response to IL-1, IL-6, TNFα, or PGE_2[147, 148].

Similarly to estrogens, androgens interact with other well-known modulators of osteoblast function. DHT increases the expression of TGF-β mRNA in human osteoblast primary cultures[46, 153]. In the human clonal osteoblast-like cell line SaOS-2, testosterone and DHT specifically inhibit the cAMP response elicited by parathyroid hormone (PTH) or parathyroid hormone-related protein, possibly via an effect on the parathyroid hormone receptor-Gs-adenylate cyclase complex[154, 155]. DHT and testosterone reduce PGE_2 production in calvarial organ cultures exposed to stimulation with PTH or IL-1[50]. Similarly, androgens have potent inhibitory effects on IL-6 production by stromal cells[105]. The effect of androgen on IL-6 production may explain to a large extent the marked increase in bone remodeling and resorption that follows orchiectomy. The effects of androgens on growth factors and cytokines production seem to be very similar to those of estrogen, which inhibits osteoclastogenesis via mechanisms that also involve IL-6 inhibition.

Recent evidence revealed that one of the major action of estrogens on bone metabolism is mediated by stimulating production of osteoprotegerin (OPG), a potent anti-osteoclastogenic factor. OPG is the soluble decoy receptor secreted by the stromal-osteoblast lineage cells, and serves to neutralize receptor activator of NF-$\kappa\beta$ ligand (RANKL), which is expressed in committed preosteoblastic cells[156]. The high-affinity binding of RANKL to RANK is apparently essential for osteoclastogenesis. RANK is expressed in osteoclast progenitors and is the final mediator of osteoclastogenesis (see chapter on RANK/RANKL). MSF, as well as IL-6, IL-11, IL-1, and TNF all exert their osteoclastogenic effects partially via stimulating the expression of RANKL. IL-6 and IL-11 may influence osteclastogenesis by stimulating the self-renewal and inhibiting the apoptosis of osteoclast progenitors[136, 157]. Estrogen loss may also increase the sensitivity of osteclasts to IL-1 by increasing the ratio of the IL-1RI over the IL-1 decoy receptor (IL1-

RII) [158]. As in the case of_IL-6, the effects of estrogen on TNF and M-CSF are mediated via protein-protein interactions between ER and other transcription factors. Because of the interdependent nature of the production of IL-1, IL-6, and TNF, a significant increase in one of them may amplify, in a cascade fashion, the effect of the others[136].

12. Effect of Sex Steroids on Bone Growth and Maturation

Sex steroids are responsible for skeletal growth and maturation and sexual dimorphism of the skeleton. Estrogens are particularly important for the regulation of epiphyseal function, and act to reduce the rate of longitudinal growth via influences on chondrocyte proliferation and action, as well as on the timing of epiphyseal closure[16]. Androgens appear to have somewhat opposite effects, and tend to promote long bone growth, chondrocyte maturation, and metaphyseal ossification. Androgen deficiency retards these processes[159]. Evidence for direct effect of androgens independent of those of estrogen comes from studies in which cultured rabbit cartilage cells eposed to testosterone or DHT increased the incorporation of [^{35}S]sulfate into proteoglycans[160] and another study showing imcorporation of calcium in a model of endochondral bone formation based on the subcutaneous implantation of demineralized bone matrix in castrate rats[161].

The exact cellular and molecular mechanisms of sex steroid on the initiation of pubertal growth spurt and the closure of the epiphyses at the end of puberty are unclear. Linear bone growth is governed by the chondrocytes of the growth plate. ER and AR are expressed on chondrocytes[109] and therefore the effects of sex steroid on pubertal growth and epiphyseal closure result from direct action of these hormones on chondrocytes.

The traditional beliefs that bone mass in men is regulated by androgens have been called into question by rare experiments of nature. Smith et al[32] reported a eunuchoid 28-year-old man with homozygous mutation of the estrogen receptor α gene associated with estrogen resistance. The patient had unfused epiphyses, low areal bone mineral density and increased bone remodeling despite normal levels of testosterone and DHT and elevated levels of estrogen. Carani et al[162] and

Bilezikian et al[163] each reported a young adult man with aromatase gene mutation and failure to convert androgens to estrogens. Both men had elevated testosterone, DHT and androstenedione levels but undetectable estrogen levels, unfused epiphyses and low areal bone mineral density. Administration of estrogen to the two patients with aromatase deficiency resulted in marked increase in areal bone mineral density and bone mass but not in the patient with estrogen resistance. These cases serve to illustrate the importance of estrogen in establishing peak bone mass in both men and women, and also highlight the possibility that subtle deficiencies in estrogen activity may contribute to low peak bone mass in some men. As treatment of male subjects with aromatase deficiency[164] with estrogen but not testosterone results in epiphyseal closure and increase in areal bone mineral density, it is likely that these processes during skeletal development are regulated by estrogen rather than androgen[165].

Bone mass accrual during puberty depends more on sexual maturation than chronological age. Skeletal size and volumetric bone mineral density (BMD) are similar in prepubertal girls and boys, however because of later onset of puberty and longer duration of growth spurts, boys acquire 10% greater body weight and 25% greater peak bone mass compared to girls. The greater bone mass in males is due to their greater bone size because testosterone promotes long bone growth, and periosteal new bone formation[166]. The excess in periosteal bone apposition over endosteal bone resorption that occurs during the pubertal growth spurt increases both the size and the volumetric BMD in growing males. A greater bone size in men confers greater mechanical strength as bone formed on the periosteal surface is biomechanically advantageous because it increases cross-sectional movement of inertia and bending strength of the long bone[167]. Male animals have larger bones and particularly thicker cortices than females[16, 61]. The effects of sex steroids on bone mass maturation can to some extent be assessed by observing the results of sex steroids withdrawal. The gender difference in size and shape of the skeleton reflects the need of the female pelvis to cater for pregnancy and delivery of the offspring. Furthermore, the gender difference is evident during fetal development[168], and short-term fetal exposure to exogenous estrogen may affect bone growth and development in postnatal life[169], suggesting the existence of an imprinting mechanism that acts on bone cell programming early in skeletal development. It is noted that both epiphyseal closure and increase in areal BMD occurred during estrogen treatment when serum testosterone was very low[164], reinforcing the concept that estrogen is the principal sex steroid involved in the final phases of skeletal maturation and mineralization. In fact, conditions that result with congenital estrogen deficiency in men are associated with eunuchoid body proportions[32, 162,163].

In rats, the nonaromatizable androgens DHT decreased biochemical markers of bone turnover and urinary calcium excretion in young rats, although it is unclear whether these effects were due to the skeletal or extraskeletal actions of the androgen[163]. In vitro, estrogen, testosterone and DHT stimulated osteoblast proliferation[170] and testosterone can prevent orchiectomy – induced bone loss in ERα knockout mice[171]. These data suggest direct action of androgens in bone growth and bone formation in males.

Androgens also play a role in bone metabolism in females. Flutamide (a specific androgen receptor antagonist) treatment is capable of inducing osteopenia in intact female rats[172]. This suggests that androgens provide crucial support to bone mass independent of estrogens. Interestingly, the characteristics of the bone loss induced by flutamide suggest that estrogen prevents bone resorption while androgens stimulate bone formation.

13. Effect of Sex Steroids in Skeletal Maintenance in Adulthood

The issue of the role of sex steroids in age-related bone loss has long been subjected to debate. Increased bone remodeling is seen following loss of sex steroids which results from upregulation of osteoblastegeneis and osteoclastogeneis. During each remodeling cycle, the rate of resorption is faster than the the rate of bone formation. The unbalanced formation and resorption rate within each remodeling cycle leads to expanded remodeling space, with osteoclasts eroding deeper cavities, resulting in removal of the entire cancellous elements and loss of connection between the remaining trabecular bone[173, 174]. While some studies reported a significant association between androgen concentrations and bone loss in older men[175-178], others failed to substantiate the relation between bone mass and androgens[179-182]. Recent epidemiology studies have demonstrated by multivariate analysis that estrogen, rather that testosterone, was the main predictor of BMD at all sites in older men, except for certain cortical bone sites in the appendicular skeleton[123, 183-190]. Szulc et al[191] showed that men with low levels of bio-17βestradiol were associated with high levels of biochemical markers of bone turnover and low BMD. Khosla et al[192] also demonstrated in a group of aging men, the rate of bone loss from the forearm correlated with bioavailable estrogen rather than bioavailable testosterone.

Falahati-Nini[193] assessed the relative contributions of estrogen and testosterone on bone turnover by rendering elderly men hypogonadal with GnRH agonist lenprolide. Conversion of androgens to estrogen was blocked by administration of the aromatase inhibitor letrozole. The subjects then received replacement doses of testosterone and estrogen in turn. The result showed that estrogen prevented the increase in markers of bone resorption whereas testosterone had only a minor effect. Bone formation markers were maintained by both testosterone and estrogen. The authors inferred that estrogen accounted for at least 70% of the effect of sex steroids on bone resorption whereas testosterone accounts for less than 30% of the effect. Using a similar study design, Khosla et al[194] observed that in elderly men treated with both GnRH agonist and aromatase inhibitor, testosterone replacement decreased osteoprotegerin (OPG) level by about 10% whereas estrogen replacement increased OPG by 18%. As mentioned in other chapters, OPG blocks the binding of RANK to RANKL and inhibits osteoblast stimulation on osteoclastogenesis. In view of the effect on OPG, Khosla et al[194] conclude that estrogen plays a more important role than androgen in inhibiting bone resorption in humans. In a similar study in younger individuals, Leder et al[195] confirmed the increase in bone resorption following aromatase inhibition, even though an additional independent role of androgens on bone resorption was also observed. Also, Taxel et al[196] demonstrated that elderly men being treated with aromatase inhibitor had significant increase in bone resorption. Collectively, these results strongly suggest that estrogen is the dominant sex steroid regulating bone resorption, but both testosterone and estrogen are important in maintaining bone formation.

The importance of estrogen in aging men is also demonstrated by experimental models of aged male rats. Orchidectomy and treatment with aromatase inhibitor in these animals produced a similar degree of bone loss[197]. In orchidectomized aged male rats, there was a reduction in cancellous bone area at the tibia and vertebra. Both increase in osteoblast surface and osteoclast number was seen, suggesting that the bone loss in these animals was due to increased bone turnover and activation frequency, which in turn stimulated bone formation. This was accompanied by increase urinary calcium excretion and N-telepeptide excretion, a marker of bone resorption[198]. Estradiol but not testosterone (total or free) was the only significant predictor of bone changes in these animals. Moreover, targeted deletion of the gene for either ERα or aromatase results in decreased BMD in male mice[199, 200].

To address the importance of endogenous estrogens in mediating the bone effect of androgens, aromatase knockout (ArKO) mice were generated[201]. Uterine weight was reduced in female ArKO mice, confirming a reduction in endogenous estrogen. In response, testosterone concentrations were elevated in both male and female ArKO mice. Femur length was reduced in male but not female ArKO mice. In contrast to ERKO models, both ArKO male and female mice had osteopenia. Female ArKO mice had reduced cortical thickness and trabecular bone volume, while male mice had reduced trabecular but not cortical bone. Bone turnover markers including serum osteocalcin and urine crosslinks were reduced in male ArKO mice, suggesting a low bone turnover state, whereas in female mice, there was increased bone turnover with elevated urinary crosslinks and reduced serum osteocalcin levels. It can be seen that this ArKO model does not reproduce the bone phenotype of ERKO mice. Whether there may be another receptor or other mechanism for estrogen action to account for the altered phenotype seen in ArKO mice is uncertain. Another study evaluated the effect of E_2 replacement in these ArKO mice and observed that treatment with E_2 completely restore bone mass in both sex[202]. In view of the fact that serum testosterone concentrations were elevated in these animals, it is conceived that androgens are not capable of reversing bone loss associated with estrogen deficiency, and estrogen may be more protective in the skeleton than androgens event in males. Putting all the results together suggested that the major action of testosterone is mediated through aromatization to estrogen and binding to the ER.

14. Relative Role of Steroid Receptors in Bone Metabolism

The relative importance of ERα, ERβ and AR in skeletal development and maintenance is tested in animal models deficit in the respective receptors by knockout techniques. Estrogen receptor α knockout (ERKOα) female mice had significantly reduced ability to superovulate[127, 203]. These mice had significantly greater body weight but their crown-rump length was similar to wildtype (WT) mice. However, the bone size of these female mice was smaller[204]. In contrast, male ERKOα mice had reduced body weight and reduced crown-rump length as well as shorter femoral length detectable after puberty. The shorter long bone length was associated with a reduced growth plate width. The cortical bone area and circumference were reduced in the male ERKOα mice[204]. Although the areal BMD measured by dual energy X-ray absorptiometry (DXA) was reduced in the total body and femur, volumetric BMD as measured by peripheral quantitative computed tomography (pQCT) and bone histomorphometry did not reveal significant changes in trabecular or cortical bone density. However, the role of ERα in these animals is difficult to assess because of the changes in body weight which may affect the mechanical loading of the skeleton. Serum IGF-I concentration was also lower than wild-type mice, suggesting the reduction in longitudinal bone growth was attributed in part to an abnormal growth hormone/IGF-I axis. Furthermore, truncated ER transcripts were detectable in these mice which might exert transactivatorial capacity.

In a second report of ERKO mice with no detectable ERα transcripts, there was no alteration in femur length in either female or male ERKOα mice[127]. The bone phenotype in the female ERKOα mice was similar as previously reported but there were differences in the male ERKOα mice. Cortical bone density and thickness was reduced in male ERKOα.

However trabecular number was increased but the trabecular thickness was reduced in these male ERKOα mice, leading to a finer network of trabeculae and more trabeculae extending into the diaphyseal region in these animals[205]. Furthermore, there was a marked reduction in bone turnover in male ERKOα mice as demonstrated by decreased flourochrome labeling of bones and reduced trabecular perimeter lined by osteoblasts and osteoclasts. Nonetheless, there was no significant change in serum osteocalcin or urinary deoxypyridinoline crosslinks, suggesting no systemic alternations in bone turnover.

Vandenput et al[171] reported another ERKOα mouse model. In this model, ERKOα mice had smaller and thinner bones, suggesting a direct role of ERα to achieve full skeletal size in male mice[171]. However, male ERKOα mice had significantly more trabecular bone, indicating that ERα is not essential to maintain cancellous bone mass in males.

Although there is general agreement about the role of ERα in mediating bone growth and maturation in both males and females from mice studies, the role of ERβ in skeletal development and maintenance of the skeleton in adults is unclear. Female mice deficit of ERβ generated by knockout technique had normal bone at prepubertal stage (4 weeks old), but there was increased cortical bone area and increased cortical bone formation in adolescent female mice which was maintained till adulthood[206]. Trabecular bone volume and BMD was unchanged. Adult female ERβ knockout (ERKOβ) mice had increased osteoblast number and function as evidenced by increased mRNA expression of alpha 1(I) collagen, alkaline phophatase and osteocalcin but osteoclast function was normal. The growth plate width of these female ERKOβ mice was unaltered. These findings suggest that ERβ probably plays a repressive function in the regulation of bone growth during adolescence, but ERβ is not required for the protective effect of estrogen on trabecular bone.

Similar findings in female ERKOβ mice were reported by Sims et al[205]. Cortical thickness and BMD as measured by pQCT was unchanged but trabecular bone volume and trabecular thickness was increased as reflected by reduction in osteoclast surfaces and urinary markers of bone resorption. If ERβ acts to limit ERα action, these results reflect the unopposed action of ERα.

The skeletal phenotype of double ERα and ERβ knockout (DERKO) mice has been described by several groups[199, 205,206]. There were similarities and differences between these mouse models. In general, the body weight of DERKO mice was similar to ERKOα mice, and the femur length of DERKO mice was intermediate between ERKOα and ERKOβ mice. Cortical and BMD in female DERKO mice was unchanged in prepubertal stage but increased in adult female mice. However trabecular BMD of the female DERKO mice was reduced. As for male mice, ERKOβ did not result in any significant bone phenotypic changes during growth and adulthood up to 1 year[205,207].

The role of AR is addressed in ARKO male mice[208]. The ARKO male mice had a female-like appearance and body weight. Apart from having smaller testes and lower serum testosterone levels, cancellous bone volumes of ARKO male mice were reduced compared with wild-type littermates. Osteoclast numbers in the femoral metaphyses were higher in ARKO than wildtype mice. Bone formation rate and mineral apposition rate were also increased in these animals, suggesting increased bone turnover in the ARKO male mice. The osteopenic bone phenotype of this ARKO mice strongly supports an important role of AR signaling in bone metabolism and the role of importance of sex steroids in skeletal health. The role of AR in female skeleton mechanism is unclear.

Interpretation of the skeletal changes of the ERKO and ARKO mice has been difficult as these models are confounded by changes in hormonal status of these animals[168]. For example, estradiol concentrations are increased in the female but not male ERKOα mice, and testosterone concentrations are increased in both female and male ERKOα mice compared to wild-type mice. Similarly, serum estradiol and testosterone levels were significantly elevated in female DERKO mice versus wild-type mice[206], and the levels of ERα increased by two fold in long bones of ERKOβ female mice[209]. In contrast, serum testosterone levels were reduced in ARKO male mice[208].

To overcome these problems, gonadectomy was performed on these KO mice. Ovariectomy performed at 13 weeks in female ERKOβ mice resulted in reduced uterine weight and BMC in long bones and vertebral[207]. The bone phenotype of these ovariectomized ERKOβ mice was similar to control female ERKβ and wild-type mice, suggesting that the ovariectomy-induced reduction in trabecular bone mass does not rely on ERβ.

As to orchidectomized ERKOα mice, reduction in both cortical and trabecular bone density was seen as similar to wild-type orchidectomized animals[209]. Orchidectomy resulted in high-turnover bone loss in predominanthy trabecular but also cortical bone. Administration of testosterone completely prevented bone loss in both WT and ERKOα mice. E_2 replacement failed to prevent cancellous bone loss in ordiectomized ERKOα mice, but was able to stimulate bone formation at the endocortical surface in these mice, suggesting that E_2 may stimulate osteoblasts through non ERα pathway. Whether E_2 works through ERβ and AR is unclear from this study, although studies from ERKOβ mouse models showed that ERβ deficiency plays no significant role in the maintenance of bone mass in adult male mice.

Unfortunately, none of these studies address the potential for androgen compensation in the female ERKOα mice. Further studies with models that can selectively regulate sex steroids and their putative compensatory factors will help clarify the skeletal impact of ERs and AR.

15. Indirect Actions of Sex Steroids on Bone Metabolism

Apart from direct action on bone cells, sex steroids affect bone metabolism indirectly through other mechanisms. One of the possible way is via modifying PTH secretion. Estradiol increases PTH secretion from bovine parathyroid cells in primary culture[210]. ER mRNA and immunoractivity are detected in parathyroid tissue[211]. In ovariectomized rats, estradiol injection increases PTH gene expression[211]. Low doses of estradiol injection increases PTH mRNA without affecting serum calcium and vitamin D levels. In humans, the periodicity of PTH secretion was lost in postmenopausal osteoporotic women[212], and estrogen-deficient women showed increased skeletal sensitivity to the resorbing actions of PTH[213]. Previous studies have found that estrogen therapy may result in increases[214-217], decrease[218], or no change[219] in serum PTH levels. The likely explanation for the inconsistent findings may be related to the time of onset of estrogen deficiency.

Estrogens have indirect actions on extraskeletal calcium homeostasis, which include reduction in intestinal calcium absorption[220] and decreasing renal calcium conservation[221]. There are two phases of bone loss that have opposing effects on parathyroid hormone (PTH) secretion. During the early phase, estrogen deficiency resulted directly in a stimulation of bone resorption, with a consequent flux of calcium from the skeleton. This in turn resulted in a suppression of PTH secretion. In the late phase, estrogens have an indirect effect on PTH secretion via extraskeletal effects of estrogen on enhancing intestinal calcium absorption and renal calcium reabsorption. This results in total body loss of calcium and a compensatory increase in PTH secretion.

Estrogens also increase IGF-I expression in osteoblasts to enhance bone formation. E_2 enhances IGF-I synthesis at the transcriptional level in rat and human osteoblast cells[222, 223]. As there is no ERE in the promoter region of IGF-I gene, it is conceived that estrogen acts through a cAMP dependent pathway to control IGF-I gene expression by interacting with C/EBP transcription factor[224].

16. Summary

Sex steroids play a major role in the regulation of skeletal growth and development. The essential role of estrogens and androgens in the maintenance of normal bone volume in vivo has been demonstrated in animal models. This chapter summarized the existing knowledge and newer data on the cellular mechanisms of estrogens and androgens on growth factor and cytokine expression in osteoblasts and osteoclasts. The genotropic and nongenotropic effect of sex steroids through different cell signaling system has recently been described. However the sex nonspecificity of sex steroids and the relative role of their receptors remain controversial. With increasing understanding of the actions of estrogens and androgens and their receptors in both males and females, this opens up new opportunities in drug development for prevention and treatment of a wide variety of bone disorders.

References

1. Michnovicz JJ and Rosenberg DW, Biochem *Pharmacol* 43 (1992).
2. Niwa T, Bradlow HL, Fishman J and Swaneck GE, *Steroids* 55 (1990).
3. Westerlind KC, Gibson KJ, Malone P, Evans GL and Turner RT, *J Bone Miner Res* 13 (1998).
4. Lim SK, Won YJ, Lee JH, Kwon SH, Lee EJ, Kim KR, Lee HC, Huh KB and Chung BC, *J Clin Endocrinol Metab* 82 (1997).
5. Kalu DN, Salerno E, Liu CC, Echon R, Ray M, Garza-Zapata M, Hollis BW, *Bone Miner* 15 (1991).
6. Lindsay R, Aitken JM, Steuterman S and Baran D, *Postgrad med J* 54 (1978).
7. Jayo MJ, Weaver DS, Adams MR and Rankin SE, *Am J Obstet Gyncol* 193 (1990).
8. Wronski TJ, Cintron M and Dann LM, *Calcif Tissue Int* 43 (1988).
9. Turner RT, Vandersteenhoven JJ and Bell NH, *J Bone Miner Res* 2 1987.
10. Goulding A and Gold E, *J Endocrinol* 121 (1989).
11. Mann DR, Gould KG and Collins DC, *J Clin endocrinol Metab* 71 (1990).
12. Wronski TJ, Yen C-F and Scott KS, *Journal of Bone and Mineral Research* 6 (1988).
13. Dahinten SL and Pucciarelli HM, *Am J Phys Anthropol* 71(1986).
14. Kalu DN, Liu CC, Hardin RR and Hollis BW, *Endocrinology* 124 (1989).
15. Turner RT, Wakley GK, and Hannon KS, *J Orthop Res* 8 (1990).
16. Turner RT, Riggs BL, Spelsberg TC, *Endocr Rev* 15 (1994).
17. Wronski TJ, Dann LM and Horner SL, *Bone* 10 (1989).
18. Jerome CP, Kimmel DB, McAlister JA and Weaver DS, *Calcif Tissue Int* 39 (1986).
19. Longcope C, Hoberg L, Steuterman S and Baran D, *Bone* 10 (1989).
20. Wronski TJ, Walsh CC and Ignaszewski LA, Bone 7 (1986).
21. Turner RT, Wakley GK, Hannon KS, Bell NH, *Endocrinology* 122 (1988).
22. Dempster DW, Birchman R, Xu R, Lindsay R and Shen U, *Bone* 16 (1995).
23. Peterziel H, Mink S, Schonert A, Becker M, Klocker H and Cato AC, *Oncogene* 18 (1999).
24. Nakano Y, Morimoto I, Ishida O, Fujihira T, Mizokami A, Tanimoto A, Yanagihara N, Izumi F, Eto S, *Bone Miner* 26 (1994).
25. Schweikert HU, Rulf W, Niederle N, Schaefer HE, Keck E, Krück F, *Acta Endocrinol* 95 (1980).
26. Vittek J, Altman K, Gordon GG, Southren AL, *Endocrinology* 94 (1974).
27. Yanase T, Suzuki S, Goto K, Nomura M, Okabe T, Takayanagi R and Nawata H, *J Steroid Biochem and Mol Biol* 86 (2003).
28. Fisher LK, Kogut MD, Moore RJ, Goebelsmann U, Weitzman JJ, Isaacs Jr H, Griffin JE, Wilson JD, *J Clin Endocrinol Metab* 47 (1978).
29. Nawata H, Tanaka S, Tanaka S, Takayanagi R, Sakai Y, Yanase T, Ikuyama S, Haji M, *J Steroid Biochem Mol.* 1–6(1995).
30. Conte, F.A., Grunbach, M.M., Ito Y, Fisher CR, Simpson ER, *J.Clin. Endocrinol. Metab.,* 78 (1994).

31. Morishima A, Grumbach MM, Simpson ER, Fisher C, Qin K, *J. Clin. Endocrinol. Metab*. 80 (1995).
32. Smith EP, Boyd J, Frank GR, Takahashi H, Cohen RM, Specker B, Williams TC, Lubahn DB, Korach KS. *N Engl J Med.* 331 (1994).
33. Buchanan JR, Hospodar P, Myers C, Leuenberger P, Demers LM. *J Clin Endocrinol Metab* 67 (1988).
34. Buchanan JR, Myers C, Lloyd T, Leuenberger P, Demers LM. *J Bone Miner Res* 3 (1988).
35. Daniel M, Martin AD, Drinkwater DT. *Calif Tissue Int* 50 (1992).
36. Leuenberger PK, Buchanan JR, Myers CA, Lloyd T, Demers LM. *Am J Clin Nutr* 50 (1989).
37. Dixon JE, Rodin A, Murby B, Chapman MG, Fogelman I. *Clin Endocrinol* 30 (1989).
38. Prezelj J and Kocijancic A, *Calcif Tissue Int* 52 (1993).
39. Ohta H, Masuzawa T, Ikeda T, Suda Y, Makita K, Nozawa S, *Bone Miner* 19 (1992).
40. Slemenda C, Hui SL, Longcope C, Johnston CC, *J Clin Invest* 80 (1987).
41. Steinberg KK, Freni-Titulaer LW, DePuey EG, Miller DT, Sgoutas DS, Coralli CH, Phillips DL, Rogers TN, Clark RV, *J Clin Endocrinol Metab* 69 (1989).
42. Gasperino J, *Clin Orthop Rel Res* 311 (1995).
43. Vanderschueren D and Bouillon R, *Calcif Tissue Int* (1995) 314-346.
44. Davis SR, McCloud P, Straus BJG, Burger H. *Maturitas* 21 (1995).
45. Kasperk CH, Wergedal JE, Farley JR, Linkhart TA, Turner RT and Baylink DJ, *Endocrinology* 124 (1989).
46. Benz DJ, Haussler MR, Thomas MA, Speelman B and Komm BS, *Endocrinology* 128 (1991).
47. Gray C, Colston KW, Mackay AG, Taylor ML, Arnett TR, *J Bone Miner Res* 7 (1992).
48. Somjen D, Weisman Y, Harell A, Berger E, Kaye AM, *Proc Natl Acad Sci* 86 (1989).
49. Canalis E and Raisz LG, *Calcif Tissue Res* 25 (1978).
50. Pilbeam CC and Raisz LG, *J Bone Miner Res* 5 (1990).
51. Hock, B., and Bunsey, M, *J. Neuroscience* 18(17) (1998).
52. Kapitola J, Kubickova J, Andrle J, *Bone* 16 (1995).
53. Saville, P. D., *J. Am. Geriatr. Soc.* 17 (1969).
54. Schoutens A, Verhas M, L'Hermite-Baleriaux M, L'Hermite M, Verschaeren A, Dourov N, Mone M, Heilporn A, Tricot A, *Acta Endocrinol* 107 (1984).
55. Vanderschueren D, Van Herck E, Geusens P, Suiker A, Visser W, Chung K, Bouillon R, *Calcif Tissue Int* (1994).
56. Ongphiphadhanakul B, Alex S, Braverman LE, Baran DT, *J Bone Miner Res* 7 (1992).
57. Rosen HN, Chen V, Cittadini A, Greenspan SL, Douglas PS, Moses AC, Beamer WG, *J Bone Miner Res* 10 (1995).
58. Turner RT, Lifrak ET, Beckner M, Wakley GK, Hannon KS, Parker LN, *Am J Physiol* 258 (1990).

59. Ornoy A, Giron S, Aner R, Goldstein M, Boyan BD, Schwartz Z, *Bone Miner* 24 (1994).
60. Turner RT, Hannon KS, Demers LM, Buchanan J, Bell NH, *J Bone Miner Res* 4 (1989).
61. Kasra M and Grynapas MD, *Bone* 17 (1995).
62. Kim Bt, Mosekilde L, Duan Y, Zhang XZ, Tornvig L, Thomsen JS, Seeman E, *J Bone Miner Res* 18 (2003).
63. Danielsen CC, Mosekilde L, Andreassen TT, *Calcif Tissue Int* 50 (1992).
64. Gunness M and Orwoll E, *J Bone Miner Res* 10 (1995).
65. Vanderschueren D, Van Herck E, Suiker AM, Allewaert K, Visser WJ, Geusens P, Bouillon R, *J Bone Miner Res* 7 (1992).
66. Vanderschueren D, Van Herck E, Suiker AM, Visser WJ, Schot LP, Bouillon R, *Endocrinology* 130 (1992).
67. Vanderschueren D, van Herck E, Rush E, Einhorn T, Geusens P, Bouillon R, *Calcif Tissue Int* 53 (1993).
68. Rosen HN, Tollin S, Balena R, Middlebrooks VL, Beamer WG, Donohue LR, Rosen C, Turner A, Holick M, Greenspan SL, *Calcif Tissue Int* 57 (1995).
69. Wakley GK, Schutte HD Jr, Hannon KS, Turner RT, *J Bone Miner Res* 6 (1991).
70. Vanderschueren D, Jans I, van Herck E, Moermans K, Verhaeghe J, Bouillon R, *Bone and Miner* 26 (1994).
71. Verhas M, Schoutens A, L'hermite-Baleriaux M, Dourov N, Verschaeren A, Mone M, Heilporn A, *Calcif Tissue Int* 39 (1986).
72. Jensen EV, Jacobson HI, *Recent Prog Hom Res* 18 (1962).
73. Kuiper GG, Enmark E, Pelto-Huikko M, Nilsson S, Gustafsson J, *Proc Natl Acad Sci USA* 93 (1996).
74. Gosden, J. R.; Middleton, P. G.; Rout, D., *Cytogenet. Cell Genet.* 43 (1986).
75. Menasce, L. P.; White, G. R. M.; Harrison, C. J.; Boyle, J. M., *Genomics* 17 (1993).
76. Enmark, E.; Pelto-Huikko, M.; Grandien, K.; Lagercrantz, S.; Lagercrantz, J.; Fried, G.; Nordenskjold, M.; Gustafsson, J.-A., *J. Clin. Endocr. Metab.* 82 (1997).
77. Hall JM, McDonnell DP, *Endocrinology* 140 (1999).
78. Delaunay F, Pettersson K, Tujague M, Gustafsson J-A, *Mol Pharmacol* 58 (2000).
79. Métivier R, Stark A, Flouriot G, Hűbner MR, Brand H, Penot G, Manu D, Denger S, Reid G, Kos M, Russell RB, Kah O, Pakdel F, Gannon F, *Mol Cell* 10 (2002).
80. Glass CK, Rosenfeld MG, *Genes Dev* 14 (2000).
81. Beckman JM, Allan GF, Tsai SY, Tsai M-J, O'Malley BW, *Mol Endocrinol* 7 (1993).
82. Fritsch M, Leary CM, Furlow JD, Ahrens H, Schuh TJ, Mueller GC, Gorski J, *Biochenistry* 31 (1992).
83. McKenna NJ, Lanz RB, O'Malley BW, *Endocr Rev* 20 (1999).
84. McKenna NJ, O'Malley BW, *Cell* 108 (2002).
85. Robyr D, Wolffe AP, Wahli W, *Mol Endocrinol* 14 (2000).
86. Hermanson O, Glass CK, Rosenfeld MG, *Trends Endocrinol Metab* 13 (2002).
87. Schulman IG, Juguilon H, Evans RM, *Mol Cell Biol* 16 (1996).
88. Razandi M, Pedram A, Greene GL, Levin ER, *Mol Endocrinol* 13 (1999).

89. Okamoto T, Schlegel A, Scherer PE, Lisanti MP, *J Biol Chem* 273 (1998).
90. Solomon KR,Danciu TE, Adolphson LD, Hecht LE, Hauschka PV, *J Bone Miner Res* 15 (2000).
91. Migeon BR, Brown TR, Axelman J and Migeon CJ, *Proc Natl Acad Sci USA* 78 (1981)
92. Brown CJ, Goss SJ, Lubahn DB, Joseph DR, Wilson EM, French FS, Willard HF, *Am J Hum Genet* 44 (1989).
93. Simental JA, Sar M, Lane MV, French FS, Wilson EM, *J Biol Chem* 266 (1991).
94. Jenster G, van der Korput HAGM, van Vroonhoven C, van der Kwast TH, Trapman J , Brinkmann AO, *Mol Endocrinol* 5 (1991).
95. Wong C, Zhou Z, Sar M, Wilson EM, *J Biol Chem* 268 (1993).
96. Moilanen A, rouleau N, Ikonen T, Palvimo JJ, Janne OA, *FEBS Lett* 412(2) (1997).
97. Wilson EM, French FS, *J Biol Chem* 251 (1976).
98. Culig Z, Hobisch A, Cronauer MV, Cato ACB, Hittmair A, Radmayr C, Eberle J, Bartsch G, Klocker H, *Mol Endocrinol* 7 (1993).
99. Denger S, Reid G, Gannon F, flouriot G, Parsch D, Brand H, Korach KS, Sonntag-Buck V, *Mol Endocrinol* 15 (2001).
100. Colvard DS, Eriksen EF, Keeting PE, Wilson EM, Lubahn DB, French FS, Riggs BL, Spelsberg TC, *Proc Natl Acad Sci USA* 86 (1989).
101. Zhuang YH, Blauer M, Pekki A, Tuohimaa P, *J Steroid Biochem Mol Biol* 41 (1992).
102. Oursler MJ, Osdoby P, Pyfferoen J, Riggs BL, Spelsberg TC, *Proc Natl Acad Sci USA* 88 (1991).
103. Davis VL, Couse JF, Gary TK and Korach KS, *J Bone Miner Res* 9 (1994).
104. Mizuno Y, Hosoi T, Inoue S, Ikegami A, Kaneki M, Akedo Y, Nakamura T, Ouchi Y, Chang C, Orimo H, *Calcif Tissue Int* 54 (1994).
105. Bellido T, Jilka RL, Boyce BF, Girasole G, Broxmeyer H, Dalrymple SA, Murray R and Manolagas SC, *J Clin Invest* 95 (1995).
106. Abu EO, Horner A, Kusec V, Triffitt JT and Compston JE, *J Clin Endocrinoly Metab* 82 (1997).
107. Bord S, Horner A, Beavan S, Compston J, *J Clin Endocrinol Metab* 86(5) (2001).
108. Kennedy JJ, Baris C, Hoyland JA, Selby PL, Freemont AJ, Braidman IP, *Bone* (1999).
109. Carrascosa A, Audi L, Ferrandez MA, Ballabriga A, *J Clin Endocrinol Metab* 70 (1990).
110. Braidman IP, Baris C, Selby PL, Adams JE, Freemont AJ and Hoyland JA, *Journal of Pathology* 192 (2000).
111. Arts J, Kuiper GGJM, Janssen JMMF, Gustafsson JA, Lowik CW, Pols HAP, van Leeuwen JP, *Endocrinol* 138 (1997).
112. Onoe Y, Miyaura C, Ohta H, Nozawa S, Suda T, *Endocrinology* 138 (1997).
113. Bodine PV, Henderson RA, Green J, Aronow M, Owen T, Stein GS, Lian JB, Komm BS, *Endocrinology* 139 (1998).
114. Wiren KM, Chapman EA and Zhang XW, *J Endocrinol* 175 (2002).
115. Pettersson K., K. Grandien, G. G. J. M. Kuiper and J.-Å., *Endocrinol.* 11 (1997).
116. Pettersson K, Delaunay F, Gustafsson JA., *Oncogene*. Oct 12 (2000).

117. Britt KL, Kerr J, O'Donnell L, Jones ME, Drummond AE, Davis SR, Simpson ER, Findlay JK, *FASEB J.* Sep; 16(11) (2002).
118. Kousteni S, Bellido T, Plotkin LI, O'Brien CA, Bodenner DL, Han L, Han K, DiGregorio GB, Katzenellenbogen JA, Katzenellenbogen BS, Roberson PK, Weinstein RS, Jilka RL and Manolagas SC, *Cell* 104 (2001).
119. Chang L and Karin Mmammalian, *Nature* 410 (2001).
120. Schlegel A, Wang C, Katzenellenbogen BS, Pestell RG, Lisanti MP, *J Biol Chem* 274 (1999).
121. Kousteni S, Chen JR, Bellido T, Han L, Ali AA, O'Brien CA, Plotkin L, Fu Q, Mancino AT, Wen Y, Vertino AM, Powers CC, Stewart SA, Ebert R, Parfitt AM, Weinstein RS, Jilka RL, Manolagas SC, *Science* 298 (2002).
122. Manolagas SC and Kousteni S, *Endocrinology* 142 (2001).
123. Khosla S, Melton LJ 3rd, Atkinson EJ, O'Fallon WM, Klee GG, Riggs BL, *J Clin Endocrinol Metab* 83 (1998).
124. Hodgin JB, Knowles JW, Kim HS, Smithies O and Maeda N, *J Clin Invest* 109 (2002).
125. Vanderschueren D, Boonen S and Bouillon R, *Bone* 23 (1998).
126. Chen JR, Kousteni S, Bellido T, Han L, diGregorio GB, Jilka RL, Manolagas SC, *J Bone Miner Res* 16 (2001) S159(Abst).
127. Dupont S, Krust A, Gansmuller A, Dierich A, Chambon P, Mark M, *Development* 127 (2000).
128. Gentile MA, Zhang H, Harada S, Rodan GA and Kimmel DB, *J Bone Minier Res* 16 S146 (Abstr) (2001).
129. Lindberg MK, Weihua Z, Andersson N, Moverare S, Gao H, Vidal O, Erlandsson M, Windahl S, Andersson G, Lubahn DB, Carlsten H, Dahlman-Wright K, Gustafsson JA, Ohlsson C, *J Endocrinol* 174 (2002).
130. Jilka RL, Hangoc G, Girasole G, Passeri G, Williams DC, Abrams JS, Boyce B, Broxmeyer H, Manolagas SC, *Science* 257 (1992).
131. Manabe N, Kawaguchi H, Chikuda H, Miyaura C, Inada M, Nagai R, Nabeshima Y, Nakamura K, Sinclair AM, Scheuermann RH, Kuro-o M, *J Immunol* 167 (2001).
132. Sato T, Shibara T, Ikeda K and Watanabe K, *J Bone Miner Res* 16 (2001).
133. Katavic V, Grcevic D, Lee SK, Kalinowski J, Jastrzebski S, Dougall W, Anderson D, Puddington L, Aguila HL, Lorenzo JA, *Bone* Jun. 32(6) (2003).
134. Kousteni S, Han L, Chen JR, Aloneida M, Plotkin LI, Bellido T, Manologas SC, *J Clin Invest* 111 (2003).
135. Manolagas SC, Kousteni S and Jilka RL, *Recent Prog Horm Res* 57 (2002).
136. Jilka RL, Takahashi K, Munshi M, Williams DC, Roberson PK, Manolagas SC,*J Clin Invest* 101 (1998).
137. Jilka RL, Weinstein RS, Takahashi K, Parfitt AM, Manolagas SC, *J Clin Invest* 97 (1996).
138. Weinstein RS, Jilka RL, Parfitt AM, Manolagas SC, *Endorinology* 138 (1997).
139. Di Gregorio GB, Yamamoto M, Ali AA, Abe E, Roberson P, Manolagas SC, Jilka RL, *J Clin Invest* 107 (2001).
140. Scheven BA, Visser JW, Nijweide PJ, *Nature* 321 (1986).

141. Kurihara N, Suda T, Miura Y, Nakauchi H, Kodama H, Hiura K, Hakeda Y, Kumegawa M, *Blood* 74 (1989).
142. Pottratz ST, Bellido T, Mocharla H, Crabb D, Manolagas SC, *J Clin Invest* 93 (1994).
143. McDonnell and Norris, *Osteoporos Int.* 7 Suppl 1 (1997).
144. Lin SC, Yamate T, Taguchi Y, Borba VZ, Girasole G, O'Brien CA, Bellido T, Abe E, Manolagas SC, *J Clin Invest* 100 (1997).
145. Manolagas SC, *Ann NY Acad Sci* 840 (1998).
146. Poli V, Balena R, Fattori E, Markatos A, Yamamoto A, Tanaka H, Ciliberto G, Rodan GA, Costantini F, *EMBO J* 13 (1994).
147. Pacifici R, *J Bone Miner Res* 11 (1996).
148. Manolagas SC, *Endocr Rev* 21 (2000).
149. Srivastava S, Weitzmann MN, Cenci S, Ross FP, Adler S and Pacifici R, *J Clin Invest* 104 (1999).
150. Cenci S, Weitzmann MN, Roggia C, Namba N, Novack D, Woodring J and Pacifici R, *J Clin Inves* 106 (2000).
151. Roggia C, Gao Y, Cenci S, Weitzmann MN, Toraldo G, Isaia G and Pacifici R, *Proc Natl Acad Sci USA* 98 (2001).
152. Horowitz MC, Xi Y, Wilson K, Kacena MA, *Cytokine Growth Factor Rev* Mar. 12(1) (2001).
153. Kasperk C, Fitzsimmons R, Strong D, Mohan S, Jennings J, Wergedal J, Baylink D, J Clin Endocrinol Metab 71 (1990).
154. Fukayama S and Tashjian AHJ, *Endocrinology* 125 (1989).
155. Gray A, Feldman HA, McKinlay JB, Longcope C, *J Clin Endocrinol Metab* 73 (1991).
156. Hofbauer LC, Khosla S, Dunstan CR, Lacey DL, Spelsberg TC, Riggs BL, *Endocrinology* Sep. 140(9) (1999).
157. Girasole G, Passeri G, Jilka RL, Manolagas SC, *J Clin Invest* Apr. 93(4) (1994).
158. Sunyer T, Lewis J, Collin-Osdoby P, Osdoby P, *J Clin Invest* 103(10) (1999).
159. Lebovitz HE and Eisenbarth GS, *Vitam Horn (NY)* 33 (1975).
160. Corvol MT, Carrascosa A, Tsagris L, Blanchard O, Rappaport R, *Endocrinology* 120 (1987).
161. Kapur SP, Reddi AH, *Calcif Tissue Int* Feb. 44(2) (1989).
162. Carani C, Qin K, Simoni M, Fanstini-Fustini M, Serpanti S, Boyd J, Korach K and Simpson ER, *New England Journal of Medicine* 337 (1997).
163. Bilezikian JP, Morishima A, Bell J, Grumbach MM, *N Engl J Med.* Aug. 339(9) (1998).
164. Maffei L, Murata Y, Rochira V, Tubert G, Aranda C, Vazquez M, Clyne CD, Davis S, Simpson ER, Carani C, *J Clin Endocrinol Metab.* Jan. 89(1) (2004).
165. Weise M, De-Levi S, Barnes KM, Gafni RI, Abad V, Baron J, *Proc Natl Acad Sci USA* 98 (2001).
166. Seeman E, *Baillieres Clin Rheumatol.* Aug. 11(3) (1997).
167. Schnoeu E, Neu CM, Rauch F, Manz F, *J Clin Endocrinol Metab* 86 (2001).
168. Couse JF, Korach KS, *Endocr Rev.* 20 (1999).

169. Migliaccio S, Newbold RR, Bullock BC, Jefferson WJ, Sutton FG Jr, McLachlan JA, Korach KS, *Endocrinology* 137 (1996).
170. Damien, E., J.S. Price, and L.E. Lanyon., *J. Bone Miner. Res.* 15 (2000).
171. Vandenput L, Ederveen AG, Erben RG, Stahr K, Swinnen JV, Van Herck Verstuyf A, Boonen S, Bouillon R, Vanderschueren D, *Biochem Biophys Res* 285(1) (2001).
172. Goulding A and Gold E, *J Bone Miner Res* 8 (1993).
173. Parfitt, A. M., *J. Cell Biochem.* 55 (1996).
174. Eriksen EF, Langdahl B, Vesterby A, Rungby J and Kassem M., *Journal of Bone and Mineral Research* 14 (1999).
175. Kelly PJ, Twomey L, Sambrook PN, Eisman JA, *J Bone Miner Res* 5 (1990).
176. McElduff A, Wilkinson M, Ward P, Posen S. *Bone* 9 (1988).
177. Murphy S, Khaw KT, Cassidy A, Compston JE. *Bone Miner* 20 (1993).
178. Rudman D, Drinka PJ, Wilson CR, Mattson DE, Scherman F, Cuisinier MC, Schultz S, *Clin Endocrinol (Oxf)* 10 (1994).
179. Barrett-Connor E, Kritz-Silverstein D, Edelstein SL. *Am J Epidemiol* 137 (1993).
180. Drinka PJ, Olson J, Bauwens S, Voeks SK, Carlson I, Wilson M. *Calcif Tissue Int* 52 (1993).
181. Meier DE, Orwoll ES, Keenan EJ, Fagerstrom RM. *J Am Geriatr Soc* 35 (1987).
182. Wishart JM, Need AG, Horowitz M, Morris HA, Nordin BE. *Clin Endocrinol (Oxf)* 42 (1995).
183. Slemenda CW, Longcope C, Zhou L, Hui SL, Peacock M, Johnston C. *J Clin Invest* 100 (1997).
184. Greendale GA, Edelstein S, Barrett-Connor E. *J Bone Miner Res* 12 (1997).
185. Tenover JS, Matsumoto AM, Plymate SR, Bremner WJ. *J Clin Endocrinol Metab* 65 (1987).
186. Centre JR, Nguyen TV, White CP, Eisman JA. *J Bone Miner Res* 12 (1997).
187. Ongphiphadhanakul B, Rajatanavin R, Chanprasertyothin S, Piaseau N, Chailurkit L. *Clin Endocrinol (Oxf)* 49 (1998).
188. van den Beld AW, de Jong FH, Grobbee DE, Pols HA, Lamberts SW. *J Clin Endocrinol Metab* 85 (2000).
189. Ho AY, Yeung SS, Kung AW, *Calcif Tissue Int.* 66(6) (2000)
190. Amin S, Zhang Y, Sawin CT, Evans SR, Hannan MT, Kiel DP, Wilson PW, Felson DT, *Ann Intern Med* 133 (2000).
191. Szulc P, Munoz F, Claustrat B, Garnero P, Marchand F. *J Clin Endocrinol Metab* 86 (2001).
192. Khosla S, Melton LJ 3rd, Atkinson EJ, O'Fallon WM. *J Clin Endocrinol Metab* 86 (2001).
193. Falahati-Nini A, Riggs BL, Atkinson EJ, O'Fallon WM, Eastell R, Khosla S. *J Clin Invest* 106 (2000).
194. Khosla SK, Atkinson EJ, Dunstan CR, O'Fallon WM, *J Clin Endocrinol Metab* 87 (2002).
195. Leder BZ, LeBlanc KM, Schoenfeld DA, Eastall R, Finkelstein JS. *J Clin Endocrinol Metab* 88 (2003).
196. Taxel P, Kennedy DG, Fall PM, Willand AK, Clive JM, Raisz LG. *J Clin Endocrinol Metab* 86 (2001).

197. Vanderschueren D, Van Herck E, Nijs J, Ederveen AG, De Coster R, Bouillon R. *Endocrinology* 138 (1997).
198. Erben RG, Eberle J, Stahr K, Goldberg M. *J Bone Miner Res* 15 (2000);15.
199. Vidal O, Lindberg MK, Hollberg K, Baylink DJ, Andersson G, Lubahn DB, Mohan S, Gustafsson JA, Ohlsson C, *Proc Natl Acad Sci U S A* 97 (2000).
200. Oz OK, Zerwekh JE, Fisher C, Graves K, Nanu L, Millsaps R, Simpson ER, *J Bone Miner Res* 15 (2000).
201. Fisher, C. R., Graves, K. H., Parlow, A. F., and Simpson, E. R. *Proc Natl Acad Sci U S A.* 95 (12) (1998).
202. Miyaura, C, K Toda, M Inada, T Ohshiba, C Matsumoto, T Okada, M Ito, Y Shizuta, and A. Ito. *Biochem Biophys Res* 280 (2001)
203. Lubahn DB, Moyer JS, Golding TS, Couse JF, Korach KS, Smithies O, *Natl Acad Sci USA* 90 (1993)
204. Vidal, O., Kindblom, L.G., and Ohlsson, C. *J. Bone Miner. Res.* 14 (1999).
205. Sims NA, Dupont S, Krust A, Clement-Lacroix P, Minet D, Resche-Rigon M, Gaillard-Kelly M, Baron R, *Bone* 30 (2002).
206. Lindberg MK, Alatalo SL, Halleen JM, Mohan S, Gustafsson J-Å, Ohlsson C, *J Endocrinol* 171 (2001).
207. Windahl SH, Vidal O, Andersson G, Gustafsson JA, Ohlsson C, *J Clin Invest* 104(7) (1999).
208. Yeh S, Tsai MY, Cu Q, Mu XM, Lardy H, Huang KE, Lin H, Yeh SD, Altuw S, Zhou X, Xing L, Boyce BF, Hung MC, Zhang S, Gan L, Chang C, Hung MC, Zhang S, Gan L, Chang C, *Proc Nati Acad Sci USA* 99(21) (2002).
209. Windahl SH, Hollberg K, Vidal O, Gustafsson JA, Ohlsson C, Andersson G, *J. Bone Miner. Res.* 16 (2001).
210. Greenberg Z, Chorev M, Muhlrad A, Shteyer A, Namdar-Attar M, Casap N, Tartakovsky A, Vidson M, Bab I, *J Clin Endocrinol Metab* 80 (1995).
211. Naveh-Many T, Almogi G, Livni N and Silver J, *J Clin Invest* 90 (1992).
212. Prank K, Nowlan SJ, Harms HM, Kloppstech M, Brabant G, Hesch RD and Sejnowski TJ, *J Clin Invest* 95 (1995).
213. Orimo H, Fujita T and Yoshikawa M, *Endocrinology* 90 (1972).
214. Prince RL, Schiff I and Neer RM, *J Clin Endocrinol Metab* 71 (1990).
215. Khosla S, Atkinson EJ, Melton LJ 3rd, Riggs BL. *J Clin Endocrinol Metab.* 82(5) (1997).
216. Eastell R, Yergey AL, Vieira NE, Cedel SL, Kumar R and Riggs BL, *J. Bone Miner.Res.* 6 (1991).
217. Zofkova I and Kancheva RL *Bone* 19 (1996).
218. Uemura H, Irahara M, Yoneda N, Yasui T, Genjida K, Miyamoto KI, Aono T and Takeda E *J.Clin.Endocrinol.Metab* 85 (2000).
219. Stock JL, Coderre JA and Posillico JT, *Clin.Chem.* 35 (1989).
220. Gennari C, Agnusdei D, Nardi P and Civitelli R, *J.Clin.Endocrinol.Metab* 71 (1990).
221. McKane WR, Khosla S, Burritt MF, Kao PC, Wilson DM, Ory SJ and Riggs BL, *J Clin Endocrinol Metab* 80 (1995).
222. Ernst M and Rodan GA, *Mol Endocrinol* 5 (1991).

223. Kassem M, Okasaki R, Harris SA, Spelsberg TC, Conover CA, Riggs BL, *Calcif Tissue Int* 62 (1998).
224. Aronica SM, Kraus WL and Katzenellenbogen BS, *Proc Natl Acad Sci USA* 91 (1994).
225. Greg Miller, ***Science*** 298 (2002).

CHAPTER 11

PHYTOESTROGENS AND BONE HEALTH: MECHANISMS OF ACTION

ZhiChao Dang

Department of Endocrinology and Metabolic Diseases (C4-R),
Leiden University Medical Center,
Albinusdreef 2, 2300 RC, Leiden,
The Netherlands
Email: ZCDANG@LUMC.NL

Interest in phytoestrogens for maintenance of bone health and delaying or preventing osteoporosis has exploded in the past decade. These substances have estrogen-like activity and are viewed as potential natural alternatives to estrogen replacement therapy. However, the evidence that support beneficial effects of phytoestrogens on bone is still not completely convincing. Data on bone formation *in vitro* show that phytoestrogens affect osteoblasts and osteoprogenitor cells in a biphasic dose-dependent way, which is estrogen receptor (ER)-dependent and ER-independent. In contrast, studies on bone resorption *in vitro* show that phytoestrogens only inhibit osteoclast formation and activity. Similar findings were also observed in some animal models *in vivo*. However, experiments in primates and clinical investigations showed that the beneficial effects of phytoestrogens on bone are not conclusive. Furthermore, peroxisome proliferator-activated receptors (PPARs) as new molecular targets of phytoestrogens, their differential effects on bone, and their interactions with ERs are discussed. Dose-dependent responses of phytoestrogens result from the balance between ERs and PPARs. This newly-proposed mechanisms of action of phytoestrogens are important for the future study to find precise beneficial doses *in vivo*. Further investigations are needed to find the critical effective doses and evaluate the long-term beneficial effects of phytoestrogens on bone.

1. Introduction

Osteoporosis, caused by estrogen deficiency in post-menopausal women, is characterized by a progressive reduction in bone and consequent increase in fracture risk. Estrogen replacement therapy is an effective prevention and treatment therapy for the post-menopausal osteoporosis. [1,2] Long-term use of estrogen, however, has been specifically related to increased risk of breast and endometrial cancers and other side effects. [3-5] So the development of well-tolerated and safe alternatives to estrogen replacement therapy should be a public health priority for our aging population. Phytoestrogens, among all the natural alternatives currently investigated, may become the most effective means of preventing bone loss. [6-8]

Phytoestrogens are plant-derived nonsteroidal compounds that bind to estrogen receptors (ERs) and have estrogen-like activity [9-11]. These substances are available as food supplements because they may have beneficial effects on many Western diseases including cardiovascular diseases, osteoporosis, diabetes and obesity, and various cancers [9,12,13]. On the other hand, phytoestrogens have been categorized as endocrine disruptors because they may cause environmental problems and have deleterious effects on reproductive systems [14-16]. Both topics have been a focus of many studies in the past decade. In this chapter, the effects of phytoestrogens on bone health will be further discussed. The issue on endocrine disruptor has been extensively reviewed elsewhere [16-18] and will not be further discussed here.

Interest in phytoestrogens for maintenance of bone health and delaying or preventing osteoporosis has exploded in the past decade [6,10,11,19]. Substantial data from epidemiological studies and nutritional intervention experiments in humans and animals support the notion that phytoestrogens affect bone in a beneficial way. However, this notion remains inconclusive [6,11,19]. Some studies showed that phytoestrogens, similar to estrogen, have ER-dependent beneficial effects on bone remodelling. Others, however, showed that phytoestrogens have no effects or even antiestrogenic effects on bone cells. These contradictory results reported in the literature may attribute to the limited doses selected in different studies. One of the striking phenomena is that phytoestrogens affect bone cells *in vitro* and *in vivo* in a biphasic dose-dependent way [10,20,21]. These biphasic dose responses are ER-dependent and ER-independent. In general, the ER-dependent or estrogenic effects of phytoestrogens occur at relative low concentrations, whereas ER-independent or antiestrogenic effects at relative high concentrations. The limited doses selected in some studies may fall into the concentration range in which the effects of phytoestrogens are only estrogenic or antiestrogenic. As a result, the precise action of phytoestrogens is often lacking due to the limited dose selection in different studies and/or the overlapping of such pleiotropic biological properties of phytoestrogens at certain concentrations.

Biphasic dose responses of phytoestrogens result from their pleiotropic biological properties [20,21]. ERs have been proposed as their major molecular targets, but the health beneficial effects of phytoestrogens cannot be solely explained by ER-mediated pathway. It has been concluded that phytoestrogens cannot be simply viewed as either agonists or antagonists of estrogen [22,23]. In addition to their estrogenic activity, phytoestrogens have biological properties that are quite separate from classic estrogen action. They inhibit enzymes like tyrosine kinase [24], topoisomerase I and II [25,26], and mitogen-activated protein kinase [27], which are critical for cell signal transduction. The different enzymes like tyrosine kinase and topoisomerase II have been referred to as molecular targets of phytoestrogens. However, this mechanism of action may only limit to phytoestrogens that have enzyme inhibiting effects. Recently, we identified peroxisome proliferator-activated receptors (PPARs), in addition to ERs, are the critical molecular targets of phytoestrogens. [20,21] PPARs are the critical targets of many Western diseases including cancer, glucose/lipid metabolisms, obesity, and inflammatory diseases [28-30] and have differential effects on bone remodelling [31]. Based on our studies, the dose-responses of phytoestrogens result from the balance among these phytoestrogen-induced pleiotropic actions [20,21]. It is these pleiotropic biological properties of phytoestrogens that often result in complex dose-dependent responses in bone cellular systems *in vitro* and *in vivo*. This chapter focuses on the current state of the mechanisms of action of phytoestrogens on bone cells.

The following topics with regards to phytoestrogens and bone health are covered in this chapter: background on phytoestrogens, effects of phytoestrogens on bone cells *in vitro*, effects of phytoestrogens on bone remodelling *in vivo*, and the molecular mechanisms of action of phytoestrogens.

2. Background of Phytoestrogens

Phytoestrogens are so named because they originate from plants and have estrogenic activity. More specifically, they are defined as any plant substance or metabolite that induces biological responses in vertebrates and can mimic or modulate the actions of endogenous estrogens usually by binding to estrogen receptors[32] Structurally, they are similar to naturally occurring most potent mammalian oestrogen 17β-oestradiol, synthetic oestrogens, and anti-oestrogens (Fig. 1). Phytoestrogens can be divided into three main classes: isoflavones, coumestans, and lignans. In

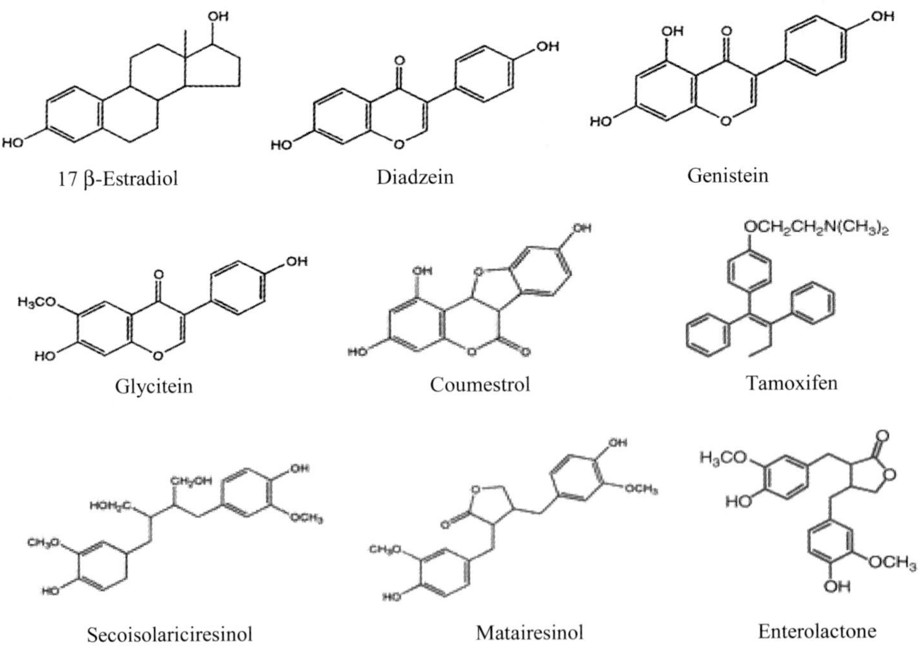

Fig. 1. Structures of 17β-estradiol, phytoestrogens and tamoxifen

some reviews, the fourth class of prenylated flavonoids is also included [11,32]. Despite of these differences in classification, phytoestrogens, except lignans, belong to a large group of substituted phenolic flavonoids that contain more than 4000 plant phytochemicals [33].

Isoflavones are the most extensively studied phytoestrogens [6,19,33]. These compounds are found in highest amounts in soybeans and soy foods, with approximately 0.2 –1.6 mg of isoflavones/g dry weight [34]. Due to extremely high amount of isoflavones present in soy and soy products, some in vivo experiments used soy products as a source of isoflavones to study their beneficial effects on bone. It has been a focus of many reviews that beneficial effects of soy are due to the soy protein component itself, its isoflavones or a combination of both [35-37]. Further discussion on this topic is beyond the scope of this review. The other dietary sources of phytoestrogens include clover, oilseeds, nuts and bluegrass [34,38]. The major isoflavones present in plant-derived foods are genistein, daidzein, glycitein, biochanin A, and formononetin [13,32,36]. Genistein, daidzein and glycitein are three main compounds from soybeans, which have received the most attention. Beer is another surprising source of isoflavones, with genistein of 0.05-1.8 μg/g, daidzein of 0.02 - 0.6 μg/g, biochanin A of 0.2 –1.4 μg/g, and formononetin of 0.05 – 4.5 μg/g [39]. It is estimated that typical daily intake of isoflavones in Asian populations are 25-50 mg, which is much higher than the level (less than 1 mg/day) of Western population [11,40]. Isoflavones are often present as unconjugated form aglycone or as glycoside conjugates glycones [36]. In plants, they exist mainly as an inactive form of glycosides. After ingestion, isoflavone glycosides such as genistin and daidzin are hydrolysed in the intestines by bacterial β-glucosidases and are coverted into bioactive aglycones like genistein and daidzein. The aglycones are absorbed from intestine and conjugated to glucuronides in the liver. Glucuronides are reexcreted through the bile and reabsorbed by enterohepatic recycling or excreted unchanged in the urine. Especially in rats, daidzein may further be metabolised to equol, a more potent estrogenic compound than daidzein and genistein [41]. Clearly, absorption of isoflavones is rapid and requires intestinal metabolism, which plays a crucial role in their bioavailability and biological activity [36,41,42]. Normally, isoflavones can be detected after 15 to 30 minutes of ingestion and reach a peak value between 3 and 7 hours. This high daily intake of isoflavones is accompanied by the high plasma concentrations and daily urinary isoflavones excretion [36]. Human plasma concentrations of isoflavones in the populations on soy-rich diet and soy-poor diet can be more than 100-fold difference, with the levels lower than 6 μmol/L or 40 nmol/L, respectively [11,36]. The levels of the daily urinary isoflavones excretion has been reported as 3412 - 8770 nmol in Japanese men and 67.5- 324 nmol in Finnish people [34]. Isoflavones can be detected in many tissues of animals and humans. For example, genistein has been detected in brain, liver, mammary, ovary, prostate, testis, thyroid and uterus when Sprague-Dawley rats exposed to dietary genistein. In addition, isoflavones have been also reported in lung, skeletal muscle, spleen, heart and kidney [34,36]. Commercial rodent diets widely used in many laboratories contain large amounts of isoflavones. Serum concentrations of isoflavones in these rodents can reach plasma level as high as 8.5 μM [43]. Therefore, caution should be taken on animal experiments especially those influenced by hormones [43,44].

Coumestans are structurally related to isoflavones but less commonly found in human diets [39]. Coumestrol is the most common form of coumestans found in clover and fresh alfalfa sprouts [34,39,40]. In addition, 4'-methoxycoumestrol is another estrogenic coumestan often mentioned in the literature [34,39]. The content of coumestrol in plant varies according to many factors like plant variety, stage of growth, the presence of diseases [38]. Soy sprouts have high levels of coumestrol of 71.1 μg/g wet weight [39].

Compared to isoflavones, lignans are widely distributed in many oilseeds, cereals, vegetables, seaweed, and fruits but have been less well studied [38,39,45]. The levels of lignans are normally low on an individual food basis. But the ubiquity of these substances in plants suggests that they may be an important source of phytoestrogens. The principal lignans identified in food are secoisolariciresinol, matairesinol, lariciresinol, and isolariciresinol [38-40]. Oilseeds like flaxseed are the richest known plant source of lignans. Flaxseed contain as high as 0.8 mg of secoisolariciresinol/g dry weight, the content is about 100 times higher than the other foods [34]. Similar to isoflavones, the absorption and metabolism of lignans occur in the gastrointestinal tract. Lignans are absorbed more slowly than isoflavones, with the peak value at 8 hours after ingestion. The plant lignans, secoisolariciresinol, matairesinol, are the precursors of the mammalian lignans, enterodiol and enterolactone, which are converted by intestinal bacteria [34,40]. It has been reported that the plasma lignan concentrations increased from a baseline level of 29 nM to 52 nM after ingestion of flaxseed in young healthy women. Urinary excretion of lignans is between 1.5 to 3.3 μmol/24h in omnivorous women and 2 to 3 fold higher in vegetarian women [46].

3. Effects of Phytoestrogens on Bone Cells *in vitro*

Effects of phytoestrogens on bone cells *in vitro* have been studied in isolated cells, cell lines, and tissue cultures [6,10]. These studies mainly focus on regulation of phytoestrogens on osteoblasts and their progenitor cells, the cytokines derived from these cells, and osteoclast formation and activity. Strikingly, data on bone formation *in vitro* show that phytoestrogens affect osteoblasts and osteoprogenitor cells in a biphasic dose-dependent way, which is ER-dependent and ER-independent [10,20,21]. It is worth noting that some studies in the literature only showed the estrogenic effects of phytoestrogens, whereas others only showed their non-estrogenic effects. These pure estrogenic or non-estrogenic results may attribute to the doses selected in these studies, the cellular systems and parameter used, and the time and duration of exposure to phytoestrogens. Definitely, these results do not rule out the possibility of biphasic dose effects on bone formation. In contrast, studies on bone resorption *in vitro* show that phytoestrogens inhibit osteoclast formation and activity, which is not biphasic dose-dependent.

Most studies on the effects of phytoestrogens on bone cells *in vitro* focus on isoflavones. However, one of the earliest studies of phytoestrogens on bone *in vitro* was used coumestrol. This study, using 9-d-old chick embryonic femur cultures, showed that coumestrol stimulated bone mineralization and inhibited bone resorption stimulated by parathyroid hormone (PTH), vitamin D, and prostaglandin E_2. The optimal concentrations of coumestrol for bone formation and resorption is 10^{-5} M, and the effects were similar to those of estrogen [47]. In another study, coumestrol and estrogen had no effects on the proliferation and alkaline phosphatase (ALP) activity of ROS 17/2.8 cells and osteoblasts from neonatal mouse calvaria. But coumestrol at tested concentration between 10^{-9} and 10^{-5} M dose-dependently inhibited mouse osteoclast-like cell formation [48]. Different from the above results, coumestrol at tested concentrations between 10^{-10} and 10^{-5} M influenced proliferation and ALP activity of preosteoblastic MC3T3-E1 cells in a biphasic way, with the optimal stimulated bone formation at the concentration of 10^{-7} M [49].

Bone forming osteoblasts are derived from bone marrow mesenchymal or stromal stem cells that are also the progenitor cells of adipocytes, chondrocytes, and myoblasts [2,50,51]. A progressive decrease in bone mass and an increase of adipocyte formation in bone marrow are associated with aging. This inverse relationship between osteoblastogenesis and adipogenesis suggests that osteoprogenitor cells play a critical role in bone formation [2]. The complex process of bone formation involving the commitment of osteoprogenitor cells and the differentiation of osteoblasts, both are regulated by phytoestrogens. It is known that the commitment of these osteoprogenitor cells to osteoblast lineage cells is mediated by transcription factors like Runt-related transcription factor (Runx)-2 or core-binding factor (Cbfa)-1 [52]. In contrast, the master transcriptional adipogenic commitment factor PPAR γ2 downregulates differentiation of these progenitor cells into osteoblasts [20,53]. PPARγ is likely to be a potential target for intervention in osteoporosis [54]. Furthermore, these two transcription factors can be regulated by p44/42 mitogen-activated protein kinases (MAPKs). Phosphorylation of Cbfa1 increases osteogenesis, whereas an inhibition of this phosphorylation decreases osteogenesis [55,56]. In contrast, phosphorylation of PPARγ downregulates adipogenesis, whereas an inhibition of the phosphorylation upregulates adipogenesis [57-59]. Some phytoestrogens like genistein, apigenin, quercetin that have enzyme inhibiting effects may influence osteogenesis and adipogenesis via inhibiting of phosphorylation of these transcriptional factors.

Estrogen is a crucial systemic factor regulating skeletal homeostasis. Osteoblasts and their progenitor cells contain ERα and ERβ and are the direct target of estrogen [1]. Recent data show that estrogen stimulates osteogenesis and inhibits adipogenesis in an ER-dependent way [60,61].

Consistent with these estrogenic effects, several reports showed that phytoestrogen genistein at relative low doses upregulated osteogenesis and concurrently downregulated adipogenesis in mouse osteoprogenitor KS483 cells, mouse and human primary bone marrow mesenchymal cells [20,62]. In these studies, osteogenesis has been shown by alkaline phosphatase (ALP) activity, calcified nodule formation, mRNA expression of Cbfa1, osteocalcin, and PTH/PTHrP-R; whereas adipogenesis by the number of adipocytes, mRNA expression of PPAR γ2, aP2 and LPL. Similarly, another phytoestrogen daidzein showed an ER-dependent stimulated osteogenesis and concurrently inhibited adipogenesis in KS483 cells, mouse and human primary bone marrow mesenchymal cells [21]. It is, however, important to note that these typical ER-dependent estrogenic effects of phytoestrogens are effective only within certain doses. The specific antiestrogen compound ICI182,780 could completely block phytoestrogen-action only within these doses [20,21,31]. These doses are normally below micromolar and may vary with the different compounds and the cellular system studied. In contrast to these estrogenic effects of phytoestrogens at low doses, they may also have antiestrogenic effects at high doses. Both genistein and daidzein downregulated osteogenesis and concurrently upregulated adipogenesis in KS483 cells, which are completely opposite to those of estrogenic actions [20,21]. The specific antiestrogen compound ICI182,780 could only partly block phytoestrogen-action at these relative high doses [20,21,31]. It is interesting to note that genistein and daidzein at some physiological doses of range showed a typical estrogenic effects on osteogenesis and adipogenesis in KS483 cells. When these phytoestrogen-stimulated cells were exposed to ICI182,780, osteogenesis of these cells were lower than controls, whereas adipogenesis higher than controls. These results indicate that estrogenic effects of isoflavones even at physiological doses are only part of story. The responses of different organs to phytoestrogens may depend on the amount of receptors present in these tissues. In summary, data on mesenchymal stem cells indicate that phytoestrogens directly act on these cells. The effects of phytoestrogens on osteogenesis and adipogenesis are dose-dependent and biphasic. At low doses, they act as estrogen, stimulating osteogenesis and inhibiting adipogenesis; At high doses, they act oppositely, inhibiting osteogenesis and stimulating adipogenesis.

Effects of isoflavones on proliferation have been studied in osteoblastic cell lines. Most studies used isoflavones at doses between 10^{-10} and 10^{-4} M and the results are biphasic. The stimulatory doses on proliferation normally are below 10 μM as evidenced [^{3}H]-thymidine incorporation or MTT cell proliferation assay, whereas the inhibitory effects are between 30 and 100 μM. [63-65]. Anabolic effects of isoflavones on osteoblasts have been reported. Despite of difference in inhibition of tyrosine kinases activity, both genistein and daidzein at the concentrations between 10^{-7} and 10^{-5} M dose-dependently increased protein contein, ALP activity, and DNA content in preosteoblast MC3T3-E1 cells [66,67]. It is suggested that isoflavones may directly activate leucyl-tRNA synthetase, a rate-limiting enzyme of the translational process of protein synthesis. Similar to estrogen, these anabolic effects of isoflavones were blocked by antiestrogen tamoxifen, indicating the effects were ER-dependent. In these studies, biphsic effects of genistein and daidzein were not found within the tested concentrations between 10^{-7} and 10^{-5} M [66-68]. In contrast, it has been shown that glycitein, similar to genistein and daidzein, increased the ALP activity and osteocalcin in MC3T3-E1 cells in a biphasic way at the concentrations between 10^{-8} and 10^{-6} M [69]. The reason for these different results from two groups is not clear, but it may be attributed to the duration of treatment. For example, daidzein at the concentrations between 2 and 100 μM influenced osteoblast viability in a biphasic way when these primary osteoblastic cells isolated from newborn Wistar rats were exposed for 6 days, whereas it increased osteocalcin in a dose-dependent way when these cells were exposed for 2 days [65]. In another study, MC3T3-E1 cells showed biphasic dose responses to genistein or daidzein at the concentrations between 10^{-10} and 10^{-5} M for 3 days, but did not response to the proliferation test when these cells exposed for 1 day (Kanno et al., 2004).

During past five years, progresses have been made on elucidating the molecular mechanisms of estrogen and phytoestrogens on bone cells and the studies have been focus on interaction of estrogen and phytoestrogens with local factors like BMPs. One of the targets identified is BMP-2, which is a potent key inducer of osteogenic differentiation *in vitro* and *in vivo*. Estrogen induced mRNA expression of BMP-2 in an ER-dependent and transcriptional mechanisms in mouse mesenchymal stem cells [70]. Similarly, both genistein and daidzein increased mRNA expression of BMP-2 in mouse bone marrow mesenchymal stem cells, in C3H10T1/2 cells and in osteoblastic cells isolated from newborn Wistar rats [65,70]. It has been shown that genistein at the concentration of 10^{-7} M stimulates BMP-2 promoter activity via ERβ, but not ERα [70]. As this study did not show the effects of other concentrations of genistein on BMP-2 promoter activity, this conclusion may have its limitation. In fact, several other reports demonstrated that the differential effects of ERβ-mediated responses can only be found at the concentrations of genistein lower than 1 μM [71,72]. It is therefore important to test in the future the effects of different doses of isoflavones on BMP-2 promoter activity.

The long-sought osteoblast-derived RANKL plays an important role in osteoclastogenesis, whereas its decoy receptor OPG inhibits osteoclastogenesis. Several reports show that isoflavones influenced RANKL, OPG and other cytokines like interleukin-6 (IL-6). Similar to estrogen that downregulated RANKL and upregulated OPG [73-75], genistein at the concentrations between 10^{-10} and 10^{-6} M increased mRNA expression of OPG and decreased IL-6 production or mRNA expression of RANKL in MC3T3-E1 cells, in hFOB cells, and in human primary bone marrow stromal cells [62,76,77]. The effects of genistein at concentrations between 10^{-10} and 10^{-6} M on mRNA expression of OPG and protein secretion in human trabecular osteoblasts were biphasic, with the peak value at 10^{-7} M [78].

In addition to suppression of bone-resorbing cytokines and stimulation of anti-resorptive factors, isoflavones have inhibitory, but not biphasic effects on bone resorption and osteoclast formation and activity. The precise molecular mechanisms of action, however, are not clear. Osteoclasts, derived from hematopoietic cells of the monocyte/macrophage linage, are responsible for bone resorption. Similar to estrogen, genistein at the concentrations between 10^{-7} and 10^{-5} M inhibited parathyroid hormone (PTH) induced bone resorption in femoral-metaphyseal tissues of elderly female rats in an ER-dependent way [79]. These inhibitory effects may result from a decrease in osteoclast differentiation and an increase in osteoclast apoptosis [80,81]. It has been further shown that the suppressive effects of genistein were partly mediated through the pathway of calcium signalling, including the inhibition of protein kinase or cyclic AMP and activation of protein tyrosine phosphatase [81,82]. In another study using isolated rat osteoclasts, however, both genistein and daizein inhibited inward rectifier K^+ current, which is independent of protein tyrosine kinases and a rise of intracellular [Ca2+] level [83]. Similar inhibitory effects of genistein at the concentrations between 10^{-9} and 10^{-4} M on osteoclast formation as shown by tartrate-resistant acid phosphatase (TRAP) staining have been observed in mouse bone marrow cultures stimulated by $1,25(OH)_2D_3$ [84] and in coculture system of mouse osteogenic stromal ST2 cells and spleen cells [85]. The mechanisms of this action, however, may be different. In the coculture system of ST2 cells and spleen cells, daidzein, E2, PD98059 (inhibitor of p42/44 MAPK), wortmannin (inhibitor of PI3 kinase) and lavendustin A (inhibitor of tyrosine kinase) did not affect the number of TRAP-positive cells, whereas the inhibitors of topoisomerase II and genistein dose-dependently decreased the number of TRAP-positive cells, which suggests that genistein inhibited osteoclast formation via inhibition of topoisomerase II activity [85]. In addition, ER-dependent down-regulation of osteoclast differentiation by daidzein has been suggested via promoting caspase-8 and caspase-3 cleavage and DNA fragmentation of monocytic bone marrow cells [86].

4. Effects of Isoflavones on Bone *in vivo*

Dose-dependent effects of isoflavones on bone *in vivo* have been studied in rodents and primates and biphasic dose-dependent responses have also been observed in these in vivo models. These results suggest that isoflavones may mainly influence bone formation other than bone resorption. In contrast, there are also data shown that isoflavones affect both bone formation and bone resorption. Due to the limited dose-selection in some experiments, time and length of exposure, and different responses in various in vivo models, it is not surprised that some experiments showed bone-sparing effects and others did not or even showed contradictory results. Clearly, finding the critical doses of isoflavones that exert beneficial effects on bone in different models *in vivo* will be essential for future study.

Ovariectomized rodent models have been widely used to induce bone loss and to test whether isoflavones can prevent bone loss. Ovariectomized, lactating rats on low-calcium diet lost about 50% of bone mineral mass over two weeks. Low doses of genistein of 0.5 mg/d increased femur bone retention, which is effective as estrogen, whereas high doses of genistein of 1.6 mg/d and of 5 mg/d were less effective [87]. These biphasic dose responses effects in vivo have been observed in other rodent studies. The doses of isoflavones used in rodent studies are normally within the range of 0.1 mg/d to 5 mg/d. Some studies showed that genistein was more effective than estrogen in bone-sparing effects, whereas others showed opposite results. It has been argued that the different potency in isoflavones compared to estrogen may be attributed to the doses selected. Both suboptimal and excessive doses could result in only partial bone-sparing effects. In addition, bone-sparing effects of genistein is more effective in ameliorating ovariectomy-induced loss of trabecular and compact bone in young rodents than in older rodents. It has been shown that genistein doses that protect against bone loss were 10-fold lower than that induce uterine hypertrophy [88]. Genistein, daidzein and glycetein have been studied in rodents on their bone protection effects. In rats, however, it seems that daidzein is more efficient than genistein in preventing ovariectomy-induced bone loss, which may be due to the metabolism of daidzein to equol. Further discussion of this issue will be out of the scope of this review. Data got from rodents *in vivo* suggest that isoflavones stimulate bone formation and inhibit bone resorption or stimulate bone formation rather than suppress bone resorption, which indicate that the mechanism for isoflavone action may be distinct from that of estrogen.

Effects of coumestrol on bone remodelling in vivo have been conducted in oophorectomized rats. This 6-week experiment showed that rats receiving coumestrol of 1.5 μmol twice per week reduced bone mineral density globally and at the spine and femur. Coumestrol reduced urine calcium excretion and the bone resorption markers pyridinoline and deoxypyridinoline. It is unknown, from this study, whether coumestrol has biphasic effects on bone *in vivo* [89].

Several studies used flaxseed as a source of lignans to test their effects on bone. However, flaxseed is also the richest known source of α-linolenic acid that decreased bone resorption by inhibiting the biosynthesis of prostaglandins [90,91]. Studies on flaxseed showed that lignans, similar to isoflavones, altered bone development in female rats receiving 50 and 100 g flaxseed/kg diet during lactation [92]. These effects may only be effective in the young but not the old rats [93]. There is no report showing biphasic dose-dependent responses on lignans so far. But it has been reported that lignans have estrogenic and antiestrogenic effects [90,91].

Ovariectomized primates models have been conducted to study the effects of isoflavones on bone-sparing effects. Different from rodent studies that concluded isoflavones are effective at reducing bone loss, two studies using ovariectomized primates showed that soy phytoestrogens do not prevent bone loss [94,95]. Due to the limited number of studies and dose-selected in these studies, the lack of bone-sparing effects may result from species difference and/or the selected-doses that were not sufficiently to exert beneficial effects on bone. The beneficial effects of isoflavones on bone remain inconclusive.

It has long been hypothesized that phytoestrogen-rich diets may have beneficial effects on human bone. This hypothesis is based on the fact that osteoporotic fracture rates are lower in Asian woman, who consume large amounts of soy products. However, this hypothesis has not yet been fully proved in human studies. So far, data from human studies are variable and conflicting [6,11,19]. The major problem of these studies is the difference in experimental design, relative small number of studied subjects, and relative short duration of study. Most studies used soy, clover, and flaxseed as sources of phytoestrogens [11,96,97]. The doses of phytoestrogens studied so far in human are below 150 mg isoflavones/d [6,11,98]. In a double-blind placebo-controlled 6-month period postmenopausal woman study, intake of 90 mg isoflavones/d increased bone mineral content and density in the lumbar spine. In contrast, bone mineral content and density in the lumbar spine remained unchanged when consumed 56 mg isoflavones/d [99]. In another study that thirty-seven postmenopausal women were given 150 mg/d of isoflavone supplements twice daily for six months, calcaneus bone mineral density was not changed [98]. Although it is difficult to pool the data from the different human studies and the doses calculated in various studies may differ, it is clear from human studies that the beneficial effects of phytoestrogens on bone are dose-dependent [6,8]. It is likely that the dose-dependent effects are also biphasic. Finding the critical doses of phytoestrogens that exert bone-sparing effects will be crucial for the future human studies. Large, randomised, and long clinical trials are needed to clarity these issues

5. Mechanisms of Action

Phytoestrogens have been reported as an agonist, partial agonist or antagonist for oestrogen [10,10,36]. Consensus exists that the estrogenic or antagonistic effects of phytoestrogens was dependent upon their concentrations and the cell types/tissues. Normally, the estrogenic effects occurred at low concentrations, whereas the antagonistic effects at high concentrations. The molecular mechanisms of estrogenic action of phytoestrogens have been related to ER-mediated actions and this notion is well accepted. In contrast, the molecular mechanisms of antiestrogenic action of phytoestrogens are the major discussion in the past years and several explanations have been proposed.

One explanation suggested that the antiestrogenic effects of phytoestrogens are dependent upon the amount of estrogen produced in the body. If the estrogen level is low, as it is in menopause, empty receptor sites can be filled with phytoestrogens that can produce estrogenic effects. If the estrogen levels are high, phytoestrogens can compete with endogenous estrogens for binding to receptors. Data from this prospective trial of healthy, menstruating young adult women suggest that postmenopausal women with low circulating levels of estrogens are more likely to benefit the bone-sparing effects of phytoestrogens. However, this explanation has its limitation and cannot be proved in the experiments both *in vitro* and *in vivo* [20,21].

Estrogen is the most important sex steroid for maintenance of bone homeostasis. Both ERα and ERβ are present in osteoblasts and their precusor cells, osteocytes, and osteoclasts. The second explanation has been proposed based on the presence of ERβ in bone cells, the dominant negative regulation of ERβ on estrogen signaling, and the differential binding affinity of isoflavones to ERα and ERβ. Because of the structural similarities between phytoestrogens and 17β-oestrodiol (E2), studies on the molecular mechanisms of action of phytoestrogens have focused on interaction with ERs. Binding affinity of phytoestrogens with both ERα and ERβ has been studied. Different from E2 that has similar binding affinity to both receptors, some phytoestrogens preferentially bind to ERβ. The rank of the affinity for ERα is E2 > coumestrol > zearalenone > genistein > apigenin > daidzein = glycitein = kaempferol > quercetin = naringenin > biochanin A = formononetin = chrysin. In contrast, the order of the affinity for ERβ is E2 = coumestrol = genistein > apigenin > zearalenone > kaempferol > daidzein = glycitein > naringenin > quercetin > biochanin A = formononetin = chrysin [100,101]. Despite of the large differences in binding affinity of phytoestrogens for ERα and ERβ, it has been shown that the binding affinity is unlikely to account entirely for the distinct transcriptional actions [72]. Based on their transcriptional activity, the order of estrogenic potency of phytoestrogens is E2 > zearalenone = coumestrol > genistein > daidzein > apigenin > biochanin A = kaempferol = naringenin = glycitein > formononetin = quercetin = chrysin for ERα and E2 > genistein = coumestrol > zearalenone > daidzein > biochanin A = apigenin = kaempferol = naringenin = glycitein > formononetin = quercetin = chrysin for ERβ. Phytoestrogens like coumestrol and genistein have a higher affinity for ERβ [102-104]. Structurely, genistein adopts an antagonist conformation rather than an agonist conformation [105]. The maximal transcriptional stimulation by phytoestrogens achieved with ERβ is only about half that of ERα [100]. Although the estrogenic potency of phytoestrogens is lower than that of E2, it is possible that some phytoestrogens like genistein at high concentrations are more potent than that of E2 [100,103,104]. The higher estrogenic potency of genistein resulted from its enzyme inhibiting effects [106]. Interestingly, antagonism of phytoestrogens could not be detected in the gene reporter assays. This result indicates that gene report assay is good method to detect estrogenic activity of phytoestrogens, but it may not detect whether phytoestrogens have estrogenic or antiestrogenic effects in cellular system. Estrogenic or antiestrogenic effects of phytoestrogens at the cellular and molecular level may depend on their concentrations, receptor status, the concentration of endogenous estrogens and the type of target organs or cells.

Consensus exists that ERα mediates most of actions of estrogen on bone cells, whereas ERβ may function as a dominant negative regulator. It has been shown, in many different cellular context including bone cells *in vitro* and *in vivo*, that ERβ antagonize the actions of ERα and this relationship is referred to yin/yang [71,107-109]. Consistent with this yin/yang relationship between ERα and ERβ, isoflavones are ERβ selective agonists and acts as a dominant regulator of estrogen signalling [71,72]. It has been further shown that isoflavones were over a 1000-fold more potent at triggering transcriptional activity with ERβ compared to ERα [72]. The yin/yang relationship between ERα and ERβ may explain the biphasic effects of phytoestrogens observed below micromolar range because the preferential activation of ERβ and yin/yang relationship between ERα and ERβ were only observed at relative low isoflavone concentrations. In contrast, the antiestrogenic effects of phytoestrogens at micromolar range cannot be solely explained by ER-mediated actions.

In addition to genomic mechanisms, non-genomic mechanisms have been proposed [10]. Non-genomic actions of phytoestrogens are often based on the inhibitory effects of genistein on enzyme activities like tyrosine kinase and topoisomerases II. Since genistein was found to inhibit tyrosine protein kinase activity in 1987 [24], there have been more than 2000 publications on this issue [22]. Furthermore, genistein inhibits DNA topoisomerases I and II [25,26], protein histidine kinase activity [110], 5α-reductase [111,112], and aromatase enzyme activity [112,113]. Some studies suggested the importance of inhibiting enzyme activity of genistein in bone resorption [10,85,114]. However, these non-genomic effects have the limitation to explain data on the effects of other phytoestrogens like daidzein and glycetein on bone formation and bone resorption. It is known that daidzein, an inactive of form of genistein, does not exhibit inhibitory activity on protein tyrosine kinase at certain concentrations and has similar effects on osteogenesis and adipogenesis [21].

The fourth explanation suggested that genistein action is via transforming growth factor-β (TGF-β) signalling pathway [22]. TGF-β is a polypeptide expressing in bone cells. TGF-β stimulates bone formation via direct actions on the differentiated function of the osteoblasts and osteoprogenitor cells. It decreases bone resorption by inducing apoptosis of osteoclasts. Some actions of phytoestrogens can be explained by TGF-β pathway. However, the TGF-β signalling pathway cannot explain the observations that biphasic dose effects of genistein on osteogenesis and adipogenesis in osteoprogenitor cells because TGF-β stimulates bone formation but inhibits adipogenesis [2]. It has been also proposed that prostaglandin and nitric oxide may implicated in the effects of phytoestrogens on bone cells [115]. However, these explanations may limit to certain compounds and have not yet been fully proved in bone cells.

Recently, we identified PPARs as molecular targets of isoflavones and proposed that the complex biological effects of isoflavones are determined by the balance of many factors mainly including ERs, PPARs and enzyme inhibiting activity. This molecular mechanism of action can explain not only the action of genistein that has enzyme-inhibiting effects but also the action of daidzein that do not have enzyme-inhibiting effects but has a similar effects on bone cells. One of the critical points for this explanation is phytoestrogens at micromolar concentrations concurrently activate ERs and PPARs, which exert distinct actions on bone cells. This means that estrogenic activity of isoflavones still exists despite their effects on bone cells are antiestrogenic. Indeed, data *in vitro* and *in vivo* show that phytoestrogens produced estrogenic activity when these substances have antiestrogenic effects in cells/tissues. For example, genistein at the concentrations above 10 μM had antiestrogenic effects on osteogenesis and adipogenesis of KS483 cells. In contrast, genistein at concentrations between 0.1 to 50 μM dose-dependent increased ERE-luc activity in KS483 cells. The estrogenic potency of genistein at the concentrations higher than 1 μM was greater than that of estrogen at the concentration of 0.01 μM [20]. In a gene expression assay using a yeast system, a dose-dependent increase of activation of ERα was in a range of 1 to 100 μM, whereas ERβ in the range of 0.01 to 100 μM [103,116]. In a human estrogen-dependent breast cancer cell line MCF-7, genistein in the concentration range between 0.1 and 10 μM dose-dependently increased mRNA levels of presenelin-2 (pS2), an estrogen-responsive gene. When MCF-7 cells were treated with genistein and E2 together, the expression of pS2 was lower than when each compound was added alone, which indicates antiestrogenic effects [117-119]. In addition, in vivo data showed that high levels of genistein increased uterine weight, suggesting that estrogenic activity of genistein still exist at high concentrations. Dose-dependent increase of uterine hypertrophy has been observed in mice exposed to genistein of 0.7, 2, and 5 mg/day [88]. The estrogenic potency of genistein at the dose of 50 mg/kg per day in neonatal mice was comparable to the diethylstilbestrol, a potent synthetic estrogen [120]. Similar observations that high amount of genistein induced estrogenic effects were also reported in immature rats [121]. Furthermore, the estrogenic activity of other isoflavones daidzein and glycetein at high concentrations have been shown when these substances exert antiestrogenic effects [21,101].

After 1990 when the term was first coined, PPARs have been intensively studied and been implicated in lipid/gulcose metabolism, cellular proliferation, differentiation, adipogenesis and inflammatory signalling [122-124]. PPARs, are ligand-activated transcription factors that belong to nuclear hormone receptor superfamily. They function as heterodimer with the retinoid X receptor α and bind to specific peroxisome proliferator response elements (PPRE) in the promoters of target genes [125-127]. PPARs, like other nuclear receptors including ERs, have a similar structural organization. The N-terminal A/B domain that allows ligand-independent activation can confer constitutive activity on the receptor and is negatively regulated by phosphorylation. This region is followed by a highly conserved C domain or DNA-binding domain. The D domain or hinge domain, link the DNA-binding domain to the ligand-binding domain. The C-terminal E/F domain or ligand-binding domain allows a ligand-dependent transactivation of this receptor [128].

PPARs can be activated via either ligand binding in C-terminal E/F domain or inhibition of phosphorylation in N-terminal A/B domain of the receptors. Both natural and synthetic PPAR ligands have been identified and some have been used as drugs. For example, fatty acids, eicosanoids, fibrates and non-steroidal anti-inflammatory drugs are ligands of PPARα. Ligands of PPARγ include prostaglandin J2, thiazolidinediones and the nonsteroidal anti-inflammatory compounds indomethacin and ibuprofen. It has been shown that PPARγ [129,130]contains a consensus mitogen-activated protein kinases (MAPK) site in the A/B domain [57,58,131]. Phosphorylation of this site by p44/42 MAPKs decreases PPARγ activity, leading to a downregulation of adipogenesis. In constrast, PD98059, an inhibitor of p44/42 MAPK pathway, increased PPARγ activity leading to an upregulation of adipogenesis [57,132].

The PPAR subfamily consists of three members, PPARα, PPARγ, and PPARδ (also known as PPARβ), which have distinct tissue distribution and physiological functions. PPARα is most abundant in tissues such as the liver, kidney, skeletal and cardiac muscle, intestine, placenta, adipose tissue, and adrenal glands [133]. PPARγ is highly expressed in adipose tissue, but also found in liver, skeletal muscle, intestine, heart, bone marrow stromal cells, and immune system. PPARδ is expressed ubiquitously in all tissues of adult mammals [128,134,135].

PPARs are critical transcriptional targets of phytoestrogens. To determine whether genistein directly binds to PPARγ, we performed a membrane-bound PPARγ binding assay. Genistein can interact directly with the PPARγ ligand-binding domain and has a measurable K_i of 5.7μM [20], which is comparable to that of the known PPARγ ligands such as indomethacin [136] and prostaglandin J2 [137]. The information on the interaction of daidzein and glycetein with PPARγ is either too low or not available. Gene reporter assays show that genistein at concentrations between 1 and 100 μM activated PPARγ in a dose-dependent way in osteoprogenitor KS483 cells, breast cancer MCF7 cells, T47D cells and MDA-MD-231 cells. Activation of PPARγ by genistein is attributed to dual activation of this receptor, i.e. binding and inhibition of phosphorylation [20]. Furthermore, we observed that genistein activates PPARα as well as PPARδ (unpublished data). Consistent with our findings, activation of PPARα and PPARγ by genistein has been reported in murine macrophage-like RAW 264.7 cells [138]. Although daidzein does not exhibit inhibitory activity on protein tyrosine kinase and has lower binding affinity than genistein, it can still activate PPARs. Transcriptional activity of PPARα was stimulated to the highest level at daidzein concentrations of 10 and 20 μM and decreased thereafter. PPARγ activity was increased in a dose dependent way. Similar to PPARα, daidzein increased PPARδ activity dose-dependently and reached the highest level at the concentrations between 20 and 40 μM and then decreased. Furthermore, daidzein-induced PPAR activity was observed in several breast cancer cell lines including MCF7, T47D and

MDA-MD-231 and the results are consistent with the data in RAW 264.7 cells. [138]. It is clear that activation of PPARs by phytoestrogens is independent of their enzyme inhibiting activity. Based on our studies and others, it is known that the other phytoestrogens that include apigenin, quercetin, morin, luteolin, chrysin, biochanin A, kaempferol, and naringenin can activate PPARs.

PPARs are present in bone tissue and involved in regulation of bone formation and bone resorption. They have been observed in bone marrow mesenchymal cells, osteoblastic MC3T3-E1 cells, and KS483 cells [21,139-141]. Activation of PPARs induces differential effects on bone. Among these three isoforms, PPARγ is the most intensively studied on bone cells. Several reports showed that activation of PPARγ downregulates osteogenesis in bone marrow mesenchymal cells, MC3T3-E1 osteoblastic cells, and KS483 cells [20,139,141]. It has been further shown that PPARγ-downregulated osteocalcin expression via suppressing both the expression of Cbfa1 and interfering with the transactivation ability of Cbfa1 [142]. The important role of PPARγ in regulating differentiation of osteoprogenitor cells has been further demonstrated in following two studies. In one study, it has been shown that overexpression of PPARγ acts as a dominant inhibitor of osteoblastogenesis in mouse bone marrow cells [53]. In contrast, another study showed that PPARγ-deficient ES cells failed to differentiate into adipocytes but spontaneously differentiated into osteoblasts [54]. Furthermore, in vivo data showed that body bone mineral density was decreased when 6-month-old nondiabetic C57BL/6 mice were treated with PPARγ agonist rosiglitazone for 7 weeks. The decreased bone mineral density is accompanied by a decrease in bone formation rate and an increase in fat content in the bone marrow, indicating that this antidiabetic drug mainly affects differentiation of osteoprogenitor cells [143]. Moreover, heterozygous PPARγ-deficient mice had high bone mass that resulted from osteoblastogenesis. The function of osteoblasts and osteoclasts in these mice, however, was normal [54]. Data *in vivo* showed that activation of PPARγ decreased osteoclast formation and bone resorption [144]. It is likely that activation of PPARγ decreased soluble osteoprotegerin ligand (sOPGL)-induced NF-κB activation and consequently blocked sOPGL-induced osteoclast formation and activity [145]. Data on the effects of PPARα on bone are not consistent. In KS 483 cells, we have shown that activation of PPARα did not alter osteogenesis and adipogenesis [21], which is consistent with the observations in bone marrow mesenchymal cells isolated from PPARα knock out mice [146], but different from those in MC3T3-E1 cells [141]. Different from PPARα and PPARγ, activation of PPARδ increased osteogenesis in MC3T3-E1 cells and in KS483 cells [21,141]. In addition, it has been shown that PPARδ represses PPARα- and PPARγ-mediated transcription and adipogenesis in NIH-3T3 cells that were double transduced by PPARγ and PPARδ [147]. It is clear that different isoforms of PPARs influence each other.

In addition to divergent actions of ERs and PPARs on bone cells, both ERs and PPARs influenced each other. For example, PPARγ as well as PPARα decreased phytoestrogen-induced estrogenic activity. In contrast, PPARδ potentiated phytoestrogen-induced estrogenic activity. These results suggest that PPARs had differential effects on estrogenic transcriptional activity. Both PPARα and PPARγ may contribute to the antiestrogenic effects of phytoestrogens. Furthermore, ERα and ERβ decreased PPARγ transcriptional activity, but had no influence on PPARα or PPARδ transcriptional activity [20,21]. The cross-talks between ERs and PPARs have also been reported in uterine Leiomyoma and MCF-7 breast cancer cells [148,149]. It has been shown that PPARα-RXR heterodimers could block ER binding to the ERE of the promoter and inhibit ER-mediated transcriptional activity. [148] Inhibition of PPAR transactivation by ER could be mediated via competition for common coactivators, or increased corepressors. But the underlying molecular mechanism remains unclear.

Phytoestrogens concurrently activated transcriptional factors, ERs and PPARs, and these transcriptional factors influenced each other. As a result, the balance of divergent actions of these transcriptional factors determine phytoestrogen-induced biological effects. As demonstrated by ICI182,780, this balance was indeed observed at all tested concentrations of daidzein [21]. Based on our biochemical and transcriptional data, we conclude that the amount of activated ERs or PPARs was determined by the dose of phytoestrogens. As a result, the biological effects of daidzein were concentration-dependent. At certain concentrations, the balance of divergent actions of ERs and PPARs determines the final biological effects.

In summary, the biphasic dose effects of phytoestrogens are due to concurrent activation of ERs and PPARs, which produce divergent actions in the same cell/tissue system. Understanding this molecular mechanism of action is important for the future study on finding the precise doses in vivo. Data on beneficial bone effects of phytoestrogens are provocative but not conclusive.

References

1. B. L. Riggs, S Khosla, L J Melton, III, *Endocr. Rev.*, 23 (2002).
2. S. Ahdjoudj, O Fromigue, P J Marie, *Histol. Histopathol.*, 19 (2004).
3. S. L. Hendrix, *Curr. Opin. Rheumatol.*, 15 (2003).
4. L. M. Newcomer, P A Newcomb, J D Potter, Y Yasui, A Trentham-Dietz, B E Storer, M P Longnecker, J A Baron, J R Daling, *Cancer Causes Control,* 14 (2003).
5. J. E. Rossouw, G L Anderson, R L Prentice, A Z LaCroix, C Kooperberg, M L Stefanick, R D Jackson, S A Beresford, B V Howard, K C Johnson, J M Kotchen, J Ockene, *JAMA*, 288 (2002).
6. K. D. Setchell, E Lydeking-Olsen, *Am. J. Clin. Nutr.*, 78 (2003).
7. L. A. Fitzpatrick, *Endocrinol. Metab Clin. North Am.*, 32 (2003).
8. S. Migliaccio, J J Anderson, *Osteoporos. Int.*, 14 (2003).
9. H. Adlercreutz, W Mazur, *Ann. Med.*, 29 (1997).
10. J. J. Anderson, S C Garner, *Baillieres Clin. Endocrinol. Metab*, 12 (1998).
11. F. Branca, *Proc. Nutr. Soc.*, 62 (2003).
12. D. M. Tham, C D Gardner, W L Haskell, *J. Clin. Endocrinol. Metab,* 83 (1998).
13. S. Barnes, B Boersma, R Patel, M Kirk, V M Darley-Usmar, H Kim, J Xu, *Biofactors*, 12 (2000).
14. J. E. Thigpen, J K Haseman, H E Saunders, K D Setchell, M G Grant, D B Forsythe, *Comp Med.*, 53 (2003).
15. J. Kanno, H Kato, T Iwata, T Inoue, *J. Agric. Food Chem.*, 50 (2002).
16. P. B. Hoyer, *Biochem. Pharmacol.*, 62 (2001).
17. D. A. Whaley, D Keyes, B Khorrami, *Drug Chem. Toxicol.*, 24 (2001).
18. R. Nilsson, *Toxicol. Pathol.*, 28 (2000).
19. A. Cassidy, J. Br. *Menopause. Soc.*, 9 (2003).
20. Z. C. Dang, V Audinot, S E Papapoulos, J A Boutin, C W Lowik, *J Biol Chem*, 278 (2003).
21. Z. C. Dang and C. W. Lowik, *J. Bone Miner. Res.* (2004).
22. H. Kim, T G Peterson, S Barnes, *Am. J. Clin. Nutr.*, 68 (1998).
23. S. Barnes, H Kim, V Darley-Usmar, R Patel, J Xu, B Boersma, M Luo, *J. Nutr.*, 130 (2000).

24. T. Akiyama, I Ishida, S Nakagawa, H Ogawara, S I Watanabe, N Itoh, M Shibuya, Y Fukami, *J Biol. Chem.*, 262 (1987).
25. A. Okura, H Arakawa, H Oka, T Yoshinari, Y Monden, *Biochem. Biophys. Res. Commun.*, 157 (1988).
26. J. Markovits, C Linassier, P Fosse, J Couprie, J Pierre, A Jacquemin-Sablon, J M Saucier, J B Le Pecq, A K Larsen, *Cancer Res.,* 49 (1989).
27. S. P. Davies, H Reddy, M Caivano, P Cohen, *Biochem. J.*, 351 (2000).
28. D. P. Kelly, *Circ. Res.*, 92 (2003).
29. G. A. Francis, J S Annicotte, J Auwerx, *Curr. Opin. Pharmacol.*, 3 (2003).
30. E. L. Schiffrin, F Amiri, K Benkirane, M Iglarz, Q N Diep, Hypertension, . (2003).
31. Z. C. Dang and C. W. Lowik, *J. Bone Miner. Res.* (2003).
32. Food Standards Agency, available at http://www. food. gov. uk/multimedia/pdfs/phytoestrogenreport. pdf (2002).
33. Middleton E Jr, C Kandaswami, T C *Theoharides, Pharmacol. Rev.*, 52 (2000).
34. M. S. Kurzer, X Xu, *Annu Rev Nutr.*, 17 (1997).
35. B. H. Arjmandi, B J Smith, *J Nutr Biochem*, 13 (2002).
36. K. D. Setchell, *Am. J. Clin. Nutr.*, 68 (1998).
37. M. J. Messina, C L Loprinzi, *J. Nutr.*, 131 (2001).
38. A. L. Ososki, E J Kennelly, *Phytother Res.,* 17 (2003).
39. D. Ibarreta, A Daxenberger, H H Meyer, *APMIS*, 109 (2001).
40. A. M. Duncan, W R Phipps, M S Kurzer, *Best Pract Res Clin Endocrinol Metab.*, 17 (2003).
41. K. D. Setchell, N M Brown, E Lydeking-Olsen, *J. Nutr.*, 132 (2002).
42. K. D. Setchell, N M Brown, L Zimmer-Nechemias, W T Brashear, B E Wolfe, A S Kirschner, J E Heubi, *Am. J. Clin. Nutr.*, 76 (2002).
43. N. M. Brown, K D Setchell, *Lab Invest*, 81 (2001).
44. L. A. Leinwand, *J. Clin. Invest*, 112 (2003).
45. A. L. Murkies, G Wilcox, S R Davis, *J. Clin. Endocrinol. Metab*, 83 (1998).
46. S. J. Bhathena, M T Velasquez, *Am. J. Clin. Nutr.*, 76 (2002).
47. N. Tsutsumi, *Biol. Pharm. Bull.*, 18 (1995).
48. K. Kawashima, T Inoue, N Tsutsumi, H Endo, *Biochem. Pharmacol.*, 51 (1996).
49. S. Kanno, S Hirano, F Kayama, *Toxicology*, 196 (2004).
50. P. Bianco, R P Gehron, *J. Clin. Invest.*, 105 (2000).
51. M. E. Nuttall, J M Gimble, *Bone*, 27 (2000).
52. A. Yamaguchi, T Komori, T Suda, *Endocr. Rev.*, 21 (2000).
53. B. Lecka-Czernik, I Gubrij, E J Moerman, O Kajkenova, D A Lipschitz, S C Manolagas, R L Jilka, *J. Cell Biochem.*, 74 (1999).
54. T. Akune, S. Ohba, S. Kamekura, M. Yamaguchi, U. Chung, N. Kubota, Y. Terauchi, Harada, Y. , Y. Azuma, K. Nakamura, T. Kadowaki, and H. Kawaguchi, *J Clin. Invest* 113 (2004).
55. G. Xiao, D Jiang, P Thomas, M D Benson, K Guan, G Karsenty, R T Franceschi, *J. Biol. Chem.*, 275 (2000).
56. C. F. Lai, L Chaudhary, A Fausto, L R Halstead, D S Ory, L V Avioli, S L Cheng, *J. Biol. Chem.*, 276 (2001).
57. H. S. Camp, S R Tafuri, *J. Biol. Chem.*, 272 (1997).

58. E. Hu, J B Kim, P Sarraf, B M Spiegelman, *Science*, 274 (1996).
59. M. Adams, C T Montague, J B Prins, J C Holder, S A Smith, L Sanders, J E Digby, C P Sewter, M A Lazar, V K Chatterjee, S O'Rahilly, *J. Clin. Invest*, 100 (1997).
60. Z. C. Dang, R L van Bezooijen, M Karperien, S E Papapoulos, C W Lowik, *J. Bone Miner. Res*., 17 (2002).
61. R. Okazaki, D Inoue, M Shibata, M Saika, S Kido, H Ooka, H Tomiyama, Y Sakamoto, T Matsumoto, *Endocrinology*, 143 (2002).
62. M. Heim, O Frank, G Kampmann, N Sochocky, T Pennimpede, P Fuchs, W Hunziker, P Weber, I Martin, I Bendik, *Endocrinology*, 145 (2004).
63. Lee Y.S, Chen X, and J. J. Anderson, *Nutr. Rev*. 21 (2001).
64. K. S. Suh, G Koh, C Y Park, J T Woo, S W Kim, J W Kim, I K Park, Y S Kim, *Phytochemistry*, 63 (2003).
65. T. L. Jia, H Z Wang, L P Xie, X Y Wang, R Q Zhang, *Biochem. Pharmacol*., 65 (2003).
66. E. Sugimoto, M Yamaguchi, *Int. J. Mol. Med*., 5 (2000).
67. E. Sugimoto, M Yamaguchi, *Biochem. Pharmacol*., 59 (2000).
68. M. Yamaguchi, E Sugimoto, *Mol. Cell Biochem*., 214 (2000).
69. H. Yoshida, T Teramoto, K Ikeda, Y Yamori, *Biosci. Biotechnol. Biochem*., 65 (2001).
70. S. Zhou, G Turgeman, S E Harris, D C Leitman, B S Komm, P V Bodine, D Gazit, *Mol. Endocrinol*., 17 (2003).
71. K. Pettersson, F Delaunay, J A Gustafsson, *Oncogene*, 19 (2000).
72. J. An, C Tzagarakis-Foster, T C Scharschmidt, N Lomri, D C Leitman, J Biol Chem, 176 (2001).
73. S. Bord, D C Ireland, S R Beavan, J E Compston, *Bone*, 32 (2003).
74. M. K. Lindberg, M Erlandsson, S L Alatalo, S Windahl, G Andersson, J M Halleen, H Carlsten, J A Gustafsson, C Ohlsson, *J. Endocrinol*., 171 (2001).
75. N. K. Shevde, A C Bendixen, K M Dienger, J W Pike, *Proc. Natl. Acad. Sci. U. S. A*, 97 (2000).
76. X. W. Chen, S C Garner, J J Anderson, *Biochem. Biophys. Res. Commun.,* 295 (2002).
77. X. Chen, S C Garner, L D Quarles, J J Anderson, *J. Nutr. Biochem*., 14 (2003).
78. V. Viereck, C Grundker, S Blaschke, H Siggelkow, G Emons, L C Hofbauer, *J. Cell Biochem*., 84 (2002).
79. M. Yamaguchi, Y H Gao, *Biochem. Pharmacol*., 55 (1998).
80. Y. H. Gao, M Yamaguchi, *Biol. Pharm. Bull*., 22 (1999).
81. Y. H. Gao, M Yamaguchi, *Biochem. Pharmacol*., 58 (1999).
82. Y. H. Gao, M Yamaguchi, *Int. J. Mol. Med*., 5 (2000).
83. F. Okamoto, K Okabe, H Kajiya, *Jpn. J. Physiol*, 51 (2001).
84. B. Li, S Yu, *Biol. Pharm. Bull*., 26 (2003).
85. T. Yamagishi, E Otsuka, H Hagiwara, *Endocrinology*, 142 (2001).
86. C. M. Rassi, M Lieberherr, G Chaumaz, A Pointillart, G Cournot, *J. Bone Miner. Res.,* 17 (2002).
87. J. J. Anderson, W W Ambrose, S C Garner, *Proc. Soc. Exp. Biol. Med*., 217 (1998).

88. Y. Ishimi, N Arai, X Wang, J Wu, K Umegaki, C Miyaura, A Takeda, S Ikegami, *Biochem. Biophys. Res. Commun.*, 274 (2000).
89. C. R. Draper, M J Edel, I M Dick, A G Randall, G B Martin, R L Prince, *J. Nutr.*, 127 (1997).
90. B. H. Arjmandi, *J. Am. Coll. Nutr.*, 20 (2001).
91. L. T. Bloedon, P O Szapary, *Nutr. Rev.*, 62 (2004).
92. W. E. Ward, Y V Yuan, A M Cheung, L U Thompson, *Br. J. Nutr.*, 86 (2001).
93. W. E. Ward, Y V Yuan, A M Cheung, L U Thompson, *J. Toxicol. Environ. Health A*, 63 (2001).
94. K. D. Ham, R F Loeser, B R Lindgren, C S Carlson, *Arthritis Rheum.*, 46 (2002).
95. T. C. Register, M J Jayo, M S Anthony, *J. Clin Endocrinol Metab*, 88 (2003).
96. D. L. Alekel, A S Germain, C T Peterson, K B Hanson, J W Stewart, T Toda, *Am. J. Clin. Nutr.*, 72 (2000).
97. J. D. Brooks, W E Ward, J E Lewis, J Hilditch, L Nickell, E Wong, L U Thompson, *Am. J. Clin. Nutr.*, 79 (2004).
98. C. S. Hsu, W W Shen, Y M Hsueh, S L Yeh, *J. Reprod. Med.*, 46 (2001).
99. S. M. Potter, J A Baum, H Teng, R J Stillman, N F Shay, J W Erdman, Jr., *Am. J. Clin. Nutr.*, 68 (1998).
100. G. G. Kuiper, J G Lemmen, B Carlsson, J C Corton, S H Safe, P T van der Saag, B B van der, J A Gustafsson, *Endocrinology*, 139 (1998).
101. T. T. Song, S Hendrich, P A Murphy, *J Agric Food Chem.*, 47 (1999).
102. T. Barkhem, B Carlsson, Y Nilsson, E Enmark, J Gustafsson, S Nilsson, *Mol. Pharmacol.*, 54 (1998).
103. K. Morito, T Hirose, J Kinjo, T Hirakawa, M Okawa, T Nohara, S Ogawa, S Inoue, M Muramatsu, Y Masamune, *Biol. Pharm. Bull.*, 24 (2001).
104. G. N. Nikov, N E Hopkins, S Boue, W L Alworth, *Environ. Health Perspect.*, 108 (2000).
105. A. C. Pike, A M Brzozowski, R E Hubbard, T Bonn, A G Thorsell, O Engstrom, J Ljunggren, J A Gustafsson, M Carlquist, *EMBO J.*, 18 (1999).
106. Z. C. Dang and C. W. Lowik, *J. Cell Biochem.* (2004).
107. Z. Weihua, S Andersson, G Cheng, E R Simpson, M Warner, J A Gustafsson, *FEBS Lett.*, 546 (2003).
108. A. S. Chagin, M K Lindberg, N Andersson, S Moverare, J A Gustafsson, L Savendahl, C Ohlsson, *J Bone Miner Res.*, 19 (2004).
109. M. K. Lindberg, S Moverare, S Skrtic, H Gao, K Dahlman-Wright, J A Gustafsson, C Ohlsson, Mol. Endocrinol., 17 (2003).
110. J. Huang, M Nasr, Y Kim, H R Matthews, *J Biol Chem*, 267 (1992).
111. B. A. Evans, K Griffiths, M S Morton, *J Endocrinol.*, 147 (1995).
112. K. S. Weber, N A Jacobson, K D Setchell, E D Lephart, *Proc. Soc. Exp. Biol. Med.*, 221 (1999).
113. D. R. Campbell, M S Kurzer, *J. Steroid Biochem. Mol. Biol.*, 46 (1993).
114. J. P. Williams, S E Jordan, S Barnes, H C Blair, *Am. J. Clin. Nutr.*, 68 (1998).
115. P. Albertazzi, *Climacteric.,* 5 (2002).
116. P. De Boever, W Demare, E Vanderperren, K Cooreman, P Bossier, W Verstraete, *Environ. Health Perspect.*, 109 (2001).

117. C. Y. Hsieh, R C Santell, S Z Haslam, W G Helferich, R C Santell, Y C Chang, M G Nair, W G Helferich, *Cancer Res.*, 58 (1998).
118. N. Sathyamoorthy, T T Wang, J M Phang, *Cancer Res.*, 54 (1994).
119. D. T. Zava, G Duwe, *Nutr Cancer*, 27 (1997).
120. R. R. Newbold, E P Banks, B Bullock, W N Jefferson, *Cancer Res.*, 61 (2001).
121. R. C. Santell, Y C Chang, M G Nair, W G Helferich, *J. Nutr.*, 127 (1997).
122. T. M. Willson, P J Brown, D D Sternbach, B R Henke, *J. Med. Chem.*, 43 (2000).
123. E. D. Rosen, B M Spiegelman, *J. Biol. Chem.*, 276 (2001).
124. G. J. Murphy, J C Holder, *Trends Pharmacol. Sci.*, 21 (2000).
125. V. Bocher, I Pineda-Torra, J C Fruchart, B Staels, *Ann. N. Y. Acad. Sci.*, 967:7-18. (2002).
126. P. Escher, W Wahli, *Mutat. Res.*, 448 (2000).
127. P. Puddu, G M Puddu, A Muscari, *Int J Cardiol.*, 90 (2003).
128. S. Kersten, W Wahli, EXS, 89:141-51. (2000).
129. E. D. Rosen, C H Hsu, X Wang, S Sakai, M W Freeman, F J Gonzalez, B M Spiegelman, *Genes Dev.*, 16 (2002).
130. D. Ren, T N Collingwood, E J Rebar, A P Wolffe, H S Camp, *Genes Dev.*, 16 (2002).
131. M. Adams, M J Reginato, D Shao, M A Lazar, V K Chatterjee, *J. Biol. Chem.*, 272 (1997).
132. G. K. Chan, R A Deckelbaum, I Bolivar, D Goltzman, A C Karaplis, *Endocrinology*, 142 (2001).
133. P. Escher, O Braissant, S Basu-Modak, L Michalik, W Wahli, B Desvergne, *Endocrinology*, 142 (2001).
134. W. Wahli, O Braissant, B Desvergne, *Chem. Biol.*, 2 (1995).
135. L. Michalik, B Desvergne, C Dreyer, M Gavillet, R N Laurini, W Wahli, *Int. J. Dev. Biol.*, 46 (2002).
136. J. M. Lehmann, J M Lenhard, B B Oliver, G M Ringold, S A Kliewer, *J. Biol. Chem.*, 272 (1997).
137. G. Ferry, V Bruneau, P Beauverger, M Goussard, M Rodriguez, V Lamamy, S Dromaint, E Canet, J P Galizzi, J A Boutin, *Eur. J. Pharmacol.*, 417 (2001).
138. O. Mezei, W J Banz, R W Steger, M R Peluso, T A Winters, N Shay, *J Nutr.*, 133 (2003).
139. J. M. Gimble, C E Robinson, X Wu, K A Kelly, B R Rodriguez, S A Kliewer, J M Lehmann, D C Morris, *Mol. Pharmacol.*, 50 (1996).
140. H. Mano, C Kimura, Y Fujisawa, T Kameda, M Watanabe-Mano, H Kaneko, T Kaneda, Y Hakeda, M Kumegawa, *J. Biol. Chem.*, 275 (2000).
141. S. M. Jackson, L L Demer, *FEBS Lett.*, 471 (2000).
142. M. J. Jeon, J A Kim, S H Kwon, S W Kim, K S Park, S W Park, S Y Kim, C S Shin, *J. Biol. Chem.*, 278 (2003).
143. S. O. Rzonca, L J Suva, D Gaddy, D C Montague, B Lecka-Czernik, *Endocrinology*, 145 (2004).
144. R. Okazaki, M Toriumi, S Fukumoto, M Miyamoto, T Fujita, K Tanaka, Y Takeuchi, *Endocrinology*, 140 (1999).

145. G. Mbalaviele, Y Abu-Amer, A Meng, R Jaiswal, S Beck, M F Pittenger, M A Thiede, D R Marshak, *J. Biol. Chem.*, 275 (2000).
146. X. Wu, J M Peters, F J Gonzalez, H S Prasad, M D Rohrer, J M Gimble, *Bone*, 26 (2000).
147. Y. Shi, M Hon, R M Evans, *Proc. Natl. Acad. Sci. U. S. A,* 99 (2002).
148. K. D. Houston, J A Copland, R R Broaddus, M M Gottardis, S M Fischer, C L Walker, *Cancer Res.*, 63 (2003).
149. X. Wang, M W Kilgore, *Mol. Cell Endocrinol.*, 194 (2002).

CHAPTER 12

REGULATION OF BONE REMODELING

Di Chen[1,2], Mo Chen[1], Ying Yan[1], Yong-jun Wang[1], and Tianhui Zhu[2]

[1]*Center for Musculoskeletal Research, University of Rochester Medical Center, Rochester, New York 14642, USA*

[2]*Medical College, Nankai University, Tianjin 300071, P.R. China*

1. Introduction

The adult skeleton is in a dynamic state, being continually broken down and reformed by the coordinated actions of osteoclasts and osteoblasts on trabecular bone surfaces and in Haversian systems. This turnover or remodeling of bone occurs in focal and discrete packets throughout the skeleton. On average, the remodeling of each packet takes about 3-4 months, with it taking longer in trabecular bone than in cortical bone. The remodeling which occurs in each packet (a bone remodeling unit)[1] is geographically and chronologically separated from other packets of remodeling. This suggests that activation of the sequence of cellular events responsible for remodeling is locally controlled, most likely by local mechanisms in the bone microenvironment.

The adult skeleton consists of cortical (or compact) bone and trabecular (or cancellous) bone. Current evidence indicates that regulatory mechanisms of cortical bone and trabecular bone may not be exactly the same. For example, the function of the PTH/PTHrP receptor (PPR) was recently investigated in transgenic mice that express one of the constitutively active receptors found in Jansen's metaphyseal chondrodysplasia (expression of the transgene is targeted to osteoblasts using the bone-specific type I collagen promoter). In these transgenic mice, osteoblast numbers and function are increased in trabecular and endosteal bones but was decreased in the periosteum. In trabecular bone of the transgenic mice, there was an increase in osteoblast precursors, as well as mature osteoblasts. Osteoblastic expression of the constitutively active PPR induces a dramatic increase in osteoclast numbers in both trabecular and compact bone in transgenic mice leading to a substantial increase in trabecular bone volume and a decrease in cortical bone thickness of the long bone[2]. These findings indicate that PPR signaling differentially regulates trabecular and cortical bone formation in mice.

2. Bone Remodeling

2.1. Bone remodeling at different parts of the skeleton

The relative amount of cortical and trabecular bone is variable at the different sites in the skeleton. Trabecular bone is relatively prominent in the vertebral column, the most common site of fracture associated with osteoporosis. In the lumbar spine, trabecular bone comprises more than 65% of the total bone. In the intertrochanteric area of the femur, bone is comprised of 50% cortical and 50% trabecular. In the neck of the femur, the bone is 75% cortical and 25% trabecular. In contrast, in the mid-radius more than 95% of the bone is cortical bone. The difference in behavior of bone at these different sites is most likely due to the unique environment presented to bone cells exposed to cortical or trabecular architecture. For example, osteoblasts and osteoclasts on trabecular bone surfaces are in intimate contact with the cells of the marrow cavity, which produce a variety of osteotropic cytokines. Comparatively, it is likely that the cells in cortical bone, which are more distant from the influences of these cytokines, are controlled more by the systemic osteotropic hormones such as parathyroid hormone and 1,25-dihydroxyvitamin D_3. As a result of these differences in foci-dependent regulation, behavior of cells in each of these microenviroments will be unique.

2.2. *Cortical bone remodeling*

Cortical bone comprises 85% of the total bone in the body, and is most abundant in the long bone shafts of the appendicular skeleton. The volume of cortical bone is regulated by the formation of periosteal bone, by remodeling within Haversian systems, and by endosteal bone resorption. Cortical bone is removed primarily by endosteal resorption and resorption within the Haversian canals. The resorption in the Haversian canals leads to increased porosity of cortical bone. However, periosteal bone formation continues to increase the thickness of cortical bone throughout life. Cortical bone loss probably begins to occur after the age of 40 and there is an acceleration of cortical bone loss that occurs for 5 to 10 years after menopause. This accelerated phase of cortical bone loss continues for 15 years and then gradually slows. There is irrefutable evidence that estrogen replacement therapy after the menopause preserves cortical bone. In later life, women with osteoporosis lose cortical bone at similar rates to those of premenopausal women. Loss of cortical bone is the major predisposing factor for fractures that occur at the hip and around the wrist. Cortical bone is particularly prone to increased resorption in patients with primary hyperparathyroidism.

2.3. *Trabecular bone remodeling*

Although trabecular bone comprises only 15% of the skeleton, the changes that occur in this type of bone after the age of 30 largely determine whether spinal osteoporotic fractures will occur. It is widely held that decline in trabecular bone mass begins in early adult life, occurring earlier than the decline in cortical bone mass[3]. Nevertheless, there are studies that have disagreed with these findings by suggested that the decline in trabecular bone mass begins later, after ovarian function ceases[4]. Regardless of the timing, loss of trabecular bone that occurs with aging is not due simply to a generalized thinning of the bone plates, but is rather due to complete perforation and fragmentation of trabeculae [5,6]. Since trabecular bone has a broad surface area, resorption may be modulated by focal osteoclastic resorption regulated by growth factors produced by cells in the bone microenvironment.

Osteoclast activation is the initial step in the remodeling process. Thus, the rate of bone remodeling is to a large degree dependent on the activation frequency of osteoclasts. The mechanism responsible for the initiation of bone remodeling is unknown. One possibility is that osteoclast precursors recognize a change in the mechanical properties of aging bone, which requires replacement with new bone for optimal structural integrity. The eventual activation of the osteoclast may occur because of interactions which occur between integral membrane proteins, integrins, on osteoclast cell membranes with proteins in bone matrix which contain RGD (arginine-glycine-asparagine) amino acid sequences (such as osteopontin)[7]. Overall, the resorptive phase of the remodeling process has been estimated to last 10 days. This period is followed by repair of the defect by a team of osteoblasts which are attracted to the site of the resorption defect and then presumably proceed to make new bone. This part of the process takes approximately three months.

2.4. Bone coupling

In most physiological and pathological circumstances, the coupling of bone formation to previous bone resorption occurs faithfully. Packets of bone which are removed during resorption are replaced by enhanced focal bone formation. The cellular and molecular mechanisms which are responsible for mediating the coupling process are still not clear. A number of theories have been proposed to account for coupling. Many investigators have favored the notion that coupling is mediated locally by growth factors, such as IGFs, TGFβs, FGFs or BMPs, released from bone matrix during the process of osteoclastic bone resorption[8]. These growth factors stimulate osteoblast activity and new bone formation. More recently, it has been suggested that coupling, or at least the bone formation phase of bone remodeling, may be systemically mediated by actions of leptin on the hypothalamus that leads to β2-adrenergic stimulation of cells in the osteoblast lineage to inhibit bone formation.

Obviously, understanding this sequence of cellular events may lead to clarification of the mechanisms responsible for the decreased osteoblast activity which occurs in age-related bone loss, and possibly the pathophysiology of osteoporosis, as well as the specific defects in osteoblast function which occur in malignancies such as myeloma and breast cancer.

3. Regulation of Bone Resorption

The first event during bone remodeling is to recruit osteoclast precursor cells, followed by osteoclast formation and activation, formation of a ruffled border, resorption and ultimately apoptosis. These steps of osteoclast formation and differentiation which ultimately lead to bone resorption are regulated by systemic hormones and local bone growth factors. For example, one potent activation mechanism involves local stimulation of the osteoclast by a factor expressed by cells in the osteoblast lineage[9,10], recently discovered to be RANK ligand (also called OPGL, TRANCE or ODF)[11]. Overall, osteoclastic resorption may be stimulated by factors which enhance proliferation of osteoclast progenitors, which cause differentiation of committed precursors into mature osteoclasts or activation of the mature multinucleated cell to resorb bone[12]. Local and systemic regulatory mechanisms are addressed below.

3.1. Systemic hormones

Parathyroid hormone (PTH)

PTH stimulates committed progenitors to fuse to form mature multinucleated osteoclasts. It also activates preformed osteoclasts to resorb bone. However, it does not increase CFU-GM, the earliest detectable cells in the osteoclast lineage. The activation of osteoclasts is indirect, likely mediated through cells in the osteoblast lineage such as the lining cells[13], which produce RANK ligand (see below). It should be noted that parathyroid hormone-related protein (PTH-rP) has identical effects to those of PTH on osteoclasts.

1,25-dihydroxyvitamin D3

1,25-dihydroxyvitamin D_3 is a potent stimulator of osteoclastic bone resorption. Like PTH, it stimulates osteoclast progenitors to differentiate and fuse[14]. It is now clear that 1,25-dihydroxyvitamin D_3 promotes osteoclast formation and function indirectly by stimulating RANKL production by osteobalsts/stromal cells. 1,25-dihydroxyvitamin D_3 inhibits T-cell proliferation and the production of the cytokine interleukin-2 and it enhances interleukin-1 production from cells with monocyte characteristics. Thus, the overall effects of 1,25-dihydroxyvitamin D_3 on bone resorption are multiple and complex.

Calcitonin

Calcitonin is a polypeptide hormone which is a potent inhibitor of osteoclast bone resorption, but its effects are only transient. Ultimately, osteoclasts escape from the inhibitory effects of calcitonin during continuous exposure[15]. Thus, patients treated for hypercalcemia with calcitonin will respond for only a limited period of time before hypercalcemia recurs (usually 48-72 hours). In Pagetic patients, the beneficial effects of calcitonin may eventually be lost with continued treatment. The "escape" phenomenon is likely due to down-regulation of the calcitonin receptor[16]. Calcitonin causes cytoplasmic contraction of the osteoclast cell membrane, which has been correlated with its capacity to inhibit bone resorption[17]. It also causes the dissolution of mature osteoclasts into mononuclear cells. However, it also blocks osteoclast formation, inhibiting both proliferation of the progenitors and differentiation of the committed precursors. The effects of calcitonin on osteoclasts are mediated by cyclic AMP.

3.2. Local factors

Local growth factors may play a more important role than systemic hormones for the initiation of physiologic bone resorption and for the normal bone remodeling. Since bone remodeling occurs in discrete and distinct packets throughout the skeleton, it seems probable that cellular events are mainly influenced by factors generated in the microenvironment of bone in combination with systemic factors. A number of potent local stimulators and inhibitors of osteoclast activity have recently been identified.

OPG/RANKL/RANK

It has become clear in the last few years that the tumor necrosis factor (TNF) ligand family member, Receptor Activator of NF-κB (RANK) ligand and its two known receptors RANK and osteoprotegerin (OPG), are the key local regulators of osteoclastic bone resorption *in vivo*.

OPG is a TNF receptor (TNFR) superfamily member that lacks a transmembrane domain and is thus secreted. When expressed, recombinant OPG inhibits both physiological and pathological bone resorption and hepatic over-expression of the *OPG* gene in mice results in severe osteopetrosis[18]. OPG has only two known ligands, RANK ligand (RANKL) and TRAIL, both of which are type II membrane-bound TNF homologs[19,20]. In contrast to most other TNF receptor family members, OPG is secreted and circulates *in vivo*[18]. *OPG*-deficient mice produced by targeted disruption of the gene exhibit profound osteoporosis from birth caused by enhanced osteoclast formation and function as well as prolonged osteoclast survival. These results indicate that OPG is a physiological regulator of osteoclast-mediated bone resorption during postnatal bone growth.

RANKL (also called TRANCE, OPG ligand and ODF) is a TNF ligand superfamily member which has been cloned by four independent groups in recent years[21-24]. The expression of this molecule is obligatory for osteoclastic resorption and normal bone remodeling. In RANKL knockout mice, typical osteopetrosis occurs with total occlusion of bone marrow space by endosteal bone. The bones of the RANKL null mutant mice lack osteoclasts, although they contain osteoclast progenitors which differentiate into functionally active osteoclasts when co-cultured with normal osteoblasts/stromal cells from wild-type littermates[25]. Administration of recombinant RANKL to mice induces osteoclast formation and increases blood ionized calcium[23, 26]. These results suggest that RANKL is absolutly required for osteoclast development.

The TNFR superfamily member RANK is the only known signaling receptor for RANKL. RANK, a type I transmembrane protein, mediates all of the signals essential for osteoclast differentiation from hematopoietic progenitors to activation of mature osteoclasts[27-29]. *Rank*-deficient mice have been generated[29,30], and as expected they exhibit severe osteopetrosis due to complete absence of osteoclasts and lack of bone resorption. These results indicate that the absolute requirement of an intact RANKL/RANK pathway for osteoclastogenesis *in vivo*.

Interleukin-1

There are two interleukin-1 molecules, interleukin-1α and β. Their effects on bone appear to be the same, and are mediated through the same receptor. Interleukin-1 is released by activated monocytes, but also by other types of cells including osteoblasts and tumor cells. It is a potent stimulator of osteoclasts. It works at all phases in the formation and activation of osteoclasts and its effects are mediated through RANK ligand.

Interleukin-1 also stimulates osteoclastic bone resorption when infused *in vivo*, leading to a substantial increase in plasma calcium levels[31,32]. It has been implicated as a potential mediator of bone resorption and increased bone turnover in osteoporosis[33]. It may be responsible for the increase in bone resorption seen in some malignancies, as well as the localized bone resorption associated with collections of chronic inflammatory cells in diseases such as rheumatoid arthritis.

Macrophage colony stimulating factor (M-CSF)

Macrophage colony stimulating factor (M-CSF), which was once thought to be specific for the monocyte-macrophage lineage, has recently been shown to be required for normal osteoclast formation in rodents during the neonatal period. In the *op/op* variant of osteopetrosis, there is impaired production of M-CSF and the consequence is osteopetrosis due to decreased normal osteoclast formation. The disease can be cured by the treatment with M-CSF[34]. M-CSF is produced by stromal cells. Cells in the osteoclast lineage contain the M-CSF receptor (a receptor tyrosine kinase). M-CSF works in conjunction with RANK ligand to cause osteoclastic bone resorption.

Interleukin-6

Interleukin-6 is a pleiotropic cytokine which has important effects on bone. It is expressed and secreted by normal bone cells in response to osteotropic hormones such as PTH, 1,25-dihydroxyvitamin D_3 and interleukin-1[35]. The osteoblast is the major cell source of interleukin-6 so far described. Interleukin-6 is a weak stimulator of osteoclast formation, and less potent than other cytokines such as interleukin-1 and tumor necrosis factor[36,37]. It has been implicated in the bone loss associated with estrogen withdrawal (ovariectomy) in the mice[37,38]. It may work through similar mechanisms to other cytokines such as interleukin-11 and leukemia inhibitory factor.

Interferons

Both interferon-beta and gamma (IFN-β and γ) are negative regulators of osteoclastogenesis. RANKL induces expression of the IFN-β gene in osteoclast precursor cells and IFN-β inhibits the osteoclast differentiation by interfering with the RANKL-induced expression of c-Fos, an essential transcription factor for the formation of osteoclasts. Mice deficient in IFN-β signaling exhibit severe osteopenia accompanied by enhanced osteoclastogenesis[39]. T-cell production of IFN-γ strongly suppresses osteoclastogenesis by interfering with the RANKL-RANK signaling pathway. IFN-γ induces rapid degradation of the RNAK adaptor protein, tumor necrosis factor receptor associated protein 6 (TRAF6), which results in strong inhibition of the RANKL-induced activation of NF-κB and JNK. This inhibition of osteoclastogenesis is rescued by over-expressing TRAF6 in osteoclast precursor cells, which indicates that TRAF6 is the target protein for the IFN-γ action[40].

Transforming growth factor β

Although TGFβ has been reported to inhibit osteoclast proliferation and differentiation in osteoclast precursor cells and to inhibit bone resorption in many *in vitro* studies[41,42], recent *in vitro* and *in vivo* evidence demonstrates that TGFβ supports osteoclast formation. In transgenic mice in which expression of a dominant-negative truncated type II TGFβ receptor is targeted to mature osteoblasts by the osteocalcin promoter or the type I collagen promoter, osteoclast formation was decreased and mice exhibit a mild osteopetrosis phenotype. In these mice, bone mineral density and bone volume are increased but bone formation rates are not significantly changed[43,44]. In Col1a1-dnTGFβR2 transgenic mice, bone marrow culture experiments show that M-CSF and RANKL-induced osteoclast formation is decreased and 1,25-dihydroxyvitmin D_3 and TNFα-induced osteoclast formation is also decreased. Co-culture experiments show that osteoblast-supported osteoclast formation is also reduced when osteoblasts from these same transgenic mice are cultured with spleen cells from wild-type mice[44]. These findings indicate that TGFβ signaling in osteoblasts supports osteoclast formation *in vivo*. It has been suggested that at least part of the mechanism for TGFβ-induced osteoclast formation is due to the up-regulation of expression of a STAT-induced factor, SOCS-3 (suppressor of cytokine signaling 3) in osteoclasts[45].

4. Regulation of Bone Formation

The specific cellular events which occur at sites of osteoclastic resorption include a series of sequential changes in cells in the osteoblast lineage, such as osteoblast chemotaxis, proliferation and differentiation, which are followed by formation of mineralized bone and cessation of osteoblast activity. During bone resorption, active TGFβ is released from the resorbed bone[46] which is followed by chemotactic attraction of osteoblasts or their precursors to the sites of the resorption lacunae. Proliferation of osteoblast precursors is an important event at the remodeling site. Likely candidates for stimulation of osteoblast proliferation include TGFβ1 and 2, IGFs and FGFs. Sequentially, the next event during the formation phase is the differentiation of osteoblast precursors into mature osteoblasts. Several of the bone-derived growth factors including BMPs enhance expression of osteoblast marker genes and stimulate mineralized bone matrix formation. The final phase of the formation process is cessation of osteoblast activity. This may involve osteoblast apoptosis induced by TGFβ, BMPs or other growth factors produced by osteoblasts. The complete sequence of cellular events which occur at the bone surface during the remodeling process have been described in detail by Baron et al.[47] in studies on the alveolar bone of the rat, and from studies by Boyce et al.[48] using the mouse calvaria as a model. The cellular events which occur in these models are similar to those in adult human bone.

4.1. Systemic hormones

Parathyroid hormone (PTH)

PTH is a potent modulator of bone metabolism. In experimental animals and patients with osteoporosis, intermittent administration of PTH increases bone mass by stimulating *de novo* bone formation[49-52]. A recent report indicates that PTH induces bone formation by enhancing expression of the transcription factor Runx2/Cbfa1 and this effect is predominantly mediated by the protein kinase A (PKA) signaling pathway[53]. In contrast, sustained administration of PTH induces bone resorption and inhibits bone formation. These findings suggest that the dual effects of PTH on bone formation and bone resorption may be determined by the temporal expression patterns of Runx2 induced by PTH and also suggest that Runx2 may play critical a role in bone remodeling.

The bone resorption induced by sustained administration of PTH is probably due to the induction of Runx2 expression which in turn activates RANKL expression in osteoblasts. Recent reports show that continuous over-expression of Runx2 in osteoblasts in transgenic mice (under control of the murine type I collagen promoter) induces progressive osteopenia and high cortical bone turnover during adulthood and aging[54, 55]. Continuous over-expression of Runx2 gene induces expression of RANKL[56] in osteoblasts which stimulates osteoclast formation and bone resorption. In addition, continuous over-expression of Runx2 may also inhibit osteoblast maturation since in Col1a1-Runx2 transgenic mice, osteopontin-positive cells are accumulated and osteocalcin-positive cells and osteocytes are decreased[54].

Leptin

It has been proposed that leptin plays a specific role in mediating the relationship between body weight and bone mass[57]. Ducy et al. has investigated the role of leptin in bone formation using two murine models of obesity, the *ob/ob* and *db/db* mice which are *Leptin* and *Leptin-receptor* deficient respectively. Both types of mutant mice have high bone mass and increased bone formation rates. Although leptin receptors cannot be detected in osteoblasts, they did find that intracerebro-ventricular infusions of leptin induce bone loss in wild-type mice, and can reverse the high bone mass phenotype in *ob/ob* mice that are leptin-deficient. Thus, they propose that leptin modulates bone mass indirectly by a central effect, and link leptin deficiency to high bone mass. More recently, they have suggested that leptin acts in the hypothalamus and its downstream effects in bone are mediated by β2-adrenergic receptors expressed in osteoblasts[58]. These findings are potentially very important, although not all investigators agree that leptin does not have direct effects on bone cells[59]. Irrespective of whether leptin has effects on bone cells indirectly or directly, these studies emphasize the potential importance of leptin as a central regulator of bone formation.

Overall it is likely that the effects of systemic factors such as leptin and PTH are superimposed on a constellation of factors released locally to coordinate formation of new bone which occurs at recent resorption sites.

Local Factors

Transforming growth factor β (TGFβ)

As mentioned earlier, TGFβ is a member of a large superfamily of proteins that are important regulators of bone cell activity. Multiple isoforms of TGFβ exist and appear to control cell proliferation and differentiation of osteoblast precursor cells and mature osteoblasts. The prototype of TGFβ isoforms, TGFβ1, is highly expressed by differentiated osteoblasts and osteoclasts and is stored in bone matrix[60] and released in active form during osteoclastic bone resorption[61]. TGFβ is chemotactic for bone cells[62] and it attracts osteoblast precursor cells to the resorption sites. TGFβ stimulates proliferation of mesenchymal cells but inhibits osteoblast terminal differentiation *in vitro*[63,64]. TGFβ also stimulates bone formation *in vivo*[65], possibly because of its ability to increase the total numbers of osteoblasts. Effects of TGFβ are mediated through interactions with specific transmembrane serine-threonine kinase receptors. TGFβ binds the TGFβ type II receptor and this complex recruits and phosphorylates the TGFβ type I receptor, which in turn initiates signal transduction mediated by the signaling molecules Smad2 and 3.

Bone morphogenetic proteins (BMPs)

BMPs are members of the TGFβ superfamily. Like TGFβ, BMPs signal through serine/threonine kinase receptors, composed of type I and type II subtypes. Type I and type II BMP receptors are both indispensable for signal transduction. After ligand binding they form an active heterotetrameric receptor complex consisting of two pairs of a type I and type II receptor complex[66]. The type I BMP receptor substrates include Smad1, 5 and 8, all of which play a central role in relaying the BMP signal from the receptor to target genes in the nucleus. Smad proteins are phosphorylated by BMP receptors in a ligand-dependent manner[67-69]. After release from the receptor, phosphorylated Smads associate with Smad4, which acts as a shared partner. This complex translocates into the nucleus and participates in gene transcription with other transcription factors.

Physiological roles of BMPs and BMP receptor signaling in normal bone formation have been investigated. Injection of BMP-2 locally over the surface of calvariae of mice induces periosteal bone formation on the surface of calvariae without a prior cartilage phase[70]. Systemic administration of recombinant human BMP-2 increases mesenchymal stem cell activity and reverses ovariectomy-induced bone loss in female ICR mice and age-related bone loss in 24-month-old male BALB/c mice[71]. Over-expression of a dominant-negative truncated BMPR-IB in osteoblast precursor 2T3 cells inhibits osteoblast-specific gene expression and mineralized bone matrix formation[72]. In the transgenic mice in which expression of a dominant-negative truncated BMPR-IB transgene is targeted to the osteoblast lineage using the osteoblast-specific type I collagen promoter, the postnatal bone formation, including bone mineral density, static bone volume and dynamic bone formation rates, is decreased[73]. These results demonstrate that BMP receptor signaling plays a necessary role in normal postnatal bone formation.

The activity of BMPs is controlled at different molecular levels: 1) A series of BMP antagonists bind BMP ligands and inhibit BMP functions; 2) Smad6 is a member of the Smad family. It binds type I BMP receptors and prevents the binding and phosphorylation of Smad1 and 5; 3) Tob is an anti-proliferative protein. It selectively binds Smad1 and 5 and inhibits BMP signaling in osteoblasts; and 4) Smad ubiquitin regulatory factor 1 (Smurf1) is an E3 ubiquitin ligase that interacts with Smad1 and 5 and mediates the degradation of these Smad proteins.

Noggin is a secreted polypeptide which binds and inactivates BMP-2, 4 and 7. Co-crystal structures of the noggin-BMP-7 complex show that noggin inhibits BMP signaling by blocking the molecular interfaces of the binding epitopes for both type I and type II BMP receptors[74], thus preventing BMP-7 from binding to BMP receptors. A transgenic mouse model has recently been established using the osteocalcin promoter to drive the noggin transgene. The animals develop severe osteoporosis evidenced by significant reductions in bone mineral density, bone volume and bone formation rates[75, 76]. Sclerostosis is a recessive inherited osteosclerotic disorder caused by mutations in the protein sclerostin. The disorder is primarily due to increased bone formation[77]. Recently it was found that sclerostin is related in sequence to the family of secreted BMP antagonists, which includes Noggin, Chordin, Gremlin and Dan. Sclerostin is expressed in osteoblasts and osteocytes and binds BMP-5, 6 and 7 with high affinity. Over-expression of sclerostin in osteoblasts under the control of the osteocalcin promoter in transgenic mice causes osteoporosis[78]. These findings suggest that activation of endogenous BMP signaling can enhance bone formation, and regulation of the amount of BMP activity in postnatal stage is required for normal bone formation.

Tob is a member of a novel anti-proliferative protein family. Tob inhibits BMP-induced, Smad-dependent transcription in osteoblasts through its association with Smad1 and 5 proteins. In Tob knockout mice, BMP-2 signaling is enhanced and the effects of BMP-2 on osteoblast proliferation and differentiation are increased. Bone volume and bone formation rates are increased in adult Tob knockout mice as well[79].

Another important regulatory mechanism by which the activity of BMP signaling proteins is modulated involves ubiquitin-mediated proteasomal degradation. The formation of ubiquitin-protein conjugates requires three enzymes that participate in a cascade of ubiquitin transfer reactions: ubiquitin-activating enzyme (E1), ubiquitin-conjugating enzyme (E2), and ubiquitin ligase (E3). The specificity of protein ubiquitination is determined by E3 ubiquitin ligases, which play a crucial role in defining substrate specificity and subsequent protein degradation by 26S proteasomes[80,81]. Smad ubiquitin regulatory factor 1 (Smurf1) was identified by a yeast two-hybrid assay by its ability to interact with Smad1 and 5[82]. To determine the role of Smurf1 in bone formation *in vivo*, we have recently generated transgenic mice (Col1a1-Smurf1) in which expression of a Smurf1 transgene is targeted to osteoblasts using the bone-specific type I collagen promoter. Trabecular bone volume and bone formation rates are decreased in Col1a1-Smurf1 transgenic mice. Osteoblast proliferation and differentiation are inhibited in these mice, suggesting that bone formation defects observed in Smurf1 transgenic mice are mainly due to decreased osteoblast proliferation and differentiation[83]. Consistent with these findings, recent studies also demonstrate that bone formation is enhanced in Smurf1 null mutant mice[84]. These results demonstrate that regulation of BMP signaling proteins may also play an important physiological role *in vivo*.

Although cumulative evidence shows that BMP signaling plays an important role in bone formation *in vivo*, the effects of BMPs and BMP signaling proteins in bone remodeling, especially in the bone formation phase during bone coupling induced by ovariectomy or bone resorption cytokines have not been demonstrated. In addition, since null mutations of most of BMP ligands, receptors and signaling molecules produce lethal phenotype perinatally, tissue-specific knock-outs of the individual BMP ligands, receptors and signaling molecules are required to gain further information about physiological functions of BMP signaling in postnatal and adult animals.

Insulin-like growth factors (IGFs)

IGFs are polypeptides and two IGFs have been characterized: IGF-I and IGF-II. These peptides are present in the systemic circulation and are synthesized by multiple tissues, including bone. Systemic IGF-I is secreted by the liver, and its synthesis is regulated by growth hormone, whereas the synthesis of IGF-I in peripheral tissues is regulated by diverse hormones[85]. In the circulation, IGFs are bound to IGF binding proteins (IGFBPs), the most abundant of which in serum is IGFBP-3, which is also regulated by growth hormone[86]. IGF-I and II have similar biological activities but IGF-I is more potent than IGF-II. IGFs have mitogenic activity for osteoblasts and stimulate function of mature osteoblasts including enhanced synthesis of bone matrix proteins[87].

IGF-I null mutant mice have decreased bone formation and transgenic mice over-expressing IGF-I under the control of the osteocalcin promoter have increased bone formation[88,89]. In contrast, Infusion of IGF-I into humans causes a generalized anabolic effect[90]. IGFs play a role in the maintenance of cortical bone in rodents, and low IGF-I levels correlate with a decrease in bone mineral density in elderly women and in patients with anorexia nervosa[90-92].

The synthesis of skeletal IGF-I is regulated by hormones and growth factors. PTH stimulates IGF-I synthesis in osteoblasts and PTH-induced bone anabolic effect is dependent on IGF-I[88]. Growth hormone also enhances IGF-I production. In growth hormone receptor null mutant mice, bone formation is decreased and administration of IGF-I into these mice reverses the decreased bone formation phenotype[93]. Glucocorticoids inhibit IGF-I expression and this inhibition may contribute to the mechanism of glucocorticoid-induced bone loss[94]. Glucocorticoids inhibit IGF-I transcription through the induction of CCAAT-enhancer binding proteins (C/EBPs)[95].

Fibroblast Growth Factors (FGFs)

Fibroblast growth factors (FGFs) make up a large family of polypeptide growth factors. FGFs have angiogenic properties and play important roles in wound healing and bone repair. FGFs stimulate osteoblast proliferation and synthesis of collagen matrix. FGF is important in the maintenance of bone mass evidenced by decreased osteoblast numbers and bone formation in FGF-2 null mutant mice[96]. There are four related FGF receptors (FGFR) and they are FGFR-1 to 4. The pattern of FGFR-1, 2 and 3 expression differs during skeletal development, but all three forms mediate the mitogenic effect in target tissues[97,98]. Conditional knockout of the FGFR2 gene in condensed mesenchymal cells, which give rise to both osteoblast and chondrocyte lineages, results in mice with skeletal dwarfism and decreased bone density. This phenotype develops because of severely affected proliferation of osteoprogenitors and function of mature osteoblasts. It should be noted that in this conditional knockout approach, the Cre recombinase transgene is driven by the Dermo1 (Twist2) gene and CRE is highly expressed in condensed mesenchymal cells[99]. Conversely, the systemic administration of FGF-2 increases the number of preosteoblasts that eventually mature and form bone[100]. These data collectively establish that FGF signaling is essential for skeletal and limb development and for chondrogenesis. Hence, mutations of FGFR-1, 2 and 3 result in serious skeletal abnormalities.

References

1. H. M. Frost, *Bone Biodynamics*, 315 (1964).
2. L. M. Calvi, N. A. Sims, J. L. Hunzelman, M.C. Knight, A. Giovannetti, J.M. Saxton, H. M. Kronenberg, R. Baron, E. Schipani E, *J Clin Invest,* 107, 277-286 (2001).
3. Riggs BL, Wahner HW, Melton LJ III, et al. *J Clin Invest*, 77, 1487-1491 (1986).
4. Genant HK, Cann CE, Ettinger B, et al. *Ann Intern Med,* 97, 699-705 (1982).
5. Parfitt AM, Mathews CHE, Villanueva AR, et al. *J Clin Invest*, 72, 1396-1409 (1983).
6. Kleerekoper M, Villanueva AR, Stanciu J, et al. *Calcif Tissue Int*, 37, 594-597 (1985).
7. Miyauchi A, Alvarez J, Greenfield EM, et al. *J Biol Chem*, 266, 20369-20374 (1991).
8. Howard GA, Bottemiller BL, Turner RT, et al*Proc Natl Acad Sci* USA, 78, 3204-3208 (1981).
9. Rodan GA, Martin TJ. *Calcif Tiss Int*, 33, 349-351 (1981).
10. McSheehy PMJ, Chambers TJ. *Endocrinology*, 118, 824-828 (1986).
11. Wong BR, Josien R, Lee SY, et al. *J Exp Med*, 186, 2075-2080 (1997).
12. Mundy GR, Roodman GD. *Bone and_Mineral Research*, V. Elsevier: Amsterdam, 209-280 (1987).
13. McSheehy PMJ, Chambers TJ. *Endocrinology*, 118, 824-828 (1986).
14. Roodman GD, Ibbotson KJ, MacDonald BR, et al. *Proc Natl Acad Sci USA*, 82, 8213-8217 (1985).
15. Wener JA, Gorton SJ, Raisz LG. *Endocrinology*, 90, 752-759 (1972).
16. Takahashi S, Goldring S, Katz M, et al. *J Clin Invest*, 95, 167-171(1995).
17. Chambers TJ, Magnus CJ. *J Pathol*, 136, 27-40 (1982).
18. Simonet WS, Lacey DL, Dunstan CR, et al. *Cell*, 89, 309-310 (1997).
19. Simonet WS, Lacey DL, Timms E, et al. *Cell*, 93, 165-176 (1998).
20. Emery JG, McDonnell P, Burke MB, et al. *J Biol Chem*, 273, 14363-14367 (1998).
21. Anderson DM, Maraskovsky E, Billingsley WL, et al. *Nature*, 390, 175-179 (1997).
22. Wong BR, Rho J, Arron J, et al. *J Biol Chem*, 272, 25190-25194 (1997).
23. Simonet WS, Lacey DL, Timms E, et al. *Cell*, 93, 165-176 (1998).
24. Yasuda H, Shima N, Nakagawa N, et al. *Endocrinology*, 139, 1329-1337 (1998).
25. Kong Y-Y, Yoshida H, Sarosi I, et al. *Nature*, 397, 315-323 (1999).
26. Burgess TL, Qian Y-X, Kaufman S, et al. *J Cell Biol*, 145, 527-538 (1999).
27. Hofbauer LC, Heufelder AE. *J Mol Med*, 79, 243-253 (2001).
28. Hsu H, DL Lacey, CR Dunstan, et al. *Proc Natl Acad Sci USA*, 96, 3540-3545 (1999).
29. Li J, Sarosi I, Yan X-Q, et al. *Proc Natl Acad Sci USA*, 97, 1566-1571 (1999).
30. Dougall B, Glaccum M, Charrier K, et al. *Genes Develop*, 13, 2412-2424 (1999).
31. Sabatini M, Boyce B, Aufdemorte T, et al. *Proc Natl Acad Sci USA*, 85, 5235-5239 (1988).

32. Boyce BF, Aufdemorte TB, Garrett IR, et al. *Endocrinology*, 123,1142-1150 (1989).
33. Pacifici R, Rifas L, McCracken R, et al. *Proc Natl Acad Sci USA*, 86, 2398-2402 (1989).
34. Felix R, Cecchini MG, Fleisch H. *Endocrinology*, 127, 2592-2594 (1990).
35. Feyen JHM, Elford P, Dipadova FE, et al. *J Bone Min Res*, 4, 633-638 (1989).
36. Black K, Garrett IR, Mundy GR. *Endocrinology*, 128, 2657-2659 (1991).
37. Ishimi Y, Miyaura C, Jin CH, et al. *J Immunol,* 145, 3297-3303 (1990).
38. Jilka RL, Hangoc G, Girasole G, et al. *Science,* 257, 88-91 (1992).
39. Takayanagi H, Kim S, Matsuo K, et al. *Nature*, 416, 744-749 (2002).
40. Takayanagi H, Ogasawara K, Hida S, et al. *Nature,* 408, 600-605 (2000).
41. Pfeilschifiter JP, Seyedin S, Mundy GR. *J Clin Invest*, 82, 680-685 (1988).
42. Chenu C, Pfeilschifter J, Mundy GR, et al. *Proc Natl Acad Sci USA*, 85, 5683-5687 (1988).
43. Filvaroff E, Erlebacher A, Ye J-Q, et al. *Development*, 126, 4267-4279 (1999).
44. Chen D, Mi Z, Zhao M, Qiao M, et al. *Bone* OR1 (2003).
45. Fox SW, Haque SJ, Lovibond AC, et al. *J Immunol*, 170, 3679-3687 (2003).
46. Hughes DE, Tiffee JC, Li HH et al. *Nature Medicine*, 7, 1132-1136 (1996).
47. Baron R, Vignery A, Horowitz M. *Bone and Mineral Research.* (Elsevier: Amsterdam), 175-243 (1984).
48. Boyce BF, Aufdemorte TB, Garrett IR et al. *Endocrinology*, 125, 1142-1150 (1989).
49. Hock JM, Fonseca GJ, and Raisz LG. *Endocrinology*, 122, 2899-2904 (1988).
50. Hock JM and Gera I. *J Bone Miner Res*, 7, 65-72 (1992).
51. Dempster DW, Cosman F, Parisien M, et al. *Endocr Rev,* 14,690-708 (1993).
52. Neer RM, Arnaud CD, Zanchetta JR, et al. *N Engl J Med*, 344, 1434-1441 (2001).
53. Krishnan V, Moore TL, Ma, YL, et al. *Mol Endocrinol*, 17, 423-435 (2003).
54. Liu W, Toyosawa S, Furuichi T, et al. *J Cell Biol,* 155, 157-166 (2001).
55. Geoffroy V, Kneissel M, Fournier B, et al. *Mol Cell Biol,* 22, 6222-6233 (2002).
56. Enomoto H, Shiojiri S, Hoshi K, et al. *J Biol Chem,* 278, 23971-23977 (2003).
57. Ducy P, Amling M, Takeda S, et al. *Cell*, 100, 197-207 (2000).
58. Takeda S, Elefteriou F, Levasseur R, et al. *Cell*, 111, 305-317 (2002).
59. Gordeladze JO, Roseland JE. *J Cell Biochem*, 88, 706-712 (2003).
60. Hauschka PV, Mavrakos AE, Iafrati MD. *J Biol Chem,* 261, 12665-12674 (1986).
61. Pfeilschifter J, Mundy GR. *Proc Natl Acad Sci,* 84, 2024-2028 (1987).
62. Pfeilschifter J, Wolf O, Naumann A, et al. *J Bone Miner Res*, 5, 825-830 (1990).
63. Harris SE, Bonewald LF, Harris MA, et al. *J Bone Miner Res,* 9, 855-863 (1994).
64. Harris SE, Sabatini M, Harris MA, et al. *J Bone Miner Res,* 9, 389-394 (1994).
65. Marcelli C, Yates AJP, Mundy GR. *J Bone Miner Res,* 5,1087-1096 (1990).
66. Moustakas A. and C.H. *Genes Dev,* 16, 67-1871 (2002)
67. Hoodless P.A., T. Haerry, S. Abdollah, et al. *Cell,* 85, 489-500 (1996).
68. Nishimura R., Y. Kato, D. Chen, et al. *J Biol Chem,* 273, 1872-1879 (1998).
69. Chen Y., A. Bhushan and W. Vale. *Proc Natl Acad Sci USA,* 94, 12938-12943 (1997).
70. Chen D., M.A. Harris, G. Rossini, et al. *Calcif Tissue Int,* 60, 283-290 (1997).

71. Turgeman G., Y. Zilberman, S. Zhou, et al. *J Cell Biochem,* 86, 461-74 (2002).
72. Chen D., X. Ji, M.A. Harris, et al. *J Cell Biol,* 142, 295-305 (1998).
73. Zhao M., S.E. Harris, et al. *J Cell Biol,* 157, 1049-1060 (2002).
74. Groppe J., J. Greenwald, E. Wiater, et al. *Nature,* 420, 636-642 (2002).
75. Wu X.B., Y. Li, A. Schneider, et al. *J Clin Invest,* 112, 924-934 (2003).
76. Devlin R.D., Z. Du, R.C. Pereira, et al. *Endocrinology,* 144, 1972-1978 (2003).
77. Stein S.A., C. Witkop, S. Hill, et al. *Neurology*, 33, 267-277 (1983).
78. Winkler D.G., M.K. Sutherland, J.C. Geoghegan, et al. *EMBO J,* 22, 6267-6276 (2003).
79. Yoshida Y., S. Tanaka, H. Umemori, et al. *Cell,* 103, 1085-1097 (2000).
80. Hershko A. *Cell,* 34, 11-12 (1983).
81. Ciechanover A., A. Orian, and A.L. Schwartz. *J Cell Biochem,* 77(S34), 40-51 (2000).
82. Zhu H., P. Kavsak, S. Abdollah,et al. *Nature,* 400, 687-693 (1999).
83. Zhao M., M. Qiao, S.E. Harris, et al. *J Biol Chem* 10.1074/jbc.M313294200 (2003).
84. Yamashita M., S. Ying, S.Y. Cheng, et al. FASEB Summer Research Conferences July 12-17, 2003, Tucson, Arizona.
85. Jones JI, Clemmons. *Endocr Rev,* 16, 3-34 (1995).
86. Hwa V, Oh Y, Rosenfeld RG. *Endocr Rev*, 20, 761-787(1999).
87. Hock JM, Centrella M, Canalis E. *Endocrinology* 122, 254-260 (1998).
88. Bikle DD, Sakata T, Leary C, et al. *J Bone Miner Res,* 17, 1570-1578 (2002).
89. Zhao G, Monier-Faugere MC, Langub MC, et al. *Endocrinology,* 141, 2674-2682 (2000).
90. Grinspoon S, Thomas L, Miller K, et al. J *Clin Endocrinol Metab,* 87, 2883-2891 (2002).
91. Barrett-Connor E, Goodman-Gruen D. *J Bone Miner Res*, 13,1343-1349 (1998).
92. Rosen CJ, Dimai HP, Vereault D,et al. *Bone,* 21 217-223 (1997).
93. Sims NA, Clement-Lacroix P, Da Ponte F,et al. *J Clin Invest,* 106, 1095-1103 (2000).
94. Canalis E, Giustina A. *J Clin Endocrinol Metab*, 86, 5681-5685 (2001).
95. Delany AM, Durant D, Canalis E. *Mol Endocrinol,* 15, 1781-1789 (2001).
96. Montero A, Okada Y, Tomita M, et al. *J Clin Invest,* 105, 1085-1093 (2000).
97. Ornitz DM. *Bioessays*, 22, 108-112 (2000).
98. Marie PJ. *Gene*, 316, 23-32 (2003).
99. Yu K, Xu J, Liu Z, et al. *Development*, 130, 3063-3074 (2003).
100. Nakamura T, Hanada K, Tamura M, et al. *Endocrinology*, 136, 1276-1284 (1995).

CHAPTER 13

TGFB IN CHONDROCYTE BIOLOGY AND CARTILAGE PATHOLOGY

Tian-Fang Li, Regis J. O'Keefe, and Di Chen

[1]*Center for Musculoskeletal Research,*
University of Rochester Medical Center, Rochester, New York 14642, USA

1. Introduction

Transforming growth factor-beta (TGFβ) signaling is involved in a wide array of cellular activities in both physiological and pathological conditions. It induces gene transcription through a canonical signaling pathway from the TGFβ type II receptor (TβRII) to type I receptor (TβRI) to Smad activation. The receptors for TGFβ superfamily have intrinsic serine/threonine kinase activity. Binding of TGF-β to TβRII causes the phosphorylation of TβRI. Seven type I receptors for TGFβ superfamily members, which are also referred to as activin receptor-like kinase (ALK), have been cloned in mammals. ALK5 predominantly transduces TGFβ signaling, while ALK3 and ALK6 specifically mediate BMP responses[1-3].

There are eight known Smad proteins in vertebrates which can be classified into three categories: receptor-activated Smads (R-Smads, Smad1/5/8 for BMP signaling, Smad2/3 for TGFβ and activin signaling); the common mediator Smad (Co-Smad, Smad4); and inhibitory Smads (I-Smads, Smad6/7). Upon TGFβ activation, Smad2 and 3 are phosphorylated by TβRI; these phosphorylated Smads form protein complexes with Smad4. These complexes then translocate into nucleus, where they regulate target gene expression. Smad2 and Smad3 have distinct roles in mediating TGFβ signaling. Smad3 binds DNA directly, whereas Smad2 regulates gene expression indirectly, because a 30 amino acid insertion in the MH1 domain encoded by exon 3 of Smad2 prevents its binding to the DNA. Smad2 is closely involved in the embryonic development; in contrast, Smad3 may play a more important role in adult life[4-8].

Felici *et al.* recently reported that TRAP-1-like protein (TLP), associates with both active and kinase-deficient TβRII, and represses Smad3-mediated transcription. TLP prevents Smad3 from forming a protein complex with Smad4, but has little effect on Smad2-mediated TGFβ signaling. TLP may thus act as a molecular switch balancing the effects of Smad3 and Smad2[9].

2. Regulation of TGFβ Signaling

TGFβ signaling cascade is tightly controlled by the feedback mechanisms at the different levels: extracellular milieu, membrane, cytoplasm, and nucleus. Normal physiological functions of Smads depend on the delicately balanced regulations. I-Smads antagonize R-Smads by inhibiting phosphorylation of R-Smads through competitive binding to the activated TβRI. In addition, I-Smads recruit E3 ubiquitin ligase Smurf1 or 2 to target the activated TβRI at the cell membrane and thus induce receptor degradation. I-Smads also repress the function of R-Smads in nucleus by recruiting transcriptional co-repressors. Smad7 preferentially inhibits TGFβ signaling and plays a role in the dephosphorylation of TβRI[10, 11]. The Ski proto-oncoprotein family members (c-Ski and SnoN) hinder the TGFβ-mediated transcriptional activation by disrupting the formation of a heteromeric Smad complex. These oncoproteins immigrate with R-Smad/Co-Smad complexes to their DNA binding sites, where they recruit histone deacetylase and repress the transcription of target genes. c-Ski also stabilizes inactive Smad complexes on DNA, thereby suppressing gene transcription[12-15]. Ubiquitin-dependent degradation of Smads by Smurf1 or Smurf2 is critical in regulating the turnover of Smads. Smurfs 1 and 2 are members of the HECT domain family of E3 ubiquitin ligases. They mediate proteasomal degradation of Smad1, Smad5, Smad2, TβRI and BMPRI proteins. Smad7 serves as an adaptor protein for Smurfs in the process of TGFβ receptor degradation[16-19]. Sumolyation is a ubiquitination-like protein conjugation process, but it does not target specific substrates for proteasomal degradation. SUMO-1 is actively involved in the regulation of many transcription factors. It modulates Smad4 activity through a p38 MAP kinase pathway. The protein inhibitor of activated STAT, PIASy, prevents TGFβ-mediated Smad3 activation by inducing its sumolyation[20, 21].

3. Smad-Dependent and Independent Pathways

In addition to TβRI, increasing numbers of cellular kinases have been shown to participate in the functional modulation of Smad proteins. The interdependent relationship of Smads and these kinases is essential for maintaining the balance between different signaling pathways and is responsible for TGFβ-mediated responses. The phosphorylation of Smads by these kinases not only activates Smads but also changes their capability for nuclear translocation and DNA binding. TGFβ elicits cellular response through TGFβ activating kinase 1 (TAK1) which is a member of MAPKKK involved in the activation of p38 and JNK and ultimately activates ATF2. A large body of evidence shows that several MAP kinases such as ERK, JNK and p38, are activated in response to TGFβ via Smad-dependent or -independent mechanisms. Treatment with TGFβ induces a biphasic JNK response: a rapid Smad-independent activation followed by a Smad-dependent activation. JNK in turn phosphorylates Smad3 and facilitates its nuclear translocation. Functional and physical interactions between Smad3/Smad4 and c-Jun/c-Fos have been observed after TGFβ stimulation, indicating that Smad and MAPK/JNK signaling converge at AP1-binding sites of the target gene[22-27]. A recent report demonstrates that TGFβ activates PKA in the absence of increased cAMP. The possible mechanism is because of the complex formation between Smad proteins and the regulatory subunit of PKA with subsequent release of the catalytic subunit. TGFβ may also stimulate the translocation of α-catalytic subunit of PKA to the nucleus[28, 29]. Calmodulin-dependent protein kinase II and PKC provides negative feedback for Smad2 and Smad3, respectively. In the presence of Ca^{++}, calmodulin interacts with Smad2, 3 and 4 through the conserved N-terminal domain of Smad proteins. PKC phosphorylates the MH1 domain of Smad3 and interferes Smad3 DNA binding[30-33]. In osteoblasts, PTH induces Smad3 expression through PKA as well as PKC pathways[34].

4. TGFβ and Other Signaling Cascades

TGFβ cooperates with Notch signaling through a protein-protein interaction between Smad3 and intracellular domain of Notch-1[35]. There is evidence for TGFβ-dependent activation of PKB/Akt through PI3 kinase and Rho-dependent signaling pathways. PKB/Akt sequentially regulates TGFβ signaling through a direct interaction with Smad3. PKB represses TGFβ-Smad3 signaling through a kinase-activity-independent mechanism. It prevents Smad3 phosphorylation, Smad3/4 complex formation and Smad3 nuclear translocation. In contrast, Smad3 does not inhibit PKB[36-39]. STAT pathway modulates Smad activity through the induction of Smad7. PIAS3, a member of the protein inhibitor of activated STAT (PIAS) family, activates TGFβ/Smad transcriptional responses. PIAS3 interacts with the MH2 domain of Smad3 and with general co-activators p300/CBP. The formation of PIAS3/Smad3/p300 complex enhances TGFβ signaling. PIASy, another member of the PIAS family, interacts strongly with Smad4 and Smad3 and inhibits TGFβ/Smad transcriptional responses through a negative feedback loop. PIASy also stimulates the sumolyation of Smad3 *in vivo*[40-43]. NF-κB signaling also affects the TGFβ-Smad pathway through regulating the expression of Smad7[44, 45].

5. Endochondral Bone Formation

Chondrogenesis begins with mesenchymal condensation, which is driven by transcription factor *Sox9* and is characterized by expression of type II collagen (*col-II*). Subsequently, other chondrocyte-specific genes such as

Col-IX and *Col-XI*, N-cadherin and aggrecan, are also expressed. The committed chondrocytes are destined to a well-coordinated process referred to as endochondral ossification. This consecutive process consists of proliferation, prehypertrophy, hypertrophy, apoptosis, and osteoblastic cells ingression together with blood vessels invasion, which culminates in the formation of major axial bones. Several molecules, e.g. Indian hedgehog (*Ihh*), BMP6, type X collagen (*Col-X*) and alkaline phosphatase, are detected in the hypertrophic chondrocytes. They are well-accepted chondrocyte maturation marker genes. Articular chondrocytes, on the other hand, do not express these marker genes. Instead, their differentiation is arrested before cells become hypertrophic[46-49].

6. Expression of TGFβ and Related Molecules in Chondrocytes

TGFβ is produced by chondrocytes as a latent, high molecular weight molecule in association with latent TGFβ binding protein (LTBP). In growth plate chondrocytes, the storage of TGFβ by LTBP is cell maturation-dependent. Plasmin, transglutaminase and MMPs help the release and activation of TGFβ[50-54]. In epiphyseal growth plate, TGFβ1 and TGFβ3 are expressed in the resting, proliferating, and hypertrophic zones in 6 to 24-week-old rats. TGFβ2 is expressed at similar areas at 6-week-old of age, but decreased during growth. The expression of TGFβs in hypertrophic chondrocytes is weak. TβRI is co-expressed with TGFβ ligands in resting, proliferating, and hypertrophic zones throughout the process of development. TβRII is detected in 6-week-old rats but decreased after that. In the growing human bone, TGFβ2 exits in all zones of endochondral ossification, with the highest expression seen in hypertrophic and mineralizing zones. TGFβ3 is expressed in the chondrocytes of proliferative and hypertrophic zones. TGFβ1 is only found in the proliferative and upper hypertrophic zones. TβRI and TβRII are expressed intensively in the hypertrophic and mineralizing zones. The expression of TGFβ-restricted Smads is correlated with that of TGFβ1 and its receptors. Smad2 and Smad3 are expressed in a somewhat complementary manner. Smad2 is strongly expressed in proliferating chondrocytes. In contrast, Smad3 is found predominantly in mature chondrocytes. Smad4 is seen basically in all zones of the growth plate. Smad6 and Smad7 are mainly detected in mature chondrocytes. Smad3 is also found in perichondrium in developing cartilage undergoing endochondral bone formation[55-58]. These findings suggest that TGFβ may play an important role in chondrocyte differentiation and maturation in epiphyseal growth plate.

Articular cartilage is structurally divided into three zones: superficial or tangential zone; transitional or intermediate zone; basal or deep zone. TGFβ1, 2 and 3 are found in all zones of articular cartilage in the 6 to 50-week-old rats. In basal zone, the expression of TGFβ2 and 3 is very weak, while moderate expression of TGFβ1 is seen. The expression of TGFβ1 and TβRII decreases in osteoarthritic cartilage[59]. Endoglin and betaglycan are also referred to as type III TGFβ receptors. Endoglin is expressed in human articular chondrocytes and forms complex with betaglycan, TβRI and TβRII. It has been postulated that endoglin is an important modulator of TGFβ signaling in articular chondrocytes[60, 61].

7. Effects of TGFβ on Chondrocyte Function

Effects of TGFβ on articular chondrocytes are controversial. Intra-articular TGFβ injection aimed to treat osteoarthritis yields contradicting results. Different doses of TGFβ, different time point of treatment and the different severity of the arthritis may be responsible for the inconsistency. The most important factor that affects the effect of TGFβ is the duration of the treatment. Short-term treatment with TGFβ induces the repair of cartilage defects, while long-term injection causes the degradation of articular cartilage[62, 63].

Although TGFβ is a well-known inhibitor of cell cycle, its overall *in vitro* effect on chondrocyte proliferation is stimulatory. The response to TGFβ may be affected by cell culture conditions, the origin and differentiation state of cells, and existence of other growth factors such as IGF and FGF[64-68]. In cultured rat articular chondrocytes, TGFβ1 stimulates cell proliferation and extracellular matrix formation. Transient expression of *c-fos* through PKC activation is required for the mitogenic effect of TGFβ. TGFβ transduces a predominant signal through a MEK-ERK-Elk1 pathway to regulate chondrocyte proliferation. The MEK-ERK pathway activated by TGFβ is negatively regulated by PKA but transactivated by PKC[69-71]. The pro-proliferative effect of TGFβ is at least partially through induction of cyclin D1[72].

Treatment with TGFβ increases cartilage matrix synthesis, especially aggrecan. Gene transfer with adenoviral or retroviral vectors expressing TGFβ augments the production of *Col-II* and proteoglycans. TGFβ increases total glycosaminglycan synthesis in immature, but not mature cartilage. It helps maintain the matrix components of cartilage in the immature states[73-77]. TGFβ induces the aggrecan expression in chondrogenic cell line through a Smad-dependent pathway. In response to TGFβ, Smad2 is rapidly phosphorylated, leading to the initial activation of aggrecan gene. This rapid phosphorylation of Smad2 is, however, not indispensable for keeping high level of aggrecan expression. In contrast, TGFβ-induced phosphorylation of MAP kinases (ERK and p38) is crucial for maintaining the elevated basal aggrecan level[78]. TGFβ simultaneously boosts the expression of aggrecanase, and thus accelerates the turnover of cartilage matrix[79]. Over-expression of Smad7 in chondrocytes completely blunts TGFβ1-mediated effects on cell proliferation and proteoglycan synthesis[80]. Despite of its protective effect against the matrix loss in articular cartilage, endogenous TGFβ may cause excessive response to injury, resulting in the formation of osteophytes[81].

TGFβ enhances chondrogenesis but inhibits the terminal differentiation of chondrocytes. It helps chondrocytes maintain in the prehypertrophic stage. TGFβ acts downstream of Ihh and upstream of PTHrP. However, it has both PTHrP-dependent and -independent effects on chondrocyte maturation. TGFβ1, 2, and 3 all stimulate PTHrP expression, at least partially through Smad-mediated signaling pathway. In response to TGFβ, ATF2 is rapidly phosphorylated via the activation of p38 MAP kinase. The phosphorylated ATF2 cooperates with Smad3 to inhibit the rate of chondrocyte maturation. Smad3 has stronger inhibitory effect on the terminal differentiation of chondrocytes than Smad2[82-87]. Oh *et al.* reported that when chondrogenesis proceeds, ERK activity is reduced, whereas p38 activity is continuously increased. Inhibition of ERK induces chondrogenesis, whereas inhibition of p38 activity represses chondrogenesis[88]. In growth plate chondrocytes, TGFβ1 stimulates PLA-2 and prostaglandin release via the induction of COX-1 on arachidonic acid. PGE2 activates the EP2 receptor, leading to G-protein-dependent activation of PKA. PKA signaling in turn upregulates PKC activity. The end result is inhibition of chondrocyte proliferation and maturation[89]. Over-expression of ALK5, a TGFβ specific type I receptor, hampers chondrocytes maturation and hypertrophy[90].

8. TGFβ and Osteoarthritis (OA)

OA is a highly prevalent medical condition and has the greatest public health implications. In US, over 20 million people suffer from OA. It affects approximately 60% of men and 70% of women older than 65-year-old of age. OA is clinically characterized by pain, tenderness, swelling and decreased joint functions. Radiological features include joint-space narrowing, marginal osteophytes, subchondral bone sclerosis and cyst formation, and deformity of bone ends. Current treatment of OA is based on the control of symptoms because the disease-modifying, chondroprotective drugs with convincing results are not available yet. It is therefore imperative to investigate the mechanism underlying OA. Although OA is a multi-factorial disease, growth factors, particularly TGFβ superfamily members, play a fundamental role in the pathogenesis of OA. In OA, articular chondrocytes recapitulate the maturation cascade seen in endochondral ossification. Loss of maturation constrains imposed by TGFβ and a shift toward BMP signaling may be one of the mechanisms responsible for the pathogenesis of OA. TGFβ stimulates MMP13 production in OA chondrocytes through coordinated effects of Smad-dependent and -independent pathways. TGFβ concomitantly upregulates the expression of tissue inhibitor of metalloproteinases (TIMPs) in articular cartilage through serine/threonine kinase and tyrosine kinase pathways[91-93].

Transgenic mice over-expressing a cytoplasmically truncated, dominant- negative form of the TβRII in cartilage develop maturation of articular cartilage resembling the events in growth plate. Articular cartilage becomes fibrillated and disorganized. As this degenerative process progresses, articular surface is replaced by bone or hypertrophic cartilage with osteophytes formation, a characteristic feature of human OA. The growth plate is also disorganized with increased expression of *Col-X* and *Ihh*. The hyperplasia and cartilaginous metaplasia of synovium is observed in different joints. The mice also display several other skeletal deformities including bifurcation of the xiphoid process and sternum and kyphoscolisosis[94].

Smad3 deficient mice have been created on the basis of a targeted disruption of exon 8 of the Smad3 gene. At birth, Smad3$^{-/-}$ mice have no developmental or skeletal abnormalities and are indistinguishable from their wild-type littermates both grossly and with histology, except for an angular distortion of the forelimb in approximately 30% mice. Skeletal deformities start to appear after three weeks. The abnormalities are characterized by premature chondrocyte maturation with increased length of the hypertrophic region, disorganization of the chondrocyte columns, and early expression of *Col-X*. Thirty days after birth, articular chondrocytes become hypertrophic and severe degenerative joint disease is developed. The characteristic features of the joint diseases in Smad3$^{-/-}$ mice is similar to human OA, e.g. progressive loss of articular cartilage, formation of large osteophytes, decreased production of proteoglycans, and an increased expression of *Col-X* in articular chondrocytes in synovial joints. Degeneration also appears to involve the vertebral discs, mimicking a spinal disorder common in the human population and a source of enormous morbidity[95]. Sternal chondrocytes from Smad3$^{-/-}$ mice have higher expression levels of *Col-X* and other chondrocyte maturation compared to wild-type mice. No compensatory changes, e.g. upregulation of TβRI, TβRII, Smad2, or Smad4, are observed in Smad3 null mutant mice. Instead, both BMP-2 and 6 are upregulated in Smad3$^{-/-}$ chondrocytes. TGFβ inhibits *Col-II* expression in wild-type cells, but increases *Col-II* expression in Smad3$^{-/-}$ chondrocytes. Smad3$^{-/-}$ cells are more sensitive to BMP treatment than wild-type cells (unpublished data).

9. TGFβ and Runx2

Runx2 is key transcription factor that plays a critical role in osteoblast differentiation as well as in chondrocyte maturation[96]. Treatment with TGFβ results in phosphorylation of Runx2 at threonine residues possibly through ERK signaling pathway. The effect of TGFβ on Runx2 expression varies in different cells. In mesenchymal cells, TGFβ induces Runx2 expression via MAPK pathway. JunB is the upstream molecule of Runx2 in response to TGFβ. Runx2 in turn regulates TGFβ-mediated responses. The promoter of TβRI gene contains several Runx2 binding sites and forced expression of Runx2 enhances TβRI promoter activity[97-100]. In osteoblasts, TGFβ inhibits the expression of Runx2 through Smad3. Over-expression of Smad2 also represses Runx2 expression in osteoblats. Physical interaction of Smad2 and Smad3 with Runx2 is essential for the collagenase 3 expression in osteoblasts and in breast cancer cells[101-103].

10. TGFβ and Wnt Signaling

TGFβ and Wnt signaling pathways independently or cooperatively regulate LEF1/TCF target genes. The cooperative enhancement of TGFβ and Wnt signalings relies not only on the physical association of transcription factors in protein level but also on DNA level at the Smad binding element (SBE) adjacent to the LEF1 binding sites. After TGFβ stimulation, Smad3 interacts with LEF1 to activate target gene transcription. Deletion of SBEs near the LEF1/TCF abrogates Smad-dependent transcription. TGFβ-dependent activation of LEF1/TCF target genes occurs independently of β-catenin. Axin, a negative regulator of Wnt signaling, modulates the effects of Smad3. It functions as an adapter for Smad3, facilitating its activation by TβRI for efficient TGFβ signaling. In human marrow stromal cells, TGFβ upregulates the expression of Wnt2, 4, 5a, 7a, 10a, and Wnt co-receptor LRP5. TGFβ also increases nuclear accumulation and stability of β-catenin. Working synergistically with Wnt signal pathways, TGFβ stimulates chondrocyte differentiation from mesenchymal cells. Wnt 7 may be a critical mediator of this process[104-106]. Interestingly, in chicken upper sternal chondrocytes, TGFβ inhibits the expression of Wnt 4, 5, 8, and 14. It also suppresses the β-catenin and TCF-induced *Col-X* expression (unpublished data). Possible explanation is that interdependent relationship between TGFβ and Wnt signaling pathways may vary at the different stage of cell differentiation.

References

1. Miyazawa K, Shinozaki M, Hara T, et al. *Genes Cells,* 7, 1191-1204 (2002).
2. Shi Y, Massague J. *Cell* ,113,685-700 (2003).
3. de Caestecker M. *Cytokine Growth Factor Rev*, 15,1-11 (2004).
4. Shi Y, Wang YF, Jayaraman L, et al. *Cell,* 94,585-594 (1998).
5. Yagi K, Goto D, Hamamoto T, et al. *J Biol Chem,* 274,703-709 (1999).
6. Heyer J, Escalante-Alcalde D, Lia M, et al. *Proc Natl Acad Sci U S A*, 96, 12595-12600 (1999).
7. Datto MB, Frederick JP, Pan L, et al. *Mol Cell Biol*, 19, 2495 2504 (1999)
8. Kretschmer A, Moepert K, Dames S, et al. *Oncogene*, 22, 6748-6763 (2003).
9. Felici A, Wurthner JU, Parks WT, et al. *EMBO J,* 22, 4465-4477 (2003).
10. Wrana JL. *Cell,* 100, 189-192 (2000).
11. Shi W, Sun C, He B, et al. *J Cell Biol,* 164, 291-300 (2004).
12. Wu JW, Krawitz AR, Chai J, et al. *Cell*, 111, 357-367 (2002).
13. Frederick JP, Wang XF. *Structure (Camb),* 10, 1607-1611 (2002).
14. Liu X, Sun Y, Weinberg RA, et al. *Cytokine Growth Factor Rev*, 12, 1-8 (2001).
15. Suzuki H, Yagi K, Kondo M, et al. *Oncogene*, 23, 5068-5076 (2004).
16. Bonni S, Wang HR, Causing CG, et al. *Nat Cell Biol,* 3, 587-595 (2001).
17. Izzi L, Attisano L. *Oncogene*, 15, 2071-2078 (2004).
18. Zhang F, Laiho M. *Exp Cell Res,* 291, 275-281 (2003).
19. Zhao M, Qiao M, Harris SE, et al. *J Biol Chem*, 279, 12854-12859 (2004).
20. Ohshima T, Shimotohno K. *J Biol Chem,* 278, 50833-50842 (2003).
21. Imoto S, Sugiyama K, Muromoto R, et al. *J Biol Chem*, 278, 34253-34258 (2003).
22. Kretzschmar M, Doody J, Massague J. *Nature*, 389, 618-622 (1997).
23. Zhang Y, Feng XH, Derynck R. *Nature*, 394, 909-913 (1998).
24. Engel ME, McDonnell MA, Law BK, et al. *J Biol Chem*, 274, 37413-37420 (1999).
25. Massague J. *Genes Dev*, 17,2993-2997 (2003).
26. Ionescu AM, Schwarz EM, Zuscik MJ, et al. *Exp Cell Res,* 288, 198-207 (2003).
27. Derynck R, Zhang YE. *Nature*, 425, 577-584 (2003).
28. Wang L, Zhu Y, Sharma K. *J Biol Chem,* 273, 8522-8527 (1998).
29. Zhang L, Duan CJ, Binkley C, et al. *Mol Cell Biol,* 24, 2169-2180 (2004).
30. Zimmerman CM, Kariapper MS, Mathews LS. *J Biol Chem*, 273, 677-680 (1998).
31. Wicks SJ, Lui S, Abdel-Wahab N, et al. *Mol Cell Biol,* 20, 8103-8111 (2000).
32. Yakymovych I, Ten Dijke P, Heldin CH, et al. *FASEB J*, 15, 553-555 (2001).
33. Sowa H, Kaji H, Iu MF, et al. *J Biol Chem*, 278, 52240-52252 (2003).
34. Blokzijl A, Dahlqvist C, Reissmann E, et al. *J Cell Biol*, 163, 723-728 (2003).
35. Bakin AV, Tomlinson AK, Bhowmick NA, et al. *J Biol Chem*, 275, 36803-36810 (2000).
36. Bhowmick NA, Ghiassi M, Bakin A, et al. *Mol Biol Cell*, 12, 27-36 (2001).

37. Remy I, Montmarquette A, Michnick SW. *Nat Cell Biol,* 6, 358-365 (2004).
38. Conery AR, Cao Y, Thompson EA, et al. *Nat Cell Biol,* 6, 366-372 (2004).
39. Ulloa L, Doody J, Massague J. *Nature*, 397, 710-713 (1999).
40. Long J, Matsuura I, He D, et al. *Proc Natl Acad Sci U S A*, 100, 9791-9796 (2003).
41. Imoto S, Sugiyama K, Muromoto R, et al. *J Biol Chem*, 278, 34253-34258 (2003).
42. Long J, Wang G, Matsuura I, et al. *Proc Natl Acad Sci U S A,* 101, 99-104 (2004).
43. Nagarajan RP, Chen F, Li W, et al. *Biochem J*, 348, 591-596 (2000).
44. Bitzer M, von Gersdorff G, Liang D, et al. *Genes Dev*, 14,187-197 (2000).
45. de Crombrugghe B, Lefebvre V, Behringer RR, et al. *Matrix Biol,* 19, 389-394 (2000).
46. DeLise AM, Fischer L, Tuan RS. *Osteoarthritis Cartilage*, 8, 309-334 (2000).
47. Ballock RT, O'Keefe RJ. *Birth Defects Res Part C Embryo Today*, 69, 123-143 (2003).
48. Ulrich-Vinther M, Maloney MD, Schwarz EM, et al. *J Am Acad Orthop Surg*, 11, 421-430 (2003).
49. Pedrozo HA, Schwartz Z, Gomez R, et al. *J Cell Physiol,* 177, 343-354(1998).
50. Pedrozo HA, Schwartz Z, Robinson M, et al. *Endocrinology,* 140, 5806-5816 (1999).
51. Rosenthal AK, Gohr CM, Henry LA, et al. *Arthritis Rheum*, 43, 1729-1733 (2000).
52. D'Angelo M, Billings PC, Pacifici M, et al. *J Biol Chem,* 276,11347-11353 (2001).
53. Maeda S, Dean DD, Gay I, et al. *J Bone Miner Res,* 16, 1281-1290 (2001).
54. Matsunaga S, Yamamoto T, Fukumura K. *Int J Oncol,* 14, 1063-1067 (1999).
55. Horner A, Kemp P, Summers C, et al. *Bone,* 23, 95-102 (1998).
56. Sakou T, Onishi T, Yamamoto T, et al. *J Bone Miner Res,* 14, 1145-1152 (1999).
57. Verdier MP, Seite S, Guntzer K, et al. *Rheumatol Int,* 2003 Nov 14 [Epub ahead of print])
58. Fukumura K, Matsunaga S, Yamamoto T, et al. *Anticancer Res,* 18, 4189-4193 (1998).
59. Letamendia A, Lastres P, Botella LM, et al. *J Biol Chem,* 273, 33011-33019 (1998).
60. Parker WL, Goldring MB, Philip A. *J Bone Miner Res,* 18, 289-302 (2003).
61. Glansbeek HL, van Beuningen HM, Vitters EL, et al. *Lab Invest,* 78, 133-142 (1998).
62. van Beuningen HM, Glansbeek HL, van der Kraan PM, et al. *Osteoarthritis Cartilage,* 8, 25-33 (2000).
63. Blanco FJ, Geng Y, Lotz M. *J Immunol,* 154, 4018-4026 (1995).
64. Boumediene K, Vivien D, Macro M, et al. *Cell Prolif,* 28, 221-234 (1995).
65. Okazaki R, Sakai A, Nakamura T, et al. *Ann Rheum Dis,* 55,181-186 (1996).
66. Nixon AJ, Lillich JT, Burton-Wurster N, et al. *J Orthop Res,* 16, 531-541 (1998).
67. de Haart M, Marijnissen WJ, van Osch GJ, et al. *Acta Orthop Scand,* 70, 55-61 (1999).
68. Yonekura A, Osaki M, Hirota Y, et al. *Endocr J,* 46, 545-553 (1999).
69. Osaki M, Tsukazaki T, Yonekura A, et al. *Endocr J,* 46, 253-261 (1999).
70. Hirota Y, Tsukazaki T, Yonekura A, et al. *Osteoarthritis Cartilage,* 8, 241-247 (2000).
71. Beier F, Ali Z, Mok D, et al. *Mol Biol Cell,* 12, 3852-3863 (2001).

72. Baragi VM, Renkiewicz RR, Jordan H, et al. *J Clin Invest,* 96, 2454-2460 (1995).
73. Arai Y, Kubo T, Kobayashi K, et al. *J Rheumatol,* 24, 1787-1795 (1997).
74. Shuler FD, Georgescu HI, Niyibizi C, et al. *J Orthop Res,* 18, 585-592 (2000).
75. Smith P, Shuler FD, Georgescu HI, et al. *Arthritis Rheum,* 43, 1156-1164 (2000).
76. Hickery MS, Bayliss MT, Dudhia J, et al. *J Biol Chem,* 278, 53063-53071 (2003).
77. Watanabe H, de Caestecker MP, Yamada Y. *J Biol Chem,* 276, 14466-14473 (2001).
78. Moulharat N, Lesur C, Thomas M, et al. et al. *Osteoarthritis Cartilage,* 12, 296-305(2004).
79. Scharstuhl A, Diepens R, Lensen J, et al. *Osteoarthritis Cartilage,* 11,773-782 (2003).
80. Scharstuhl A, Glansbeek HL, van Beuningen HM, et al. *J Immunol,* 169, 507-514 (2002).
81. Zhang X, Ziran N, Goater JJ, et al. *Bone,* 34, 809-817 (2004).
82. Alvarez J, Sohn P, Zeng X, et al. *Development,* 129,1913-1924 (2002).
83. Serra R, Karaplis A, Sohn P. *J Cell Biol,* 145, 783-794 (1999).
84. Ionescu AM, Schwarz EM, Zuscik MJ, et al. *Exp Cell Res,* 288, 198-207 (2003).
85. Pateder DB, Rosier RN, Schwarz EM, et al. *Exp Cell Res,* 256, 555-562 (2000).
86. Ferguson CM, Schwarz EM, Reynolds PR, et al. *Endocrinology,* 141,4728-4735 (2000).
87. Oh CD, Chang SH, Yoon YM, et al. *J Biol Chem,* 275, 5613-5619 (2000).
88. Rosado E, Schwartz Z, Sylvia VL, et al. *Biochim Biophys Acta,* 1590, 1-15 (2002).
89. Valcourt U, Gouttenoire J, Moustakas A, et al. *J Biol Chem,* 277, 33545-33558 (2002).
90. Tardif G, Reboul P, Dupuis M, et al. *J Rheumatol,* 28, 1631-1639 (2001).
91. Su S, DiBattista JA, Sun Y, et al. *J Cell Biochem,* 70, 517-527 (1998).
92. Goldring MB. *Arthritis Rheum,* 43,:1916-1926 (2000).
93. Serra R, Johnson M, Filvaroff EH, et al. *J Cell Biol,* 139, 541-552 (1997).
94. Yang X, Chen L, Xu X, et al. *J Cell Biol,* 153, 35-46 (2001).
95. Yang X, Karsenty G. *Trends Mol Med,* 8, 340-345 (2002).
96. Chang DJ, Ji C, Kim KK, et al. *J Biol Chem,* 273, 4892-4896 (1998).
97. Ji C, Casinghino S, Chang DJ, et al. *J Cell Biochem,* 69, 353-363 (1998).
98. Lee KS, Kim HJ, Li QL, et al. *Mol Cell Biol,* 20, 8783-8792 (2000).
99. Lee KS, Hong SH, Bae SC. *Oncogene,* 21, 7156-7163 (2002).
100. Alliston T, Choy L, Ducy P, et al. *EMBO J,* 20, 2254-2272 (2001).
101. Li J, Tsuji K, Komori T, et al. *J Biol Chem,* 273, 31009-31015 (1998).
102. Selvamurugan N, Kwok S, Alliston T, et al. *J Biol Chem,* 279, 19327-19334 (2004).
103. Labbe E, Letamendia A, Attisano L. *Proc Natl Acad Sci U S A,* 97, 8358-8363 (2000).
104. Letamendia A, Labbe E, Attisano L. *J Bone Joint Surg Am,* 83A, S31-39 (2001).
105. Furuhashi M, Yagi K, Yamamoto H, et al. *Mol Cell Biol,* 21, 5132-5141 (2001).
106. Zhou S, Eid K, Glowacki J. *J Bone Miner Res,* 19, 463-470 (2004).

CHAPTER 14

BONE HEALTH IN CHILDREN AND ADOLESCENTS

Joan M Lappe, Ph.D., R.N., F.A.A.N.
Professor of Nursing, Professor of Medicine
Director of Clinical and Pediatric Studies

Osteoporosis Research Center, Creighton University
Omaha, NE 68131
Tel: 402-280-4646
Email: jmlappe@creighton.edu

Osteoporosis is a paediatric disease.
Charles Dent, 1972 [47]

1. Introduction

Optimum development of the skeleton during childhood is critical for prevention of osteoporosis in adulthood.[1;30;98;166;180;196;198] Achievement of the full genetic potential for peak bone mass protects against osteoporotic fracture. The relative risk of fracture increases by as much as 2.6 for each decrease in bone mass of 1 standard deviation.[42] It has been estimated that increasing the bone mass of the elderly population by 5% would decrease hip fracture incidence by 50%[156]. Although peak mass is not achieved until about age 30, at least 90% of the adult mass is accrued by the age of 19 or 20.[27;146;149;150;186;201;212]

The relationship between skeletal health and bone strength is found throughout the lifespan. Not only older adults, but also children and adolescents with low bone mass have been found to be at greater risk of fracture than their peers with higher bone mass.[75;76;121] While 70-80% of peak bone mass is genetically determined, strong evidence suggests that lifestyle factors have an impact on bone mass accrual. [51;81;110;149;181;182]

Thus, there is considerable interest in methods of maximizing bone mass accrual during childhood and adolescence so that individuals achieve their full potential for peak bone mass. During the last decade, there has been a considerable increase in research directed toward understanding and improving the bone health of children and adolescents. The purpose of this chapter is to review findings from that research and to provide a foundation for applying those findings to pediatric populations.

2. Growth and Development

From birth to young adulthood, skeletal mass increases from about 70-95 grams to 2400-3300 grams in women and men, respectively.[214] During this time, longitudinal growth occurs along with changes in skeletal size and shape. Although skeletal growth during infancy is rapid, it slows during childhood and then increases again at puberty. Bone mass accretion accelerates during puberty, corresponding with the rapid gain in longitudinal growth.[14] In fact, about 45% of the adult skeleton is built and enlarged during adolescence. [150]

Modeling is the primary mode of bone turnover in the growing skeleton.[173]Modeling involves the persistence of bone formation or resorption for a period of time at the same location during growth. The resultant changes in the relative locations of bone surfaces result in shaping the bones. After the cessation of longitudinal growth, the process of bone **remodeling** is predominant and is characterized by the continuous formation and resorption of bone tissue with minimal change in bone volume or shape. (Modeling and remodeling are more thoroughly discussed in Chapter 21.)

3. Hormonal Status

The accelerated bone accretion during puberty is influenced by the sex steroids. [14;118;212] Bone accrual plateaus as full pubertal development and final adult height are achieved. While the sequence of sexual maturation and physical growth associated with puberty are predictable, the ages at which these changes occur and the rate of developmental progression vary considerably among individuals. [208;227] The earliest signs of puberty in girls appear at 9-13 years of age and in boys at 9.5-14 years of age. [227] It is estimated that 20-25% of adult height is achieved during 2-3 years of puberty, the adolescent growth spurt. It is further estimated that about 25% of adult bone density is acquired during the two years surrounding peak height velocity. [19] The maximum height increase in both genders, occurs at Tanner stages 2-3, while the maximal bone mass gain occurs at stages 3-4. [60] Thus, adolescence is a critical time for attention to bone healthy behaviors.

Many, but not all, cross-sectional studies have found an inverse relationship between age at menarche and adult bone mineral density (BMD). [61;99;100;139;188] There are also a few reports of increased risk of osteoporotic fracture in women who experienced late menarche. [106;170]

Heaney et al [86]pointed out that interpretation of the data surrounding peak bone mass and age at menarche is very difficult. One reason for this difficulty is failure of published reports to distinguish between BMD and bone mineral content (BMC). Initiation of estrogen production at puberty arrests longitudinal growth and increases true trabecular density. [70] Thus, in a girl who experiences early menarche, growth is arrested early, and she has smaller bones compared to a girl with later menarche. Girls with later menarche have larger bones, and they almost always have lower BMD even though their BMC is equal to or greater than the girls with smaller bones. The mineral in the larger skeletons is spread over a greater area.

Another difficulty in interpreting the BMD data in adolescent girls is that cross-sectional studies often do not take into consideration the time since menarche. Heaney et al suggest that in order to determine the effect of late menarche on the skeleton, bone strength and/or BMC should be evaluated either at the adult peak or at a constant number of years after menarche. [86] Few studies have evaluated the effects of late menarche in this way.

The strongest correlate of age of menarche is age of maternal menarche; thus, late menarche in a girl whose mother experienced a similar late menarche is probably not of concern. However, primary amenorrhea, delay in menarche beyond the age of 16, may be due to low body weight or body fat, poor nutrition, dieting behavior, psychological stress, or excessive physical training. [223] Most of these factors are also associated with osteopenia and/or osteoporosis. Thus, girls with late menarche need careful evaluation to assure that their bone mass accrual is not compromised. Of greatest concern is late menarche in competitive athletes, who often exhibit a number of factors that affect bone mass. This is further discussed later in this chapter.

4. Tracking of Bone Mass

Ideally, we would target bone building interventions towards children who are at risk of low peak bone mass. Targeting children who currently have low bone mass is based on the assumption that an individual's bone mass tracks throughout life, i.e. remains in the same position relative to percentiles of the referent population. In fact, limited evidence suggests that bone mass does track. Ferrari et al [58] followed eight-year-old girls for two years and found that during that time BMC, BMD, and bone area increased dramatically, while Z-scores of these bone traits changed minimally (<0.5 Z-score). A longitudinal study using computed tomography (CT) found that trabecular bone density and cross-sectional areas of the vertebrae and cross-sectional and cortical bone areas of the femora in early puberty predicted values at sexual maturity. [142] Baseline values of the bone traits were divided into quartiles and plotted by Tanner stage. For each trait, the regression lines differed among the quartiles and ran parallel to each other without overlapping. Thus, bone size and volumetric density tracked. This provides a basis for identifying and targeting children who are prone to develop suboptimal peak bone mass.

5. Genetics

Estimates are that genetic factors account for about 70% to 80% of the variance in peak bone mass.[107;159;178] However, there are limited reports of the relationship between specific genes and bone mass in children. Studies of polymorphisms in the VDR gene are those most often reported in children and young adults; however, the findings are equivocal. The polymorphisms routinely studied include BsmI, ApaI, TaqI and FokI. For BsmI, the BB genotype has been associated with higher BMD in males and females ranging from prepubertal to young adult. [141;191;195;220] Conversely, others have found that BB genotype is associated with lower BMD in adolescents and young men. [55;57] Gunnes [79] found no association between the Bsm1 polymorphisms and forearm BMD gain in boys and girls between ages 8 and 17.

For ApaI, the AA genotype has been associated with lower femoral BMD in Mexican-American girls [195], while the A allele has been associated with lower distal radius BMD in late adolescent women. [115] In one study, the aa genotype was associated with lower femoral neck BMD in young women [191], while in a second study no association was found between the Apa1 polymorphism and BMD. [79] For TaqI, the T allele has been associated with higher spine BMD in seven-year-old girls [211] and with lower femoral neck BMD in young adult females.[191] Gunnes et al[191] found no associations.

Studies of the association of FokI with BMD in children have also been equivocal. In children aged 7-12, the F allele was associated with higher BMD and higher calcium absorption. [10] The FF genotype was associated with higher spine BMD in adolescent boys [209] but it was not associated with BMD in prepubertal girls. [56] Thus, association studies of genotype and bone mass in youth have not provided clear relationships. It is highly likely that a multitude of genes are involved in the hereditability of peak bone mass. The genetics of osteoporosis are more thoroughly discussed in Chapters 24 and 26.

6. Gender Differences

Ample evidence supports the gender difference in bone mass and skeletal strength, i.e. that men have more massive skeletons than women. A number of studies suggest that bone mass is similar in males and females before puberty.[71;162;164] However, CT studies indicate that prepubertal females have smaller vertebral cross-sectional area than males, even after controlling for body size.[26] This difference increases during growth and stabilizes at skeletal maturity at which time the vertebral cross-sectional area is about 35% smaller in women than in men. This may account for the higher incidence of vertebral fractures in females than in males.

7. Ethnicity/Race

It is well-established that there are ethnic/racial differences in bone mass in children.[15;23;72;134;162] In the U.S., blacks have higher bone mass than other races at early ages. However, black Gambian children have a lower bone mass than British children,[180] and South African blacks have bone mass similar to white children.[174] Bone mass of Asian and Hispanic children and adolescents in the U.S. is considerably lower than that of black but similar to that of white youth.[15;96] In Canada, prepubertal Asian boys were found to have significantly lower femoral neck BMC and BMD than white males. [155]

In their studies using CT in children, Gilsanz et al found that race has significant, differential effects on the density and size of bones in both the axial and appendicular skeleton. [71] Although cancellous bone density in the vertebral bodies of prepubertal children does not differ between blacks and whites, a difference is seen during the late stages of puberty. Cancellous bone density increases in all adolescents, but the magnitude of the increase is substantially greater in black than in white youth (34% vs 11%, respectively). However, there are no differences in cross-sectional area of the vertebral bodies between black and white children. In the appendicular skeleton, black children have a greater femoral cross-sectional area but similar cortical bone area and density. At sexual maturity, the vertebral bone density and the femoral cross-sectional area were 10.75% and 5.7% higher, respectively, in black than in white children. These racial differences likely contribute to the lower risk of fracture observed in black adults.[161]

8. Fractures during Childhood

It has long been recognized that a sharp increase in incidence of fractures occurs during puberty. [8;114] The peak frequency is seen at ages that coincide with the ages of peak height velocity. To explain this, it has been hypothesized that a delay occurs in consolidation of bone mass during rapid longitudinal growth. [173] This results in a transient state of relatively low bone mass while the skeleton is adapting to increased loading as body size and weight increase with growth. Studies have found that children who fracture during this phase have bone mass that is lower than age- and gender-matched controls.[75;76;121] These studies suggest that interventions to improve bone health in children with lower than average bone mass might reduce the incidence of childhood fractures as well as decrease the risk for osteoporosis in old age.

9. Maximizing Peak Bone Mass

Peak bone mass and optimal bone health are determined by a combination of factors: genetic and environmental. While genetic factors account for about 70% to 80% of the variance in peak bone mass, the remaining 20 to 30% is accounted for by environmental factors, such as nutrition, physical activity, medications, diseases, and lifestyle factors such as smoking cigarettes and drinking alcohol. [33;35;38;46;49;65;69;85;101;107;169;204;205;217] Optimizing these environmental influence can account for substantial differences in peak bone mass.

9.1. Nutrition and peak bone mass

A number of nutrients are important to bone health.

Calcium

Evidence suggests that optimal calcium nutrition in childhood may positively affect peak bone mass by as much as 5-10%[149], an amount equal to about 0.5 - 1.0 standard deviation in peak skeletal mass. This difference is enough to decrease the risk of hip fractures later in life by 25-50%. [86]Growing children and adolescents are depositing large amounts of calcium in their skeletons. Thus, calcium deficiency during formation of the skeleton likely decreases the level of peak bone mass.

Calcium is a threshold nutrient, i.e. there exists an intake level below which bone accretion varies with intake and above which bone accretion appears to be constant, regardless of intake. Above the threshold, the amount of bone accrued is dependent on genetic programming along with mechanical loading. However, below the threshold of intake, the skeleton does not retain enough calcium to build the amount of bone for which it was programmed. Thus, children and adolescents whoe calcium intake is below the threshold level will be unable to achieve their genetic potential for peak bone mass. [86]

Matkovic and Heaney[148] completed a meta-analysis of published calcium balances to arrive at threshold intakes for children and adolescents, 1390 and 1480 mg/day respectively. More recently, Jackman [102] has conducted a prospective balance study that arrived at a similar threshold level for adolescents. Abrams et al [4] found a significant increase in absorption of calcium correlated with the onset of puberty. In children of Tanner stages 2-4, calcium absorption increased across the range of 750-1350 mg/day, which is congruent with the established threshold. It is important to note that studies suggest that calcium absorption efficiency is different in African-Americans [5] and in Chinese [125] than in whites; accordingly, thresholds are most likely different. Further research is needed to establish threshold intake levels in various populations.

The threshold intakes are somewhat higher than the Adequate Intake (AI) for calcium established by the Food and Nutrition Board of the Institute of Medicine. (2386} (See Table 1.) The thresholds are better estimates of calcium requirements since they are predicated on the level of dietary calcium above which balance no longer rises with intake, while the AI is a value based on experimentally derived intake approximations of observed mean nutrient intakes by a group of healthy people. [160] An AI is set when data are insufficient to establish a Recommended Dietary Allowance (RDA). Clinical trials of calcium supplementation in children and adolescents suggest that bone mass accrual is greater when calcium intake is closer to the published threshold levels.

Table 1. Recommendations for Dietary Calcium Intake (mg/day)

Age (yrs)	Matkovic	AI
3-8	1390	800
9-17	1480	1300

Unfortunately, many children and adolescents do not consume the recommended levels of dietary calcium. For example, in the U.S. a recent national survey showed that the mean intake for girls ages 9-13 was 869 mg/day, while for those ages 14-18 the mean intake was 777 mg/day.[68] The mean intake for boys was closer to the recommended level. For ages 9-13, it was 1085 mg/day and for ages 14-18 it was 1172 mg/day. According to the China National Nutrition Survey in 1992, [36;67] the average calcium intake in children and teenagers is about 40% of the RDA. (The RDA for children and adolescents is 800 mg/day and 1000 mg/day, respectively.) In urban areas of China, about 72%-79% of the children and teenagers have calcium intake less than 50% of the RDA. Only 3.7%-5.8% of them have calcium intakes that meet the RDA. Calcium intake is even lower in rural than urban areas.

Support for the long-term effect of childhood calcium intake is provided from epidemiological studies conducted in Croatia and China. Lifelong inhabitants of districts with high calcium intakes had higher bone mass [97;151] and decreased risk of hip fracture. [151] Retrospective studies demonstrate that long-term increased calcium intake can result in approximately 5% higher mineral content of the mature skeleton. [177;196;202]. Results of cross-sectional studies in children are conflicting, with some finding an association between BMD or BMC and calcium intake [29;34;107;128;190;192] and others finding no relationship. [153;219] Reasons for conflicting outcomes might be the difficulty of assessing calcium intake accurately or the lack of variation in calcium intake in the population studied. Since calcium is a threshold nutrient, finding a relationship between calcium and bone mass in populations with a high calcium intake might not be possible.

Solid evidence of the effect of calcium in childhood skeletons is provided by prospective studies that have found that high calcium intake in children and adolescents is associated with a more positive calcium balance [147] and greater BMD and/or BMC than in controls. [33;33;35;35;69;69;101;107;169;204;205] For example, Lee et al supplemented seven-year-old children in the Guangdong Province with 300 mg/day of calcium. [129]The baseline intake was very low (279 mg/day). After 18 months of intervention, the treatment group demonstrated significantly greater gains in radial BMC than the control group (16.5% vs 14.0%,

P=0.02). In a subsequent study, Lee et al [127]supplemented children in Hong Kong with 300 mg/day of calcium. In this case, baseline calcium intake was 560 mg/day. After 18 months, the intervention group had a greater increase in spine BMC but not in radial BMD or BMC than the controls.

Wosje and Specker [228] recently published a review of the role of calcium intake on bone mineralization. They included a critical analysis of eight pediatric calcium supplementation trials and bone health, in which they determined annualized percent changes in BMD. They found that increases in BMD in calcium supplemented children occur primarily in cortical bone sites and are most pronounced in study populations with low baseline calcium intakes. Also, increases in lumbar BMD are greater in pubertal children than in prepubertal children. Review of studies conducted in infants indicated that a high mineral intake is positively associated with short-term gains in BMC.

From this comprehensive review, the authors concluded that in pediatric populations the increases in BMD and BMC associated with high calcium intake do not persist after the supplementation has been stopped. Evidence is provided from several studies. [107;126;126;137] In the Lee study of Hong Kong children, follow up measurements 18 months after cessation of the supplementation showed that the benefits of calcium supplementation were no longer apparent. [126] Other randomized controlled trials of calcium interventions in children show that high calcium diets increase bone mass accrual, but that after completion of the intervention, the control group catches up.[107;137] Thus, the approach to calcium nutrition in children is to help them establish the behavior of ingesting optimal calcium in their diets. The window of opportunity for encouraging calcium-rich diets is in the pre-pubertal stage since this is the age at which children are formulating their lifelong health behaviors.[9] Children who avoid dairy products in childhood are unlikely to change this behavior in adolescence.

Food sources of calcium are preferable to calcium supplements. The primary reason for this is that diets low in one nutrient are probably low in other essential nutrients, and single nutrient supplementation can do nothing to correct the other nutritional inadequacies. Recently, a study in which 2000 Israeli girls (mean age 14.5 years) completed a semiquantitative food frequency questionnaire found that a large percentage of diets with calcium intakes below 800 mg/day were deficient in phosphorus, magnesium, iron and zinc.[189] Chan [35] reported that when the dairy food intake was increased in pubertal females, the intake of calcium, as well as the intake of phosphorus, vitamin D, and protein, increased significantly over those of control subjects. Lappe et al [122] found that prepubertal girls who consumed a calcium-rich diet (mean calcium intake 1656 mg/day) significantly improved their overall nutritional intake. During two years, these girls increased their intake of 8 of the 14 nutrients measured (calcium, protein, vitamins A and D, phosphorus, potassium and magnesium). Girls in thc control group, who remained on their customary diets (mean calcium intake 961 mg/day), increased their intake of only two nutrients, iron and zinc. Intake of calcium, vitamin D, and phosphorus, which are critical for optimal bone health of growing girls, was inadequate. Thus, considerable evidence supports the overall nutritional value of a calcium-rich diet.

Concern exists that diets high in dairy foods, as a source of calcium, place individuals at risk for excessive weight gain. In fact, reports indicate that adolescent girls avoid dairy foods due to concern about weight gain and beliefs that milk and other dairy foods are fattening.[2;20;163;171] However, considerable evidence exists to refute this concern. For example, in a study of the effect of dietary calcium on bone mass accrual, Lappe et al (unpublished data) found that adolescent girls who had been on a calcium-rich diet (mean calcium intake 1694 mg/day) for five and one half years gained no more weight that the girls in the control group (mean calcium intake 875 mg/day), even though the calcium-rich group consumed on average 150 kcal/day more than the control group., There were no statistically significant differences between groups in increase in weight, body mass index or body fat composition as measured by DXA. This study corroborates the findings of others who have reported that supplementation with calcium-rich foods does not increase weight gain in pubertal girls over periods of 12-24 months.[16;33;35;69;157] Considerable evidence demonstrates that diets high in dairy food help with weight loss and maintenance in adults, although this has not been confirmed in children. [45;103;135;231] However, high calcium diets can be recommended for children and adolescents without concern about excessive weight gain.

Vitamin D

Vitamin D plays a critical role in skeletal health of children as well as of adults because it is necessary for the active transport of calcium across the intestinal mucosa. Without adequate vitamin D, adults absorb only about 10-15% of dietary calcium, while with adequate vitamin D, they absorb about 30% of ingested calcium.[92] If dietary calcium is inadequate, 1,25-dihydroxyvitamin D ($1,25(OH)_2D$), along with parathyroid hormone (PTH), stimulates physiologic activities that mobilize calcium from the skeleton.[92;105;193]

Children who have severe vitamin D deficiency are at high risk of rickets. Milder vitamin D deficiency is asymptomatic but can lead to decreased calcium absorption and thereby interfere with maximal acquisition of peak bone mass. Since adolescents accrue bone mass at extremely high rates and absorb calcium with higher efficiency than individuals in any other age group, optimal levels of vitamin D are particularly important during the teen years.[6;18;145;224]

Hollick has pointed out that vitamin D deficiency is a "major unrecognized epidemic in the older adult population."[93] One reason that this epidemic is unrecognized is that lack of clarity exists about what constitutes vitamin D deficiency. Historically, vitamin D deficiency was defined as the level below which rickets or osteomalacia developed. Based on this definition, the minimal acceptable level for serum 25-hydroxyvitamin D [25(OH)D] for clinical use is typically about 10-12 ng/mL, based on the referent population of individual laboratories. Use of local referents is confusing since the "normal" values in locations such as southern Spain differ considerably from those in locations such as northern Germany. Based on studies that have evaluated PTH and 25(OH)D levels, it is likely that the minimum acceptable level of 25(OH)D for preventing secondary hyperparathyroidism and maintaining bone health is closer to 30 ng/mL (75 nmol/L).[37;136;213]

There are many published reports of vitamin D deficiency in otherwise healthy children and adolescents, especially during winter months.[25;48;50;78;109;130;140;207;218] Du et al [50] conducted a cross-sectional study in a random sample of 1248 Beijing girls ages 12-14 years. They found a prevalence of subclinical vitamin D deficiency (serum 25OHD < 12.5 nmol/L) of 45.2% in winter and 6.7% in summer. In a study of children living in northern Spain (43° north latitude), vitamin D metabolites and PTH serum levels were measured in March. Thirty-one percent of the children had 25(OH)D levels below 12 ng/ml, and 80% had levels lower then 20 ng/ml. Serum 25(OH)D was inversely associated with serum PTH. In a study of eight-year-olds in Tasmania (42 ° south latitude), bone density was inversely associated with winter sunlight exposure. Thus, suboptimal serum vitamin D levels are prevalent in children, and evidence suggests that this is harmful to their bone health.

Very few foods naturally contain vitamin D. Variable amounts can be obtained from egg yolks, fatty fish, and fish liver oils. Likewise, few foods are fortified with vitamin D. The primary source in the U.S. is vitamin D-fortified milk. However, the amount of vitamin D in milk is highly variable; some samples taken from vitamin D-fortified milk contained scant amounts of this nutrient.[94]

Much of the required vitamin D for children [91;93;206] is obtained from exposure to sunlight. Ultraviolet B radiation converts 7-dehydrocholesterol in the skin to pre-vitamin D, which then is converted to the more stable vitamin D_3.[93] The amount of sun exposure necessary depends on latitude, time of year, amount of melanin pigment in the skin, age of the individual, body mass index, and type of clothing worn. For example, individuals with large amounts of melanin pigment in their skin have considerably less cutaneous production of vitamin D_3 than those with lighter skin tones. In fact, this largely accounts for the high incidence of vitamin D deficiency in black breast-fed infants.[25;84] Due to the oblique zenith angle of the sun during the winter, very little ultraviolet B radiation reaches the earth during these months. Thus, during the winter little cutaneous vitamin D is produced in individuals who dwell at latitudes above and below -35°.[225] Even during summer months, individuals may obtain less than the needed amount of vitamin D because of sunscreen use. The amount of vitamin D produced by the skin is reduced by 97.5% when a sunscreen with a skin protection factor of 8 is worn.[152]

The National Academy of Sciences [160] and the American Academy of Pediatrics [3] recommend that in order to prevent rickets, all children, including those who are exclusively breastfed, have a minimum intake of 200 IU of vitamin D per day beginning during the first two months of life and continued throughout childhood and adolescence. These groups do not currently have recommendations for the vitamin D intake that is needed for children to optimize bone mass, but 200 IU/day is most likely too low. Further research needs to be done in this area.

Overall nutrition

Phosphorus, protein and several trace nutrients, such as magnesium, zinc, fluoride, copper and vitamin C are important for skeletal health. However, literature reporting deficiencies of these nutrients and the relationship to bone health in children is scant. Furthermore, estimates of intake of trace nutrients are often underestimated because these nutrients are not available in food nutrient databases.[86] If, as indicated earlier, children are encouraged to adopt calcium-rich diets, they are more likely to also ingest a diet rich in other nutrients needed to reach their genetic potential for peak bone mass.

9.2. Physical activity

Bone adapts to varying levels of mechanical loading, i.e., physical activity. This is the skeleton's attempt to maintain bone mass and architecture able to support the needs for mechanical usage without sustaining damage to bone tissue. [185] Modeling, the process that alters the shape and mass of the bones, is a response to mechanical loading.[173] As mentioned earlier, in growing children **modeling** is the dominant mode by which recruitment and function of bone cells are coordinated to increase bone mass and strength.

Cells respond differently to mechanical loading in modeling than they do in remodeling. According to Parfitt, [173] exercise increases the rate of osteoblast recruitment. In the modeling mode, recruitment of osteoblasts continues at the same location for many years until bone strength is sufficient to resist bending (strain). In the remodeling mode, osteoclast activity may be suppressed and osteoblast activity increased to prevent bone loss in turnover. [173] These differences in osteoblast recruitment account for the varied responses to physical activity, resulting in enhanced bone gain during growth and retarded bone loss in aging. [59;80] During modeling, bone formation or resorption persists for a long time at one location, and, as a result, changes or "drifts" occur in the relative location of bone. At high-load sites bone strength is increased by adding mass and improving internal architecture.

Thus, during growth the ability of bone to adapt to mechanical loading is significantly greater than after maturity has been reached. [59;62;66;173;173] This allows the individual to develop a skeleton able to support his/her adult level of physical activity. [173] Children who have little physical activity will develop bone strength sufficient to sustain that sedentary level of activity and may not be able to make up for this lack of bone strength in later years. Therefore, physical activity is an important modifiable environmental influence that can account for substantial differences in peak bone mass.

A considerable body of evidence supports the positive effect of physical activity on bone mass accrual in children and adolescents. Longitudinal studies demonstrate that high physical activity in childhood and adolescence is associated with high adult bone mass. [21;40;46;111;113;183;217;226] Child athletes in various sports have been found to have greater BMD at weight-bearing sites than do controls[39;77;81;154;168;221]. Studies of childhood activities that preferentially stress one side of the body over the other, such as tennis, find significantly greater BMD in the dominant arms and/or legs. [22;63;81;116;117]High-impact activities have the greatest osteogenic stimulus. For example, numerous studies have found that gymnasts have higher bone mineral density than controls. [54;88;120;131;165;230] Gymnasts have higher BMD than runners while the runners have higher BMD than controls.[104;131]

Many questions remain about the optimal type, duration and frequency of exercise to provide the greatest skeletal effect. In mechanical loading studies of the rat tibia, Robling et al [187] found that more frequent, shorter loading sessions (60 cycles x 6 sessions or 90 cycles x 4 sessions) provided more of an osteogenic stimulus than a single bout of 360 cycles over three minutes. Relative (loaded minus non-loaded limb) values for endocortical lamellar bone formation rate, mineralizing surface and mineral apposition rate were 65-94% greater in the two divided applications than in the single loading bout. The authors concluded that a saturation curve exists for bone cell mechanosensitivity and suggested that a rest period between loading bouts increases the ostegenic effectiveness of loading. In another animal study, skeletal loading three times a week was as effective in stimulating bone formation as loading every day. [184] A strong body of evidence supports the site specificity of loading. [39;77;81;111;154;168;203;210;221].

Recent studies have found that short-term interventions with high-impact loading can have a significant effect on bone mass accrual in children. For example, a nine-month school-based prospective step-aerobic program with two 50-minute sessions per week resulted in a significantly higher increase in spine and femoral neck BMC compared to controls (8.6% vs 5.3% and 9.3% vs 5.3%, respectively) in premenarcheal females. [87] In another school-based study, prepubescent children who performed 100 jumps off 61-cm boxes three times per week for seven months had significantly greater increases in spine and hip BMC than controls.[64] These studies also demonstrate the feasibility of implementing bone loading activity in the physical education classes of schools.

Exercise has more of an effect if started in early age. For example, a cross-sectional study of Finnish female tennis and squash players found that among different age groups of athletes, group differences in BMC were significant, with the group means lower in those who were older when they began participating in the sport. [117]Furthermore, the benefit of playing was about 2 times greater if females started playing at or before menarche rather than after it. A 5-year follow up revealed that, despite reduced training, the exercise-induced bone gain was well maintained. [116] Duppe et al [51] also found that BMD effects of training are maintained. They found that former competitive soccer players, who had ceased playing on average 9.7 years previously, had retained their femur and total body BMD advantage over controls.

The skeletons of children are unique from those of adults in that young skeletons adapt to mechanical loading by modifying bone geometry as well as bone mass. Prepubertal children assigned to performing 100 jumps three times a week had greater bone area at the femoral neck than the control group after seven months of intervention. [64] In another study, pre- and early-pubertal children were enrolled into a school-based, jumping intervention (12 minutes, 3 times/week). [143;144;176] After seven months of intervention, hip structural analysis (HSA) was evaluated in the girls. In the early-pubertal girls (Tanner stages 2 and 3), there was increased bone cross-sectional area and reduced endosteal expansion at the femoral neck, which improved section modulus (bone bending strength) (+4.0%, P=0.04) [176] Changes in HSA in the boys were evaluated after 20 months. [144] The treatment group had non-significantly greater bone expansion at the narrow femoral neck on both the endosteal (+2.6%, P=0.1) and periosteal (+2.7%, P=0.2) surfaces than did controls. This resulted in significantly greater increases in section modulus in the treatment group than in the controls (+7.5%, P=0.02).

These findings are consistent with those of Haapasalo et al [82] who used peripheral quantitative computed tomography to evaluate bone differences in the dominant vs non-dominant forearm of tennis players. He found a greater increase in BMC but not in volumetric bone density in the dominant compared to the non-dominant forearm. The higher BMC in the dominant arm was due to differences in the total area, cross-sectional area of the marrow cavity, and cortical area. In other words, the enhanced bone mineral was mainly due to increased bone size. In an earlier study by this group [83], those who started playing tennis in childhood demonstrated a greater increase in forearm diameter than those who started playing tennis in adolescence. Thus, active children can develop changes in bone geometry that provide for lifetime stronger bones.

The effect of exercise on bone in children is enhanced in the presence of adequate calcium intake. In a study of pre-pubertal and pubertal girls, participants were assigned to one of four groups: moderate impact exercise with or without calcium supplementation or low-impact exercise with or without calcium. Those assigned to moderate impact exercise and calcium supplementation experienced a greater increase in BMC of the femur than the other three groups [101] Specker et al [205] conducted a randomized study of gross motor vs. fine motor activity in infants and found that the effect of activity on total body bone mineral content (TBBMC) accrual was dependent upon the infant's calcium intake. Infants with a high calcium intake who were assigned to bone loading activities had a greater increase in TBBMC than infants with low or moderate calcium intake. In randomized study of calcium supplementation and physical activity in children ages three to five years [204], there was a significant interaction between physical activity and calcium supplementation. Positive effects on changes in leg bone mass occurred only in children consuming high calcium intakes (≥1100 mg/day). However, there were increases in periosteal and endosteal tibial circumferences in the physical activity group independent of calcium intake. The authors concluded that physical activity stimulates bone growth in diameter, but that both physical activity and calcium intake are needed for increase in bone mineralization. The findings of these studies are not surprising since calcium is the major substrate needed for formation of new bones. Further research is needed to determine the optimum levels of physical activity and calcium intake.

Due to modern technology children and adolescents are decreasingly active. Because of the automobile, they are less likely to walk or ride their bicycles. Also, during leisure time they tend to participate in sedentary activities such as watching television or playing on the computer. In the US, a 1997 survey found that only 64% of high school students spend at least 20 minutes three times a week in vigorous physical activity. [216] On average, girls are less active than boys.

Thus, strong evidence supports the efficacy of physical activity in development of strong bones during childhood and adolescence. Weight-bearing exercise should be initiated in childhood and maintained throughout adolescence to maximize peak bone mass and to enhance bone geometry for greater strength. Frequent short exercise sessions are effective and should include loading of various skeletal sites. However, physical activity is suboptimal in many young persons. It is critical that health professionals, school officials, parents and policy makers work together to help children develop and maintain physically active lifestyles.

10. Measurement of Bone Mass in Children

Dual energy x-ray absorptiometry (DXA) is the method most commonly used to assess bone mass in children due to its low cost and ease of use. Interpretation of DXA measurements in children is problematic for many reasons. One problem is that BMD is not a true volumetric density measurement. BMD is calculated as BMC/bone area (grams/cm^2), and this erroneously assumes that BMC and area are directly proportional. In fact, during adolescence, there is dissociation between linear growth and BMC accrual. Peak linear growth precedes peak BMC accretion by 8-11 months.[19] Similarly, bone area, as measured by DXA, peaks before BMC. This confounds the comparison of BMD values between similar age children of different sizes and the comparison of values in the same child across time. BMC has been proposed as a better assessment of skeletal health in children. However, there is not agreement about the factors that should be used to adjust BMC, such as bone size, height, pubertal stage, body composition, etc. [17;132;228]

There is an increasing need for DXA screening protocols in children. Many children with chronic disorders, such as Crohn's disease and juvenile rheumatoid arthritis, have been found to have low bone mass compared to age- and gender-matched controls. [11-13;89;90;95;108;112;158;175;199] In addition, medications such as anticonvulsants and corticosteroids have been found to decrease bone mineralization. [31;175;200] Treatment protocols including bisphosphonates, calcitonin, vitamin D, calcium, thyroid, and growth hormone have been proposed for children with low bone mass. [24;73;158;167;194] Thus, it is vital that children with low bone mass are identified so that interventions to maximize their bone mass accrual are implemented as early as possible.

In pediatric clinical practice, BMD/BMC is interpreted in terms of the number of standard deviations (SD's) above or below the age-specific mean among healthy individuals (the Z-score). In order to identify children accurately with low bone mass (osteopenia), the distribution of BMD/BMC in the reference population must predict not only the age-specific mean value of BMD/BMC but also the SD.[133] Leonard et al. recently measured spine BMD in 95 children with childhood disorders and compared the values with five different reference data sets.[133] BMD values from the 95 children were converted to age-specific Z-scores for each of the reference data sets. The percentage of children that were osteopenic (Z-score <-2.0) ranged between 11% and 30%, depending on which reference data set was used. The prevalence of osteopenia was significantly different across the five reference sets (P=0.005). The inconsistencies were greatest when the reference data were not gender-specific. It is widely recognized that population-based pediatric reference bases are sorely needed. In the US, a large NIH study is currently underway to document the normal pattern of skeletal development in a multiethnic sample of healthy US children and adolescents.

The International Society for Clinical Densitometry (ISCD) recently published a position statement on Diagnosis of Osteoporosis and Osteopenia in Children (Individuals less than age 20).[132] (http://www.iscd.org/visitors/official.cfm) The main points of the position are:

- T-scores should not be used in children; Z-scores should be used instead.
- The diagnosis of osteoporosis in children should not be made on the basis of densitometric criteria alone.
- Terminology such as "low bone density for chronologic age" may be used if the Z-score is below -2.0.
- Z-scores must be interpreted in light of the best available pediatric databases of age-matched controls. The reference data base should be cited in the report.
- Spine and total body are the preferred sites for measurement.
- There is no agreement on standards for adjusting BMD or BMC composition. If adjustments are made, they should be clearly stated in the report.
- Serial BMD studies should be done on the same machine using the same scanning mode, software and analysis when appropriate. Changes may be required with growth of the child.
- Any deviation from standard adult acquisition protocols, such as use of low-density software and manual adjustment of region of interest, should be stated in the report.

11. Factors Interfering with Bone Mass Accrual

A few factors that can adversely affect skeleton health in children and adolescents deserve attention. These include but are not limited to cigarette smoking, use of certain contraceptives, and the female triad.

11.1. Smoking

It is widely accepted that cigarette smoking is a risk factor for osteoporosis. Ward et al [222] recently published a meta-analysis of the effects of smoking on BMD. They pooled data across 86 studies, enrolling 40,753 subjects. They found that smokers had significantly lower bone mass than non-smokers. Hip bone mass deficits were greater than deficits at other skeletal sites. In prospective studies, smokers lost more bone over time than did non-smokers. Twin studies, which control for age, gender, and genetic composition, found that a dose effect, i.e. greater exposure to smoking was associated with lower bone mass. Measures of exposure included pack-years, cigarettes per day, and number of years smoked. Bone loss was greatest for current smokers compared with never smokers and intermediate for current smokers compared with former smokers. The least difference occurred between former smokers compared with never smokers. From the foregoing, the authors concluded that smoking cessation may have a positive impact on bone mass.

Based on the meta-analysis, the authors estimated that smoking increases the lifetime risk of having a vertebral fracture by 13% in women and 32% in men. They estimated that the smoking-related increase in the lifetime risk of hip fracture is 31% and 40% in women and men, respectively.

Scant data are available regarding the effect of smoking on adolescent skeletons. Some, but not all, studies of cigarette smoking in adolescents and young adults have found a negative association with bone mass. [7;28;53;215;217;229] However, since bone loss is associated with years of smoking, it is likely that the smoking effect on bone is gradual and difficult to find in young persons who have a short history of smoking. Thus, the effects of smoking on peak bone mass are not known, but they are most likely negative. It is known that most persons smoke their first cigarette before the age of 18 and that those that start smoking at an early age are more likely to continue smoking. [216]Thus, efforts to prevent children and adolescents from initiation of smoking are an important component of bone health promotion.

11.2. Contraceptives

Contraceptives that contain hormones have the potential to affect peak bone mass. Evidence suggests that depomedroxyprogesterone acetate (DMPA) and ultra-low dose oral contraceptives (containing 20μg ethinyl estradiol) negatively affect bone mass accrual in young females.

DMPA exerts its contraceptive effect by inhibiting pituitary gonadotropin secretion, which in turn suppresses ovulation and production of estrogen by the ovaries. This is supported by a study of Brusen et al who found that 17β estradiol levels in adolescent girls receiving DMPA were similar to levels typically seen in perimenopausal women who are losing bone. [32]

In a cross-sectional study, Scholes et al [197] found that spine BMD was 9.4% lower in young women 18-21 years of age who were treated with DMPA than age-matched women not receiving DMPA. Another cross-sectional study reported that in women ages 18-54, BMD was significantly lower in DMPA users than in non-users. Furthermore, those who had started using DMPA before the age of 21 had lower BMD compared to age-matched controls than those who started when they were older. [44]

In three prospective studies of girls ranging from 12-18 years of age, mean loss of spine BMD ranged from -1.59% to -3.52% during one year of DMPA use and from -3.44% to -5.99% during two years of use. [32;52;124] On the other hand, untreated adolescent girls in the control groups experienced mean BMD increases of 1.20% to 2.45% over one year and as much as 5.89% over two years. Loss of bone mass in girls who are at the age when they should be experiencing peak bone mass velocity is a grave concern.

There are scant data associating DMPA use with increased risk of fracture. In a study of female Army recruits with a mean age of 21.1±3.7

years, Lappe et al found that, in non-Hispanic white women only, use of DMPA increased the risk of stress fracture during basic training.(RR 1.71, CI 1.01-2.90; p=0.04). [123] However, when adjusted for speed of sound measured with quantitative ultrasound of the heel, DMPA use was no longer a statistically significant risk factor. This further suggests that use of DMPA negatively affects bone mass.

There is some evidence that BMD is recovered after DMPA is discontinued. In a three-year prospective study of DMPA users aged 18-39 at baseline, discontinuers had significant increases in BMD compared to continuing users. By 30 months post-discontinuation, mean BMD for discontinuers was similar to that of non-users. However, in the subgroup of women aged 18-21, discontinuers still had lower BMD than non-users after 30 months. In this age group, the women on DMPA also had lower baseline BMD than the untreated women of the same age. Thus, concern exists about whether women who start taking DMPA at a young age and continue its use for a considerable period of time might ever reach their full genetic potential for peak bone mass.

Although it is well-established that low doses of estrogen in oral contraceptives are able to maintain bone mass in women in their 30's and 40's, [119] there is scant research to indicate the levels of estrogen needed for women in their teens and 20's. What evidence there is suggests that oral contraceptives containing 30-40 μg of ethinyl estradiol have no negative and perhaps a slight positive effect on BMD in adolescents. [124;138;179] However, there are some preliminary data that suggest that ultra-low doses (containing 20μg ethinyl estradiol) may have a negative impact on bone mass. [41;179] For example, in a study of healthy adult women ages 19-22 using a 20μg oral contraceptive, there was no change in spine BMD over five years. [179] However, mean spine BMD in the control group increased by 7.8% over the observation period. The concern is that endogenous production of estrogen is suppressed and that the exogenous dose of estrogen is inadequate to promote bone mass accrual. Further research is needed to determine the effect of ultra-low doses of oral contraceptives on bone mass accrual.

It should be emphasized that teenage pregnancy often has adverse consequences, including a negative impact on the growing skeleton of the mother. Thus, the risks and benefits of each type of contraceptive need to be weighed. DMPA is a very effective contraceptive used throughout the world. Administered as an injection given quarterly, DMPA is especially useful for adolescents who do not reliably take their pills. In fact, DMPA has been credited for contributing greatly to the decrease in teen pregnancy rates that has occurred during the past decade. [41] At least one study has demonstrated that giving oral estrogen therapy along with DMPA arrested the bone loss associated with use of DMPA in adult women. [43] Thus, concomitant use of oral estrogen along with DMPA may provide for both reliable contraception and protection of bone mass accrual. For teens that choose an oral contraceptive, it may be prudent to use doses of 30-40 μg of ethinyl estradiol until evidence demonstrates that lower doses are safe for the skeletons of growing females.

Those who use DMPA or ultra-low dose oral contraception should be strongly encouraged to adopt a bone healthy lifestyle –optimize calcium intake, assure adequate vitamin D, engage in regular weight-bearing physical activity and avoid smoking. However, it should be noted that such behaviors cannot totally compensate for the bone effect of the absence of estrogen. Young females whose contraceptive use may negatively affect their bone should be monitored with DXA measurements.

11.3. The Female Triad

The female athlete triad is a syndrome consisting of eating disorders/ disordered eating behavior, amenorrhea/oligomenorrhea, and decreased bone mineral density (osteoporosis and osteopenia). [172] The disorder is seen in adolescent and young adult females and is increasing in prevalence as more women are participating in competitive sports/dance. It is seen most frequently in activities that emphasize a lean appearance, such as gymnastics or ballet, and most often these women are underweight. When coupled with inadequate energy intake, the high caloric expenditure of rigorous exercise results in a sustained negative caloric balance that is sensed by the hypothalamus. This initiates a neuroendocrine adaptive cascade associated with changes in the hypothalamic-pituitary-ovarian axis and resulting in decreased estrogen levels. Subsequently, the individual may develop amenorrhea, oligomenorrhea or anovulation. The low estrogen levels may lead to impaired bone accretion as well as increased risk of fractures. It has been shown that in athletes the most severe loss of BMD occurs in women who are amenorrheic, while there is less severe loss in women with oligomenorrha, and still less loss in those with normal menstrual function and normal body weight. [49] Since the precipitating factor of the female triad is increased energy expenditure with inadequate caloric intake, optimizing the nutritional status of female athletes would serve to prevent the syndrome.

Early recognition of this syndrome and prompt intervention will help to minimize both the short-term and the long-term consequences. The treatment strategy includes reducing the intensity of training until menses resume, increasing caloric intake, ensuring optimal calcium and vitamin D, encouraging weight-bearing activity as is appropriate, and considering hormone replacement therapy. [74]

12. Summary

In summary, optimization of peak bone mass is critical for prevention of osteoporosis. Strong evidence supports the efficacy of bone healthy behaviors, particularly calcium-rich diets, optimal vitamin D, and weight-bearing exercise, in assuring that the skeleton achieves its full genetic potential. Unfortunately, many children and adolescents are not engaged in bone healthy behaviors. Attention of policy makers, school administrators, parents, health professionals, and other groups that influence children is needed to establish environments that strongly encourage young persons to adopt and maintain bone healthy behaviors. Only then will the epidemic of osteoporosis be stopped.

References

1. Consensus development conference: Diagnosis, prophylaxis, and treatment of osteoporosis. 1-10. 1993. Ref Type: Report
2. Assessing women's perceived benefits, barriers, and stage of change for meeting milk product consumption recommendations. *J Am Diet Assoc* 2001;**101**:1354-7.

3. Prevention of rickets and vitamin D deficiency: New guidelines for vitamin D intake. *Pediatrics* 2003;**111**:908-10.
4. Abrams S, Grusak M, Stuff J, O'Brien K. Calcium and magnesium balance in 9-14-y-old children. *Am J Clin Nutr* 1997;**66**:1172-7.
5. Abrams S, O'Brien K, Liang L, Stuff J. Differences in calcium absorption and kinetics between Black and White girls aged 5-16 years. *J.Bone Miner.Res.* 1995;**10**:829-33.
6. Abrams S,.Stuff J. Calcium metabolism in girls: Current dietary intakes lead to low rates of calcium absorption and retention during puberty. *Am.Soc.Clin.Nutr.* 1994;**60**:739-43.
7. Afghani A, Xie B, Wiswell R, Gong J, Li Y, Johnson C. Bone mass of asian adolescents in china: influence of physical activity and smoking. *Med Sci Sports Exerc* 2003;**35**:720-9.
8. Alffram P,.Bauer G. Epidemiology of fracture of the forearm. *J Bone Jt Surg* 1962;**44A**:105.
9. Allensworth D. Health education: State of the art. *J.Sch.Health* 1993;**63**:14-20.
10. Ames S, Ellis K, Gunn S, Copeland K, Abrams S. Vitamin D receptor gene fok1 polymorphism predicts calcium absorption and bone mineral density in children. *J Bone Miner Res* 1999;**14**:740-6.
11. Arikoski P, Komulainen J, Riikonen P, Parviainen M, Jurvelin J, Voutilainen R *et al.* Impaired development of bone mineral density during chemotherapy: A prospective analysis of 46 children newly diagnosed with cancer. *J Bone Miner Res* 1999;**14**:2002-9.
12. Arikoski P, Komulainen J, Rikonen P, Voutilainen R, Knip M, Kroger H. Alterations in bone turnover and impaired development of bone mineral density in newly diagnosed children with cancer: a 1-year prospective study. *J Clin Endocrinol Metab* 1999;**84**:3174-81.
13. Atkinson S, Halton J, Bradley C, Wu B, Barr R. Bone and mineral abnormalities in childhood acute lymphoblastic leukemia: Influence of disease, drugs and nutrition. *Int J Cancer* 1998;**Supplement 11**:35-9.
14. Bachrach L. Bone acquisition in childhood and adolescence. In Marcus R, Feldman D, Kelsey J, eds. *Osteoporosis*, pp 69-106. San Diego: Academic Press, 1996.
15. Bachrach L, Hastie T, Wang M, Balasubramanian N, Marcus B. Bone mineral acquisition in healthy Asian, Hispanic, Black, and Caucasian youth: A longitudinal study. *J Clin Endocrinol Metab* 1999;**84**:4702-12.
16. Badenhop, N., Ilcih, J., Skugor, M., Landoll, J., and Matkovic, V. Changes in body composition and serum leptin in young females with high vs. low dairy intake. J Bone Miner Res 12 (Suppl 1), s487. 1997.

Ref Type: Abstract

17. Bailey D, Baxter-Jones A, Mirwald R, Faulkner R. Bone growth and exercise studies: The complications of maturation. *J Musculoskei Neuron Interact* 2003;**3**:335-7.

18. Bailey, D., Faulkner, R., and McKay, H. Growth, physical activity, and bone mineral acquisition. J.Holloszy. (8), 233-266. 1996. Baltimore, Williams & Wilkins. Exercise and Sport Sciences Reviews.

Ref Type: Serial (Book,Monograph)

19. Bailey DA, Martin A, McKay H, Whiting S, Mirwald R. Calcium accretion in girls and boys during puberty: A longitudinal analysis. *J Bone Miner Res* 2000;**15**:2245-50.
20. Barr S. Dieting attitudes and behavior in urban high school students: Implications for calcium intake. *J Adolesc Health* 2004;**16**:458-64.
21. Bass S, Pearce G, Bradney M, Hendrich E, Delmas P, Harding A *et al.* Exercise before puberty may confer residual benefits in bone density in adulthood: studies in active prepubertal and retired female gymnasts. *J Bone Miner Res* 1998;**13**:500-7.
22. Bass S, Saxon L, Daly R, Turner C, Robling A, Seeman E *et al.* The effect of mechanical loading on the size and shape of bone in pre-, peri-, and postpubertal girls: A study in tennis players. *J Bone Miner Res* 2002;**17**:2274-80.
23. Bell N, Shary J, Stevens J, Garza M, Gordon L, Edwards J. Demonstration that bone mass is greater in black than in white children. *J.Bone Miner.Res.* 1991;**6**:719-23.
24. Bianchil M, Cimaz R, Bardare M, Zulian F, Lepore L, Boncompagni A *et al.* Efficacy and safety of alendronate for the treatment of osteoporosis in diffuse connective tissue diseases in children. *Arthritis Rheum.* 2000;**43**:1960-6.
25. Biser-Rohrbaugh A,.Haddley-Miller N. Vitamin D deficiency in breast-fed toddlers. *J Pediatr Orthop* 2001;**21**:508-11.
26. Boechat M, Mora S, Loro M, Roe TGW, Gilsanz V. Gender differences in vertebral bone mass in children and adolescents. *10th International Bone Densitometry Workshop* 1993.
27. Bonjour J, Theintz G, Buchs B, Slosman D, Rizoli R. Critical years and stages of puberty for spinal and femoral bone mass accumulation during adolescence. *J.Clin.Endocrinol.Metabol.* 1991;**73**:555-63.
28. Boot A, de Ridder M, Pols H, Krenning E, Keizer-Schrama S. Bone mineral density in children and adolescents: relation to puberty, calcium intake, and physical activity. *J.Clin.Endocrinol.Metab.* 1997;**82**:57-62.
29. Boot A, de Ridder M, Pols H, Krenning E, Keizer-Schrama S. Bone mineral density in children and adolescents: relation to puberty, calcium intake, and physical activity*. *J Clin Endocrinol Metab* 1997;**82**:57-62.
30. Bouillon R, Burckhardt P, Christiansen C, Fleisch H, Fujita T, Gennari C *et al.* Consensus development conference: Prophylaxis and treatment of osteoporosis. *Osteoporosis Int.* 1991;**1**:114-7.
31. Brik R, Keidar Z, Schapira D, Israel O. Bone mineral density and turnover in children with systemic juvenile chronic arthritis. *J of Rheumaology* 1998;**25**:990-2.
32. Busen N, Britt R, Rianon N. Bone mineral density in a cohort of adolescent women using depot medroxyprogesterone acetate for one to two years. *J Adolesc Health* 2003;**32**:257-9.

33. Cadogan J, Eastell R, Jones N, Barker ME. Milk intake and bone mineral acquisition in adolescent girls: randomized, controlled intervention trial. *BMJ* 1997;**315**:1255-60.
34. Chan G, Hoffman K, McMurray M. The effect of dietary calcium supplementation on pubertal girls' growth and bone mineral status. <*None Specified*> 1995.
35. Chan G, Hoffman K, McMurry M. Effects of dairy products on bone and body composition in pubertal girls. *J Pediatr* 1995;**126**:551-6.
36. Chang S, Ge K, Zhai F, Jia F. Dietary intake of some micronutrients in different provinces of China. *Wei Sheng Yan Jiu* 1999;**28**:364-6.
37. Chapuy M, Prezoisi P, Maamer M. Prevalence of vitamin D insufficiency in an adult normal population. *Osteoporos Int.* 1997;**7**:439-43.
38. Clark, K. and Sowers, M. Alcohol dependence, smoking status, reproductive characteristics and bone mineral density in premenopausal women. 2001.

Ref Type: Unpublished Work

39. Conroy B, Kraemer W, Maresh C, Fleck S, Stone M, Fry A *et al.* Bone mineral density in elite junior Olympic weightlifters. *Med.Sci.Sports Exerc.* 1993;**25**:1103-9.
40. Cooper C, Cawley M, Bhalla A, Egger P, Ring F, Morton L *et al.* Childhood growth, physical activity, and peak bone mass in women. *J.Bone Miner.Res.* 1995;**10**:940-7.
41. Cromer B. Bone mineral density in adolescent and yound adult women on injectable or oral contraception. *Curr Opin Obstet Gynecol* 2003;**15**:353-7.
42. Cummings C, Black D, Nevitt M, Browner W, Cauley J, Ensurd K *et al.* bone density at various sites for prediction of hip fractures. *Lancet* 1993;**341**:72-5.
43. Cundy T, Ames R, Horne A, Clearwater J, Roberts H, Gamble G *et al.* A randomized controlled trial of estrogen replacement therapy in long-term users of depot medroxyprogesterone acetate. *J Clin Endocrinol Metab* 2003;**88**:78-81.
44. Cundy T, Cornish J, Roberts H, Elder H, Reid I. Spinal bone density in women using depot medroxyprogesterone contraception. *Obstet.Gynecol.* 1998;**92**:569-73.
45. Davies K, Heaney R, Recker R, Lappe J, Barger-Lux J, Rafferty K *et al.* Calcium intake and body weight. *J Clin Endocrinol Metab* 2000;**85**:4635-8.
46. Delvaux K, Lefevre J, Philippaerts R, Dequeker J, Thomis M, Vanreusel B *et al.* Bone mass and lifetime physical activity in Flemish males: a 27-year follow-up study. *Med Sci Sports Exerc* 2001;1868-75.
47. Dent, C. Problems in metabolic bone disease. 1-7. 1972. Detroit, Henry Ford Hospital. Clinical aspects of metabolic bone disease. Proceedings of the international symposium on clinical aspects of metabolic bone disease. Frame, B., Parfitt, A., and Duncan, H. Ref Type: Conference Proceeding
48. Docio S, Riancho J, Perez A, Olmos J, Amado J, Gonzalez-Macias J. Seasonal deficiency of vitamin D in children: A potential target for osteoporosis-preventing strategies? *J Bone Miner Res* 1998;**13**:544-8.
49. Drinkwater B, Bruemner B, Chesnut C. Menstrual history as a determinant of current bone density in young athletes. *J.A.M.A.* 1990;**263**:545-8.

50. Du X, Greenfield H, Fraser D, Ge K, Trube A, Wang Y. Vitamin D deficiency and associated factors in adolescent girls in Beijing. *Am J Clin Nutr* 2001;**74**:494-500.
51. Duppe H, Gardsell P, Johnell O, Ornstein E. Bone mineral density in female junior, senior and former football players. *Osteoporosis Int* 1996;**6**:437-41.
52. Edwards, C., Hertweck, S., and Perlman, S. A prospective study evaluating the effects of Depo Provera on bone mineral density in adolescent females: a preliminary report. J Pediat Adol Gynecol 11, 201. 1998. Ref Type: Abstract
53. Elgan C, Samsioe G, Dykes A-K. Influence of smoking and oral contraceptives on bone mineral density and bone remodeling in young women: a 2-year study. *Contraception* 2003;**67**:439-47.
54. Faulkner R, Forwood M, Beck T, Mafukidze J, Russell K, Wallace W. Strength indices of the proximal femur and shaft in prepubertal female gymnasts. *Med Sci Sports Exerc* 2002;513-8.
55. Ferrari S, Manen D, Bonjour J, Slosman D, Rizzoli R. Bone mineral mass and calcium and phosphate metabolism in young men: Relationships with vitamin D receptor allelic polymorphisms. *J Clin Endocrinol Metab* 1999;**84**:2043-8.
56. Ferrari S, Rizzoli R, Manen D, Slosman D, Bonjour J. Vitamin D receptor gene start codon polymorphisms (FokI) and bone mineral density: interaction with age, dietary calcium, and 3'-end region polymorphisms. *J Bone Miner Res* 1998;**13**:925-30.
57. Ferrari S, Rizzoli R, Slosman D, Bonjour J. Do dietary calcium and age explain the controversy surrounding the relationship between bone mineral density and vitamin D receptor gene polymorphisms? *J Bone Miner Res* 1998;**13**:363-70.
58. Ferrari S, Rizzoli R, Slosman D, Bonjour J. Familial resemblance for bone mineral mass is expressed before puberty. *J Clin Endocrinol Metab* 1998;**83**:358-61.
59. Forwood M,.Burr D. Physical activity and bone mass: Exercises in futility? *Bone Miner.* 1993;**21**:89-112.
60. Fournier P, Rizzoli R, Slosman D, Theintz G, Bonjour J. Asynchrony between the rates of standing height gain and bone mass accumulation during puberty. *Osteoporosis Int* 1997;**7**:525-32.
61. Fox K, Magaziner J, Sherwin R, Scott J, Plato C, Nevitt M *et al.* Reproductive correlates of bone mass in elderly women. *J.Bone Miner.Res.* 1993;**8**:901-8.
62. Frost H. The role of changes in mechanical usage set points in the pathogenesis of osteoporosis. *J.Bone Miner.Res.* 1992;**7**:253-61.
63. Frost H. Why do long-distance runners not have more bone? A vital biomechanical explanation and an estrogen effect. *J Bone Miner Metab* 1997;**15**:9-16.
64. Fuchs R, Bauer JJ, Snow CM. Jumping improves hip and lumbar spine bone mass in prepubescent children: A randomized controlled trial. *Bone and Mineral Research* 2001;**16**:148-56.
65. Gabrielsson B, Carmignac D, Flavell D, et al. Steroid regulation of growth hormone (GH) receptor and GH-binding protein messenger ribonucleic acids in the rat. *Endocrinology* 1995;**136**:209-17.
66. Garn S. The earlier gain and the later loss of cortical bone. Springfield, IL: C. Thomas, 1970.

67. Ge K,.Chang S. Dietary intake of some essential micronutrients in China. *Biomed Environ Sci* 2001;**14**:318-24.
68. Geller G, Botkin J, Green MJ, Press N, Biesecker BB, Wilfond B *et al.* Genetic testing for susceptibility to adult-onset cancer. *JAMA* 1997;**277**:1467-74.
69. Gilchrist, N., Smart, E., Turner, J., Hooke, E., March, R., and Frampton, C. Dairy food supplementation in teenage girls: Effects on bone density, biochemistry and dietary profile. Osteoporosis Int 6(Supp. 1), 217. 1996. Ref Type: Abstract
70. Gilsanz V, Gibbens D, Roe T, Carlson M, Senac M, Boechat M *et al.* Vertebral bone density in children: Effect of puberty. *Radiology* 1988;**166**:847-50.
71. Gilsanz V, Skaggs D, Kovanlikaya A, Sayre J, Loro M, Kaufman F *et al.* Differential effect of race on the axial and appendicular skeletons of children. *J Clin Endocrinol Metab* 1998;**83**:1420-7.
72. Gilzanz V, Roe T, Stefano M, Costin G, Goodman W. Changes in vertebral bone density in black girls and white girls during childhood and puberty. *New Engl.J.Med.* 1991;**325**:1597-600.
73. Glorieux F, Bishop N, Plotkin H, Chabot G, Lanoue G, Travers R. Cyclic administration of pamidronate in children with severe osteogenesis imperfecta. *New Engl.J.Med.* 1998;**339**:947-52.
74. Golden N. A review of the female athlete triad (amenorrhea, osteoporosis and disordered eating). *Int J Adolesc Med Health* 2002;**14**:9-17.
75. Goulding A, Cannan R, Williams SM, Gold EJ, Taylor RW, Lewis-Barned N. Bone mineral density in girls with forearm fractures. *J Bone Miner Res* 1998;**13**:143-8.
76. Goulding A, Jones L, Taylor R, Manning P, Williams SM. More broken bones: A 4-year double cohort study of young girls with and without distal forearm fractures. *J Bone Miner Res* 2000;**15**:20112018.
77. Grimston S, Willows N, Hanley D. Mechanical loading regime and its relationship to bone mineral density in children. *Off.J.Am.Coll.Sports Med.* 1993;1203-9.
78. Guillemant J, Taupin P, Le H, Taright N, Allemandou A, Peres G *et al.* Vitamin D status during puberty in french healthy male adolescents. *Osteo Int.* 1999;**10**:222-5.
79. Gunnes M, Berg J, Halse J, Lehmann E. Lack of relationship between vitamin D receptor genotype and forearm bone gain in healthy children, adolescents and young adults. *J Clin Endocrinol Metab* 1997;**83**:851-5.
80. Gutin B,.Kasper M. Can vigorous exercise play a role in osteoporosis prevention? A review. *Osteoporosis Int.* 1992;**2**:55-69.
81. Haapasalo H, Kannus P, Sievanen H, Heinonen A, Oja P, Vuori I. Long-term unilateral loading and bone mineral density and content in female squash players. *Calcif.Tissue Int.* 1994;**54**:249-55.
82. Haapasalo H, Kontulainen S, Sievanen H, Kannus P, Jarvinen M, Vuori I. Exercise-induced bone gain is due to enlargement in bone size without a change in volumetric bone density: A peripheral quantitative computed tomography study of the upper arms of male tennis players. *Bone* 2000;**27**:351-7.

83. Haapasalo H, Sievanen H, Kannus P, Oja P, Vuori I. Dimensions and estimated mechanical characteristics of the humerus after long-term tennis loading. *J Bone Miner Res* 1996;**11**:864-72.
84. Hartman J. Vitamin Deficiency rickets in children: prevalence and need for community education. *Orthop Nurs* 2000;**19**:63-8.
85. Heaney RP. Calcium, dairy products and osteoporosis. *J Am Coll Nutr* 2000;**19**:83S-99S.
86. Heaney R, Abrams S, Dawson-Houghes B, Looker A, Marcus B, Matkovic V *et al*. Peak Bone Mass. *Osteoporos Int.* 2000;**11**:985-1009.
87. Heinonen A, Sievanen H, Kannus P, Pasanen M, Vuori I. High-impact exercise and bones of growing girls: A 9-month controlled trial. *Osteoporos Int.* 2000;**11**:1010-7.
88. Helge E,.Kanstrup I. Bone density in female elite gymnasts:: impact of muscle strength and sex hormones. *Med Sci Sports Exerc* 2002;174-80.
89. Henderson C, Cawkwell G, Specker B, Sierra R, Wilmott R, Campaigne B *et al.* Predictors of total body bone mineral density in non-corticosteroid-treated prepubertal children with juvenile rheumatoid arthritis. *Arthritis Rheum.* 1997;**40**:1967-75.
90. Henderson C, Specker B, Siervogel R, Campaigne B, Lovell D. Total-body bone mineral content in non-corticosteroid-treated postpubertal females with juvenile rheumatoid arthritis. *Arthritis Rheum.* 2000;**43**:531-40.
91. Ho M, Yen H, Tsang R. Randomized study of sunshine exposure and serum 25-OHD in breast-fed infants in Beijing, China. *J Pediatr* 1995;**107**:928-31.
92. Holick M. Vitamin D: photobiology, metabolism, mechanism of action, and clinical applicatons. In M.Favus, ed. *Primer on the Metabolic Bone Diseases and Disorders of Bone Metabolism*, pp 92-8. Philadelphia: Lippincott-Raven, 1999.
93. Holick M. Vitamin D: the underappreciated D-lightful hormone that isimportant for skeletal and cellular health. *Current Opinion in Endocrinology & Diabetes* 2002;**9**:87-98.
94. Holick M, Shao Q, Liu W. The vitamin D content of fortified milk and infant formula. *N Engl J Med* 1992;**326**:1178-81.
95. Hopp R, Degan J, Biven R, Kinberg K, Gallagher C. Longitudinal assessment of bone mineral density in children with chronic asthma. *Ann of Allergy, Asthma,& Immunology* 1995;**75**:143-8.
96. Horlick M, Rosenbaum M, Nicolson M, Levine L, Fedun B, Wang J *et al.* Effect of puberty on the relationship between circulating leptin and body composition. *J Clin Endocrinol Metab* 2000;**85**:2509-18.
97. Hu J, Zhao X, Jia J, Parpia B, Campbell T. Dietary calcium and bone density among middle-aged and elderly women in China. *Am.J.Clin.Nutr.* 1993;**58**:219-27.
98. Hui S, Slemenda C, Johnston C. Age and bone mass as predictors of fracture in a prospective study. *J.Clin.Invest.* 1988;**81**:1804-9.
99. Ilich J, Skugor M, Badenhop N, Landoll J, Matkovic V. Time since menarche is positively related to bone mass of total body and radius in premenopausal women. *J Bone Miner Res* 1997;**12**:S252.

100. Ito M, Yamada M, Hayashi K, Ohki M, Uetani M, Nakamura T. Relation of early menarche to high bone mineral density. *Calcif.Tissue Int.* 1995;**57**:11-4.
101. Iuliano-Burns S, Saxon L, Naughton G, Gibbons K, Bass S. Regional specificity of exercise and calcium during skeletal growth in girls: a randomized controlled trial. *J Bone Miner Res* 2003;**18**:156-62.
102. Jackman L, Millane S, Martin B, Wood O, McCabe G, Peacock M *et al.* Calcium retention in relation to calcium intake and post menarcheal age in adolescent females. *Am J Clin Nutr* 1997;**66**:327-33.
103. Jacqmain M, Doucet E, Despres J, Bouchard C, Tremblay A. Calcium intake, body composition, and lipoprotein-lipid concentrations in adults. *Am J Clin Nutr* 2003;**77**:1448-52.
104. Jaffe C, Snow-Harter C, Connolly D, Robinson T, Marcus R. Differential effects of swimming versus weight-bearing activity on bone mineral status of eumenorrheic athletes. *J Bone Miner Res* 1995;**10**:586-93.
105. Jimi E, Nakamura I, Amano H. Osteoclast function is activated by osteoblastic cells through a mechanism involving cell-to-cell contact. *Endocrinology* 1996;**137**:2187-90.
106. Johnell O, Gullberg B, Kanis J, Allander E, Elffors L, Dequeker J *et al.* Risk factors for hip fracture in European women: the MEDOS study. *J.Bone Miner.Res.* 1995;**10**:1802-15.
107. Johnston C, Miller J, Slemenda C, Reister T, Hui S, Christian J *et al.* Calcium supplementation and increases in bone mineral density in children. *New Engl.J.Med.* 1992;**327**:82-7.
108. Joiner T, Foster C, Shope T. The many faces of vitamin D deficiency rickets. *Ped in Review* 2000;**21**:296-302.
109. Jones G,.Dwyer T. Bone mass in prepubertal children: gender differences and the role of physical activity and sunlight exposure. *J Clin Endocrinol Metab* 1998;**83**:4274-9.
110. Kanders B, Dempster D, Lindsay R. Interaction of calcium nutrition and physical activity on bone mass in young women. *J.Bone Miner.Res.* 1988;**3**:145-9.
111. Kannus P, Haapasalo H, Sankelo M, Sievanen H, Pasanen M, Heinonen A *et al.* Effect of starting age of physical activity on bone mass in the dominant arm of tennis and squash players. *Ann.Intern.Med.* 1995;**123**:27-31.
112. Kaste S, Chesney R, Hudson R, Rose S, Carbone L. Editorial: Bone mineral status during and after therapy of childhood cancer: An increasing population wtih multiple risk factors for impaired bone health. *J Bone Miner Res* 2000;**14**:2010-4.
113. Kemper H, Twisk J, Mechelen W, Post G, Roos J, Lips P. A fifteen-year longitudinal study in young adults on the relation of physical activity and fitness with the development of the bone mass: the Amsterdam growth and health longitudinal study. *Bone* 2000;**27**:847-53.
114. Khosla S, Melton III L, Dekutoski M, Achenback S, Oberg A, Riggs B. Incidence of childhood distal forearm fractures over 30 years. *JAMA* 2003;1479-85.
115. Kitagawa L, Kitagawa Y, Kawase Y, Nagaya T, Tokudome S. Advanced onset of menarche and higher bone mineral density depending on vitamin D receptor gene polymorphism. *Eur J Endocrinol* 1998;**139**:522-7.

116. Kontulainen S, Kannus P, Haapasalo H, Sievanen H, Pasanen M, Heinonen A *et al*. Good maintenance of exercise-induced bone gain with decreased training of female tennis and squash players: A prospective 5-year follow-up study of young and old starters and controls. *J Bone Miner Res* 2001;**16**:195-201.
117. Kontulainen S, Sievanen S, Kannus P, Pasanen M, Vuori I. Effect of long-term impact-loading on mass, size, and estimated strength of humerus and radius of female racquet-sports players: A peripheral quantitative computed tomography study between young and old starters and controls. *J Bone Miner Res* 2002;**17**:2281-9.
118. Krabbe S, Hummer L, Christiansen C. Longitudinal study of calcium metabolism in male puberty. *Act Paediatr Scand* 1984;**73**:750-5.
119. Kuohung W, Borgatta L, Stubblefield P. Low-dose oral contraceptives and bone mineral density: an evidence-based analysis. *Contraception* 2000;**61**:77-82.
120. Laing E, Massoni J, Nickols-Richardson S, Modlesky C, O'Connor P, Lewis R. A prospective study of bone mass and body composition in female adolescent gymnasts. *J Pediatr* 2002;**141**:211-6.
121. Landin L,.Nilsson B. Bone mineral content in children with fractures. *Clin.Orthop.* 1983;**178**:292-6.
122. Lappe J, Rafferty K, Lypaczewski G, Haynatzki G, Davies M. Girls on a high calcium diet gain weight at the same rate as girls on a normal diet: A pilot study. *J of Amer Dietetic Asso.* 2004;**In press**.
123. Lappe J, Stegman M, Recker R. The impact of lifestyle factors on stress fractures in female Army recruits. *Osteoporosis Int* 2001;**12**:35-42.
124. Lara-Toxe, E., Edwards, C., Pelman, S., and Hetweile, P. Prospective study evaluating bone mineral density in adolescent females using Depo-Provera (DMPA). 56. 2003.
Ref Type: Conference Proceeding
125. Lee W, Cheng J, Jiang J, Hu P, Hu X, Roberts D. Calcium absorption measured by stable calcium isotopes (^{42} Ca & ^{44} Ca) among northern Chinese adolescents with low vitamin D status. *J Orthop Surg* 2002;**10**:61-6.
126. Lee W, Leung S, Leung D, Cheng J. A follow-up study on the effect of calcium-supplement withdrawal and puberty on bone acquisition in children. *Am J Clin Nutr* 1996;**64**:71-7.
127. Lee W, Leung S, Leung D, Tsang H, Lau J, Cheng J. A randomized doubel-blind controlled calcium supplementation trial, and bone mineral accretion in children accustomed to a low-calcium diet. *Br J Nutr* 1995;**74**:125-39.
128. Lee W, Leung S, Lui S. Relationship between long-term calcium intake and bone mineral content of chiildren from birth to 5 years. *Br J Nutr* 1993;**70**:235-48.
129. Lee W, Leung S, Wang S, Xu Y, Zeng W, Lau J *et al*. Double-blind, controlled calcium supplementation and bone mineral accretion in children accustomed to a low-calcium diet. *Am.J.Clin.Nutr.* 1994;**60**:744-50.
130. Lehtonen-Veromaa M, Mottonen T, Nuotio I, Irjala K, Leino A, Viikari J. Vitamin D and attainment of peak bone mass among peripubertal Finnish girls: a 3-y prospective study. *Am J Clin Nutr* 2002;**76**:1446-53.

131. Lehtonen-Veromaa M, Mottonen T, Svedstrom E, Hakola P, Heinonen O, Viikari J. Physical activity and bone mineral acquisition in peripubertal girls. *Scand J Med Sci Sports* 2000;**10**:236-43.

132. Leib, E., Lewiecki, E., Binkley, N., and Hamdy, R. Official Positions: The International Society for Clinical Densitometry. The International Society for Clinical Densitometry 7. 2004.

Ref Type: Journal (Full)

133. Leonard M, Propert K, Zemel B, Stallings V, Feldman B. Discrepancies in pediatric bone mineral density reference data: Potential for misdiagnosis of osteopenia. *J Pediatr* 1999;**135**:182-8.

134. Li J, Specker B, Ho M, Tsang R. Bone mineral content in black and white children 1 to 6 years of age. Early appearance of race and sex differences. *Am J Dis Child* 1989;**143**:1346-9.

135. Lin Y, Lyle R, McCabe L, McCabe G, Weaver C, Teegarden D. Dairy calcium is related to changes in body composition during a two-year exercise intervention in young women. *J Am Coll Nutr* 2000;**19**:754-60.

136. Lips P. Vitamin D deficiency and secondary hyperparathyroidism in the elderly: Consequences for bone loss and fractures and therapeutic implications. *Endocrine Reviews* 2001;**22**:477-501.

137. Lloyd T, Rollings N, Andon M, Eggli D, Maugeri D, Chinchilli V. Enhanced bone gain in early adolescence due to calcium supplementation does not persist in late adolescence. *J Bone Miner Res* 1996;**11**:S154.

138. Lloyd T, Taylor D, Lin H, Matthews A, Eggli D, Legro R. Oral contraceptive use by teenage women does not affect peak bone mass: a longitudinal study. *Fertil.Steril.* 2000;**74**:734-8.

139. Lofman O, Larsson L, Ross I, Toss G, Berglund K. Bone mineral density in normal Swedish women. *Bone* 1997;**20**:167-74.

140. Looker A, Dawson-Hughes B, Calvo M, Gunter E, Sahyoun N. Serum 25-hydroxyvitamin D status of adolescents and adults in two seasonal subpopulations from NHANES III. *Bone* 2002;**30**:771-7.

141. Lorentzon M, Lorentzon R, Nordstrom P. Vitamin D receptor gene polymorphism is related to bone density, circulating osteocalcin, and parathyroid hormone in healthy adolescent girls. *Bone Miner Metab* 2001;**19**:302-7.

142. Loro M, Sayre J, Roe T, Goran M, Kaufman F, Gilsanz V. Early identification of children predisposed to low peak bone mass and osteoporosis later in life. *J Clin Endocrinol Metab* 2000;**85**:3908-18.

143. MacKelvie K, Khan K, Petit MA, Janssen P, McKay H. A school-based exercise intervention elicits substantial bone health benefits: a 2-year randomized controlled trial in girls. *Pediatrics* 2003;**112**:e447-e452.

144. MacKelvie K, Petit MA, Khan K, Beck T, McKay H. Bone mass and structure are enhanced following a 2-year randomized controlled trial of exercise in prepubertal boys. *Bone* 2004;**34**:755-64.

145. Matkovic V. Calcium metabolism and calcium requirements during skeletal modeling and consolidation of bone mass. *Am.J.Clin.Nutr.* 1991;**54**:245S-60S.

146. Matkovic V. Calcium intake and peak bone mass. *New Engl.J.Med.* 1992;**327**:119-20.
147. Matkovic V, Fontana D, Tominac L, Prem G, and Chesnut C. Factors that influence peak bone mass formation: A study of calcium balance and the inheritance of bone mass in adolescent females. *Am.J.Clin.Nutr.* 1990;**52**:878-88.
148. Matkovic V,.Heaney R. Calcium balance during human growth: evidence for threshold behavior. *Am.J.Clin.Nutr.* 1992;**55**:992-6.
149. Matkovic V, Ilich J, Skugor M. Calcium intake and skeletal formation. In P.Burchhardt, R.Heaney, eds. *Nutritional aspects of osteoporosis '94*, pp 129-45. Rome, Italy: Ares-Serono Symposia Pub., 1995.
150. Matkovic V, Jelic T, Wardlaw G, Ilich J, Goel P, Wright J *et al.* Timing of peak bone mass in Caucasian females and its implication for the prevention of osteoporosis: Inference from a cross-sectional model. *J.Clin.Invest.* 1994;**93**:799-808.
151. Matkovic V, Kostial K, Simonovic I, Buzina R, Brodarec A, Nordin B. Bone status and fracture rates in two regions of Yugoslavia. *Am.J.Clin.Nutr.* 1979;**32**:540-9.
152. Matsuoka L, Ide L, Wortsman J. Sunscreens suppress cutaneous vitamin D3 synthesis. *J Clin Endocrinol Metab* 1987;**64**:1165-8.
153. McCulloch R, Bailey D, Houston C, Dodd B. Effects of physical activity, dietary calcium intake and selected lifestyle factors on bone density in young women. *Can.Med.Assoc.J.* 1990;**142**:221-7.
154. McCulloch R, Bailey D, Whalen R, Houston C, Gaulkner R, Craven B. Bone density and mineral content of adolescent soccer athletes and competitive swimmers. *Pediatr.Exer.Sci.* 1992;**4**:319-30.
155. McKay H, Petit MA, Khan KM, Schutz RW. Lifestyle determinants of bone mineral: A comparison between prepubertal Asian- and Caucasian-Candian bodys and girls. *Calcif Tissue Int* 2000;**66**:320-4.
156. Melton, J. Osteoporosis: A worldwide problem. In the Proceedings of the Third International Symposium: Osteoporosis--Research advances and clinical applications. 23. 1994.
Ref Type: Conference Proceeding
157. Merrilees M, Smart E, Gilchrist N, Frampton C, Turner J, Hooke E *et al.* Effects of dairy food supplements on bone mineral density in teenage girls. *Eur J Nutr* 2000;**39**:256-62.
158. Mora S, Weber G, Marenzi K, Signorini E, Rovelli R, Proverbio M *et al.* Longitudinal changes of bone density and bone resorption in hyperthyroid girls during treatment. *J Bone Miner Res* 1999;**14**:1971-7.
159. Morrison N, Cheng QI J, Tokita A, Kelly P, Crofts L, Nguyen T *et al.* Prediction of bone density from vitamin D receptor alleles. *Nature* 1994;**367**.
160. National Academy of Science. Dietary reference intakes for calcium, magnesium, phosphorus, vitamin D, and fluoride. Washington, D.C.: Food and Nutrition Board, Institute of Medicine. National Academy Press, 1999.

161. Nelson D, Jacobsen G, Barondess D, Parfitt A. Ethnic differences in regional bone density, hip axis length, and lifestyle variables among healthy Black and White men. *J.Bone Miner.Res.* 1995;**10**:782-7.
162. Nelson D, Simpson PM, Johnson CC, Barondess D, Kleerekoper M. The accumulation of whole body skeletal mass in third- and fourth-grade children: effects of age, gender, ethnicity, and body composition. *Bone* 1997;**20**:73-8.
163. Neumark-Sztainer D, Story M, Dixon L, Resnick M, Blum R. Correlates of inadequate consumption of dairy products among adolescents. *JNE* 1997;**29**:12-20.
164. Nguyen T, Maynard L, Towne B, Roche A, Wisemandle W, Li J *et al.* Sex differences in bone mass acquisition during growth. *J Clin Dens* 2001;**4**:147-57.
165. Nichols D, Sanborn C, Bonnick S, Gench B, DiMarco N. Relationship of regional body composition to bone mineral density in college females. *Med.Sci.Sports Exerc.* 1995;**27**:178-82.
166. NIH Consensus Development Panel on Optimal Calcium Intake. Optimal calcium intake. *J.A.M.A.* 1994;**272**:1942-8.
167. Nishioka T, Kurayama H, Yasuda T, Udagawa JMcNH. Nasal administration of salmon calcitonin for prevention of glucocorticoid-induced osteoporosis in children with nephrosis. *J Pediatr* 1991;**118**:703-8.
168. Nordstrom P,.Lorentzon R. Site-specific bone mass differences of the lower extremities in 17-year-old ice hockey players. *Calcif Tissue Int* 1996;**59**:443-8.
169. Nowson C, Green R, Hopper J, Sherwin A, Young D, Kaymakci B *et al.* A co-twin study of the effect of calcium supplementation on bone density during adolescence. *Osteoporosis Int.* 1997;**7**:219-25.
170. O'Neill T, Silman A, Naves Diaz M, Cooper C, Kanis J, Felsenberg D *et al.* Influence of hormonal and reproductive factors on the risk of vertebral density in european women. *Osteoporosis Int* 1997;**7**:72-8.
171. Packard P,.Krogstrand K. Half of rural girls aged 8 to 17 years report weight concerns and dietary changes, with both more prevalent with increased age. *J Am Diet Assoc* 2002;**102**:672-7.
172. Papanek P. The female athlete triad: an emerging role for physical therapy. *J Orthop Sports Phys Ther* 2003;**33**:594-614.
173. Parfitt A. The two faces of growth: Benefits and risks to bone integrity. *Osteoporosis Int.* 1994;**4**:382-98.
174. Patel D, Pettifor J, Becker P, Grieve C, Leschner K. The effect of ethnic group on appendicular bone mass in children. *J Bone Miner Res* 1992;**7**:263-72.
175. Perez M, Abrams S, Loddeke L, Shypailo R, Ellis K. Effects of rheumatic disease and corticosteroid treatment on calcium metabolism and bone density in children assessed one year after diagnosis, using stable isotopes and dual energy x-ray absorptiometry. *J of Rheumaology* 2000;**27**:38-43.
176. Petit MA, McKay H, MacKelvie K, Heinonen A, Khan K, Beck T. A randomized school-based jumping intervention confers site and maturity-specific benefits on bone structural properties in girls: a hip structural analysis study. *J Bone Miner Res* 2002;**17**:363-72.

177. Picard D, Ste-Marie L, Coutu D. Premenopausal bone mineral content relates to height, weight, and calcium intake during early adulthood. *Bone Miner.* 1988;**4**:299-309.
178. Pocock N, Eisman J, Hopper J. Genetic determinants of bone mass in adults: A twin study. *J.Clin.Invest.* 1987;**80**:706-10.
179. Polatti F, Perotti F, Filippa N, Gallina D, Nappi R. Bone mass and long-term monophasic oral contraceptive treatment in young women. *Contraception* 1995;**51**:221-4.
180. Prentice A, Laskey M, Shaw J, Cole T, Fraser D. Bone mineral content of Gambian and British children aged 0-36 months. *Bone Miner.* 1990;**10**:211-24.
181. Prince R, Devine A, Dick I, Criddle A, Kerr D, Kent N *et al.* The effects of calcium supplementaion (milk powder or tablets) and exercise on bone density in post menopausal women. *J.Bone Miner.Res.* 1995;**10**:1068-75.
182. Prince R, Smith M, Dick I, Price R, Webb P, Henderson K *et al.* Prevention of postmenopausal osteoporosis. *N Engl J Med* 1991;**325**:1189-95.
183. Puntila E, Kroger H, Lakka T, Honkanen R, Tuppurainen M. Physical activity in adolescence and bone density in peri- and postmenopausal women: a population-based study. *Bone* 1997;**21**:363-7.
184. Raab-Cullen D, Akhter M, Kimmel D, Recker R. Bone response to alternate-day mechanical loading of the rat tibia. *J.Bone Miner.Res.* 1994;**9**:203-11.
185. Recker R. Embryology, anatomy, and microstructure of bone. In Coe F, Favus M, eds. *Disorders of bone and mineral metabolism*, pp 219-40. 1992.
186. Recker R, Davies M, Hinders S, Heaney R, Stegman M, Kimmel D. Bone gain in young adult women. *JAMA* 1992;**268**:2403-8.
187. Robling A, Burr D, Turner C. Partitioning a daily mechanical stimulus into decrete loading bouts improves the osteogenic response to loading. *Journal of Bone and Mineral Research* 2000;**15**:1596-602.
188. Rosenthal D, Mayo-Smith W, Hayes C, Khurana J, Biller B, Neer R *et al.* Age and bone mass in premenopausal women. *J.Bone Miner.Res.* 1989;**4**:533-8.
189. Rozen G, Rennert G, Rennert H, Diab G, Ish-Shalom S. Calcium intake and bone mass development among Israeli adolescent girls. *J Am Coll Nutr* 2001;**20**:219-24.
190. Rubin K, Schirduan V, Gendreau P, Sarfarazi M, Mendola R, Dalsky G. Predictors of axial and peripheral bone mineral density in healthy children and adolescents, with special attention to the role of puberty. *J.Pediatr.* 1993;**123**:863-70.
191. Rubin L, Hawker G, Peltekova V, Fielding L, Ridout R, Cole D. Determinants of peak bone mass: clinical and genetic analyses in a young female Canadian cohort. *J Bone Miner Res* 1999;**14**:633-43.
192. Ruiz J,.Mandel C. Influence of spontaneous calcium intake and physical exercise on the vertebral and femoral bone mineral density of children and adolescents. *J.Bone Miner.Res.* 1995;**10**:675-81.
193. Ryder K,.Duncan R. Parathyroid hormone modulates the response of osteoblast-like cells to mechanical stimulation. *Calcif Tissue Int* 2000;**67**:241-6.

194. Saggese G, Baroncelli G, Bertelloni S, Perri G. Long-term growth hormone treatment in children with renal hypophosphatemic rickets: effects on growth, mineral metabolism and bone density. *J Pediatr* 1995;**127**:395-402.
195. Sainz J, Van Tornout J, Loro M, Sayre J, Roe T, Gilsanz V. Vitamin D-receptor gene polymorphisms and bone density in prepubertal American girls of Mexican descent. *N Engl J Med* 1997;**337**:77-82.
196. Sandler R, Slemenda C, LaPorte R, Cauley J, Schramm M, Barresi M *et al.* Postmenopausal bone density and milk consumption in childhood and adolescence. *Am.J.Clin.Nutr.* 1985;**42**:270-4.
197. Scholes D, LaCroix A, Ott S, Ichikawa L, Barlow W. Bone mineral density in women using depot medroxyprogesterone acetate for contraception. *Obstet.Gynecol.* 1999;**93**:223-8.
198. Seeman E, Hopper J, Back L, Cooper M, Parkinson E, McKay J *et al.* Reduced bone mass in daughters of women with osteoporosis. *New Engl.J.Med.* 1989;**320**:554-8.
199. Semeao E, Jawad A, Stouffer N, Zemel B, Piccoli D, Stallings V. Risk factors for low bone mineral density in children and young adults with crohn's disease. *J Pediatr* 1999;**135**:593-600.
200. Sempos C, Johnson N, Smith E, Gilligan C. Effects of intraindividual and interindividual variation in repeated dietary records. *Am.J.Epidemiol.* 1984;**120**:281-5.
201. Slemenda C. Development of vertebral skeletal mass during childhood and adolescence. *Spine* 1994;**8**:83-9.
202. Slemenda, C. The development and significance of peak bone mass. In the Proceedings of the Third International Symposium--Osteoporosis research advances and clinical applications. 27-28. 1994.
Ref Type: Conference Proceeding
203. Snow-Harter C, Bouxsein M, Lewis B, Carter D, and Marcus R. Effects of resistance and endurance exercise on bone mineral status of young women: A randomized intervention trial. *J.Bone Miner.Res.* 1992;**7**:761-9.
204. Specker B,.Binkley T. Randomized trial of physical activity and calcium supplementation on bone mineral content in 3- to 5-year-old children. *J Bone Miner Res* 2003;**18**:885-92.
205. Specker B, Mulligan L, Ho M. Longitudinal study of calcium intake, physical activity, and bone mineral content in infants 6-18 months of age. *J Bone Miner Res* 1999;**14**:569-76.
206. Specker B,.Tsang R. Cyclical serum 25-hydroxyvitamin D concentrations paralleling sunshine exposure in exclusively breast-fed infants. *J Pediatr* 1987;**110**:744-7.
207. Stephons W, Lkimiuk P, Warrington S, Taylor J, Berry J, Mawer E. Observations on the natural history of vitamin D deficiency amongst Asian immigrants. *Q J Med* 1982;**51**:171-88.
208. Story M, Alton I. Becoming a woman: Nutrition in adolescence. In Krummel D, Kris-Etherton P, eds. *Nutrition in women's health*, pp 1-34. Gaithersburg, MD: Aspen, 1995.

209. Strandberg S, Nordstrom P, Lorentzon R, Lorentzon M. Vitamin D receptor start codon polymorphism (FokI) is related to bone mineral density in healthy adolescent boys. *J Bone Miner Metab* 2004;**21**:109-13.
210. Taaffe D, Robinson T, Snow C, Marcus R. High-impact exercise promotes bone gain in well-trained female athletes. *Bone and Mineral Research* 1997;**12**:255-60.
211. Tao C, Yu T, Garnett S, Briody J, Knight J, Wookhead H *et al.* Vitamin D receptor alleles predict growth and bone density in girls. *Archives of Disease in Childhood* 1998;**79**:488-94.
212. Theintz G, Buchs B, Rizzoli R, Slosman D, Clavien H, Sizonenko P *et al.* Longitudinal monitoring of bone mass accumulation in healthy adolescents: Evidence for a marked reduction after 16 years of age at the levels of lumbar spine and femoral neck in female subjects. *J.Clin.Endocrinol.Metabol.* 1992;**75**:1060-5.
213. Thomas M, Lloyd-Jones D, Thadhani R, Shaw A, Deraska D, Kitch B *et al.* Hypovitaminosis D in medical inpatients. *New Engl.J.Med.* 1998;**338**:777-83.
214. Trotter M,.Hixon B. Sequential changes in weight, density, and percentage ash weight of human skeletons from an early fetal period through old age. *Anat Rec* 1974;**179**:1-18.
215. Turner J, Gilchrist N, Ayling E, Hassall A, Hooke E, Sadler W. Factors affecting bone mineral density in high school girls. *NZ Med J* 1992;**105**:95-6.
216. U.S.Department of Health and Human Services. Healthy People 2010: Partnerships for the New Millennium. Washington, DC: 2000.
217. Valimaki M, Karkkainen M, Lamberg-Allardt C, Laitinen K, Alhava E, Heikkinen J *et al.* Exercise, smoking, and calcium intake during adolescence and early adulthood as determinants of peak bone mass. *BMJ* 1994;**309**:230-4.
218. Valimaki V-V, Alfthan H, Lehmuskallio E, Loyttyniemi E, Sahi T, Stenman U-H *et al.* Vitamin D status as a determinant of peak bone mass in young finnish men. *J Clin Endocrinol Metab* 2004;**89**:76-80.
219. VandenBergh M, DeMan S, Witteman C, Hofman A, Trouerback W, Grobbee D. Physical activity, calcium intake, and bone mineral content in children in the Netherlands. *J Epidemiol Community Health* 1995;**49**:299-304.
220. Viitanen A, Karkkainen M, Laitinen K, Lamberg-Allardt C, Kainulainen K, Rasanen L *et al.* Common polymorphism of the vitamin D receptor gene is assoiciated iwth variation of peak bone mass in young Finns. *Calcif Tissue Int* 1996;**59**:231-4.
221. Virvidakis K, Georgious A, Korkotsidis A, Ntalles K, Proukakis C. Bone mineral content of junior competitive weightlifters. *Int.J.Sports Med.* 1990;**11**:244-6.
222. Ward K,.Klesges R. A meta-analysis of the effects of cigarette smoking on bone mineral density. *Calcif Tissue Int* 2001;**68**:259-70.
223. Warren M,.Stiehl A. Exercise and female adolescents: effects on the reproductive and skeletal systems. *JAMWA* 1999;**54**:115-20.
224. Weaver C, Martin B, Plawecki K, Peacock M, Wood O, Smith D *et al.* Differences in calcium metabolism between adolescent and adult females. *Am.J.Clin.Nutr.* 1995;**61**:577-81.
225. Webb A, Kline L, Holick M. Influence of season and latitude on the cutaneous synthesis of vitamin D3: Exposure to winter sunlight in Boston and Edmonton

will not promote vitamin D3 synthesis in human skin. *J Clin Endocrinol Metab* 1988;**67**:373-8.

226. Welten D, Kemper H, Post G, VanMechelen W, Twisk J, Lips P *et al.* Weight-bearing activity during youth is a more important factor for peak bone mass than calcium intake. *J.Bone Miner.Res.* 1994;**9**:1089-95.
227. Wong D, Perry S. Maternal Child Nursing Care. St. Louis: Mosby, 1998.
228. Wosje K,.Specker B. Role of calcium in bone health during childhood. *Nutrition reviews* 2000;**58**:253-68.
229. Young D, Hopper J, Nowson C, Green R, Sherwin A, Kaymakci B *et al.* Determinants of bone mass in 10- to 26-year-old females: a twin study. *J.Bone Miner.Res.* 1995;**10**:558-67.
230. Zanker C, Gannon L, Cooke C, Gee K, Oldroyd B, Truscott J. Differences in bone density, body composition, physical activity, and diet between child gymnasts and untrained children 7-8 years of age. *J Bone Miner Res* 2003;**18**:1043-50.
231. Zemel M, Shi H, Greer B, Dirienzo D, Zemel P. Regulation of adiposity by dietary calcium. *FASEB J.* 2000;**14**:1132-8.

CHAPTER 15

THE MECHANOSTAT HYPOTHESIS FOR BONES AND OTHER SKELETAL ORGANS

Harold M. Frost, B.A., M.D., Dr. Sc.

Department of Orthopaedic Surgery, Southern Colorado Clinic
Adjunct Professor of Anatomy, Purdue University
Adjunct Professor of Radiobiology, University of Utah
Professor of Surgery, Guangjong Medical University
American Academy of Orthopaedic Surgeons
Association of Bone and Joint Surgeons
International Society for Musculoskeletal and Neuronal Interactions
Correspondence: Dr. Frost. FAX (719) 561-4458

Our load-bearing bones, fascia, growth plates, ligaments, tendons and joints have enough strength to keep typical peak voluntary loads on them from muscle forces from breaking them suddenly or in fatigue. The tissue-level biologic machinery that achieves that relationship between an organ's loads and its strength could be named the mechanostat. This article explains in general terms how mechanostats could achieve that relationship for our load-bearing skeletal organs. Bone's mechanostat can provide the illustrative case; analogs of it presumably apply to extraosseous load-bearing organs.

1. Introduction

The mechanostat hypothesis depends on several elementary observations. To wit:

#1. Throughout life most of our bones, joints, fascias, ligaments and tendons carry typical peak voluntary loads from muscle forces without breaking, rupturing, or developing arthroses (Note A). An elegant way to make them strong enough to do that would make the typical peak loads on them determine their strengths. Cybernetic consideration [13] suggest that should require at least five things: *(i)* Ways to monitor and evaluate the relationship between the strength of such organs and the kinds and sizes of the loads on them; *(ii)* special criteria for too little and too much strength of those organs relative to the typical peak loads on them; *(iii)* biologic mechanisms that could increase or decrease the strengths of such organs; *(iv)* ways to let the criteria in (ii) turn the mechanisms in (iii) on and off when and where needed; *(v)* and feedback that connected those functions to each other.

Combining those features could form tissue-level mechanisms for load-bearig skeletal organs which one could call *mechanostats*. In effect they would make the typical peak voluntary loads on load-bearing skeletal organs determine the strengths of those organs.

#2. *A)* Load-bearing skeletal organs can fail structurally and mechanically in one of two modes. Typical peak voluntary muscle forces can cause nontraumatic ("spontaneous") fractures of bones and ruptures of fascia, ligaments, and tendons, while forces caused by trauma (injuries) can cause traumatic failures. When such failures are expressed as annual numbers per 10,000 such structures in 100,000 or more people (each human body contains well over 500 such structures), traumatic failures far exceed nontraumatic ones (Note A). *B)* Ergo, the development and design of our load-bearing skeletal organs preferentially protects them from nontraumatic failures, so a curious mind would wonder how that is done. Part II below explains how bones do it, and Part III suggests how extraosseous load-bearing organs do it (fascias, ligaments, growth plates, tendons, synovial joints).

#3. "Connecting the dots" between evidence and ideas from many lines of inquiry shows that postnatal voluntary loads on load-bearig bones do determine most of their postnatal strength [4, 7]. Related evidence and ideas strongly suggest that an analogous stratagem determines most of the postnatal strength of load-bearing extraosseous organs (fascias, ligaments, tendons, growth plates, synovial joints; Note A).

If so, mechanisms one could name mechanostats would normally determine most of the postnatal strengths of load-bearing skeletal organs (i.e, bones, fascias, ligaments, tendons, growth plates, synovial joints).

#4. A four-step strategy can help to understand the physiology of any organ or organ system [5]. That powerful strategy descends the ladder of biologic organization. It helped to reveal the Utah paradigm of skeletal physiology [5,12] that does at least two things. *A)* It injects tissue-level realities into the former "knowledge gap" between organ-level and cell-level realities in skeletal physiology, *B),* and it adds some roles of muscle and biomechanics in that physiology.

Understanding/explaining renal physiology can illustrate that four-step strategy. Step #1 would describe functions the kidney (the organ) provides to the body. Step #2 would describe how functions and other features of tissue-level nephorns contribute to the organ's functions. Step #3 would describe how cell and molecular-biologic realities (varied kinds of cells, intercellular materials, genes, receptors, cytokines, RNA, ultrastructure, apoptosis, etc) directly support the nephron's Step #2 functions, and only indirectly support organ-level Step #1 or renal functions. With all that information Step #4 could *A)*, describe the pathogenesis of known renal disorders, *B)* or predict still unrecognized disorders.

#5. Unfortunately, how the skeleton's Step #3 matters support its Step #2 features including its mechanostats remained little-studied and nearly unknown in 2003. If opinions abounded, proof did not. Future research must fill that "knowledge gap" in skeletal physiology, and the Utah paradigm concerns that gap.

So said, Part II of this article summarizes some general features of bone's mechanostat. While devils still lie in some details, those general features comprise mostly tissue-level Step #2 matters that Step #3 matters *must* support. Nevertheless, and to repeat, how the skeleton's Step #3 matters support Step #2 features remained little-studied and nearly unknown in 2003.

Table 1 defines frequently-used abbreviations in the following text.

2. Bone's Mechanostat: A Summary

Twelve general features of bone's mechanostat follow.

1) By the time of birth gene expression patterns *in utero* have created our body's "baseline conditions", including its basic bony anatomy and anatomical relationships, its basic neuromuscular anatomy and physiology, and the biologic "machinery" that will adapt load-bearing bones (LBBs) to their postnatal typical peak voluntary loads (TPVMLs). *2)* In the mechanical sense human bones come in two kinds. *a)* They include LBBs such as tibias, femurs, radii, scapulae, mandibles, etc (so our LBBs are not limited to weight-bearing bones), which are designed mainly to have enough strength to keep the TPVMLs on them from breaking them suddenly or in fatigue. *b)* A few bones, including the cranial vault, turbinates and inner-ear ossicles, presumably have other chief functions.

3) In each LBB a multicellular mechanism called *modeling* by formation and resorption drifts [6] can increase its strength [5, 8, 12]. *4)* Each hollow LBB has another multicellular mechanism called disuse-mode BMU-based *remodeling* which can decrease a hollow LBB's strength by removing bone next to or close to marrow [5, 6]. Changing a bone's strength describes a Step #2 function provided to the organ by those two nephron-equivalent functions.

Hence (iii) in Section #1 of Part I Above.

5) All loads on LBBs cause strain-dependent bone signals that can both monitor and reveal the *relationship* between each LBB's strength and the size and kinds of the TPVMLs on it [7, 8].

Hence (i) in Section #1, Part I above.

6) When bone strains rise to or above a range one could name the modeling threshold (MESm), mechanically-controlled modeling turns on; otherwise it remains off [5, 8, 12]. (MES signifies the minimum effective strain or other stimulus). When bone strains stay in or below a range one could name the disuse-mode remodeling threshold (MESr), mechanically-controlled disuse-mode remodeling removes bone next to or close to marrow at its maximal rate in hollow bones. When strains exceed that range those net bone losses begin to decrease [5]. Presumably genetics determines those ranges. Aided by dedicated signaling systems [2, 7, 9], those ranges do two things: They help to turn those two bone-strength functions on and off, and they can distinguish between too little, enough and too much strength of a LBB *relative* to the TPVMLs on it.

Hence (ii, iv) in Section #1 of Part I above.

7) Each LBB can develop microscopic fatigue damage (microdamage, MDx), which has its own operational threshold range in bone (MESp) [1, 4, 9]. Normally each LBB can detect and repair the small amounts of MDx caused by strains that stay below the MESp range [5]. Carter's group found that bone MDx increases over 500x when strain and unit-load magnitudes approach or exceed bone's MESp [11], which seems to coincide with bone's yield point region [8].

8) Combining such things with feedback between them would construct a tissue-level "nephron-equivalent" negative feedback system called bone's *mechanostat* [5]. AM Parfitt recently called it the "....most important....problem in bone physiology." [10]. One could view it as a LBB's Step #2 analog of the kidney's nephron (woven bone physiology might have a somewhat different MST). It would already exist at the time of birth as part of a bone's biologic machinery and as a part of its baseline conditions.

Hence (v) in Section #1 of Part I above.

9) On earth bone's mechanostat adapts a LBB's strength and architecture chiefly to postnatal *muscle strength* (and power?). Why not to body weight? On earth, lever arm and gravitational effects make muscles put by far the largest TPVMLs on LBBs, including on weight-bearing bones like the *femur* and tibia [2,8]. In principle three organ-level feature should follow. *(i)* Chronically strong muscles should associate with stronger bones, and chronically weak muscles should usually associate with weaker bones. Both associations do occur. *(ii)* Ignoring general body growth and longitudinal bone growth, a postnatal LBB's strength (X) should sum a baseline conditions part (Y), plus any adaptations (Z) added to "Y" after birth, so X=Y+Z (Note A). If so, after total and permanent paralysis a LBB in a paralyzed limb should never disappear completely. That is true, and perhaps its "Y" part persists. *(iii)* Since bones cannot foresee one-time loads from injuries or from rare activities such as jumping from great heights, bones could not adapt their strengths to them [2].

10) Most nonmechanical influences previously thought to dominate control of bone physiology, and by implication to dominate control of whole-bone strength too, would act as permissive agents the mechanostat needs in order to work, but not ones that "guide" the mechanostat in time and anatomical space (see Section #2 in comments; "whole bone" distinguishes bones as organs from bone as a tissue or material).

11) Bone's mechanostat functions abnormally in current cell, tissue and organ culture systems [4, 5], so its *in vivo* properties would need study in live-animal research.

12) The "*general biomechanical relation*" (*GBR) for healthy LBBs* [4, 5]. Let MESr denote the strain or stress range in and below which bone's maximal disuse-mode remodeling function of decreasing a hollow LBB's strength turns on, and above which it begins to decrease or turn off. Let "E" denote the typical peak postnatal strains or stresses of a normally-adapted LBB caused by its TPVMLs. Let MESm denote the strain of stress range in and above which bone's mechanically-controlled modeling function of increasing a LBB's strength turns on. Let MESp denote the strain or stress range in and above which unrepaired bone MDx begins to accumulate. Let Fx denote a bone's ultimate or facture strength. Then the GBR for healthy postnatal young-adult mammalian LBBs made with lamellar bone could express the laddered magnitudes of those things thus: MESr < "E" < MESm < MESp < Fx [5].

One could express the GBR's entries in the corresponding strain, stress or unit-load terms in Table 2. Instead of step functions, those entries are ranges with unknown breadths, so in a first approximation the centers of their ranges could define their "set points". Note that maintaining the GBR's relationships would achieve the objective expressed in Section #2, B of Part I above.

3. Extraosseous Mechanostats

The skeleton's extraosseous mechanostats are discussed elsewhere [5, 12]. Suffice it here to note two things. A) In addition to analogs of the above 12 features of bone's mechanostat, the extraosseous mechanostats include provisions to detect and correct limited amounts of irreversible plastic flow in tension in collagenous and chondral tissues, B) and in shear in chondral tissues (Note A).

Physiologists studying extraousseous load-bearing skeletal organs seldom mention their mechanostats or their other tissue-level components mechanisms. In those respects bone physiologists advanced far beyond people who study load-bearing organs made from cartilage or collagenous tissue.

4. Comments

4.1. On mechanostat roles in skeletal physiology

In 2003 trying to explain renal physiology and disorders without accounting for nephrons would be incredibly naïve. May I suggest the skeleton's mechanostats should have the same relationship to skeletal physiology and disorders as nephrons do to renal physiology and disorders? If so, trying to explain skeletal physiology and disorders without accounting for the skeleton's mechanostats could be equally naïve.

4.2. On permissive roles

Many nonmechanical agents (hormones, minerals, vitamins, drugs, etc) formerly thought to dominate the health of bones, joints, ligaments and tendons could instead have permissive roles in the functions of the skeleton's mechanostats [5]. To explain, a car has hundreds of parts (gears, bolts, radiators, fuel, motors, brakes, etc) that are needed for it to be driven, but they do not decide if it drives to Berlin or into some ditch. For the function of being driven most of a car's parts are permissive ones.

It seems that things like vitamin D, calcium, growth and parathyroid hormones, androgens, estrogen, and other humoral and nonmechanical agents, have mainly permissive roles in the functions of our skeleton's mechanostats, and chiefly mechanical instead of nonmechanical factors guide and limit those functions in time and anatomical space.

Permissive humoral and local agents have a revealing behavioral property. Their deficiencies can cause big health problems, but their excesses in healthy subjects have small or no effects, or different kinds of effects including toxicity [4]. Thus vitamin c deficiency causes scurvy but this vitamin's excesses have little effect on healthy bodies. Vitamin D and thyroxine deficiencies cause short stature, yet their excesses do no cause giantism but can cause toxicity. Growth hormone may mainly permit whole-bone strength to increase during adaptations to larger bone loads. A clever Australian study showed that lacking such loads the hormone does not increase that strength [3], so the hormone could have a permissive role in that activity. Analogous effects may apply to androgen effects on whole-bone strength and on muscle strength etc.

4.3. On defining the health of load-bearing skeletal organs

Parts I and II above suggest that one could define the health of any load-bearing bone, ligament, tendon, growth plate or joint in terms of the three-way *relationship* between its strength, the size and kinds of TPVMLs on it, and any nontraumatic fracture, rupture or arthrosis caused by those loads, whether suddenly or in fatigue. That would define the organ's health in terms of its chief mechanical function instead of in terms of its "mass", size, shape, strength or other physical or anatomical feature. One could say that organs that satisfied that definition – and that satisfied the GBR too – were "mechanically competent".

One could define a load-bearing skeletal organ's mechanical competence thus: The organ's biologic machinery (meaning its mechanostat) would make healthy load-bearing organs strong enough to keep typical peak voluntary mechanical loads on them from breaking or rupturing them suddenly or in fatigue, or from causing an arthrosis, whether those loads are chronically subnormal, normal or supranormal, or chronically small or large. That statement could define the mechanical competence of such organs equally well in the skeletons of mice, humans and elephants, and probably in dinosaurs too. Elsewhere and in brevity's interest "Proposition #1" signified the mechanical competence of load-bearing skeletal organs [5].

4.4. Implications

Other publications [4, 5] discuss the many implications of the mechanostat hypothesis for the nature, diagnosis, study, prevention and management of numerous disorders of load-bearing skeletal organs. Many of those implications question currently "accepted wisdom", so time and more work must resolve any resulting questions and controversies to the satisfaction of the general skeletal science and clinical communities.

In conclusion please let me note three things. *A)* Controversies fuelled progress in all science in the past, *B)* reasonable people can usually find more than one explanation for a given collection of facts, *C)* and younger people than this octogenarian must resolve any controversies associated with the mechanostat hypothesis.

4.5. Note A

An observation by the author of a feature that others must have noted too. Yet in the past it did not seem important enough to deserve formal study and report.

Table 1 Meanings of Abbreviations and Symbols in this Text

BMU:	basic multicellular unit of bone remodeling
Fx:	the ultimate (fracture) strength healthy young adult bone
LBB:	a load-bearing bone (femur, vertebra, tibia, phalanx, sesamoid, etc).
MDx:	microdamage, microscopic fatigue damage.
MESm:	the strain range in and above which the modeling function of increasing a load-bearing organ's strength turns on (the modeling threshold). "MES" stands for Minimum Effective Strain or other Stimulus.
MESp:	the strain range in and above which MDx increases beyond the ability to repair it so it accumulates.
MESr:	the strain region in and below which a load-bearing skeletal organ's disuse-mode activity acts maximally, and above which it begins to decrease (the disuse-mode remodeling threshold).
TPVMLs:	typical peak voluntary mechanical loads on a skeletal organ, so it implies voluntary muscle forces. It excludes load from injuries or from rare kinds of very strenuous activities like jumping from great heights.
$\approx$:	approximately, or approximately equals.
$<$:	less than.

Table 2 Set Point Values for Bone's Thresholds and Ultimate Strength **(in microstrain, stress and unit load terms)***

MESr: 50-100 microstrain; ≈ 1-2 mpa, or ≈ 0.1 kg/mm^2 (one can argue for a value of ≈ 400 microstrain).

MESm: 1000-1500 microstrain; ≈ 20 mpa, or ≈ 2 kg/mm^2.

MESp: ≈ 3000 microstrain; ≈60 mpa, or ≈ 6 kg/mm^2. Whether by coincidence or not, bone's MESp seems to coincide with bone's yield point [9].

Fx: the ultimate (fracture) strength healthy young adult bone of ≈ 25,000 microstrain in healthy young adults (a bit more in children and less in aging adults); ≈ 120 mpa, or ≈ 12 kg/mm^2.

*: values for analogous features for extraosseous load-bearing organs are currently unknown. The bone values apply to cortical lamellar bone in healthy young adults, and depend on information available to the author in 2002. kg/mm^2 means kilograms of force per mm squared, where one kg of force equals 9.8 Newtons.

References

1. Burr DB, Milgrom C (2000) Musculoskeletal Fatigue and Stress Fractures (Eds). CRC Press, Boca Raton, Fl.
2. Currey JD 2003 How well are bones designed to resist fracture? J Bone Min Res 18:593-598
3. Forwood MR, Li L, Kelly WL, Bennett MB (2001) Growth hormone is permissive for skeletal adaptation to mechanical loading. J Bone Min Res 16:2284-2290
4. Frost HM (2003) Bone's mechanostat: A 2003 update. Anat Rec (in press).
5. Frost HM (2003) The Utah Paradigm of Skeletal Physiology. Vols I, II. Hylenome (in press).
6. Jee WSS (2001) Intergrated bone tissue physiology: Anatomy and physiology. In: Bone Mechanics Handbook (2nd ed). SC Cowin (Ed). CRC Press, Boca Raton, pp 1-68.
7. Lanyon L, Skerry T (2001) Postmenopausal osteoporosis as a failure of bone's adaptation to functional loading: A hypothesis. J Bone Min Res 16:1937-1947.
8. Martin RB, Burr DB, Sharkey NA (1998) Skeletal Tissue Mechanics. Springer-Verlag, New York.
9. Martin RB (2003) Fatigue microdamage as an essential element of bone mechanics and biology. Calc Tiss Int 73:101-107
10. Parfitt AM (2000) Osteoporosis: 50 years of change, mostly in the right direction. In: Osteoporosis and Bone Biology. J Compston and S Ralston (Eds.) International Medical Press, pp. 1-13.

11. Pattin CA, Caler, WE, Carter DR (1996) Cyclic mechanical property degradation during fatigue loading of cortical bone. J Biomech 29:69-79.
12. Takahashi HE (1995) Spinal Disorders in Growth and Aging (Ed). Springer-Verlag, Tokyo.
13. Wiener N (1964). Cyberne+
14. tics. MIT Press Cambridge.

CHAPTER 16

MECHANOTRANSDUCTION AND ITS ROLE IN BONE ADAPTATION

Yixian Qin and Clinton Rubin

Department of Biomedical Engineering
Stony Brook University, NY 11794, USA
E-mail: Yi-Xian.Qin@sunysb.edu

1. Introduction

The ability of bone to adapt to its physical environment has been studied for well over a century. Bone mass and morphology accommodates changes in mechanical demands by regulating the site-specific remodeling processes which consist of resorption of bone, typically followed by bone formation. While this adaptive process is essential to achieve the structural competence of the skeleton, the specific components of the mechanical environment which control these processes are not yet fully identified. The difficulties in defining the mechanical parameters which regulate bone adaptation have confounded our ability to predict orthopaedic failure/healing or influence the etiology of metabolic bone diseases. An improved understanding of this adaptive process, e.g., if the constitutive mechanical regime could be identified, will ultimately serve to benefit treatment regimens for musculoskeletal diseases and orthopaedic interventions.

Further, the adaptive properties of bone enable the skeleton to withstand the extremes of functional load-bearing, yet optimize this structure by the strategic removal of "unnecessary" bone tissue. Bone tissue's adaptive responses can also have grave consequences, as illustrated by fractures of the hip and spine potentiated by senile or post-menopausal osteopenia, resorption of the calcar following femoral joint replacement, stress fractures in athletes and military recruits, or microgravity induced bone loss in astronauts. It is imperative that we improve our understanding of how mechanical stimulation regulates bone morphology, and yet fails to inhibit the pathogenesis of osteopenia. Understanding of the biologic and engineering basis of the mechanical modulation of bone could ultimately benefit orthopaedic outcomes and improve the treatment of many musculoskeletal disorders.

The ability of bone to respond to changes in its functional milieu is one of the most intriguing aspects of this living tissue, and certainly contributes to its success as a structure. Bone's ability to rapidly accommodate changes in its functional environment ensures that sufficient skeletal mass is appropriately placed to withstand the rigors of functional activity, an attribute described as Wolff's Law [147,148]. This adaptive capability of bone suggests that biophysical stimuli may be able to provide a site-specific, exogenous treatment for controlling both bone mass and morphology. The premise of a mechanical influence on bone morphology has become a basic tenet of bone physiology [22,23,24,81,131]. Absence of functional loading results in the loss of bone mass [20,22,40,109,120 136], while exercise or increased activity results in increased bone mass [48,57,65,91,132]. To define the formal relationship between the mechanical milieu and the adaptive response will prove instrumental in devising a mechanical intervention for skeletal disorders such as osteoporosis, designing biomechanical means to accelerate fracture healing, and promoting bony ingrowth.

High physical activity level has been associated with high bone mass and low fracture risk and is therefore recommended to reduce fractures in old age. As a direct consequence of exposure to microgravity astronauts experience a number of physiological changes, which can have serious medical complications. High physical activity level has been associated with high bone mass and low fracture risk and is therefore recommended to reduce fractures in old age [70,69,74,25,31]. Most immediate and significant are the musculoskeletal implications in bone and muscles [15,26,59,86]. Results of the joint Russian/US studies of the effect of microgravity on bone tissue from 4.5- to 14.5-month long missions are demonstrated that bone mineral density (BMD, g/cm^2) and mineral content (BMC, g) are decayed in the whole body of the astronauts [38]. The greatest BMD losses have been observed in the skeleton of the lower body, i.e., in pelvic bones (-11.99±1.22%) and in the femoral neck (-8.17±1.24%), while there was no evitable decay founded in the skull region. Overall changes in bone mass of the whole skeleton of male cosmonauts during the period of about 6 months on mission made up -1.41±0.41% and suggest the mean balance of calcium over flight equal to -227±62.8 mg/day. In average, the magnitude and rate of the loss is staggering; astronauts lose bone mineral in the lower appendicular skeleton at a rate approaching 2% per month with muscle atrophy [71,73,72,75,69,70]. In simulated or actual microgravity, human postural muscles undergo substantial atrophy: after about 270 days, the muscle mass attains a constant value of about 70% of the initial one. Most animal studies reported preferential atrophy of slow twitch fibers whose mechanical properties change towards the fast type. After microgravity, the maximal force of several muscle groups showed a substantial decrease (6-25% of pre-flight values) [15,59,14,126,82,68]. The mechanism that explains both muscle and bone decays in the function disuse environment is still unclear. In recent years, considerable attention has focused on identifying particular parameters and exercise paradigms to ameliorate the deficits of muscle atrophy and bone density. Perhaps microcirculation and interstitial fluid flow that link with exercise and muscle contraction can identify the interrelationship between muscle and bone flow in response to loading and disuse environment. In fact, head-ward shift of body fluids and the removal of gravitational loading from bone and muscles have led to progressive changes in the musculoskeletal systems. The underlying factor producing these changes may be primarily due to the fluid flow and circulations in both muscular and bone tissues.

2. Tissue and Cellular Mechanotransduction

2.1. Mechanical loading induced adaptation

Researchers have worked very hard over the past several decades in an attempt to understand the effect of physical stimuli on bone adaptation. Many strain or stress parameters have been suggested as governing components in bone cell response to mechanical loading. For example, some investigators proposed a relationship between "invariant" parameters such as those whose magnitudes are independent of a coordinate system, i.e., strain energy density [30,43]. This is consistent with a concept that bone was "self-regulating." The theories of bone adaptation range from surface modeling as a function of strain magnitude [29118], to a time-dependent modeling and remodeling [3,4,29]. As the models use a number of different parameters as the driving function for the optimization process, i.e., strain/stress magnitude, cycle number, occurrences of events, strain tensor, and strain energy density, it is very possible that these parameters are interdependent. Indeed, it is difficult to distinguish amongst them and identify the specific components responsible for regulating bone adaptation [7]. When exploring any such mechanical hypotheses, one must identify the mechanism whereby *the cells respond directly or indirectly to these mechanical parameters*. At this point, there is little evidence that peak strains or stresses directly correlates to bone's morphological response [7,7,40]. The specific mechanical stimulus that triggers and terminates the cellular adaptive response remains to be further understood.

2.2. The role of dynamic and temporal mechanical signals

A recent finding may link the mechanism of how bone responses to mechanical stimulation at the level of the cells, by concentrating on the temporal components of the stimulus, such as strain rate, loading frequency and the strain gradients. Given equivalent strain magnitudes, higher strain rates elicit a more pronounced adaptive response than low strain rates [33,57,92,115]. Similarly, loading waveforms with substantial frequency components between 15 and 60 Hz are more osteogenic than signals whose essential frequency is closer to 1 Hz [83,110]. Using sinusoidal loading of constant duration (10 min/day), but varying frequency (1 to 60 Hz), it has been demonstrated that substantially different "thresholds" for maintaining bone mass exist. While a loading at 1 Hz requires peak induced longitudinal normal strains greater than 700 με to maintain cortical bone mass, the strain necessary to maintain cortical bone mass decreases to 400 με at 30 Hz. At 60 Hz, over a ten-minute period, peak normal strains of only 270 με were found capable of maintaining cortical bone. This frequency sensitivity remodeling response is found to correlate with intracortical fluid flow, which closely mediated with loading frequency (R=0.8) [108]. Applying bending loads to the tibiae of adult rats to create equivalent peak strains in the bone tissue but with varied rates of strain, Turner et al.[138] reported that bone formation was significantly increased and the amount of new bone formation was directly proportional to the rate of strain in the bone tissue. Moreover, in the attempt to identify mechanical parameters correlated with adaptation induced by an external loading regimen, the likely parameters that predict periosteal new bone formation are the strain gradients [40,56,58]. Indeed, while many in vivo experiments have supported and shown that bone is sensitive to, and responds anabolically to, dynamic loading adaptation should be driven by signals derived from dynamic rather than static loading[13,111,116,122,138], these findings suggest that temporal, dynamic, and potentially stress-generated fluid flow are required to stimulate new bone formation and remodeling.

2.3. Mechanotransduction and interstitial fluid flow

The temporal components of strain are particularly interesting, given their relation to mechanocoupling between matrix and strain-generated fluid flow within bone. It is proposed that mechanical loading results in deformation of bone matrix and the substantial interstitial fluid space, which generates pressure gradients and further induces interstitial fluid flow [100]. For instance, strain gradients are the source, in a solid phase, which induce fluid pressure gradients within bone, which, in turn, generate fluid flow in the tissue. Fluid-induced canalicular shear stresses have been proposed as a mechanism by which bone perceives mechanical stimuli of the tissue [114,137,142,143,149]. Cortical bone is composed of a solid matrix phase and an interstitial fluid phase [103]. In contrast to the soft tissues, i.e., cartilage, approximately 80%~85% of bone volume consists of solid matrix, such as mineral and collagen, while the rest consists of fluid phase such as interstitial fluid in the porous medium [9,87,124,123,150]. This fluid fills in the various spaces and channels in bone, including the micropores, lacunae, canaliculi, Haversian canals, and Volkmann's canals [103,150]. The motion of pore fluid under mechanical loading may play an important role in the mechanism of bone cell sensing, signaling and responding to physical stimuli, as well as nutrient transport. This intracortical fluid flow is considered a critical mediator of bone mass and morphology [60,87,124,123,127]. We believe it represents a critical mechanism to explain load, frequency, strain, strain rate and strain gradient, regulated bone formation, remodeling, and maintenance.

2.4. Cellular mechanotransduction and fluid flow

In the aspects of cellular mechanotransduction response to mechanical stimuli, a number of studies demonstrate proliferative expressive response of osteoblasts-like cells to dynamic strains in vitro [8,89,90]. Even though the strain values in these studies were done above the gross strain failure of bone, these studies have been unable to demonstrate matrix protein responses of in vitro mature osteoblasts-like cells to strains high above anabolic levels [6]However, many studies have shown that bone cells in vitro do respond to fluid flow induced anabolic shear stresses. Osteoblast-like cells responded to constant and/or intermittent hydrostatic loads [10,112] and/or responded to fluid-induced shear stress at the physiological ranges of 5-100 dynes/cm^2 [6,11,44,63,98]. Reich and Frangos [113] demonstrated that Prostaglandin (PGE_2) release is stimulated by fluid flow in a dose-dependent manner. This PGE_2 expression was maintained after termination of flow after 2 h of flow loading. While the lack of responsiveness of bone cells, in culture, to strain levels up to the high physiological range, bone cells do respond to anabolic mechanical stimuli in vivo [12,18]. Ultrastructural examination revealed most of the cells covering the trabecular bone surface were flat bone lining cells. After mechanical stimulation, the trabecular bone surface cells developed ultrastructural features of osteoblastic differentiation and activity. It is likely that bone cells do not respond to direct mechanical strain, but do respond to load-induced fluid flow.

2.5. *Fluid convection enhanced metabolic exchange*

In addition to the mechanical function of fluid flow in bone, a necessary pathway for metabolism in the living tissues is an adequate supply of nutrients for anabolic activity, and a proper disposal of waste products generated from catabolic activities through fluid channels. In soft tissue, molecular diffusion is considered to be the major pathway for transportation of metabolism [96,105]. However, because of the relatively dense structure of cortical bone, a diffusive mechanism may in fact not be sufficient to play an adequate role in transporting metabolic fluid between osteocytes and the surrounding vascular canals. Mechanical loading induced convectional fluid flow may enhance the transportation from the blood supply to osteocytes through this connective mechanism [100,135].

2.6. *Noninvasive dynamic fluid flow stimuli as a mean of countermeasure*

Previous studies have demonstrated that increased venous pressure can promote the formation of periosteal new bone in growing dogs [61]. The results of paired comparisons between experimental and control tibias, showed an increase in venous pressure, and an increase in periosteal new bone formation on the side of increased venous pressure. The data suggest that an increase in venous pressure results in an increase in passage of fluid from capillary to bone matrix. Increased extravascular perfusion could be a factor in increasing periosteal bone formation. In a rat tail suspension model, applying constant venous ligation at femurs increased Intramedullary pressure relative to the sham-operated control femurs (27.8 mmHg vs. 16.4 mmHg, $p < 0.05$), suggesting venous ligation increased interstitial fluid flow proportional to the pressure drop across the bone[5]. Bone mineral content (BMC) increased significantly in the venous-ligated femurs relative to control limbs (115.9 ± 15.6% vs. 103.8±13.2%, p<0.001) for a period of 19 days. Trabecular density was significantly higher in the femurs with venous ligation (351±12 g/cm^3 vs. 329±11 g/cm^3, p<0.05). These results suggest that fluid flow can directly influence bone adaptation independent of mechanical loading, and venous pressure is directly related to ImP and interstitial fluid flow. This implies that adaptive response can be initiated and accelerated by altering venous pressure by a muscular pump effect, which serve as a mean of noninvasive countermeasure. We will use a novel dynamic venous pressure noninvasive intervention to increase rate-dependent muscle tension and venous pressure, which can improve nutrient vessel pressure, ImP, and interstitial fluid flow in the HLS model, and ultimately trigger bone remodeling and inhibit bone loss.

2.7. The relationship between bone interstitial fluid flow and physical exercise

Exercise plays important role in regulating fluid flow in the microcirculation in the muscular tissues. Various mechanisms that are thought to cause blood flow to rise during rhythmic exercise [52,55,54,68,82]. Mechanisms including the muscle pump, substances released by skeletal muscle, substances transported by blood, and factors released by nerves have been postulated to contribute to the rise in muscle blood flow during exercise. Additionally, the factors that initiate the dilation may not be those which sustain it. Although there is normally a close relationship between contractile activity, metabolic rate, and muscle blood flow, this relationship can be disrupted under a variety of circumstances and the active skeletal muscle overperfused. The mechanism of muscle pump as a driving source for capillary filtration and bone interstitial fluid flow is proposed as coupling factor between muscle contraction and fluid flow through bone. Pressure wave from muscle pump contractions aided by increased blood pressure during exercise coupled with temporary occlusion of arteries leading to and veins from the bone, increase hydraulic pressure in cortical bone capillaries so as to amplify capillary filtration [144,145,96]. We propose that understanding the mechanisms responsible for the dynamic patterns of muscle pump, i.e., frequency, rest-insertion, pressure magnitude and duration, can provide new insight into the mechanisms which govern exercise hyperemia and its relation with bone fluid flow.

2.8. *Summary*

It has been demonstrated that bone fluid flow is an important player in triggering and signaling bone formation and bone remodeling [47,77,80,103,124,123,140,139,143,150]. The sources driven this interstitial flow may include mechanical force in skeleton, musculo-dynamics, i.e., dynamic muscle pump, and induced pressure gradients. In clinically related studies, the fracture healing process is highly dependent on visualization and venous pressure, suggesting that osteoblast-like cells respond to fluid flow. In addition, bone trace experiments have demonstrated that interstitial fluid flow does indeed occur in bone. Indeed, our recent data suggests that intracortical bone fluid flow induced by dynamic intramedullary pressure, applied at the minimal strain level, can regulate a site-specific adaptive response, and potentially achieve this goal. Using pressurized oscillatory hydraulic loading in the marrow cavity, fluid loading was found to stimulate new bone formation and reduce intracortical bone porosities caused by disuse, even in the absence direct tissue strain. While disuse alone resulted in significant bone loss ($5.7 \pm 1.5\%$, $p<0.05$), adding fluid flow to disuse resulted in a significant increase in bone mass at the mid-diaphysis ($18.3 \pm 7.6\%$, $p<0.05$), achieved by both periosteal and endosteal new bone formation. A strong correlation ($r = 0.75$, $p = 0.01$) was found between the transcortical fluid pressure gradient and total new bone formation. It has been demonstrated that bone fluid flow is an important player in triggering and signaling bone formation and bone remodeling [47,77,80,103,124,123,140,139,143,150]. Microgravity may result in interstitial flow reduction via several pathways. (1) Reduced physical activities (e.g., high rate impacts) of the skeleton result in decreased load-generated flow in bone, particularly in the weight-bearing skeletons. (2) Fluid mass shift from lower extremities to the head and upper body due to the lack of normal gravitational force [35,34,37,36], which substantially redistributes the hydraulic marrow pressure. (3) Reduced muscle activities result in decreased 'muscle pump' effects [146,145], which substantially reduce venous pressure and ImP. These effects will lead to a reduction in fluid perfusion and convection of the bones with a subsequent reduction in interstitial fluid flow and fluid shear stress on the cells [42]. In comparison, there is an increase in vascular perfusion of the head with subsequent increase in fluid flow.

The goals of this chapter is to overview the role of fluid flow in bone porous structure, establish the fluid flow pathway induced by mechanical loading, and determine the potentials of fluid flow in bone adaptation. Recent studies shown in this chapter include (1) fluid flow induced bone adaptation and correlation with pressure gradients; (2) Identifying fluid flow pathways in response to loading; (3) nonlinear relationship among strain magnitude, loading frequency and cycle number in their ability to maintain bone mass; (4) determining the interdependence of interstitial fluid flow, loading frequency and magnitude in a poroelastic model.

3. Fluid Channels and Porosity in Bone Structure

Like many biological tissues, bone has a functional unit [78,81] (Fig. 1). Mature cortical bone is organized structurally into lamellae, which are

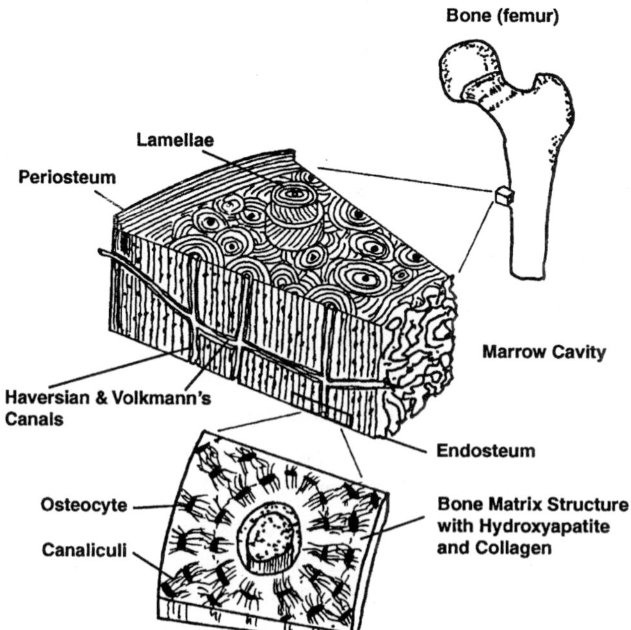

Fig. 1 Cortical bone structure.

arranged concentrically around the Haversian canal in long bones. Haversian canals run longitudinally through the bone cortex, and are transversely inter-connected through Volkmann's canals. Both canals are rich in blood vessels, supplied via the medullary cavity (two thirds) and the periosteal (one third) surfaces. Osteocytes, bone cells which reside in lacunae within the lamellae, interconnect via cell processes tunneling in canaliculi, and communicate via gap junctions. Each of these concentric bone unit, comprised intoto by the capillary canal, the concentric lamellae plates, the lacunae and the canaliculi, is called an osteon.

The Haversian canals, canaliculi and lacunae occupy 13.3% of the volume of the cortical bone [29]. The rest solid portion is the matrix, which is occupied by mineral material containing the hydroxyapatite crystal and collagen fiber. The bone solid matrix of the bone contains pores of the order of 0.01 to 0.1 micron in diameter [100,99,16]. The spatial characteristics for various ultrastructures are listed on the Table 1.

Table 1. Dimensional parameters of a human femur and its components

Structure	Effective diameter
Length	44 cm
Width at midsection	3 cm
Marrow cavity	1 cm
Trabeculae space	1 mm
Nutrient Artery	0.1~1 mm
Osteon	200~300 micron
Volkmann's & Haversian canals	50~100 micron
Osteonal vessels	~15 micron
Osteocyte lacunae	5~10 micron
Cement line thickness	1~5 micron
Canaliculi	< 1 micron
Hydroxyapatite crystal	200~600 Å

The primary constituents of an osteon are collagen (organic), hydroxyapatite (inorganic), and fluid. Fluid flows through the various microstructural spaces (Table 1) to transfer metabolites to the osteocytes, ensuring that bone tissue remain viable. This continuous perfusion allows the remodeling processes to continue in perpetuity. The requisite nutrition of the cells and essential disposal of their waste products are carried by this dynamic fluid stream. Because of the intricate microstructure of cortical bone, transportation between cells would be very poor if it relied solely simple diffusion of a fluid. Indeed, a complex network of very thin (~ 1 micron) and quite long (up to 100 micron) canaliculi must employ a more active policy of nutrient allocation and signal dispersion [100,99,16].

Regarding the potential mechanotransduction in bone induced by fluid flow, Cowin [16] and Hillsley & Frangos [42] summarized that there are three primary levels of bone porosities within cortical microstructure which might mediate fluid flow. Firstly, an osteon, the basic structure of bone with dominant cylindrical structures of 100~150 μm radii running primarily along the long axes of bone, contains at its center an *osteonal canal* (OC), including blood vessels, a nerve and surrounding fluid. There are cells attached on the walls of the OC. At the level of the osteon, fluid transduction from the Haversian canal to the cement line or its surrounding micropores, whether via perfusion or convection, appears to be significantly influenced by the pathway of the fluid flow. Secondly, an osteon contains lacunae structure of 3~10 μm in radius surrounding the OC, which connect to OC via canaliculi. Osteocytes (~2 μm radii) are contained within the lacunae. Thirdly, canaliculi, a capillary with 0.1~0.5 μm in radius, run radially, and surround and connect the lacunae, OC and cement lines together. In addition, the microstructures of collagen-apatite porosity (100~300 Å) contain fluid in the collagen matrix. The cortical vascular supply begins primarily at the marrow cavity and passes through the endosteum, accommodating approximately 2/3 of the vascular supply in the cortex. These levels of porosities participate in and interact with each other in fluid diffusion/perfusion, and fluid pathways may include vascular canals, the lacunar-canalicular spaces, and collagen-apatite spaces.

4. Fluid Flow Pathway in Bone — Defined by *in vivo* Intramedullary Pressure and Streaming Potentials Measurements [105]

In considering bone fluid flow as a regulatory factor of bone remodeling and a substantial mediator for bone formation and inhibition of bone loss, we hypothesize that load-induced bone fluid flow will play a critical role in bone adaptation and be mediated with physiological fluid intensity at specific loading frequency. It has been suggested that bone should more appropriately be viewed as a fluid saturated tissue, capable of sensing and responding to its physical environments through interstitial fluid flow. This point of view emphasizes the temporal response characteristics of the stimuli rather than the static magnitude of mechanical components. The motion of interstitial fluid within bone, which arises as a result of functional load bearing, is hypothesized to be a critical mediator in the perception and response of skeletal tissue to mechanical stimuli[60] [100,123] [16,45,87,27]. It is important to consider that bone is a highly structured composite material comprised of a variety of fluid channels which may contribute to fluid flow in bone, i.e., vasculature, collagen-hydroxyapatite matrix and a hierarchical network lacunae-canaliculi channels. These tunnels permit interstitial flow of fluid through the tiny microporosities [100,17,142], and thus "by-products" of load, such as the change in fluid velocities or pressures, represent a means by which a physical signal could be translated to the cell[60]; [16,45,80,87,114,124,123]. The pathway for intracortical fluid flow in response to a step-load was identified *in-vivo* using intramedullary pressure (ImP) and streaming potential (SP) measurements simultaneously, and allowed the development of a load-induced flow mechanism which considers mechanotransduction and mechano-electro-tranduction phenomena.

4.1. Experiment design

An avian model was used for monitoring, ImP and SP under axial loading which generated peaks of approximately 600 $\mu\varepsilon$. Following the similar osteotomy procedure, via this tube, a 50-psi pressure transducer was connected into the medullary canal, thus permitting measurement of the intramedullary pressure during animal rest and applying external mechanical loading. SPs were measured via two periosteal electrodes and a ground electrode implanted within the marrow cavity. The periosteal locations for the electrodes, both at the midshaft, were dissected free of soft tissue and the bone surface was always kept moist. The relaxation of

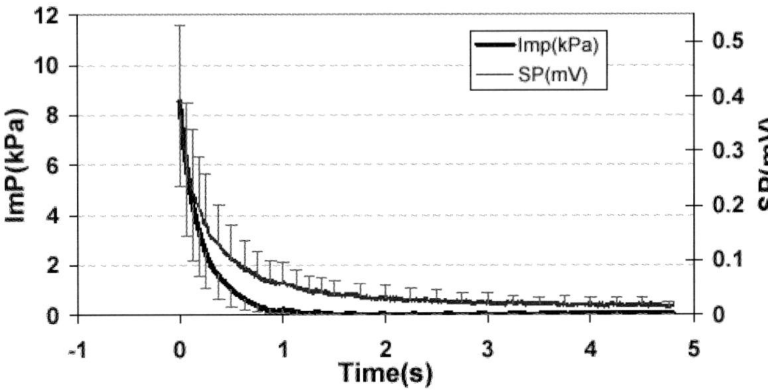

Fig. 2. Mean ImP and SP response to imposed step-load, peak of approximately 600 microstrain, versus time after imposition of load for the time period of 0~5 sec. Loading induced 8.6±3.4 kPa (mean±s.d.) peak ImP and 0.39±0.14 mV (mean±s.d.) peak magnitude of in vivo SP. The ImP decayed much faster than that of SP. (From Qin, Y. et al., J Biomech, 2003. With permission.)

both SPs and ImP response to the step loading was plotted and analyzed versus time.

Results: Step loading applied to the *in vivo* bone preparation generated both SP and ImP in the axial compression of bone (Fig. 2). The peak amplitude of the measured SP was on the order of 0.39±0.14 mV (mean±s.d.) with peak strain approaching in 600 με. A typical peak magnitude of the measured physiologic ImP was on the order of 8.6±3.4 kPa (mean±s.d.) with approximately 600 με longitudinal peak longitudinal strains. ImP response to step-load measured at same strains was decayed much more quickly than SP ($p<0.0001$). Immediately after the loading, ImP quickly relaxed to 70% of its peak value less than 180ms, while SP retain 50% of its peak magnitude at the same period.

It appears that the decay of ImP is indicative of resistive fluid flow occurring primarily in the vasculature and other intraosseous channels such as lacunar-canalicular pores, and that SP represents the fluid flow in the smaller porosities, i.e., lacunar-canalicular system or even microspores. These results suggest that SP and ImP decays are determined by a hierarchical interdependent system of multiple porosities, and that the temporal dynamics of load-bearing define the manner in which the fluid patterns and pressures are distributed.

Implication of this result: (1) It has shown that multiple porosity or distribution of multi-size fluid channels exist in bone, which affect fluid flow distribution generated by loading. (2) If a peak strain of 500~800 microstrain, exposed to an axial loading, is considered a physiological magnitude, which then substantially generates ImP on the order of 8.6 kPa (~65 mmHg). This load-induced ImP magnitude is important to determine anabolic intensity of fluid flow stimulus in the fluid movement in bone.

5. Formation of Bone Induced by Fluid Pressure Gradients[105]

While fluid flow driven by functional loading has been proposed to be a critical regulator of skeletal mass and morphology, we have tested the potential role of flow stimuli in initiating an adaptive response, using oscillatory ImP under physiological fluid pressure and at a relative high frequency achieved without the deformation of the bone tissue.

5.1. Experimental design

All surgical and experimental procedures were approved by the University's Lab Animal Use Committee. Under general halothane anesthesia, the left ulnae of 12 (including two animals used as one-day calibration) adult, one year old, skeletally mature male turkeys were functionally isolated via transverse epiphyseal osteotomies [71,83120]. The metaphyseal ends of the ulna were covered with a pair of stainless steel caps and fully sealed with 6 ml of polymethylmethacrylate. Two Steinmann pins, 4 mm in diameter and 92 mm in length, were placed through the predrilled holes in the bone and cap unit, to prevent mechanical forces to be applied during daily activity, effectively serving as functional disuse of the isolated bone. A 4-mm diameter hole was drilled through the cortex at the dorsal side, approximately 1.2 cm from the proximal cap. The hole was tapped and a special designed fluid loading device, with an inside fluid chamber approximately 0.6 cm^3 in volume, was firmly connected to the bone with an O-ring seal (Fig. 3) A diaphragm

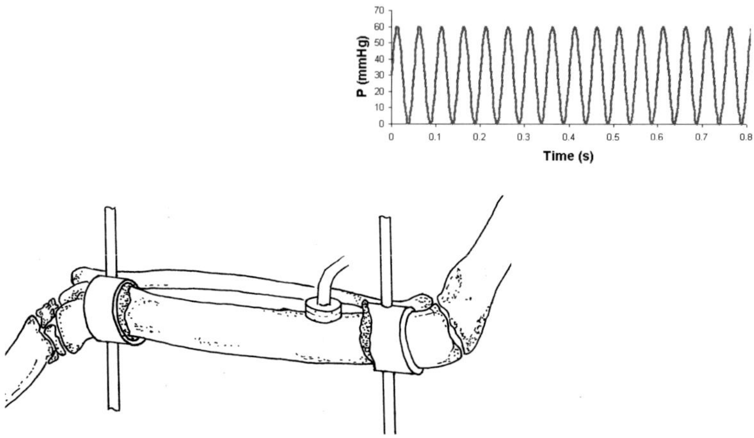

Fig. 3 A diagram of the functionally isolated turkey ulna preparation, such that oscillations in fluid flow can be achieved in the absence of matrix strain. A specially designed device connected into marrow cavity achieved oscillatory fluid loading. A low magnitude and high frequency (60 mmHg, 20 Hz) fluid pressure signal was imposed 10 min daily for 4 weeks in the disuse plus loading.

was included in the center of the chamber dividing the internal marrow fluid from the external oscillatory loading flow. A surgical plastic tube (2-mm inner diameter), which was connected to the device and passed through the skin, served as fluid coupling between the device fluid chamber and the external fluid oscillatory loading unit. An injection plug was connected to the external end of the tube to facilitate fluid flow loading. With the diaphragm and the injection plug, the bone marrow and oscillatory flow media were fully isolated from the external environment to prevent any infection. To monitor the bone remodeling response, all animals were labeled weekly using tetracycline solution (15mg-Kg^{-1}). The contralateral ulna served as control. The left ulnae were exposed a sinusoidal fluid pressure signal (60 mmHg, 20 Hz) imposed for 10 min/day.

5.2. Quantification

Following a ten-minute period of daily loading for 4 weeks, the animals were euthanized via a bolus IV injection of saturated barbiturate. Static and dynamic histomorphometry analyses were performed in determining surface modeling and intracortical remodeling response at the mid-diaphyses of the experimental and contralateral control ulnae. In addition to total new bone formation, adaptive changes were approximated using sector analysis in which the bone cross-section was divided into twelve equal angle (30°) pie sectors through the centroid of the bone section. The transcortical fluid pressure gradient was then calculated for each sector as the difference between fluid pressures at the endosteal and periosteal surfaces, where fluid pressure at the periosteal surface was considered zero.

5.3. *Results*

Disuse alone resulted in significant bone loss (5.7 ± 1.5%, $p<0.05$), achieved by thinning of the cortex via endosteal resorption and increased intracortical porosity. Adding fluid flow to disuse resulted in a significant increase in bone mass at the mid-diaphysis (18.3 ± 7.6%, $p<0.05$), achieved by both periosteal and endosteal new bone formation (Fig. 4) The spatial distribution of the transcortical fluid pressure gradients (∇P_r) (a parameter closely related to cortical fluid velocity and initiation of fluid shear stress) was then evaluated in 12 equal sectors of the mid-diaphyseal cross-section. A strong correlation ($r = 0.75$, $p = 0.01$) was demonstrated between the transcortical fluid pressure gradient and total new bone formation (Fig. 5) In the animals subject to disuse but exposed to ImP loading, a negative correlation

Fig. 4 Microradiographs of (a) animal subject to 4-week disuse resulted in significant bone loss by increase of intracortical porosity; (b) contralateral control of disuse ulna; (c) 4-week fluid flow loading resulted in significantly new bone formation in periosteal and endosteal surfaces, yielding total of 18% new bone formation as compared to control.

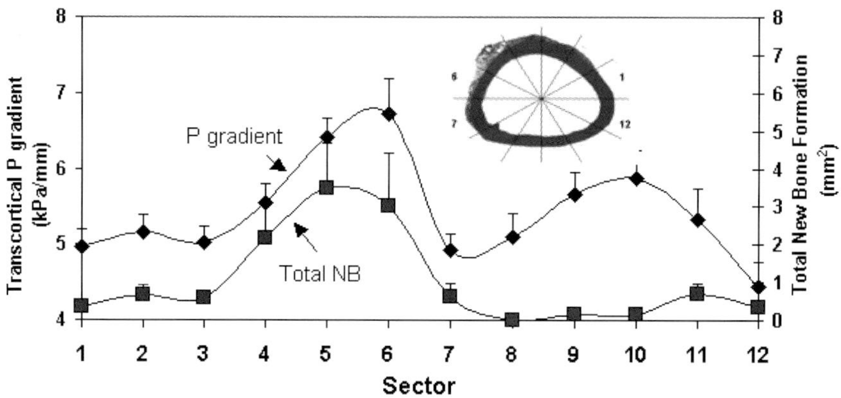

Fig. 5 Mean value (±s.e.) of transcortical fluid pressure gradient distributions for each of 12 sectors in the animals subject to fluid flow loading. Maximum pressure gradient was observed in sector 6, which corresponded to the site of maximal new bone formation. Minimum pressure gradients were located corresponding to least new bone formation sectors, i.e., sector 12 & 7. (From Qin, Y. et al., J Biomech, 2002. With permission.)

($r = -0.75$, $p = 0.01$) was demonstrated between ∇P_r. and the areas of increased intracortical porosity, indicating that a sufficient fluid flow was necessary to maintain bone mass against the challenge of disuse. These data suggest that while intramedullary pressure increases uniformly in the marrow cavity, that distinct parameters of fluid flow vary substantially, due to geometry and ultrastructure, which ultimately contribute to the spatial non-uniformity of the adaptive process. The results also indicate that the fluid flow which arises by functional loading is an important mediator in retaining bone quality and quantity, and that small fluctuations in fluid flow, achieved via pressure differentials, has a potential for therapeutic applications against skeletal disorders even in the absence of mechanical strain.

These results demonstrated that pressurizing the marrow cavity can substantially induce fluid flow in bone. While the ImP was applied at a physiological magnitude, fluid parameters may be directly involved in site-specific tissue and cellular modeling response.

6. Interstitial Fluid Flow Response to Dynamic Components of Loading [107]

Bone is considered a fluid saturated porous material. Temporal components of applied loads are particularly relevant given their direct influence on coupling between matrix deformation and strain generated fluid flow. In contrast to soft tissue (e.g., cartilage), approximately 85%-90% of bone volume consists of solid matrix, such as mineral and collagen, while the rest consists of a fluid phase such as interstitial fluid in the porous medium [9,87,124,123,150]. This fluid fills a multitude of spaces and channels in bone, including micropores, lacunae, canaliculi, Haversian canals, and Volkmann's canals [103,150,79,130,139]. The characteristics of porous medium in bone directly contributes to the temporal response of intracortical fluid flow depending on Biot's theory and Darcy's law [16,150]. While mechanical loading results in the deformation of bone matrix and the substantial interstitial fluid space, these loads also generate significant and loading rate dependent pressure gradients which further augment interstitial fluid flow [100]. Interaction between solid and fluid phases in bone is particularly important, in which loading induced fluid flows, e.g., fluid-induced canalicular shear stresses, have been proposed as a mechanism by which bone perceives mechanical stimuli in the tissue [114,137,142,143,149,27,42]. The motion of pore fluid under mechanical loading, therefore, may be critical in the regulation of bone cell activity, including the signaling and responding to physical stimuli, as well as nutrient transport [62,64,129]. Thus, not only is fluid flow in bone critical to bone remodeling, it is central to maintain bone quality.

In an attempt to evaluate, *in vivo*, the role of load-induced fluid flow it is necessary to identify three-dimensional (3-D) fluid flow behaviors as a result of dynamic loading. The macroscopic fluid parameters (i.e., pressure and pressure gradients), through computational modeling, can then be calculated as a function of variable loading parameters (e.g., frequency and rate). Thus, the principal goal of this study is to identify intracortical fluid flow distributions due to fluid pressure and pressure gradients induced by matrix strain and intramedullary pressure. In addition, the objective of this study is to evaluate macroscopic fluid flow in response to frequency and manner of loading, e.g., axial and bending loads, which have been shown to be influential in anabolic activity of bone. Thus, these entail developing a 3-D poroelastic finite element analysis (FEA) to identify the spatial distribution of fluid flow incorporated with loading frequency, and to estimate bone permeability under *in vivo* conditions using streaming potential (SP) measurements.

6.1. Experimental Design

Under a general halothane anesthesia, the left ulnae of eight adult, one year old, skeletally mature male turkeys were functionally isolated via transverse epiphyseal osteotomies [110,121] (Fig. 6). Following the osteotomy procedure, the cranial cortex of the proximal metaphysis was drilled and tapped, and a saline filled 2-cm long plastic tube was screwed into the marrow cavity. Via this tube, a 50-psi pressure transducer (Entran EPX-10IW) was inserted into the medullary canal, thus permitting measurement of the intramedullary pressure (ImP) during rest as well as during externally applied mechanical loading (Fig. 6). SPs were measured through two periosteal surface electrodes relative to a ground electrode implanted in the marrow cavity.

The animals were maintained under light anesthesia and monitored for stable vital signs, (e.g., normal pulse, O_2 concentration, and body

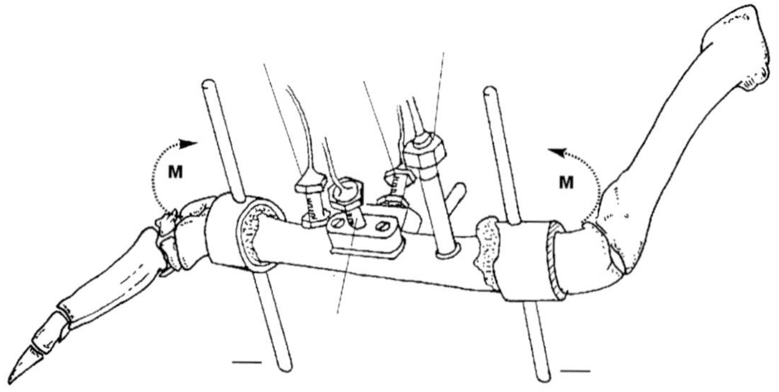

Fig. 6. A diagram of the functionally isolated turkey ulna preparation, and its loading conditions. When pins was constrained, thanscortical bone fluid flow can be generated by ImP, which detected by SP.

temperature). The ulna was loaded using load-control feedback on a modified universal mechanical testing machine (Instron) so as to generate peak normal longitudinal strains of approximately 600με. ImP, load feedback and SP signals were digitized at 800 Hz, with 16-bit resolution. Dynamic loadings were performed in both axial and bending loads, in which axial loading was applied by parallel compressing the loading pins, and bending load was achieved by compressing the loading pins only at the dorsal side resulting a bending moment in bone. Total 17 individual sinusoidal loading signals was applied for 10 second each, including frequencies of 0.1, 0.2, 0.5, 1.0, 1.5, 2.0, 2.5, 4.0, 6.0, 10, 15, 20, 25, 30, 40, 50, and 60 Hz, covering various of physiological stimuli used in previous in vivo studies [110]. For both axial and bending loading conditions, the measurements were repeated three times at each frequency.

Finally, the load feedback, ImP, SPs, and surface strains, were analyzed both in the time and in the frequency domains using Fast Fourier Transform (FFT) using custom written programs (PV-WAVE, VNI Inc., Boulder, CO). For each animal, the ImP, SP, and strain magnitudes from each frequency were averaged through the three loading trials. The mean value and the standard deviation of all measured data were calculated through the entire loading frequency range from 0.1 to 60 Hz. The ImP and SP data were normalized to the peak strain magnitude as determined by measured surface strains.

FEM development: A three-dimensional (3-D), poroelastic finite element model (FEM) was developed to simulate intracortical fluid flow induced by ImP in the avian ulna model. The geometry of the model was constructed from high-resolution CT images. The FEM consisted of 768 20-node quadratic 3-D elements in the mid-diaphyses based on the convergence study. The model was considered isotropic and assumed homogeneous with a constant 5% voids ratio to bone. An average elastic moduli and Poisson's ratio were used, with the value of $E = 12.17$ GPa and $\nu = 0.29$. The bone's hydraulic permeability, with a value of $1.8\sim2.0\times10^{-14}$ m^2 had been determined previously by correlating numerical solutions to *in vivo* streaming potential measurements. Finally, the intracortical fluid pressure (P) and pressure gradients (∇P) were calculated from FEM.

Porous Materials and Constitutive Relations of Cortical Bone: The model considers both the solid and fluid phases to be compressible. It was built using quadratic displacement and linear pore pressure elements. The constitutive analysis was based on mixture theory of biphasic materials to model solid matrix and liquid voids [1]. The mechanical behavior of the porous media consisted of the relative motion of the liquid and solid phases in response to local pressure. Assuming that the constitutive response of the poroelastic materials consists of simple bulk elasticity relationships for the fluid and the solid phases together with a constitutive theory for the pore matrix, the effective stress (σ) can be defined as a function of the strain history, temperature φ and other variables, e.g., Poisson's ratio υ: $\sigma = \sigma$ (history of $\varepsilon, \varphi, \upsilon$). The pore fluid pressure and the pore fluid flow rate are related to Darcy's law, $q_i = (k/\mu)\, \partial p/\partial x_i$, where i=1, 2, 3, q_i is the flow rate, p is the fluid pressure, k is the special permeability and μ is viscosity, which a common term often used when referring to permeability, i.e., $K = k/\mu$. This general permeability was used in this study.

Permeability of Bone Porous Structure: In this study, bone fluid flow was simulated in a 3-D poroelastic fluid model with parameters similar to the *in vivo* condition, e.g., permeability. However, due to the limited available experimental data regarding the permeabilities in various of bone porosities (e.g., Haversian canals, lacuna-canaliculi space) the bone tissue was modeled as an isotropic material, and was assumed to have an averaged 5% void ratio evenly distributed through the cortical bone [88,150]. Since there was lack of *in vivo* values for cortical permeability, the initial estimated parameter for permeability was based on previously reported *in vitro* data. For example, the permeability data varied from 1×10^{-14} m^2 to 5×10^{-14} m^2 in bovine bone [2,46]. We then narrowed down the permeability values to the range close to the frequency response of SP data. Finally, an in vivo permeability value of avian ulna was determined by the correlation between FEA and SP.

Results

Intramedullary pressure: Using a peak of 600 $\mu\varepsilon$, 1 Hz sinusoidal axial loading induced an ImP of 23 kPa (~175 mmHg), which corresponds to an approximately 10-fold increase over intramedullary pressure generated by the circulatory blood pressure alone (2.4 kPa or ~18 mmHg. Fig. 3). ImP generated by bending loads were on the order of 8 kPa (~60 mmHg) for a 1 Hz load.

SP measurement: SP magnitude rose approximately five-fold as the frequency increased from 0.1 to 30 Hz for axial loading, yet, over the same frequency range, this increase was only 2-3 fold under bending. Axial loading at 30 Hz resulted in normalized SPs of 0.64 (± 0.14) μV/με (mean ± s.d.) at the caudal site, and 0.53 (± 0.05) μV/με at the cranial site. Thirty Hz bending resulted in normalized SPs of 0.32 (± 0.06) μV/με and 0.23 (± 0.09) μV/με for the caudal and cranial sites, respectively.

It should be noted that axial and bending loads resulted in significantly different longitudinal normal strain distributions at corresponding cranial and caudal sites. Yet the transcortical SP were not significantly different in the corresponding sites. This suggests that SP may be not necessarily directly coupled with longitudinal normal strain *per se*, rather it is influenced by the local gradients of strain and fluid flow.

The value of the SPs indicated directional signals, in which the negative value represented an inward flow from the periosteal to endosteal surfaces, and a positive signal indicated an outward flow from endosteal to periosteal. While increasing ImP generated a positive SP, load-induced intracortical fluid flow was dominated by the matrix strain, i.e., axial loading generated a negative SP during loading at cranial site.

Frequency dependence of intracortical fluid pressure and pressure gradient: The poroelastic FEA demonstrated the capability to simulate fluid parameters in bone in response to dynamic loading (i.e., 0.1 to 100 Hz). In bending, the magnitude of maximum pore pressure rose 2.6 fold, from 1.49 ± 0.30 kPa/με (mean ± s.d.) at 0.1 Hz, to 3.93 ± 0.30 kPa/με at 1 Hz, and 3.5 fold to 5.21 ± 0.30 kPa/με at 30 Hz. Similar results were obtained for the axial loading, in which the pore pressure rose 3.1 fold, from 1.35 ± 0.26 kPa/με at 0.1 Hz to 4.25 ± 0.59 kPa/με at 1 Hz, and 3.9 fold to 5.28 ± 0.66 kPa/με at 30 Hz. In both manners of loading (e.g., axial and bending loads), the fluid pressure increased more quickly in the lower frequencies (0.1 to 1 Hz), than over the higher frequency

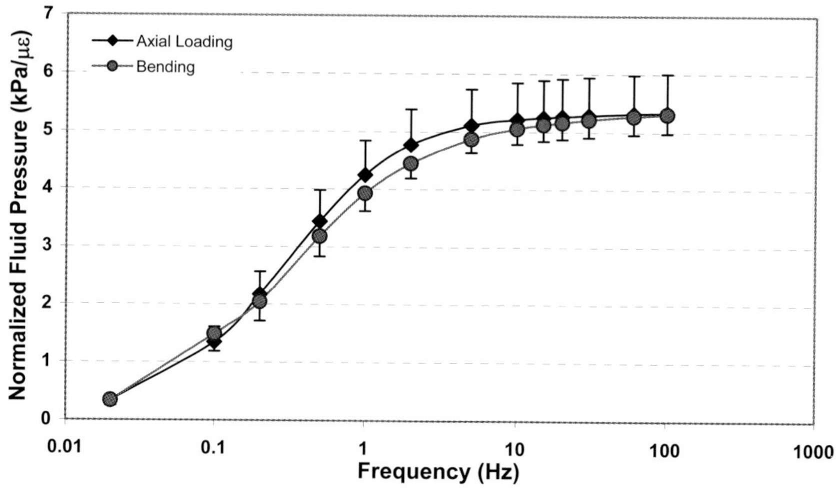

Fig. 8. Simulated in a poroelastic FEA, peak pore fluid pressure increased as loading frequency increased from 0.02 to 100 Hz. At lower frequencies (i.e., between 0.1 and 1 Hz), fluid pressure increased 3-fold. The pressure is then increased to plateau at approximately 20 Hz, and slightly reaching a threshold at approximately 60 Hz, which agreed with the SP measurement. There is no significant difference in normalized peak pore pressures generated by axial and bending loads.

range (1 to 30 Hz) (Figs. 7). A strong correlation was observed between the FEA calculated fluid pressure parameter and measured SPs in the frequencies ranging from 0.1 Hz to 30 Hz ($R^2 = 0.98$).

This work has demonstrated that intracortical fluid flow rises significantly through physiologic ImP, independent of matrix strain. This suggests that oscillation of ImP can influence the convection of fluid flow perfusion in bone tissue in many ways. For example, ImP induced by circulation alone is on the order of 18 mmHg (2.38 kPa), which will provide basic nutritional supply and fluid pressure gradients to bone. The dynamics of intracortical fluid flow in response to applied loading has been evaluated, *in vivo*, as a function of frequency and manner of loading, through a 3-D poroelastic model and validated with empirical SP measurements. It is apparent that both fluid pressure and pressure gradients are significantly influenced by the loading frequency and/or rate as indicated both empirically and theoretically [110,142,150]. However, only the fluid pressure gradient shows a strong correlation with streaming potential product, which indicates the intracortical fluid flow is strongly dependent on fluid pressure gradients, loading frequency/rate, and the surface boundary conditions. These results suggest that intracortical fluid flow is a product of not only of matrix strain gradients, but intramedullary pressure, resulting in complex spatial patterns of fluid flow. With the relationship between mechanical loading and fluid flow determined, at least in part, for bone *in vivo*, the challenge lies in correlating fluid movement under anabolic mechanical conditions to sites of new bone formation. Certainly, the fluid flow induced by ∇P may provide spatially specific, critical, and regulatory signals to the bone cell population, and certainly, it appears that the temporal aspects of the mechanical signal (i.e., frequency/rate) may play a central role in mechanotransduction.

7. Non-linear Dependence of Loading Intensity, Frequency and Cycle Number in the Maintenance of Bone Mass and Morphology[110]

As a temporal component of mechanical stimuli of bone, loading frequency has been demonstrated to be significant on bone remodeling [110,84,85,110,138]. The ability of a relatively high frequency (30 Hz) and moderate duration (60 min) loading regimen, to maintain bone mass in a turkey ulna model of disuse osteopenia was evaluated by correlating the applied strain distributions to site-specific remodeling activity. Changes in morphology were investigated following eight weeks of disuse versus disuse plus daily exposure to applied loading sufficient to induce peak strains of approximately 100 $\mu\varepsilon$. The results confirm the strong anti-resorptive influence of mechanical loading, and identify a threshold near 70 $\mu\varepsilon$ for a daily loading regimen at 30 Hz. These results suggest that the frequency or strain rate associated with the loading stimulus must play a critical role in the mechanism by which bone responds to mechanical strain (Table 2).

The adaptive sensitivity of bone to loading frequency could be explained when we consider the mechanotransduction mechanisms of the hydrodynamic interstitial fluid flow at a cellular level. Although cortical bone appears as a rigid tissue, the bone cells are tethered to the matrix in

Table 2. Mechanical stimulus resulting in equivalent strain magnitude to bone mass maintenance [110].

Model	**Strain (με)**	**Daily cycle number**
Rooster ulna	2,000 (0.5Hz) (Rubin & Lanyon, 1984)	4
Turkey ulna	1,000 (1Hz) (Rubin & Lanyon, 1985)	100
Human tibia	850 (Lanyon et al., 1975)	1,000
Human tibia	400 (Lanyon et al., 1975)	10,000
Turkey ulna	700 (1Hz) (McLeod & Rubin, 1992)	600
Turkey ulna	400 (30Hz) (McLeod & Rubin, 1992)	18,000
Turkey ulna	270 (60Hz) (McLeod & Rubin, 1992)	36,000
Turkey ulna	100 (30Hz) (Qin et al., 1998)	108,000

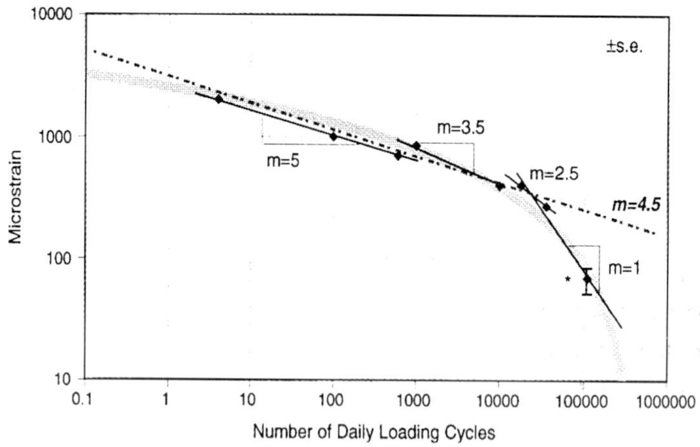

Fig. 9. Strain "threshold" (με) required to maintain bone mass as a function of daily loading cycle number. The regression lines to predict m values were determined using experimental results, following the equation of $y = 10^{2.28}(5.6 - \log_{10}x)^{1.5}$. A curve fit to the data permits extrapolation to daily loading cycle numbers less than 1 cpd and greater than 100 k cpd, with which the strain necessary to maintain bone mass will decrease as the daily loading cycle number increases. (From Qin, Y. et al., J Orthop Res, 1998. With permission.)

a bath of extracellular fluid. Clearly, there are two distinct phases in bone: the mineralized matrix (80~90%) and an interstitial fluid component. Bone cells could sense and respond to mechanical strain through this fluid layer by means of fluid pressure, pressure gradients and its generated fluid shear stresses, enhancement of nutritional supply, or mechano-electrical effects. Increasing loading frequency could substantially increase fluid pressure and pressure gradients, which may directly stimulate cellular activity. These data also support the interdependent role of loading frequency and cycle number, demonstrating that the inhibition of bone resorption and intracortical turnover to be far more sensitive to high frequency, high cycle number loading than to an equal time dose of lower frequency (and thus low cycle number) loading, even though it is of much greater magnitude.

Implication to the proposed study: Considering the strong anabolic potentials of high frequency mechanical stimulation, its potential adaptive role may be explained through bone fluid flow mechanism because interstitial fluid flow is critically sensitive to temporal components of mechanical loading. To test both fluid loading rate/frequency and intensity in the in vivo model may yield insight for understanding cellular mechanotransduction of bone adaptation.

8. Discussion

Given the porous nature of bone, the fluid filled spaces invariably generate a flow upon mechanical loading. In general, load-induced flow and its associated matrix strain are usually coupled. Therefore, segregating the regulatory potential of matrix strain from the anabolic potential of fluid flow becomes inherently difficult. Matrix strain, as a general parameter of bone receiving mechanical loading, is commonly used in describing bone tissue deformation. If bone fluid flow is indeed a key mediator for bone modeling and remodeling, then it is important to test the accommodation of tissues and cells to a customary flow loading environment.

8.1. *Bone fluid flow contributed from multiple channels of porosity — indicated by relaxation*

Complex porous structure provides a fluid pathway for mechanotransduction in bone when it subjects to loading. The data demonstrated complex fluid relaxation patterns, in which no single exponential curve could be simply fit into either ImP or SP decay curves. Multiple time constants were observed in both SP and ImP measurements. These imply that multiple fluid channels and porosities, as well as permeabilities, influence the cortical fluid movement. While the initial response of the relaxation observed in ImP was on a similar order of SPs, it decayed more quickly, yielding time constants of 0.1s and 0.2s respectively. Also the decay of ImP was approximately five times faster than the relaxation time of SP over the remaining period. The initial response of fluid relaxation appears to be a product of the vascular porosity, e.g., contributing approximately 70% of the fluid relaxation in the first 0.2s, when considering the anatomic connections between bone's vascular system and its linked marrow cavity. The ImP decayed much more quickly than the SP after the initial decay period. However, the SP dominates the remaining period of the fluid relaxation curve. This suggests that smaller porosities may have a large influence on the fluid flow in bone, and most probably may be lacunar-canalicular porosity or even the smaller systems, e.g., micropore. It must be noted that when the analysis applies to a longer tail of the relaxation curve, e.g., expanding to 2s or longer after loading, the relaxation time varies in the *slow decay track*, particularly in SP measurement, as reflected by an even smaller fluid compartment or porosity such as microporosity.

It is well demonstrated that deformation of bone matrix will produce SPs in both *in vitro* and *in vivo* conditions. The typical relaxation time for SP measurement was reported to be on the order of 0.1-1.0 s in several animal models, e.g., canine and bovine [93,123,94,39,47]. Using a canine tibia model, Otter et al. (1992) compared streaming potentials under *in vivo* and *in vitro* conditions, in which the typical relaxation times were observed around 0.45s (*in vivo*) and 0.81s (*in vitro*). The value of experimentally measured SPs, falling in the range of 0.1s and 1.0s, are dependent on the experimental conditions and animal models.

Therefore, the measured SP in this work, using the turkey ulna model, appeared to be consistent with those previously reported. Further, in this study, ImP and SP decay times were measured simultaneously, demonstrated to be similar to the rate of the initial decay period, though the ImP decayed more quickly than the SPs.

There are several possible explanations for the time constant observed in the relaxation of fluid flow in bone. First, it implies that *in vivo* these pores contain vessels and cells. Even *in vitro* these pores may contain cellular debris which would impede fluid flow. Second, the Haversian systems could be completely clogged by induced cortical pore and medullary pressure (like a "pressure chamber"), which would greatly reduce the permeability of the bone. Third, there may exist an electromechanical drag force generated by fluid movement in bone which prevents a faster decay [77]. Finally, the decay time is due to the relaxation of fluid flow in other pore structures, i.e., lacunar-canalicular systems, which may be even partially clogged during the loading and the remains of the cell processes. The first phenomena may further relate to the bone effective size of the Haversian systems which *in vivo* vasculature canal size will be smaller because of the presence of blood vessels and cells. As described by Johnson et al. (1982)[47], the fluid in the Haversian systems will be blood and extracellular fluid, both of which will be more viscous than water as used in their mathematical model. This viscosity will act to increase the time constant for the relaxation of the fluid pressure, e.g., ImP and intracortical pressure, *in vivo*. In the related fluid frequency response study[107], it demonstrates that the ImP plateau occurs at frequencies between 1 and 10 Hz, which is 2~3 orders of magnitude smaller than a model which uses a damping frequency based on water as a medium [47]. This may be important to explain the time constant observed by *in vivo* ImP and SP measurements.

8.2. Bone fluid flow induced by dynamic intramedullary pressure

In addition to the flow generated by matrix deformation in bone, this study showed a new mechanism of load induced fluid flow, i.e., induced intramedullary pressures can influence fluid movement in bone. The influence of dynamic strains on fluid stresses arises through two distinct pathways: mechanical deformation of the medullary canal and deformation of the cortical bone. The marrow cavity of bone, similar to a fluid chamber with capillaries, is normally pressurized by the blood flow into the bone. Blood flow is pulsatile, though the pressures recorded within the marrow cavity are only slightly elevated from venous pressures. In the turkey ulna, the average marrow pressures were found to be 1 to 2.4 kPa (approximately 15 mmHg) [96,107]. However, under the presence of mechanical loading, and in particular axial compression, marrow pressure rises dramatically. Our data show that the influence of mechanical deformation on intramedullary fluid pressures is highly kinematic, i.e., frequency dependent [107]. At lower frequencies, there are numerous outlets for fluid flow as pressure increases, but with more rapid pressure changes, the viscosity of the marrow fluid precludes rapid flow and pressures rapidly build. Indeed, step loading of ulna, simulating an impact strike of the stance phase, resulted in increase of marrow pressure as high as 10 kPa with approximately 600 $\mu\varepsilon$ in this particular case. The results suggest that intracortical porosities can serve as the relaxation pathways for load-generated fluid pressures.

An increase in intramedullary pressure can influence bone's fluid pathways through several coupling mechanisms. First, as the pressure in the medullary cavity increases towards the arterial blood pressure, fluid flow into the marrow cavity will be greatly inhibited and perhaps even stopped. However, the increase in fluid velocity out of the medullary canal will relieve much of the pressure buildup associated with dynamic loading, i.e., step loading. Importantly, the outward flow does not provide complete compensation for this relief mechanism, and thus high marrow cavity pressures arise immediately after step loading. Second, in the same manner that deformation of the bone results in compression of the porous space with the cortical bone, induced pore pressure is predicted at least one order of magnitude higher than load induced marrow cavity pressure by the finite element analysis [106]. This results in the kinematic loading induced fluid pressure buildup in cortical bone being much greater than that achieved within the marrow cavity, though net fluid motion may be far less. This implies the existence of a fluid-related coupling mechanism. While the ability of fluid to leave the intracortical pores is restricted because of the deformation of the pore space under loading, because of the pressure differences, the marrow cavity will still provide a pathway for relief of the intracortical pressure. Thus, while induced marrow cavity pressure can potentially reduce the fluid flow, the fluid relaxation pathways are still active, especially during unloading, in both venous and intracortical pores.

Moreover, regardless of the evidence of the dynamic fluid movement through bone, arguments remain in the level of fluid compartments involved in fluid movement and their connections, i.e., the fluid flow may be impermeable beyond the porosity of lacunae-canaliculi. For example, bone permeability studies have shown the changes which occur, as bone matures, in the interstitial pathway [76], such as degrees of mineralizing of porosity. Loading induced bone fluid flow has also been studied using stress-generated potentials (SGP) to map fluid movement [39,94,95,97,101,102,124,123]. These investigators have suggested that the site of the SGP was in part due to the collagen-apatite porosity (order 10 nm in radius), because smaller pores of approximately 16 nm radius were consistent with the relaxation time in the streaming potential measurements. To predict the contributions of both lacunae-canaliculi and collagen-apatite compartments, a theoretical model was suggested based upon the experimentally observed SGP [77,79,95,124]. However, Cowin et al. [17] have suggested that the streaming potential data [95,124] could also be consistent with the larger pore size (100 nm), i.e., canaliculi, if the effects of hydraulic drag and electrokinetic were taken into account.

8.3. Potentials of bone fluid flow in cellular stimulation

In light of the importance of lacunar-canalicular porosity in fluid flow and mechanosensory effects in bone, i.e., its mechanotransduction role associated with osteocytes and the osteocytes process, several theoretical models based on bone micro structure were presented to simulate bone fluid flow at this level of porosity [17,21,77,133,135,134,140,142,150]. Weinbaum et al. [142] and Cowin et al. [17] have suggested that fluid movement in the region between the cell membrane of the osteocyte and the matrix wall of the lacunae and canaliculi was critical to the mechanosensory process. Regarding the actual sites of bone fluid flow, using an analytical model Cowin et al. [17] and Weinbaum et al. [142] have shown that the results of *in vitro* and *in vivo* SPs [95,123,125] indicate that the lacunar-canalicular (100 nm) anatomical space is indeed the responsible sites for the strain-generated potentials, if the contributions of both the electrokinetic effect and hydraulic drag force are considered as bone fluid travels through the glycocalyx filled pore space. Current study indicates that at least two relaxation patterns of ImP and SP exist, supporting the idea that fluid flow is associated with an electrokinetic as well as a mechanotransduction mechanism. The reported study using both mechanotransduction and mechano-electronics measurement, i.e., ImP and SP, supports the argument that multiple porosity permeability structures should be considered to model bone fluid flow. The parameters associated with different pore sizes, e.g., permeability, should reflect this complex microarchitecture of bone. However, if the vascular and lacunar-canalicular pores are considered, the pore structure of bone fluid flow pathway, then two to three orders of size difference between vascular (~100μm) and lacunar-canalicular (~100nm) spaces should be considered in these parameters. This suggests that a pressure peak in the vasculature will relax 1000 times faster than a pressure peak in the lacunae-canaliculi. If the ImP is considered equivalent with vascular pressure in the osteon, the relaxation time reflects the fluid decay in the cortical bone. Nevertheless, fluid movement through cortical channels effects fluid relaxation. These data may be further used for theoretical estimation [128] and numerical simulation for determining the poroelastic parameters in cortical bone.

8.4. Fluid pressure gradients as a driving source for fluid movement in bone and initiating bone adaptation

The study tried to separate matrix strain and convective fluid flow by dynamically pressurizing the marrow cavity which drives interstitial fluid to flow. The fluid magnitude for such a flow remained in the physiological range generated in the marrow cavity by an animal's normal activity [28]. It is difficult to envision a physical mechanism by which ImP loading would result in new bone formation that could be generated by such small matrix strain, particularly in light of the strong *in vitro* evidence that fluid flow can perturb the biological response of bone cells. This experiment suggests that fluid flow can, in and of itself, influence parameters of bone formation and resorption.

The sites of greatest osteogenic response correlated with the greatest gradient of transcortical fluid pressures. The strong correlation between new bone formation and fluid flow suggests that fluid components, i.e., pressure gradients (which drive fluid velocity and fluid shear stress), may directly influence the response of bone cells to mechanical stimulation. In addition, the correlation between minimal intracortical porosity and elevated fluid pressure gradients implies that a basal level of convectional bone fluid flow is critical in preserving cortical mass against disuse, such as conditions of bed rest and microgravity. At the very least, it is clear that extremely low-level perturbations of fluid flow, as induced by high frequency oscillations, are providing necessary signals to inhibit intracortical porosity and stimulate new bone formation. Given the anabolic potential of these high frequency signals [117,117], and the rapid rise in fluid velocities that occur because of high frequencies even in conditions of very low strain [141,143], it is certainly possible that signaling the cells responsible for orchestrating bone adaptation is achieved not by subtle changes in matrix strain, but by changes in fluid flow.

8.5. Fluid pressure gradients and fluid shear stress

The strong correlation between distinct fluid flow components, i.e., pressure gradients driven by ImP, is interesting because of its potential to impose fluid shear stress in the cellular environment. A number of theoretical models have been proposed to describe a potential mechanism of fluid pressure and fluid shear stress in bone [17,19,79,142], which have been supported by mounting *in vitro* experimental work [9,27,45]. The effects of an increase in fluid flow induced by oscillatory ImP can potentially influence bone cell activities through several coupling mechanisms. First, raising the ImP can result in a corresponding increase of outward fluid flow through various fluid pathways, which include the vascular system and the extensive lacunar-canalicular spaces in which the bone cell population resides. Increased fluid velocities can produce fluid shear stresses on the endothelial lining cells of vessels [32] and on the bone cells in the lacunar spaces [142], where the oscillatory ImP can alter the fluid shear stresses on the cell population and trigger a cellular response. Second, a nutrient pathway for metabolism and the proper disposal of waste products generated from catabolic activities occur through fluid channels. In the soft tissue, molecular diffusion is considered the major pathway for transportation of metabolites [96]. Because of the relatively dense structure of cortical bone, however, the diffusive mechanism may, in fact, be insufficient to play an adequate role in transporting metabolic constituents between osteocytes and the surrounding vascular canals. An imposed dynamic ImP will enhance this fluid transportation from the blood supply to osteocytes through this convective perfusion mechanism [100,135,134,139], where the greatest exchange occurs at sites of greatest pressure gradients.

8.6. *Static vs. dynamic loading*

While physiologic fluid flow showed the potential to initiate the modeling and remodeling process, dynamic components of this fluid may also play an important role in the regulation of adaptation. It is recognized that bone tissues respond very differently to static vs. dynamic load environments, and results in an adapted structure which demonstrates similar peak strain magnitudes during vigorous activity [66,120,67,120]. These regulatory "temporal" components may include strain rate, strain frequency, and strain gradients [40,92,110,119,120,119,122,138]. These temporal components result not only in local matrix deformation, but also in fluid flow, streaming potentials, and other physical phenomena, which also influence cell responses. For example, in the case of the turkey ulna, 10 minutes of loading per day at 1 Hz requires a peak induced longitudinal normal strain greater than 700 $\mu\varepsilon$ to maintain bone mass, while a relatively high frequency (30Hz) loading regimen reduces this threshold to 70 $\mu\varepsilon$ [110]. The stimulatory effects of fluid flow, driven at physiological magnitude and high frequency but with minimal matrix strain, may depend on the cellular response due to (1) intermittent rather than static flow constant velocity, (2) direct fluid shear stress perturbation, (3) the cumulative effect of small local fluid movements resulting in cells accommodating to large flow cycles, and (4) even an "amplified" effect on the bone cell which could result in pressurization and/or fluid shear stress on the cell [143]. Again, these data, while not intended to diminish the role of bone strain, imply that anabolic fluid flows, applied in a dynamic manner, can have a tremendous influence on bone mass and morphology even under conditions of extremely low matrix deformations.

That fluid flow results in periosteal expansion in response to intramedullary pressure and transcortical pressure gradients help identify a physical mechanism for the response. Since the periosteum is often referred to as an impermeable layer for fluid perfusion, it is understandable that periosteal modeling requires fluid exchange and/or flow to initiate such an adaptive process. Fluid flow resulted in periosteal bone formation in this study, and thus implies that oscillations of ImP influence bone fluid perfusion and convection in many ways. While the endosteal surface provides an open circulation between marrow pressure and intracortical flow, the interstitial fluid flow in bone must flow out of the mineral to the periosteal surface through a variety of fluid pathways [88,135,139]. Since the loading pattern used in these experiments was oscillatory, it may not be necessary that fluid physically flowed out of the periosteal surface but, instead, the oscillation itself may serve as a stimulatory signal. Under oscillatory fluid stimulation, however, a local fluid pressure gradient may be built up with the semi-permeable periosteal boundary condition which will create a flow at the periosteal surface. The spatial distributions of such fluid flow patterns ultimately is dependent on the fluid pressure gradients, defined somewhat by the geometry, ultrastructure and fluid pathways of the bone.

8.7. Cortical bone permeability

The work presented here provides a unique *in vivo* approach (e.g., combined experimental measurements and numerical analyses) to define the permeability, *in vivo*, of cortical bone, a fundamental parameter which governs intracortical fluid flow. Thus providing insight into the applicability of determining permeability values *in situ* [105,128]. The result of permeability prediction is based on the frequency characteristics of the response to load of both SPs and the FEA calculated fluid pressure buildup and decay. While at lower frequencies, a lower fluid pressure provides sufficient time to relax from the cumulated pressure, the rapid increase in fluid pressure which results from high frequency loading provides limited time for fluid dispersion. Thus, there was a direct proportionality between bone fluid pressure and loading frequency as evidence by pressure relaxation time, and was inversely proportional to the cortical permeability. The value of permeability determined in this work fell in the range of $4.0\text{-}5.0 \times 10^{-14}$ m^2, and closely represents the *in vivo* cortical permeability of adult, secondarily remodeled, lamellar bone. It is important to emphasize, however, that this value may only reflect the bulk cortical permeability, while its contribution in relation to bone's microstructure (i.e., values for the lacunae-canaliculi system) remains unclear.

There are a number of factors that may influence bone permeability. Using a bovine model, *in situ*, Johnson [46] estimated values for permeability to fall in a range from 1 to 5×10^{-14} m^2. These values are clearly sensitive to changes in porosity, age and the sites of bone, as it has been reported in a canine model [76] that the permeability of bone from young dogs is 6x greater than adult dog bone, while the porosity was only 3.5 times higher. The average value of permeability for the adult canine tibiae was in a range of 3.35×10^{-14} m^2. It is also important to emphasize that the permeability of bone will be affected not only by the number and size of porosities (including microporosity), but the status of vascular channels, the stature of lacunae and the canaliculi which contain osteocytes and whether these cells are alive [104]. As evidence of the sensitivity of such measurements, Johnson emphasizes that cortical permeability, as measured *in situ*, was strongly dependent on the viability of the vascular channels, and particularly whether these channels were obfuscated by clotted blood and other "debris" [46,128]. The permeability may be affected by a number of porosities in bone which include: i) vascular channels; ii) the lacunae and the canaliculi that contain osteocytes; and iii) microporosity in the bone matrix. Indeed, multiple time constants of fluid relaxation were observed in bone's response to a step loading, *in vivo*, using ImP and SP measurements [105]. This suggests that multiple fluid channels and porosities, as well as permeabilities, may influence the cortical fluid movement. The ImP relaxation was shown to decay much more quickly than the SP. However, the SP still dominated the majority of fluid relaxation in bone. If ImP relaxation is considered to be associated with vasculature flow, this implies that smaller porosities, e.g., lacunae-canaliculi, must have a large influence on the fluid flow in bone. In addition, these results suggest that SP and ImP decays are determined by a hierarchical *interdependent* system of multiple porosities, which suggest that an averaged permeability may still be valid for fluid flow transportation if bone porosities are considered in the macro tissue level. The measured in vitro permeability can only represent the flow through the vascular system, because of the clotted change from in vivo to in vitro conditions. Clearly, when considering the permeability of bone, great measures must be taken to ensure that the measurements are reflective of the physiologic state, i.e., under *in vivo* condition.

8.8. Summary and significance

Understanding microcirculation regulated capillary filtration and interstitial fluid flow in musculo-skeletal tissues and the physiological mechanisms that regulate osteonal new bone formation and skeletal remodeling may lead to the understanding of bone-related and muscle-influenced diseases. Musculoskeletal complications such as osteoporosis, muscle atrophy, the loosening of joint replacements, and the delayed healing of fractures are major health problems. For example, when osteopenia, the progressive loss of bone as a function of age, becomes osteoporosis, this loss of bone can lead to crippling fractures. Annually, some 20 million women suffer from osteoporosis in this country alone, with an estimated annual cost to our health programs of over 15 billion dollars. We believe a unique strategy for the prevention and/or treatment of such skeletal complications is to harness bone tissue's sensitivity to its physical environment. More specifically, intracortical fluid flow, should it prove a key mediator of the osteogenic response, may open up unique interventional approaches for the treatment of musculoskeletal disorders, e.g., delayed-union of bone fracture, prevention of osteoporosis, and orthopedic implant fixation.

There is strong evidence that osteopenia occurs due to a lack of physical activity in the skeleton, e.g., in conditions of microgravity, in which muscle activities are strongly correlated to the degree of bone mineral density. It is also strongly demonstrated that bone fluid flow may be a potential mechanism for regulating bone adaptation and turnover. Understanding interstitial fluid flow in bone induced by dynamic muscle pump and the physiological mechanisms that regulate osteonal adaptation may lead to the understanding of musculo-skeletal diseases. If proposed research proven promising, the data will provide new insight on the role of fluid flow in understanding the mechanism of muscle-bone adaptation from tissue to cellular level in vivo. Long term, the results could lead to new approaches and modalities to treat degenerative bone loss and muscle atrophy.

This chapter overviews the capability of mechanotransduction induced by loading in enhancing microcirculatory flow, regulating marrow pressure and initiating bone adaptation. This study has evaluated the role of dynamic and temporal components of the stimuli in adaptive response. Such a designed model has provided a unique and rigorous model to quantify how adaptation responds to applied fluid stimuli. Further understanding of the mechanical control of musculo-skeleton can help for the mechanism of bone's adaptive response. Further, identification of the dynamic patterns of the specific mechanical milieu which control the adaptation may provide a possible strategy for the prevention of osteoporosis, muscle atrophy, and acceleration of fracture and injury healing.

Acknowledgements

This work is kindly supported by The Whitaker Foundation (99-0024), The US Army Medical and Materiel Command (DAMD-17-02-1-0218), and NIH #AR-49286. The authors wish to thank Marilynn Cute, Dr. Wei Lin, Tamara Kaplan, Anita Saldanha and other people in the Orthopaedic Biomechanics and Bioinstrumentation Laboratory for the assistance and support.

Corresponding Author:

Yi-Xian Qin, Ph.D.
Dept. of Biomedical Engineering
Stony Brook University
350 Psychology-A Bldg.
Stony Brook, NY 11794-2580
Voice: 631-632-1481
FAX: 631-632-8577
Email: Yi-Xian.Qin@sunysb.edu

References

1. ABAQUS Solution. Finite Element Analysis Theory Manual. Hibbitt, Karlsson & Sorensen, Inc. Version 5.7~6.0:1997.
2. Ashman, R. B., S. C. Cowin, W. C. Van Buskirk, and J. C. Rice. A continuous wave technique for the measurement of the elastic properties of cortical bone. J.Biomech. 17:349-361,1984.
3. Beaupre, G. S., T. E. Orr, and D. R. Carter. An approach for time-dependent bone modeling and remodeling- application: a preliminary remodeling simulation. J.Orthop.Res. 8:662-670,1990.
4. Beaupre, G. S., T. E. Orr, and D. R. Carter. An approach for time-dependent bone modeling and remodeling-- theoretical development. J.Orthop.Res. 8:651-661,1990.
5. Bergula, A. P., W. Huang, and J. A. Frangos. Femoral vein ligation increases bone mass in the hindlimb suspended rat. Bone. 24:171-177,1999.
6. Brand, R. A., C. M. Stanford, and D. P. Nicolella. Primary adult human bone cells do not respond to tissue (continuum) level strains. J.Orthop.Sci. 6:295-301,2001.
7. Brown, T. D., D. R. Pedersen, M. L. Gray, R. A. Brand, and C. T. Rubin. Toward an identification of mechanical parameters initiating periosteal remodeling: a combined experimental and analytic approach. J.Biomech. 23:893-905,1990.
8. Buckley, M. J., A. J. Banes, and R. D. Jordan. The effects of mechanical strain on osteoblasts in vitro. J.Oral Maxillofac.Surg. 48:276-282,1990.
9. Burger, E. H. and J. Klein-Nulend. Mechanotransduction in bone--role of the lacuno-canalicular network. FASEB J. 13 Suppl:S101-S112,1999.
10. Burger, E. H. and J. P. Veldhuijzen. Influence of Mechanical Factors on Bone Formation, resorption,Growth inVitro. 1993.
11. Chen, N. X., K. D. Ryder, F. M. Pavalko, C. H. Turner, D. B. Burr, J. Qiu, and R. L. Duncan. Ca(2+) regulates fluid shear-induced cytoskeletal reorganization and gene expression in osteoblasts. Am.J.Physiol Cell Physiol. 278:C989-C997,2000.

12. Chow, J. W. M., A. J. Wilson, T. J. Chambers, and S. W. Fox. Mechanical loading stimulates bone formation by reactivation of bone lining cells in 13-week-old rats. Journal of Bone and Mineral Research. 13:1760-1767,1998.
13. Churches, A. E. and C. R. Howlett. Functional adaptation of bone in response to sinusoidally varying controlled compressive loading of the ovine metacarpus. Clin.Orthop. 265-280,1982.
14. Convertino, V. A. Mechanisms of microgravity induced orthostatic intolerance: implications for effective countermeasures. J Gravit.Physiol. 9:1-13,2002.
15. Convertino, V. A. and H. Sandler. Exercise countermeasures for spaceflight. Acta Astronaut. 35:253-270,1995.
16. Cowin, S. C. Bone poroelasticity. J.Biomech. 32:217-238,1999.
17. Cowin, S. C., S. Weinbaum, and Y. Zeng. A case for bone canaliculi as the anatomical site of strain generated potentials. J.Biomech. 28:1281-1297,1995.
18. Cullen, D. M., R. T. Smith, and M. P. Akhter. Bone-loading response varies with strain magnitude and cycle number. J.Appl.Physiol. 91:1971-1976,2001.
19. Dillaman, R. M., R. D. Roer, and D. M. Gay. Fluid movement in bone: theoretical and empirical. J.Biomech. 24 Suppl 1:163-177,1991.
20. Donaldson, C. L., S. B. Hulley, J. M. Vogel, R. S. Hattner, J. H. Bayers, and D. E. McMillan. Effect of prolonged bed rest on bone mineral. Metabolism. 19:1071-1084,1970.
21. Doty, S. D. and B. H. Schofield. Metabolic and structural changes with osteocytes of rat bone. In: Calcium, Prarathyroid Hormone and the Calcitonins, Edited by Talmage & Munson, Excerpta Medica, Amsterdam. 353-364,1972.
22. Evans, F. G. The mechanical properties of bone. Artif.Limbs. 13:37-48,1969.
23. Evans, F. G. Factors affecting the mechanical properties of bone. Bull.N.Y.Acad.Med. 49:751-764,1973.
24. Evans, F. G. and R. Vincentelli. Relations of the compressive properties of human cortical bone to histological structure and calcification. J.Biomech. 7:1-10,1974.
25. Fiorotto, M. L., H. P. Sheng, H. J. Evans, A. D. LeBlanc, P. C. Johnson, and B. L. Nichols. Specific effects of weight loss, protein deficiency and energy deprivation on the water and electrolyte composition of young rats. J Nutr. 117:933-940,1987.
26. Fluckey, J. D., E. E. Dupont-Versteegden, D. C. Montague, M. Knox, P. Tesch, C. A. Peterson, and D. Gaddy-Kurten. A rat resistance exercise regimen attenuates losses of musculoskeletal mass during hindlimb suspension. Acta Physiol Scand. 176:293-300,2002.
27. Frangos, J. A., T. Y. Huang, and C. B. Clark. Steady shear and step changes in shear stimulate endothelium via independent mechanisms--superposition of transient and sustained nitric oxide production. Biochem.Biophys.Res.Commun. 224:660-665,7-25-1996.
28. Fritton, S. P., K. J. McLeod, and C. T. Rubin. Quantifying the strain history of bone: spatial uniformity and self- similarity of low-magnitude strains. J.Biomech. 33:317-325,2000.
29. Frost, H. M. Intermediary Organization of the Skeleton. Boca Raton:1986.
30. Fyhrie, D. P. and D. R. Carter. A unifying principle relating stress to trabecular bone morphology. J.Orthop.Res. 4:304-317,1986.

31. Gerdhem, P., K. A. Ringsberg, K. Akesson, and K. J. Obrant. Influence of muscle strength, physical activity and weight on bone mass in a population-based sample of 1004 elderly women. Osteoporos.Int. 14:768-772,2003.
32. Girard, P. R. and R. M. Nerem. Shear stress modulates endothelial cell morphology and F-actin organization through the regulation of focal adhesion-associated proteins. J.Cell Physiol. 163:179-193,1995.
33. Goldstein, S. A., L. S. Matthews, J. L. Kuhn, and S. J. Hollister. Trabecular bone remodeling: an experimental model. J.Biomech. 24 Suppl 1:135-150,1991.
34. Greenleaf, J. E., E. M. Bernauer, W. C. Adams, and L. Juhos. Fluid-electrolyte shifts and VO2max in man at simulated altitude (2,287 m). J Appl.Physiol. 44:652-658,1978.
35. Greenleaf, J. E., B. L. Castle, and D. H. Card. Blood electrolytes and temperature regulation during exercise in man. Acta Physiol Pol. 25:397-410,1974.
36. Greenleaf, J. E. and R. D. Reese. Exercise thermoregulation after 14 days of bed rest. J Appl.Physiol. 48:72-78,1980.
37. Greenleaf, J. E., E. Shvartz, S. Kravik, and I. C. Keil. Fluid shifts and endocrine responses during chair rest and water immersion in man. J Appl.Physiol. 48:79-88,1980.
38. Grigoriev, A., B. Morukov, G. Stupakov, and E. Bobrovnik. Influence of bisphosphonates on calcium metabolism and bone tissue during simulation of the physiological effects of microgravity. J.Gravit.Physiol. 5:69-70,1998.
39. Gross, D. and W. S. Williams. Streaming potential and the electromechanical response of physiologically-moist bone. J.Biomech. 15:277-295,1982.
40. Gross, T. S., J. L. Edwards, K. J. McLeod, and C. T. Rubin. Strain gradients correlate with sites of periosteal bone formation. J.Bone Miner.Res. 12:982-988,1997.
41. Hicks, A., S. McGill, and R. L. Hughson. Tissue oxygenation by near-infrared spectroscopy and muscle blood flow during isometric contractions of the forearm. Can.J Appl.Physiol. 24:216-230,1999.
42. Hillsley, M. V. and J. A. Frangos. Review: bone tissue engineering: the role of interstitial fluid flow. Biotechnology and Bioengineering. 43:573-581,1994.
43. Huiskes, R., H. Weinans, H. J. Grootenboer, M. Dalstra, B. Fudala, and T. J. Slooff. Adaptive bone-remodeling theory applied to prosthetic-design analysis. J.Biomech. 20:1135-1150,1987.
44. Hung, C. T., S. R. Pollack, T. M. Reilly, and C. T. Brighton. Real-time calcium response of cultured bone cells to fluid flow. Clin.Orthop. 256-269,1995.
45. Jacobs, C. R., C. E. Yellowley, B. R. Davis, Z. Zhou, J. M. Cimbala, and H. J. Donahue. Differential effect of steady versus oscillating flow on bone cells. J.Biomech. 31:969-976,1998.
46. Johnson, M. W. Behavior of fluid in stressed bone and cellular stimulation. Calcif.Tissue Int. 36 Suppl 1:S72-S76,1984.
47. Johnson, M. W., D. A. Chakkalakal, R. A. Harper, J. L. Katz, and S. W. Rouhana. Fluid flow in bone in vitro. J.Biomech. 15:881-885,1982.
48. Jones, H. H., J. D. Priest, W. C. Hayes, C. C. Tichenor, and D. A. Nagel. Humeral hypertrophy in response to exercise. J.Bone Joint Surg.Am. 59:204-208,1977.

49. Joyner, M. J. Does the pressor response to ischemic exercise improve blood flow to contracting muscles in humans? J Appl.Physiol. 71:1496-1501,1991.
50. Joyner, M. J. Blood pressure and exercise: failing the acid test. J Physiol. 537:331-12-1-2001.
51. Joyner, M. J. and N. M. Dietz. Nitric oxide and vasodilation in human limbs. J Appl.Physiol. 83:1785-1796,1997.
52. Joyner, M. J., R. L. Lennon, D. J. Wedel, S. H. Rose, and J. T. Shepherd. Blood flow to contracting human muscles: influence of increased sympathetic activity. J Appl.Physiol. 68:1453-1457,1990.
53. Joyner, M. J., L. A. Nauss, M. A. Warner, and D. O. Warner. Sympathetic modulation of blood flow and O2 uptake in rhythmically contracting human forearm muscles. Am.J Physiol. 263:H1078-H1083,1992.
54. Joyner, M. J. and D. N. Proctor. Muscle blood flow during exercise: the limits of reductionism. Med.Sci.Sports Exerc. 31:1036-1040,1999.
55. Joyner, M. J. and W. Wieling. Increased muscle perfusion reduces muscle sympathetic nerve activity during handgripping. J Appl.Physiol. 75:2450-2455,1993.
56. Judex, S., T. S. Gross, and R. F. Zernicke. Strain gradients correlate with sites of exercise-induced bone-forming surfaces in the adult skeleton. J.Bone Miner.Res. 12:1737-1745,1997.
57. Judex, S. and R. F. Zernicke. High-impact exercise and growing bone: relation between high strain rates and enhanced bone formation. J.Appl.Physiol. 88:2183-2191,2000.
58. Judex, S. and R. F. Zernicke. High-impact exercise and growing bone: relation between high strain rates and enhanced bone formation. J.Appl.Physiol. 88:2183-2191,2000.
59. Keller, T. S., A. M. Strauss, and M. Szpalski. Prevention of bone loss and muscle atrophy during manned space flight. Microgravity.Q. 2:89-102,1992.
60. Kelly, P. J., K. N. An, E. Y. S. Chao, and J. A. Rand. Fracture healing: biomechanical, fluid dynamic and eletrical considerations. Bone Mineral Research, Elsevier, New York. 295-319,1985.
61. Kelly, P. J. and J. T. Bronk. Venous pressure and bone formation. Microvasc.Res. 39:364-375,1990.
62. Kelly, P. J., T. Yipintsoi, and J. B. Bassingthwaighte. Blood flow in canine tibial diaphysis estimated by iodoantipy. J.Appl.Physiol. 31:38-47,1971.
63. Klein-Nulend, J., E. H. Burger, C. M. Semeins, L. G. Raisz, and C. C. Pilbeam. Pulsating fluid flow stimulates prostaglandin release and inducible prostaglandin G/H synthase mRNA expression in primary mouse bone cells. J.Bone Miner.Res. 12:45-51,1997.
64. KNOTHE TATE, M. L., R. STECK, M. R. Forwood, and P. Niederer. In vivo demonstration of load-induced fluid flow in the rat tibia and its potential implications for processes associated with functional adaptation. J Exp.Biol. 203 Pt 18:2737-2745,2000.

65. Krolner, B., B. Toft, N. S. Pors, and E. Tondevold. Physical exercise as prophylaxis against involutional vertebral bone loss: a controlled trial. Clin.Sci.(Lond). 64:541-546,1983.
66. Lanyon, L. E. Functional strain as a determinant for bone remodeling. Calcif.Tissue Int. 36 Suppl 1:S56-S61,1984.
67. Lanyon, L. E. and C. T. Rubin. Static vs dynamic loads as an influence on bone remodelling. J.Biomech. 17:897-905,1984.
68. Larina, I. M., I. A. Tcheglova, B. S. Shenkman, and T. L. Nemirovskaya. Muscle atrophy and hormonal regulation in women in 120 day bed rest. J Gravit.Physiol. 4:121-122,1997.
69. LeBlanc, A. Summary of research issues in human studies. Bone. 22:117S-118S,1998.
70. LeBlanc, A., C. Lin, L. Shackelford, V. Sinitsyn, H. Evans, O. Belichenko, B. Schenkman, I. Kozlovskaya, V. Oganov, A. Bakulin, T. Hedrick, and D. Feeback. Muscle volume, MRI relaxation times (T2), and body composition after spaceflight. J Appl.Physiol. 89:2158-2164,2000.
71. LeBlanc, A., C. Marsh, H. Evans, P. Johnson, V. Schneider, and S. Jhingran. Bone and muscle atrophy with suspension of the rat. J Appl.Physiol. 58:1669-1675,1985.
72. LeBlanc, A., R. Rowe, V. Schneider, H. Evans, and T. Hedrick. Regional muscle loss after short duration spaceflight. Aviat.Space Environ.Med. 66:1151-1154,1995.
73. LeBlanc, A. and V. Schneider. Countermeasures against space flight related bone loss. Acta Astronaut. 27:89-92,1992.
74. LeBlanc, A., V. Schneider, J. Krebs, H. Evans, S. Jhingran, and P. Johnson. Spinal bone mineral after 5 weeks of bed rest. Calcif.Tissue Int. 41:259-261,1987.
75. LeBlanc, A., L. Shackelford, and V. Schneider. Future human bone research in space. Bone. 22:113S-116S,1998.
76. Li, G. P., J. T. Bronk, K. N. An, and P. J. Kelly. Permeability of cortical bone of canine tibiae. Microvasc.Res. 34:302-310,1987.
77. Mak, A. F., D. T. Huang, J. D. Zhang, and P. Tong. Deformation-induced hierarchical flows and drag forces in bone canaliculi and matrix microporosity. J.Biomech. 30:11-18,1997.
78. Mak, A. F., L. Huang, and Q. Wang. A biphasic poroelastic analysis of the flow dependent subcutaneous tissue pressure and compaction due to epidermal loadings: issues in pressure sore. J.Biomech.Eng. 116:421-429,1994.
79. Mak, A. F., L. Qin, L. K. Hung, C. W. Cheng, and C. F. Tin. A histomorphometric observation of flows in cortical bone under dynamic loading. Microvasc.Res. 59:290-300,2000.
80. Mak, A. F. and J. D. Zhang. Numerical simulation of streaming potentials due to deformation-induced hierarchical flows in cortical bone. J.Biomech.Eng. 123:66-70,2001.
81. Martin, R. B. and D. B. Burr. Structure, Function and Adaptation of Compact Bone. New York:1989.

82. Mayet-Sornay, M. H., H. Hoppeler, B. S. Shenkman, and D. Desplanches. Structural changes in arm muscles after microgravity. J Gravit.Physiol. 7:S43-S44,2000.
83. McLeod, K. and C. T. Rubin. Sensitivity of the Bone Remodeling Response to the Frequency of Applied Strain. trans ortho.res.soc. 17:533-1992.
84. McLeod, K. J. and C. T. Rubin. Frequency specific modulation of bone adaptation by induced electric fields. J.Theor.Biol. 145:385-396,8-9-1990.
85. McLeod, K. J. and C. T. Rubin. The effect of low-frequency electrical fields on osteogenesis. J.Bone Joint Surg.Am. 74:920-929,1992.
86. McPhee, J. C. and R. J. White. Physiology, medicine, long-duration space flight and the NSBRI. Acta Astronaut. 53:239-248,2003.
87. Montgomery, R. J., B. D. Sutker, J. T. Bronk, S. R. Smith, and P. J. Kelly. Interstitial fluid flow in cortical bone. Microvasc.Res. 35:295-307,1988.
88. Morris, M. A., J. A. Lopez-Curto, S. P. Hughes, K. N. An, J. B. Bassingthwaighte, and P. J. Kelly. Fluid spaces in canine bone and marrow. Microvasc.Res. 23:188-200,1982.
89. Murray, D. W. and N. Rushton. The effect of strain on bone cell prostaglandin E2 release: a new experimental method. Calcif.Tissue Int. 47:35-39,1990.
90. Neidlinger-Wilke, C., H. J. Wilke, and L. Claes. Cyclic stretching of human osteoblasts affects proliferation and metabolism: a new experimental method and its application. J.Orthop.Res. 12:70-78,1994.
91. Nilsson, B. E. and N. E. Westlin. Bone density in athletes. Clin.Orthop. 77:179-182,1971.
92. O'Connor, J. A., L. E. Lanyon, and H. MacFie. The influence of strain rate on adaptive bone remodelling. J.Biomech. 15:767-781,1982.
93. Otter, M. W. and G. V. Cochran. Comments on 'fluid movement in bone: theoretical and empirical' [letter; comment]. J.Biomech. 25:1495-1992.
94. Otter, M. W., V. R. Palmieri, and G. V. Cochran. Transcortical streaming potentials are generated by circulatory pressure gradients in living canine tibia. J.Orthop.Res. 8:119-126,1990.
95. Otter, M. W., V. R. Palmieri, D. D. Wu, K. G. Seiz, L. A. MacGinitie, and G. V. Cochran. A comparative analysis of streaming potentials in vivo and in vitro. J.Orthop.Res. 10:710-719,1992.
96. Otter, M. W., Y. X. Qin, C. T. Rubin, and K. J. McLeod. Does bone perfusion/reperfusion initiate bone remodeling and the stress fracture syndrome? Med.Hypotheses. 53:363-368,1999.
97. Otter, M. W., D. D. Wu, W. A. Bieber, and G. V. Cochran. Intraarterial protamine sulfate reduces the magnitude of streaming potentials in living canine tibia. Calcif.Tissue Int. 53:411-415,1993.
98. Owan, I., D. B. Burr, C. H. Turner, J. Qiu, Y. Tu, J. E. Onyia, and R. L. Duncan. Mechanotransduction in bone: osteoblasts are more responsive to fluid forces than mechanical strain. Am.J.Physiol. 273:C810-C815,1997.
99. Piekarski, K., D. Demetriades, and A. Mackenzie. Osteogenetic stimulation by externally applied dc current. Acta Orthop.Scand. 49:113-120,1978.

100. Piekarski, K. and M. Munro. Transport mechanism operating between blood supply and osteocytes in long bones. Nature. 269:80-82,9-1-1977.
101. Pienkowski, D. and S. R. Pollack. The origin of stress-generated potentials in fluid-saturated bone. J.Orthop.Res. 1:30-41,1983.
102. Pollack, S. R., N. Petrov, R. Salzstein, G. Brankov, and R. Blagoeva. An anatomical model for streaming potentials in osteons. J.Biomech. 17:627-636,1984.
103. Pollack, S. R., R. Salzstein, and D. Pienkowski. Streaming potential in fluid filled bone. Ferroelectrics. 60:297-309,1984.
104. Prasad, M. Velocity-permeability relations within hydraulic units. Geophysics. 68:108-117,2003.
105. Qin, Y. X., W. Lin, and C. T. Rubin. Load-Induced Bone Fluid Flow Pathway as Definded by In-vivo Intramedullary Pressure and Streaming Potentials Measurements. Ann.Biomed.Eng. 30:693-702,2002.
106. Qin, Y. X., K. McLeod, M. W. Otter, and C. T. Rubin. The Interdependent Role of Loading Frequency, Intracortical Fluid Pressure and Pressure Gradients in Guiding SiteSpecific Bone Adaptation. 44th Ann Mtg Orthop Res Soc. 23:544-44,1998.
107. Qin, Y. X., K. McLeod, and C. T. Rubin. Intracortical fluid flow is induced by dynamic intramedullary pressure independent of matrix deformation. 46th Ann Mtg Orth Res Soc. 25:740-40,2000.
108. Qin, Y. X., K. J. McLeod, M. W. Otter, and C. T. Rubin. Patterns of Loading Induced Fluid Flow in Cortical Bone. Ann.Biomed.Eng. 2001.
109. Qin, Y. X., M. W. Otter, C. T. Rubin, and K. J. McLeod. The Influence of Intramedullary Hydrostatic Pressure on Transcortical Fluid Flow Patterns In Bone. Trans.Ortho.Res.Soc. 22:885-1997.
110. Qin, Y. X., C. T. Rubin, and K. J. McLeod. Nonlinear dependence of loading intensity and cycle number in the maintenance of bone mass and morphology. J.Orthop.Res. 16:482-489,1998.
111. Raab-Cullen, D. M., M. P. Akhter, D. B. Kimmel, and R. R. Recker. Periosteal bone formation stimulated by externally induced bending strains. J.Bone Miner.Res. 9:1143-1152,1994.
112. Reed, C., D. A. Athanasiou, and G. Constantinides. Influence of Constant Hydrostatic Pressure on Osteoblast-like Cells. J.Dent.Res. 72:276-1993.
113. Reich, K. M. and J. A. Frangos. Effect of flow on prostaglandin E2 and inositol trisphosphate levels in osteoblasts. Am.J.Physiol. 261:C428-C432,1991.
114. Reich, K. M., C. V. Gay, and J. A. Frangos. Fluid shear stress as a mediator of osteoblast cyclic adenosine monophosphate production. J.Cell Physiol. 143:100-104,1990.
115. Richards, M., K. M. Kozloff, J. A. Goulet, and S. A. Goldstein. Increased distraction rates influence precursor tissue composition without affecting bone regeneration. J.Bone Miner.Res. 15:982-989,2000.
116. Robling.A.G., J. V. Duijvelaar, J. V. Geevers, Ohashi.N., and C. H. Turner. Modulation of appositional and longitudinal bone growth in the rat ulna by applied static and dynamic force. Bone. 29:105-113,2001.

117. Rubin, C., A. S. Turner, R. Muller, E. Mittra, K. McLeod, W. Lin, and Y. X. Qin. Quantity and quality of trabecular bone in the femur are enhanced by a strongly anabolic, noninvasive mechanical intervention. J.Bone Miner.Res. 17:349-357,2002.
118. Rubin, C. T., T. S. Gross, K. J. McLeod, and S. D. Bain. Morphologic stages in lamellar bone formation stimulated by a potent mechanical stimulus. J.Bone Miner.Res. 10:488-495,1995.
119. Rubin, C. T. and L. E. Lanyon. Dynamic strain similarity in vertebrates; an alternative to allometric limb bone scaling. J.Theor.Biol. 107:321-327,3-21-1984.
120. Rubin, C. T. and L. E. Lanyon. Regulation of bone formation by applied dynamic loads. J.Bone Joint Surg.[Am.]. 66:397-402,1984.
121. Rubin, C. T. and L. E. Lanyon. Kappa Delta Award paper. Osteoregulatory nature of mechanical stimuli: function as a determinant for adaptive remodeling in bone. J.Orthop.Res. 5:300-310,1987.
122. Rubin, C. T. and K. J. McLeod. Promotion of bony ingrowth by frequency-specific, low-amplitude mechanical strain. Clin.Orthop. 165-174,1994.
123. Salzstein, R. A. and S. R. Pollack. Electromechanical potentials in cortical bone--II. Experimental analysis. J.Biomech. 20:271-280,1987.
124. Salzstein, R. A., S. R. Pollack, A. F. Mak, and N. Petrov. Electromechanical potentials in cortical bone--I. A continuum approach. J.Biomech. 20:261-270,1987.
125. Scott, G. C. and E. Korostoff. Oscillatory and step response electromechanical phenomena in human and bovine bone. J.Biomech. 23:127-143,1990.
126. Serova, L. V. Microgravity and aging of animals. J Gravit.Physiol. 8:137-138,2001.
127. Skripitz, R. and P. Aspenberg. Pressure-induced periprosthetic osteolysis: a rat model. J.Orthop.Res. 18:481-484,2000.
128. Smit, T. H., J. M. Huyghe, and S. C. Cowin. Estimation of the poroelastic parameters of cortical bone. J.Biomech. 35:829-835,2002.
129. Srinivasan, S. and T. S. Gross. Canalicular fluid flow induced by bending of a long bone. Med Eng Phys. 22:127-133,2000.
130. STECK, R., P. Niederer, and M. L. KNOTHE TATE. A Finite Element Analysis for the Prediction of Load-induced Fluid Flow and Mechanochemical Transduction in Bone. J Theor.Biol. 220:249-259,1-21-2003.
131. Stokes, I. A. Analysis of symmetry of vertebral body loading consequent to lateral spinal curvature. Spine. 22:2495-2503,11-1-1997.
132. Stokes, I. A., D. D. Aronsson, H. Spence, and J. C. Iatridis. Mechanical modulation of intervertebral disc thickness in growing rat tails. J.Spinal Disord. 11:261-265,1998.
133. Tanaka, T. and A. Sakano. Differences in permeability of microperoxidase and horseradish peroxidase into the alveolar bone of developing rats. J.Dent.Res. 64:870-876,1985.
134. Tate, M. L. and U. Knothe. An ex vivo model to study transport processes and fluid flow in loaded bone. J.Biomech. 33:247-254,2000.

135. Tate, M. L., P. Niederer, and U. Knothe. In vivo tracer transport through the lacunocanalicular system of rat bone in an environment devoid of mechanical loading. Bone. 22:107-117,1998.
136. Turner, C. H. Site-specific skeletal effects of exercise: importance of interstitial fluid pressure. Bone. 24:161-162,1999.
137. Turner, C. H., M. R. Forwood, and M. W. Otter. Mechanotransduction in bone: do bone cells act as sensors of fluid flow? FASEB J. 8:875-878,1994.
138. Turner, C. H., I. Owan, and Y. Takano. Mechanotransduction in bone: role of strain rate. Am.J.Physiol. 269:E438-E442,1995.
139. Wang, L., S. C. Cowin, S. Weinbaum, and S. P. Fritton. Modeling tracer transport in an osteon under cyclic loading. Ann.Biomed.Eng. 28:1200-1209,2000.
140. Wang, L., S. P. Fritton, S. C. Cowin, and S. Weinbaum. Fluid pressure relaxation depends upon osteonal microstructure: modeling an oscillatory bending experiment. J.Biomech. 32:663-672,1999.
141. Weinbaum, S. 1997 Whitaker Distinguished Lecture: Models to solve mysteries in biomechanics at the cellular level; a new view of fiber matrix layers. Ann.Biomed.Eng. 26:627-643,1998.
142. Weinbaum, S., S. C. Cowin, and Y. Zeng. A model for the excitation of osteocytes by mechanical loading-induced bone fluid shear stresses. J.Biomech. 27:339-360,1994.
143. Weinbaum, S., P. Guo, and L. You. A new view of mechanotransduction and strain amplification in cells with microvilli and cell processes. Biorheology. 38:119-142,2001.
144. Winet, H. The role of microvasculature in normal and perturbed bone healing as revealed by intravital microscopy. Bone. 19:39S-57S,1996.
145. Winet, H. A bone fluid flow hypothesis for muscle pump-driven capillary filtration: II. Proposed role for exercise in erodible scaffold implant incorporation. Eur.Cell Mater. 6:1-10,10-1-2003.
146. Winet, H., A. Hsieh, and J. Y. Bao. Approaches to study of ischemia in bone. J Biomed Mater.Res. 43:410-421,1998.
147. Wolff J. Das Gesetz Der Transformation Der Knochen. Berlin:1892.
148. Wolff J. The Law Of Bone Remodeling. Berlin:1986.
149. You, J., C. E. Yellowley, H. J. Donahue, Y. Zhang, Q. Chen, and C. R. Jacobs. Substrate deformation levels associated with routine physical activity are less stimulatory to bone cells relative to loading-induced oscillatory fluid flow. J.Biomech.Eng. 122:387-393,2000.
150. Zhang, D. and S. C. Cowin. Oscillatory bending of a poroelastic beam. J.Mech.Phys.Solid. 42:1575-1579,1994.

CHAPTER 17

BIO-PATHOLOGY OF BONE TUMORS

Lin Huang[1], Jiake Xu[2], and Ming Hao Zheng[2]

[1] *Department of Orthopaedics and Traumatology, The Chinese University of Hong Kong, Hong Kong*

[2] *Molecular Orthopaedic Laboratory, School of Surgery and Pathology, The University of Western Australia, QE II Medical Centre, M Block, Nedlands, Western Australia, 6009 Australia*

1. Introduction

Bone tumors represent a heterogeneous group of mesenchymal lesions, which account for approximately 1% of all malignancies. This group includes a wide range of different tumor types, characterized by very different clinical, radiological and pathological features. The pathogenesis of these tumors is based on various inherited and environmental factors. A pathological role for genetic determinants has been defined in the promotion of some bone neoplasms. With the improvement of our knowledge about the molecular events responsible for the development and progression of bone neoplasms, links between the functions of tumor suppressor genes, oncogenes, growth factors and their receptors to the behavior of the tumors have been widely explored. The mutation of several tumor suppressor genes and overexpression of several oncogenes have been reported to implicate in the development of some bone tumors; the growth factors and their receptors have also been implied in the regulation of the phenotype of bone tumors. In this chapter, WHO classification of bone tumors will be presented. Recent advances in cell biology and molecular pathology of bone tumors will be reviewed. The role of osteotropic factors including RANKL/OPG, PTHrP etc in promoting tumor-mediated osteoclastic bone resorption will also be discussed, as this is an area of particular current interest.

2. Classification of Bone Neoplasms

The most widely accepted classification of bone neoplasms is that of the World Health Organization (WHO), which was first published in 1972 and revised in 1993 and 2002 [1,2]. Histologic and cytologic characteristics are the basis for the classification, although some tumors of uncertain histogenesis have undergone reassessment as a result of findings from newer techniques of immunohistochemistry, molecular pathology and cytogenetics.

Based on the histologic criteria, including immunohistochemical studies, and observed patterns of biologic behavior, conventional approach separated primary neoplasms of bone into those forming bone, cartilage, or vessels; giant cell tumor; neoplasms that affect primarily the bone marrow; and a miscellaneous group of connective tissue, notochordal, and epithelial neoplasms (**Table 1**). The classification also recognizes a category of tumor-like lesions that are important to consider in the differential diagnosis of bone tumors. In addition, secondary bone neoplasms that metastasize from the primary site of breast, prostate, thyroid, lung, and kidney, etc., are also recognized.

Bone-forming neoplasms are divided into benign (osteoma, osteoid osteoma, osteoblastoma) and malignant (osteosarcoma) groups, and the later group characterized by bone or osteoid formation by the malignant cells. Aggressive osteoblastoma has been introduced to describe a rare, clinically locally aggressive osteoblastic lesion with characteristic histologic features but that does not show the degree of cytologic atypia associated with osteosarcoma and does not metastasize.

Cartilage-forming neoplasms now included dedifferentiated chondrosarcoma (in which low-grade and anaplastic components are juxtaposed) and clear cell chondrosarcoma (a low-grade malignancy with a typical epiphyseal location).

Giant cell tumor of bone forms a separate group in recognition of its uncertain histogenesis and biological behavior. It is generally accepted that the giant cells are indeed reactive osteoclast-like cells.

The group of "marrow tumor" includes neoplasms of diverse phenotype that collectively fall into the differential diagnosis of round cell tumors. Immunohistochemical and molecular genetic investigations of these lesions have greatly facilitated the distinction between the Ewing's sarcoma, lymphomas, and multiple myeloma.

Vascular tumors are rare, accounting for 2.5% of all primary bone neoplasms.

In the group of fibrous neoplasms, the 1993 classification includes benign and malignant fibrous histiocytomas using similar criteria to those used in soft tissue pathology. Many tumors now regarded as malignant fibrous histiocytomas would have formerly been diagnosed as fibrosarcoma, malignant giant cell tumor, or unclassified sarcoma.

The multipotential nature of the bone marrow stromal cells explains why such a diversity of types of tumors, as classified by the nature of their matrix, occurs in bone and why such anomalities as chondroblastic osteosarcoma, in which more than one type of matrix is synthesizes, exist. The presence of two stromal precursor cell pools, one in the marrow and one in the periosteum, explains the occurrence of medullary and periosteal neoplasms. And the vasculature and marrow microenvironment explain the relative differences in the behavior of these different neoplasm.

The most common neoplasms of bone are secondary carcinomas. Although incompletely understood, the nature of the bone vasculature and, probably, the expression of adhesion molecules on the vascular endothelium influence the process of metastasis.

Table 1 WHO histologic classification of bone neoplasms and tumor-like lesions

Bone-Forming Tumors

Benign

Osteoma
Osteoid osteoma
Osteoblastoma

Intermediate

Aggressive (malignant) osteoblastoma

Malignant

Central osteosarcoma
- Conventional central osteosarcoma
- Telangiectatic osteosarcoma
- Intraosseous well-differentiated (low-grade) osteosarcoma
- Round (small) cell osteosarcoma

Surface osteosarcoma
- Parosteal (juxtacortical) osteosarcoma
- Periosteal osteosarcoma
- High-grade surface osteosarcoma

Cartilage-Forming Tumors

Benign

Chondroma
- Enchodroma
- Periosteal (juxtacortical) chondroma

Osteochondroma
Chondroblastoma
Chondromyxoid fibroma

Malignant

Typical central chondrosarcoma
Juxtacortical (periosteal) chondrosarcoma
Mesenchymal chondrosarcoma
Dedifferentiated chondrosarcoma
Clear cell chondrosarcoma
Malignant chondroblastoma

Giant Cell Tumor

Marrow Tumors (Round Cell Tumors)

Ewing's sarcoma

Primitive neuroectodermal tumor of bone

Malignant lymphoma of bone
Myeloma

Vascular Tumors

Benign

Hemangioma
Lymphangioma
Glomus tumor (glomangioma)

Intermediate or Indeterminate

Hemangioendothelioma
Hemangiopericytoma

Malignant

Angiosarcoma
Malignant hemangiopericytoma

Other Connective Tissue Tumors

Benign

Benign fibrous histocytoma
Lipoma

Intermediate

Desmoplastic fibroma

Malignant

Fibrosarcoma
Malignant fibrous histocytoma
Liposarcoma
Malignant mesenchymoma
Leiomyosarcoma
Undifferentiated sarcoma

Other Tumors

Chordoma
Adamantinoma of long bones
Neurilemoma and neurofibroma

Unclassified Tumors

Tumor-like Lesions

Solitary bone cyst
Aneurysmal bone cyst
Eosinophilic granuloma (Langerhans' cell histiocytosis)
Fibrous dysplasia and fibro-osseous dysplasia
Myositis ossificans (heterotopic ossification)
Brown tumor of hyperparathyroidism
Giant cell (reparative) granuloma

Secondary Bone Tumors

Modified from Helliwell TR. Pathology of Bone and Joint Neoplasms. Vol.37. Chapter 1. [3]

3. Major genetic events in bone tumor development

Like other tumors, bone tumors results from subversion of the normal processes that regulate proliferation and apoptosis of cells. The loss of normal control mechanisms arises from the acquisition of genetic alterations in two major categories of genes: transforming genes and tumor suppressor genes. In addition, microsatellite instability is valuable genetic marker for the altered phenotypes seen in many cancers, and DNA methylation has been found to be an important part of gene regulation. Recently, progress has been made in our understandings on these genetic events of bone tumor development.

3.1. Transforming genes

c-MYC

The c-MYC gene encodes a transcriptional factor that regulates cell differentiation, DNA replication, and organogenesis. Amplification and/ or overexpression of MYC commonly occurs in a wide range of tumors including osteosarcoma, chondrosarcoma and MFH [4,5,6]. A high and uniform expression of c-myc protein and mRNA was found in relapsed patients of Ewing's sarcoma compared to disease-free patients [7].

FOS

FOS is a member of a multigene family that includes the FOS related nuclear transcription factors (FOSB, FRA1, and FRA2). FOS proteins form heterodimers with specific JUN proteins and interact with AP-1 transcription complex. These molecules are involved in cell proliferation, differentiation, transformation, and bone metabolism [8]. Overexpression of c-fos is frequently found in human osteosarcoma and with increased frequency in tumor promotion and progression of human osteosarcoma [9,10]. Further more, c-myc and c-fos were found overexpressed in a high percentage of the relapsed osteosarcoma, and overexpression of both c-myc and c-fos in the same tumor was strongly correlated to the development of metastases [11].

ERBB2/ HER2/neu/c-erbB-2

The ERBB2 gene (also known as HER2/neu or c-erbB-2) encodes a protein with structural similarity to the epidermal growth factor receptor. ERBB2 expression has been reported in osteosarcoma [12]. ErbB-2 expression is correlated with poor prognosis for patients with osteosarcoma in which expression of ErbB-2 was strongly correlated with early pulmonary metastasis and poor survival rate for the patient [13]. However other studies have found that the increased expression levels of ErbB2 in tumor cells is associated with a significantly increased probability of overall survival [14].

MET and HGF

The MET, a proto-oncogene that encodes a transmembrane tyrosine kinase, has been associated with tumor progression in different human carcinomas. It acts as the receptor for the hepatocyte growth factor (HGF). The Met/HGF receptor is expressed by epithelial cells but its ligand by cells of mesenchymal origin. The Met/HGF receptor was not detectable in the majority of bone tumors, but is over-expressed in 60% of human osteosarcomas probably by either a paracrine or an autocrine circuit [15]. One study has found that c-MET expression was frequently detected in cartilaginous tumors, such as chondroblastoma (62.5%), enchondroma (66.7%), and osteochondroma (71.4%), but no expression was observed in giant cell tumors of bone or any other benign tumors or tumor-like lesions [16]. Recent studies have suggested that HGF promote tumor malignant behavior via activating both the mitogen and motogen pathways in osteosarcoma cells [17].

3.2. Tumor-suppressor genes

Retinoblastoma (RB1)

The RB protein is a signal transducer regulating G1 to S cell cycle progression by binding to E2F protein and inhibiting the trans-activation function of E2F that is required for S phase. During G1 and S phase transition, RB becomes phosphorylated, resulting in the release and activation of the E2F family of transcription factors and a transition from the G1 to the S phase of the cell cycle. Abnormalities in RB can result in loss of this checkpoint function. Early studies have found that somatic loss of the region of human chromosome 13 that includes the RB1 locus, is associated with the development of osteosarcomas [18]. The loss of RB1 gene (locus in 13q14) may be also involved in the development of radiation-induced osteosarcoma [19]. Further studies have shown that RB1 gene alteration is present in high percentage of osteosarcomas [20], but hypermethylation of the RB1 promoter is not involved during the development of osteosarcoma [21]. More recent studies have suggested that the presence of altered RB1 gene might be regarded as a poor prognostic factor for pediatric osteosarcoma [22]. Overall, alterations of the RB1 gene, loss of heterozygosity at the RB1 gene locus on chromosome 13, structural rearrangements and point mutations are found in up to 70%, 60-70%, 30% and10% of the osteosarcoma respectively [23].

The phosphorylation of RB is regulated by its upstream molecules cyclin-dependent kinase (CDK), which in turn is regulated by a series of CDK inhibitors (CDKIs). Therefore alteration of these genes could result in functional inactivation of the RB signaling pathway. Studies have found that deletions of both p15INK4B and p16INK4 genes were detected in five of eight osteosarcoma cell lines [24]. Alterations of p19INK4D gene has been found in a small but significant number of osteosarcomas [25], whereas over-expression of p15INK4b, p16INK4a and p21CIP1/WAF1 genes has been shown to mediate growth arrest in human osteosarcoma cell lines [26].

p53 pathways

TP53, a tumor suppressor gene located at 17p13, is thought to be important in the development of osteosarcoma. The p53 protein acts as a nuclear transcription factor that inhibits cell proliferation by activation of the gene encoding p21 protein, which causes some human cells to arrest at the G1 stage of the cell cycle [27]. Mutations generating defective p53 may represent early steps of carcinogenesis in bone tumors or determine behavior of the developing tumor. Recent review has suggest that TP53 point mutations, gross gene rearrangements and allelic loss count for 20–30%, 10%–20% and 75–80% of osteosarcoma respectively [23].

3.3. Microsatellite instability

Microsatellites are interspersed tandem repeats of nucleotide base pairs scattered throughout the human genome, which exhibit length polymorphism [28]. They are very short DNA sequence repeats designated (CA)n, which the range of n being 15-30. Some neoplastic cells show allelic size alteration within the microsatellite regions compared with normal cells. These alterations, either amplification or deletion of base pairs, is termed microsatellite instability. Microsatellite instability is a valuable genetic marker for the altered phenotypes seen in many cancers. It has been reported in sporadic carcinomas (11-34%), skeletal and soft sarcomas (44%) and chordomas (50%) [29,30,31]. In giant cell tumor of bone (GCT), Scheiner et al [32] examined six microsatellites on chromosome arms 5q, 18q, 15q, 17p, 19q and 11p for their instability. These loci were chosen because their genetic instability are previously reported in other tumors, specifically colon, or because they are common sites for telomeric association in GCT. No microsatellite instability at the DNA sequences studied was shown in GCT. It has been suggested that microsatellite analysis may have a prognostic role in the future if larger studies confirm current preliminary findings in bone tumors, while, additional studies are need to further characterize the biological aggressiveness and status of microsatellite instability in malignancies.

3.4. DNA methylation

Methylation of genomic DNA is an important part of gene regulation. Its patterns are based on clonal inheritance that occurs in the early stages of embryogenesis. There is overwhelming evidence that DNA methylation patterns are altered in cancer, aberrations of which include global hypomethylation, regional hypermethylation and deregulated level of expression of DNA methyltransferases (DNMT) [33,34]. Both hypo- and hyper-methylation can co-exist in the same tumor cells. Multiple genes, such as tumor suppressors, adhesion molecules, inhibitors of angiogenesis and repair enzymes, which confer selective advantage upon cancer cells, are often transcriptional silenced through hypermethylation in tumor cells. In parallel, tumor cell genomes are, in general, seen globally less methylated than their normal tissue counterparts. It has been now well established by numerous studies that hypermethylation of certain tumor suppressor genes promotes tumorigenesis, while hypomethylation of DNA stimulates tumor invasion and metastasis [33,34,35].

The vast majority changes in the DNA methylation levels leading to cell differentiation and growth disorders occur in the promoter region of the genes. Cytosine methylation at CpG dinucleotides is thought to cause more than one-third of all transition mutations responsible for human genetic disease and cancer. Methylation at the CpG island in p16 (INK4A), p15 (INK4B) and p14 (ARF) genes have been demonstrated in osteosarcomas and Ewing family tumors [36,37,38]. Hypermethylation of the p73 gene leading to its transcriptional silencing was also observed in several leukemias and lymphomas [39]. Consequently, altered expression of these gene products were found contributing significantly to tumor pathogenesis and development in these tumors.

4. Mediators of Tumor-associated Bone Resorption

Osteolysis is a hallmark of various benign and malignant bone diseases. The clinical signs and symptoms include pathologic fractures, bone deformities, pain, and hypercalcemia. There is no evidence that tumor cells directly cause bone resorption. Instead, induction of osteoclast formation and activation of osteoclastic bone resorption by tumor cells are the cause of bone destruction [40, 41]. The cellular mechanisms by which tumor cells influence bone cell function and vice versa are still unclear. However, a number of humoral factors produced by tumor cells and inflammatory cells within the bone marrow microenvironment have been identified as osteotropic factors. Of these factors, parathyroid hormone-related protein (PTHrP), interleukin (IL)-6, and transforming growth factor-β (TGF-β) have received much attention because they are produced by various tumor cells, act as autocrine tumor growth factors, and are capable of modulating bone resorption [40,42,43]. With the identification and characterization of the RANKL/RANK/OPG cytokine system (receptor activator of nuclear factor-κB ligand [RANKL], its specific receptor --- receptor activator of nuclear factor-κB [RANK], and its decoy receptor --- osteoprotegerin [OPG]), increasing data have implicated RANKL and OPG as the essential cytokine system that regulates tumor-bone interactions (osteolytic bone metastasis, humoral hypercalcemia of malignancy), and onto which many other cytokine systems may converge [44,45,46]. In subsequent paragraphs, the role of above-mentioned osteotropic factors on tumor-associated bone resorption will be discussed.

4.1. PTHrP

PTHrP is a member of the parathyroid hormone family. It binds to the PTH receptor and can cause hypercalcemia, osteoclast-mediated bone destruction, and increased renal reabsorption of calcium and excretion of phosphate. It has been established that PTHrP is one of the key factors responsible for the development of hypercalcemia of malignancy [47,48,49], and more recent evidence showed its key role in the establishment and maintenance of bone metastases [50,51]. It is the leading candidate for an osteoclast stimulatory factor produced by breast cancer cells. Approximately 50% of human primary breast cancers express PTHrP, and more than 90% of the breast cancer metastases to bone express PTHrP [52]. Circulating PTHrP levels are markedly increased in the majority of patients with myeloma and osteolytic bone metastases [47]. Elegant experiments showed that tumor cells that overexpressed PTHrP were more likely to metastasize to bone than cancer cells that did not express PTHrP in a murine animal model [53]. The administration of antibodies to PTHrP to animals injected with PTHrP-expressing tumor cells significantly reduced the number and size of osteolytic bone lesions [53]. However, PTHrP by itself cannot directly stimulate osteoclast formation in the absence of marrow stromal cells. PTHrP induces production of RANKL, which then stimulates osteoclast formation.

4.2. IL-6

IL-6 can stimulate human osteoclast formation *in vitro* [54] and *in vivo* [55], and enhance the effects of other osteoclastogenic factors such as PTH and PTHrP [56]. IL-6 is the major growth factor for myeloma cells and is present in marrow plasma samples from myeloma patients. It has been established that adhesive interactions between marrow stromal cells and myeloma cells result in secretion of IL-6 [57]. Thomas et al [58] examined the effects of two myeloma cell lines, U266 and ARH-77, on IL-6 production by bone marrow stromal cells in a coculture system. Both cell lines strongly stimulated IL-6 production, and IL-6 production was in part dependent on physical contact between myeloma cells and stromal cells. IL-6 levels have been correlated with the clinical features of myeloma, such as osteolytic bone disease or hypercalcemia, by some investigators [59]. However, others have been unable to find a correlation between IL-6 and bone disease activity [60,61].

4.3. TGF-β

TGF-β has complex and multiple effects on bone cell function. It is present in particularly high concentrations in bone matrix, where it may act as a coupling factor between the process of bone formation and resorption. TGF-β is also produced by tumor cells [62,63]. The stimulatory effects of TGF-β on osteoclast activity appear to be mediated by prostaglandins, and the role of TGF-β in malignant bone resorption is uncertain. TGF-β may alter the behavior of many tumor cells, particularly breast carcinoma cells, to enhance the production of PTHrP from the tumor cells to further facilitate local bone resorption [63,64].

4.4. RANKL, RANK and OPG system

RANKL is a recently described osteoclast stimulatory factor that appears to mediate the effects of most osteotropic factors, such as IL-1, 1,25-$(OH)_2D_3$, PGE_2, and IL-11 etc., on osteoclast formation [65]. RANKL/RANK/OPG system has been demonstrated to be capable of regulating all aspects of osteoclast functions, including proliferation, differentiation, fusion, activation, and apoptosis of osteoclasts

[44,45,46,65]. Abnormalities of the RANKL/OPG system have been detected in various benign metabolic bone diseases, such as postmenopausal osteoporosis, Paget disease, and hyperparathyroidism, etc [66,67,68]. In these disorders, locally or systemically enhanced osteoclast activity and bone resorption are characterized, and associated with an enhanced RANKL-to-OPG ratio within the bone marrow microenvironment. Increased or decreased RANKL-to-OPG ratio has also been detected in patients with malignant bone diseases (humoral hypercalcemia of malignancy, osteolytic bone metastasis) (**Figure 1**) [69]. RANKL is either produced directly by tumor cells, or its production by osteoblastic/stromal cells or T lymphocytes is induced indirectly by tumor cells through secretion of PTHrP and other cytokines. By contrast, production of OPG is either inhibited or inappropriately low to compensate for the increase in RANKL. The exogenous administration of OPG to tumor-bearing animals corrects the increased RANKL-to-OPG ratio, and reverses the skeletal complications of malignancies.

4.5. The role of mediators in common bone tumors

Giant cell tumor of bone

Giant cell tumor of bone (GCT) is a benign primary neoplasm of bone characterized by expansile and well-delineated lytic lesions. Histologically, GCT is comprised of a mesenchymal tumor stroma surrounding areas of large multinucleated, osteoclast-like giant cells. Tumor cells of GCT have been demonstrated to produce TGF-β, IL-1, -6, -11, -17, and -18, M-CSF and PTHrP, which stimulate recruitment of reactive osteoclasts. Recently, others and we have also demonstrated that tumor cells of GCT are also a rich source of RANKL mRNA [70,71]. Moreover, osteoclast-like giant cells produce excessive levels of RANK mRNA compared with normal osteoclasts. Thus RANKL may act as an osteoclastogenesis-stimulating factor linked to interaction between tumor stroma and osteoclast progenitors in GCT. GCT can be viewed as a disease process of autonomous and unregulated overexpression of RANKL and RANK with a subsequent increase in osteoclast activity, resulting in extensive local bone destruction.

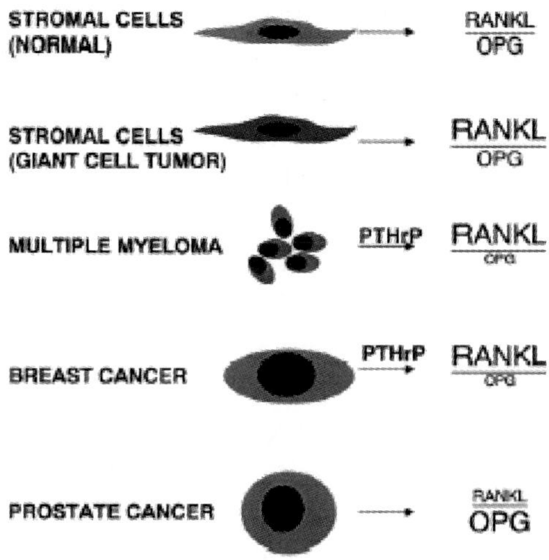

Figure 1
RANKL-to-OPG ratio in patients with malignant bone diseases. Normal stromal cells produce the stable RANKL-to-OPG ratio required for normal bone remodelling. Stromal cells derived from giant cell tumors overexpress RANKL, which results in an increased RANKL-to-OPG ratio with the subsequent excessive development of large multinucleated osteoclasts. Myeloma and some forms of breast carcinoma cells produce parathyroid hormone-related peptide (PTHrP), which induces RANKL and inhibits OPG production, thus resulting in an increased RANKL-to-OPG ratio that favors osteolysis and humoral hypercalcemia of malignancy. By contrast, decreased RANKL production in prostate carcinoma results in a reduced RANKL-to-OPG ratio and may favor an osteoblastic tumor growth pattern.

(Refer to Hofbauer LC et al. Cancer 2001, 92:460-470) [69]

Osteosarcoma

Osteosarcoma is a malignant tumor that is derived from osteoblastic lineage cells and most frequently grows as an expansive, non-osteolytic tumor. While, local growth of osteosarcoma involves destruction of host bone by proteolytic mechanisms and/or host osteoclast activation. RANKL has been demonstrated to present as a membrane-bound form in both the human osteosarcoma cell lines MG-63, HOS and SaOS-2, and mouse osteoblastic cell line MC3T3-E1 [72,73]. In addition, high OPG mRNA levels have also been detected in the cell line MG-63, and shown to be up-regulated by IL-1α and TNFs [74,75]. It is suggested that the

RANKL-to-OPG ratio in some forms of osteosarcoma might, at least in part, account for the expansive tumor growth at the expense of osteoclastic bone resorption [69], however, studies that correlate biological characteristics of osteosarcoma with changes in the RANKL/OPG system, to our knowledge, are not yet available.

Sarcomatous transformation is most often seen in severe, long-standing Paget's disease. Recent work has shown evidence of linkage between chromosome 18q21-22 locus (which encodes RANK) and familial Paget disease, some forms of osteosarcoma, as well as familial expansile osteolysis (FEO) [76,77,78]. Insertion mutations of the TNFRSF11A gene (encoding RANK) have been found to be responsible for FEO and rare cases of early onset familial Paget's disease [79]. Loss of heterozygosity (LOH) affecting the PDB/FEO critical region has also been described in osteosarcomas [79]. It is possible that TNFRSF11A might be involved in the development of osteosarcoma. However, further studies are required to assess directly the potential role of RANK in the process of transition from familial Paget disease to osteosarcoma [69].

Multiple myeloma

The skeletal effects of multiple myeloma include local osteolytic lesions, pathologic fractures and profound hypercalcemia. These complications are believed to result from interactions of myeloma cells, bone marrow stromal cells and osteoclasts, and to be mediated through myeloma-derived cytokines such as IL-6 and PTHrP [80,81,82]. In fact, IL-6 and PTHrP have been demonstrated to induce RANKL production and inhibit OPG production [83,84]. In addition to secreting certain factors that can enhance expression of RANKL on the surface of marrow stromal cells, human multiple myeloma cell lines (EJM, Karpas 707) have also been shown to directly express RANKL mRNA [69]. Moreover, multiple myeloma cells have been demonstrated to produce and to shed syndecan-1 (CD138), which inactivates OPG produced by other cell types within the confines of bone [85]. All of these distinct levels of regulation result in an increased RANKL-to-OPG ratio, which may explain the capacity of multiple myeloma cells to promote osteoclast differentiation and activation.

Skeletal metastases

Several solid tumors, most notably breast carcinoma, lung carcinoma, and prostate carcinoma, commonly metastasis to bone in patients with advanced disease, where they cause osteolysis and associated pain, hypercalcemia, and fracture. In osteolytic metastases, it has been shown that tumor cells direct osteoclastic bone resorption through a vicious cycle [86,87]: in particular, tumor cell-produced PTHrP facilitates bone resorption, and as a consequence, TGF-β is released from the bone matrix and promotes the progression of bone metastases by further inducing PTHrP production from tumor cells. Breast carcinoma cells are known to secret high levels of PTHrP. In addition, recent studies provide a molecular link between skeletal metastases and constitutive RANKL and OPG mRNA production [88,89]. An increased RANKL-to-OPG ratio in bone marrow stromal cells or osteoblasts was elicited in bone invasion model of breast cancer, which may result from breast carcinoma cell-derived PTHrP [90]. Moreover, constitutive RANKL expression at both mRNA and protein levels were found in metastatic tumor cells in lesions of breast, lung, prostate, and thyroid adenocarcinoma [91]. More direct evidence is that tumor cells of prostate cancer were found capable of inducing osteoclastogenesis in vitro, directly through the production of soluble RANKL [92]. On the other hand, RANKL mRNA expression was found to be low in tumor xenografts established using prostate carcinoma of the LnCaP cell line, which lacks the capacity to induce osteolytic metastases and grows as an osteoblastic tumor, and high in xenografts of the PC-3 cell line, which has the capacity to induce osteolytic metastases [89]. Constitutive OPG mRNA steady state levels were also found threefold to fourfold higher in prostate carcinoma cells compared with healthy prostate tissue [89]. These findings are similar to those obtained in osteosarcoma cells, which similar to prostate carcinoma metastases grow as osteoblastic tumors in bone.

References

1. Schajowitz F. Histological typing of bone tumours. 2nd ed. Berlin, Springer-Verlag, 1993.
2. Fletcher CDM, Unni KK, Mertens F. Pathology and genetics of tumours of soft tissue and bone. Lyon, IARCPress, 2002.
3. Helliwell TR. Pathology of Bone and Joint Neoplasms. Volume 37 in the series Major Problems In Pathology. Chapter 1, 1999.
4. Bogenmann E, Moghadam H, DeClerck YA, Mock A (1987): c-myc amplification and expression in newly established human osteosarcoma cell lines. Cancer Res 47:3808-14.
5. Schwab M, Ramsay G, Alitalo K, Varmus HE, Bishop JM, Martinsson T, Levan G, Levan A (1985): Amplification and enhanced expression of the c-myc oncogene in mouse SEWA tumour cells. Nature 315:345-7.
6. Brinck U, Stachura J, Kellner S, Schauer A. Prognostic relevance of Ki-67 antigen and c-myc oncoprotein in malignant fibrous histocytoma. Anticancer Res. 1995;15(5B):1915-9.
7. Sollazzo MR, Benassi MS, Magagnoli G, Gamberi G, Molendini L, Ragazzini P, Merli M, Ferrari C, Balladelli A, Picci P (1999): Increased c-myc oncogene expression in Ewing's sarcoma: correlation with Ki67 proliferation index. Tumori 85:167-73.
8. Wang ZQ, Ovitt C, Grigoriadis AE, Mohle-Steinlein U, Ruther U, Wagner EF (1992): Bone and haematopoietic defects in mice lacking c-fos. Nature 360:741-5.
9. van den Berg S, Rahmsdorf HJ, Herrlich P, Kaina B (1993): Overexpression of c-fos increases recombination frequency in human osteosarcoma cells. Carcinogenesis 14:925-8.
10. Wu JX, Carpenter PM, Gresens C, Keh R, Niman H, Morris JW, Mercola D (1990): The proto-oncogene c-fos is over-expressed in the majority of human osteosarcomas. Oncogene 5:989-1000.
11. Gamberi G, Benassi MS, Bohling T, Ragazzini P, Molendini L, Sollazzo MR, Pompetti F, Merli M, Magagnoli G, Balladelli A, Picci P (1998): C-myc and c-fos in human osteosarcoma: prognostic value of mRNA and protein expression. Oncology 55:556-63.
12. Valabrega G, Fagioli F, Corso S, Madon E, Brach del Prever A, Biasin E, Linari A, Aglietta M, Giordano S (2003): ErbB2 and bone sialoprotein as markers for metastatic osteosarcoma cells. Br J Cancer 88:396-400.
13. Onda M, Matsuda S, Higaki S, Iijima T, Fukushima J, Yokokura A, Kojima T, Horiuchi H, Kurokawa T, Yamamoto T (1996): ErbB-2 expression is correlated with poor prognosis for patients with osteosarcoma. Cancer 77:71-8.
14. Akatsuka T, Wada T, Kokai Y, Kawaguchi S, Isu K, Yamashiro K, Yamashita T, Sawada N, Yamawaki S, Ishii S (2002): ErbB2 expression is correlated with increased survival of patients with osteosarcoma. Cancer 94:1397-404.
15. Ferracini R, Di Renzo MF, Scotlandi K, Baldini N, Olivero M, Lollini P, Cremona O, Campanacci M, Comoglio PM (1995): The Met/HGF receptor is over-expressed in human osteosarcomas and is activated by either a paracrine or an autocrine circuit. Oncogene 10:739-49.

16. Naka T, Iwamoto Y, Shinohara N, Ushijima M, Chuman H, Tsuneyoshi M (1997): Expression of c-met proto-oncogene product (c-MET) in benign and malignant bone tumors. Mod Pathol 10:832-8.
17. Coltella N, Manara MC, Cerisano V, Trusolino L, Di Renzo MF, Scotlandi K, Ferracini R (2003): Role of the MET/HGF receptor in proliferation and invasive behavior of osteosarcoma. Faseb J 17:1162-4.
18. Hansen MF, Koufos A, Gallie BL, Phillips RA, Fodstad O, Brogger A, Gedde-Dahl T, Cavenee WK (1985): Osteosarcoma and retinoblastoma: a shared chromosomal mechanism revealing recessive predisposition. Proc Natl Acad Sci U S A 82:6216-20.
19. Ozisik YY, Meloni AM, Zalupski MM, Ryan JR, Qureshi F, Sandberg AA (1993): Deletion of chromosome 13 in osteosarcoma secondary to irradiation. Cancer Genet Cytogenet 69:35-7.
20. Scholz RB, Kabisch H, Weber B, Roser K, Delling G, Winkler K (1992): Studies of the RB1 gene and the p53 gene in human osteosarcomas. Pediatr Hematol Oncol 9:125-37.
21. Otte C, Scholz R, Werner M, Weber B, Delling G, Kabisch H (2004): Methylation status of the retinoblastoma gene (RB1) in osteosarcoma: no evidence for hypermethylation. Pediatr Hematol Oncol 21:57-65.
22. Patino-Garcia A, Pineiro ES, Diez MZ, Iturriagagoitia LG, Klussmann FA, Ariznabarreta LS (2003): Genetic and epigenetic alterations of the cell cycle regulators and tumor suppressor genes in pediatric osteosarcomas. J Pediatr Hematol Oncol 25:362-7.
23. Sandberg AA, Bridge JA (2003): Updates on the cytogenetics and molecular genetics of bone and soft tissue tumors: osteosarcoma and related tumors. Cancer Genet Cytogenet 145:1-30.
24. Miller CW, Aslo A, Campbell MJ, Kawamata N, Lampkin BC, Koeffler HP (1996): Alterations of the p15, p16,and p18 genes in osteosarcoma. Cancer Genet Cytogenet 86:136-42.
25. Miller CW, Yeon C, Aslo A, Mendoza S, Aytac U, Koeffler HP (1997): The p19INK4D cyclin dependent kinase inhibitor gene is altered in osteosarcoma. Oncogene 15:231-5.
26. Agiostratidou G, Derventzi A, Gonos ES (2001): Over-expression of CDKIs p15INK4b, p16INK4a and p21CIP1/WAF1 genes mediate growth arrest in human osteosarcoma cell lines. In Vivo 15:443-6.
27. Harper JW, Adami GR, Wei N, Keyomarsi K, Elledge SJ. The p21 Cdk-interacting protein Cip1 is a potent inhibitor of G1 cyclin-dependent kinases. Cell. 1993;75(4):805-16.
28. Weissenbach J, Gyapay G, Dib C, Vignal A, Morissette J, Millasseau P, Vaysseix G, Lathrop M. A second-generation linkage map of the human genome. Nature. 1992 Oct 29;359(6398):794-801.
29. Arzimanoglou II, Gilbert F, Barber HR. Microsatellite instability in human solid tumors. Cancer. 1998 May 15;82(10):1808-20. Review.
30. Martin SS, Hurt WG, Hedges LK, Butler MG, Schwartz HS. Microsatellite instability in sarcomas. Ann Surg Oncol. 1998 Jun;5(4):356-60.

31. Klingler L, Shooks J, Fiedler PN, Marney A, Butler MG, Schwartz HS. Microsatellite instability in sacral chordoma. J Surg Oncol. 2000 Feb;73(2):100-3.
32. Scheiner M, Hedges L, Schwartz HS, Butler MG. Lack of microsatellite instability in giant cell tumor of bone. Cancer Genet Cytogenet. 1996 May;88(1):35-8.
33. Szyf M. Targeting DNA methylation in cancer. Ageing Res Rev. 2003;2(3):299-328. Review.
34. Popiela A, Keith G, Borzecki A, Popiela G, Manowiec M, Gabrys M. The meaning of the methylation of genomic DNA in the regulation of gene expression levels. Eur J Gynaecol Oncol. 2004;25(2):145-9. Review.
35. Szyf M. DNA methylation and cancer therapy. Drug Resist Updat. 2003;6(6):341-53. Review.
36. Tsuchiya T, Sekine K, Hinohara S, Namiki T, Nobori T, Kaneko Y. Analysis of the p16INK4, p14ARF, p15, TP53, and MDM2 genes and their prognostic implications in osteosarcoma and Ewing sarcoma. Cancer Genet Cytogenet. 2000;120(2):91-8.
37. Lopez-Guerrero JA, Pellin A, Noguera R, Carda C, Llombart-Bosch A. Molecular analysis of the 9p21 locus and p53 genes in Ewing family tumors. Lab Invest. 2001;81(6):803-14.
38. Park YB, Park MJ, Kimura K, Shimizu K, Lee SH, Yokota J. Alterations in the INK4a/ARF locus and their effects on the growth of human osteosarcoma cell lines. Cancer Genet Cytogenet. 2002;133(2):105-11.
39. Salimath B, Marme D, Finkenzeller G. Expression of the vascular endothelial growth factor gene is inhibited by p73. Oncogene. 2000;19(31):3470-6.
40. Guise TA. Molecular mechanisms of osteolytic bone metastases. Cancer. 2000;88(12 Suppl):2892-8. Review.
41. Zheng MH, Robbins P, Xu J, Huang L, Wood DJ, Papadimitriou JM. The histogenesis of giant cell tumour of bone: a model of interaction between neoplastic cells and osteoclasts. Histol Histopathol. 2001;16(1):297-307. Review.
42. Strewler GJ. The parathyroid hormone-related protein. Endocrinol Metab Clin North Am. 2000;29(3):629-45. Review.
43. Manolagas SC. Role of cytokines in bone resorption. Bone. 1995;17(2 Suppl):63S-67S. Review.
44. Suda T, Takahashi N, Udagawa N, Jimi E, Gillepsie MT, Martin TJ. Modulation of osteoclast differentiation and function by the new members of the tumor necrosis factor receptor and ligand families. *Endocr Rev* 1999;20:345-57.
45. Dunstan CR. Osteoprotegerin and osteoprotegerin ligand mediate the local regulation of bone resorption. *Endocrinologist* 2000;10:18-26.
46. Hofbauer LC, Khosla S, Lacey DL, Dunstan CR, Boyle WJ, Riggs BL. The roles of osteoprotegerin and osteoprotegerin ligand in the paracrine regulation of bone resorption. *J Bone Miner Res* 2000;15:2–12.
47. Firkin F, Seymour JF, Watson AM, Grill V, Martin TJ. Parathyroid hormone-related protein in hypercalcaemia associated with haematological malignancy. Br J Haematol. 1996;94(3):486-92.
48. Rankin W, Grill V, Martin TJ. Parathyroid hormone-related protein and hypercalcemia. Cancer. 1997;80(8 Suppl):1564-71. Review.

49. Tsujimura H, Nagamura F, Iseki T, Kanazawa S, Saisho H. Significance of parathyroid hormone-related protein as a factor stimulating bone resorption and causing hypercalcemia in myeloma. Am J Hematol. 1998;59(2):168-70.
50. Bundred NJ, Walker RA, Ratcliffe WA, Warwick J, Morrison JM, Ratcliffe JG. Parathyroid hormone related protein and skeletal morbidity in breast cancer. Eur J Cancer. 1992;28(2-3):690-2.
51. Hiraki A, Ueoka H, Bessho A, Segawa Y, Takigawa N, Kiura K, Eguchi K, Yoneda T, Tanimoto M, Harada M. Parathyroid hormone-related protein measured at the time of first visit is an indicator of bone metastases and survival in lung carcinoma patients with hypercalcemia. Cancer. 2002;95(8):1706-13.
52. Southby J, Kissin MW, Danks JA, Hayman JA, Moseley JM, Henderson MA, Bennett RC, Martin TJ. Immunohistochemical localization of parathyroid hormone-related protein in human breast cancer. Cancer Res. 1990;50(23):7710-6.
53. Guise TA, Yin JJ, Taylor SD, Kumagai Y, Dallas M, Boyce BF, Yoneda T, Mundy GR. Evidence for a causal role of parathyroid hormone-related protein in the pathogenesis of human breast cancer-mediated osteolysis. J Clin Invest. 1996;98(7):1544-9.
54. Kurihara N, Bertolini D, Suda T, Akiyama Y, Roodman GD. IL-6 stimulates osteoclast-like multinucleated cell formation in long term human marrow cultures by inducing IL-1 release. J Immunol. 1990;144(11):4226-30.
55. Black K, Garrett IR, Mundy GR. Chinese hamster ovarian cells transfected with the murine interleukin-6 gene cause hypercalcemia as well as cachexia, leukocytosis and thrombocytosis in tumor-bearing nude mice. Endocrinology. 1991;128(5):2657-9.
56. De La Mata J, Uy HL, Guise TA, Story B, Boyce BF, Mundy GR, Roodman GD. Interleukin-6 enhances hypercalcemia and bone resorption mediated by parathyroid hormone-related protein in vivo. J Clin Invest. 1995 Jun;95(6):2846-52.
57. Teoh G, Anderson KC. Interaction of tumor and host cells with adhesion and extracellular matrix molecules in the development of multiple myeloma. Hematol Oncol Clin North Am. 1997;11(1):27-42. Review.
58. Thomas X, Anglaret B, Magaud JP, Epstein J, Archimbaud E. Interdependence between cytokines and cell adhesion molecules to induce interleukin-6 production by stromal cells in myeloma. Leuk Lymphoma. 1998;32(1-2):107-19.
59. Iwasaki T, Hamano T, Ogata A, Kakishita E. Clinical significance of interleukin-6 gene expression in the bone marrow of patients with multiple myeloma. Int J Hematol. 1999;70(3):163-8.
60. Ballester OF, Moscinski LC, Lyman GH, Chaney JV, Saba HI, Spiers AS, Klein C. High levels of interleukin-6 are associated with low tumor burden and low growth fraction in multiple myeloma. Blood. 1994;83(7):1903-8.
61. Sati HI, Greaves M, Apperley JF, Russell RG, Croucher PI. Expression of interleukin-1beta and tumour necrosis factor-alpha in plasma cells from patients with multiple myeloma. Br J Haematol. 1999;104(2):350-7.
62. Zheng MH, Fan Y, Wysocki SJ, Lau AT, Robertson T, Beilharz M, Wood DJ, Papadimitriou JM. Gene expression of transforming growth factor-beta 1 and its

type II receptor in giant cell tumors of bone. Possible involvement in osteoclast-like cell migration. Am J Pathol. 1994;145(5):1095-104.
63. Guise TA, Chirgwin JM. Transforming growth factor-beta in osteolytic breast cancer bone metastases. Clin Orthop. 2003;(415 Suppl):S32-8. Review.
64. Kakonen SM, Selander KS, Chirgwin JM, Yin JJ, Burns S, Rankin WA, Grubbs BG, Dallas M, Cui Y, Guise TA. Transforming growth factor-beta stimulates parathyroid hormone-related protein and osteolytic metastases via Smad and mitogen-activated protein kinase signaling pathways. J Biol Chem. 2002;277(27):24571-8.
65. Lacey DL, Timms E, Tan HL, Kelley MJ, Dunstan CR, Burgess T, Elliott R, Colombero A, Elliott G, Scully S, Hsu H, Sullivan J, Hawkins N, Davy E, Capparelli C, Eli A, Qian YX, Kaufman S, Sarosi I, Shalhoub V, Senaldi G, Guo J, Delaney J, Boyle WJ. Osteoprotegerin ligand is a cytokine that regulates osteoclast differentiation and activation. Cell. 1998;93(2):165-76.
66. Hofbauer LC, Heufelder AE. Clinical review 114: hot topic. The role of receptor activator of nuclear factor-kappaB ligand and osteoprotegerin in the pathogenesis and treatment of metabolic bone diseases. J Clin Endocrinol Metab. 2000;85(7):2355-63. Review.
67. Menaa C, Reddy SV, Kurihara N, Maeda H, Anderson D, Cundy T, Cornish J, Singer FR, Bruder JM, Roodman GD. Enhanced RANK ligand expression and responsivity of bone marrow cells in Paget's disease of bone. J Clin Invest. 2000;105(12):1833-8.
68. Lee SK, Lorenzo JA. Parathyroid hormone stimulates TRANCE and inhibits osteoprotegerin messenger ribonucleic acid expression in murine bone marrow cultures: correlation with osteoclast-like cell formation. Endocrinology. 1999;140(8):3552-61.
69. Hofbauer LC, Neubauer A, Heufelder AE. Receptor activator of nuclear factor-kappaB ligand and osteoprotegerin: potential implications for the pathogenesis and treatment of malignant bone diseases. Cancer. 2001;92(3):460-70. Review.
70. Huang L, Xu J, Wood DJ, Zheng MH. Gene expression of osteoprotegerin ligand, osteoprotegerin, and receptor activator of NF-kappaB in giant cell tumor of bone: possible involvement in tumor cell-induced osteoclast-like cell formation. Am J Pathol. 2000;156(3):761-7.
71. Atkins GJ, Haynes DR, Graves SE, Evdokiou A, Hay S, Bouralexis S, Findlay DM. Expression of osteoclast differentiation signals by stromal elements of giant cell tumors. J Bone Miner Res. 2000;15(4):640-9.
72. Kinpara K, Mogi M, Kuzushima M, Togari A. Osteoclast differentiation factor in human osteosarcoma cell line. J Immunoassay. 2000;21(4):327-40.
73. Miyamoto N, Higuchi Y, Mori K, Ito M, Tsurudome M, Nishio M, Yamada H, Sudo A, Kato K, Uchida A, Ito Y. Human osteosarcoma-derived cell lines produce soluble factor(s) that induces differentiation of blood monocytes to osteoclast-like cells. Int Immunopharmacol. 2002;2(1):25-38.
74. Brandstrom H, Jonsson KB, Vidal O, Ljunghall S, Ohlsson C, Ljunggren O.Biochem Tumor necrosis factor-alpha and -beta upregulate the levels of osteoprotegerin mRNA in human osteosarcoma MG-63 cells. Biophys Res Commun. 1998;248(3):454-7.

75. Vidal ON, Sjogren K, Eriksson BI, Ljunggren O, Ohlsson C. Osteoprotegerin mRNA is increased by interleukin-1 alpha in the human osteosarcoma cell line MG-63 and in human osteoblast-like cells. Biochem Biophys Res Commun. 1998;248(3):696-700.
76. Nellissery MJ, Padalecki SS, Brkanac Z, Singer FR, Roodman GD, Unni KK, Leach RJ, Hansen MF. Evidence for a novel osteosarcoma tumor-suppressor gene in the chromosome 18 region genetically linked with Paget disease of bone. Am J Hum Genet. 1998;63(3):817-24.
77. Haslam SI, Van Hul W, Morales-Piga A, Balemans W, San-Millan JL, Nakatsuka K, Willems P, Haites NE, Ralston SH. Paget's disease of bone: evidence for a susceptibility locus on chromosome 18q and for genetic heterogeneity. J Bone Miner Res. 1998;13(6):911-7.
78. Hansen MF, Nellissery MJ, Bhatia P. Common mechanisms of osteosarcoma and Paget's disease. J Bone Miner Res. 1999;14 Suppl 2:39-44. Review.
79. Sparks AB, Peterson SN, Bell C, Loftus BJ, Hocking L, Cahill DP, Frassica FJ, Streeten EA, Levine MA, Fraser CM, Adams MD, Broder S, Venter JC, Kinzler KW, Vogelstein B, Ralston SH. Mutation screening of the TNFRSF11A gene encoding receptor activator of NF kappa B (RANK) in familial and sporadic Paget's disease of bone and osteosarcoma. Calcif Tissue Int. 2001;68(3):151-5.
80. Roodman GD. Mechanisms of bone lesions in multiple myeloma and lymphoma. Cancer. 1997;80(8 Suppl):1557-63. Review.
81. Firkin F, Schneider H, Grill V. Parathyroid hormone-related protein in hypercalcemia associated with hematological malignancy. Leuk Lymphoma. 1998;29(5-6):499-506. Review.
82. Schneider HG, Kartsogiannis V, Zhou H, Chou ST, Martin TJ, Grill V. Parathyroid hormone-related protein mRNA and protein expression in multiple myeloma: a case report. J Bone Miner Res. 1998;13(10):1640-3.
83. O'Brien CA, Gubrij I, Lin SC, Saylors RL, Manolagas SC. STAT3 activation in stromal/osteoblastic cells is required for induction of the receptor activator of NF-kappaB ligand and stimulation of osteoclastogenesis by gp130-utilizing cytokines or interleukin-1 but not 1,25-dihydroxyvitamin D3 or parathyroid hormone. J Biol Chem. 1999;274(27):19301-8.
84. Thomas RJ, Guise TA, Yin JJ, Elliott J, Horwood NJ, Martin TJ, Gillespie MT. Breast cancer cells interact with osteoblasts to support osteoclast formation. Endocrinology. 1999;140(10):4451-8.
85. Tricot G. New insights into role of microenvironment in multiple myeloma. Lancet. 2000;355(9200):248-50. Review.
86. Roodman GD. Biology of osteoclast activation in cancer. J Clin Oncol 2001;19:3562-71
87. Chirgwin JM, Guise TA. Molecular mechanisms of tumor-bone interactions in osteolytic metastases. Crit Rev Eukaryot Gene Expr 2000; 10:159-78.
88. Chikatsu N, Takeuchi Y, Tamura Y, Fukumoto S, Yano K, Tsuda E, Ogata E, Fujita T. Interactions between cancer and bone marrow cells induce osteoclast differentiation factor expression and osteoclast-like cell formation in vitro. Biochem Biophys Res Commun. 2000;267(2):632-7.

89. Brown JM, Quinn JE, Buhler KR, Vessella RL. Osteoprotegerin and TRANCE are differentially expressed by prostate cancer cells in vitro amd in vivo. J Bone Miner Res 1999;14(Suppl 1):1085.
90. Kitazawa S, Mori H, Maeda S, Kitazawa R. Osteoclast differentiation factor (ODF/OPGL/RANKL/TRANCE) expression in bone invasion model of breast cancer. J Bone Miner Res 1999;14(Suppl 1):1083.
91. Huang L, Cheng YY, Chow LT, Zheng MH, Kumta SM. Tumour cells produce receptor activator of NF-kappaB ligand (RANKL) in skeletal metastases. J Clin Pathol. 2002;55(11):877-8.
92. Zhang J, Dai J, Qi Y, Lin DL, Smith P, Strayhorn C, Mizokami A, Fu Z, Westman J, Keller ET. Osteoprotegerin inhibits prostate cancer-induced osteoclastogenesis and prevents prostate tumor growth in the bone. J Clin Invest. 2001 May;107(10):1235-44.

CHAPTER 18

BONE TISSUE ENGINEERING

Xuebin Yang[1,2] M.D., MS.c., Ph.D. and Richard O.C. Oreffo[1] DPhil

[1]*Bone & Joint Research Group MP817, University of Southampton. Southampton SO16 6YD, U.K.*

[2]*Leeds Dental Institute, University of Leeds, Leeds LS16 5AF, U.K.*
E-mail:denxy@leeds.ac.uk

The ability to generate new bone for skeletal use is a major clinical need. Biomimetic scaffolds that interact and promote human osteoprogenitor differentiation and osteogenesis offer a promising approach to generate skeletal tissue to resolve this major healthcare issue. Although autogenous or allogenic bone grafts have been used for many years, a number of disadvantages have limited their use, which leads to the emerging of an attractive new field - tissue engineering. Bone tissue engineering requires three basic elements: 1) a source of osteogenic cells, 2) a source of osteoinductive agents and, 3) an osteoconductive scaffold. Bone formation comprises a complex and temporal sequence of events that begins with the recruitment and proliferation of osteoprogenitors from mesenchymal stem cells.[1-4] Central in this process is a material or scaffold for the migration, attachment and growth of stem/progenitor cells from the surrounding tissue. Mesenchymal cells including osteoblasts, chondroblasts, adipocytes, myoblasts and fibroblasts appear to be derived from a common multipotential mesenchymal stem/progenitor cell residing within bone marrow. A number of studies indicated the ability to generate osteoinductive and osteoconductive biomimetic scaffolds in combination with osteogenic growth factors and gene therapy to create a biomimetic microenvironment for osteoprogenitor growth and, significantly, the potential to generate templates for the development of a living bone substitute for clinical application.

1. Introduction

1.1. Skeletal repair — clinical need

Over one million orthopaedic operations annually involve bone repair as a consequence of replacement surgery, trauma, cancer, congenital abnormalities or skeletal deficiency.[5-8] Thus the ability to generate new bone for skeletal use is a major clinical need. Although a number of different methods have been developed to meet such a clinical requirement, to date, the most common procedures still rely on bone grafts.[9] Fresh autogenous and allogenic bone grafts, both cancellous and cortical provide a source of osteoprogenitor cells, osteoinductive growth factors and a structural scaffold for new bone formation. Furthermore, the three-dimensional framework of both autografts and allografts can function as mechanical supports for angiogenesis and the invasion of osteoprogenitor cells into the bone grafts ('osteoconduction'). However, fresh allografts can induce both local and systemic immune responses that diminish or destroy the osteoinductive and conductive processes.[9] To circumvent this issue, approaches have included the use of freezing or freeze-drying allografts to improve the usability.[9,10]

Although autogenous or allogenic bone grafts have been used for many years, a number of disadvantages have limited their use including 1) these methods are inappropriate in cases of large bone deficiency, 2) the requirement for surgery from multiple areas, 3) the loss of normal bone structure from donor areas, 4) the risk of infection and secondary deformities at the donor site [11-13] and, 5) allogenic bone graft carries potential risks of cell-mediated immune responses and pathogen transmission.[12,14-16] In addition, cancellous bone grafts are completely replaced, in time, by creeping substitution while cortical bone grafts remain an admixture of necrotic and viable bone over time.[9] The above limitations in the use of autogenous or allogenic bone graft has resulted in the search for alternative bone substitutes. Synthetic grafting materials eliminate many of the aforementioned risks and these materials do not transfer osteoinductive or osteogenic elements to the host site. However, the advantages of autograft and allograft can be considered using a composite graft for clinical application. Skeletal tissue engineering has emerged as an alternative approach to bone regeneration.[6,17-23]

1.2. Bone regeneration — a role for bone tissue engineering

Tissue engineering can be defined as the application of scientific engineering principles to the design, construction, modification, growth and maintenance of living tissue and organs from native or synthetic sources for the human body to restore function based on principles of molecular developmental biology, cell biology and biomaterial sciences.[24] Current tissue engineering programmes include skin, cornea, liver, pancreas, kidney, urinary bladder, digestive tract, vessel, muscle, nerve, ligament, bone and cartilage and many other tissues. Among the many tissues in the human body, bone has considerable powers for regeneration and is a prototype model for tissue engineering. Significantly, bone tissue engineering is gradually developing towards a clinical reality.[23,25-27]

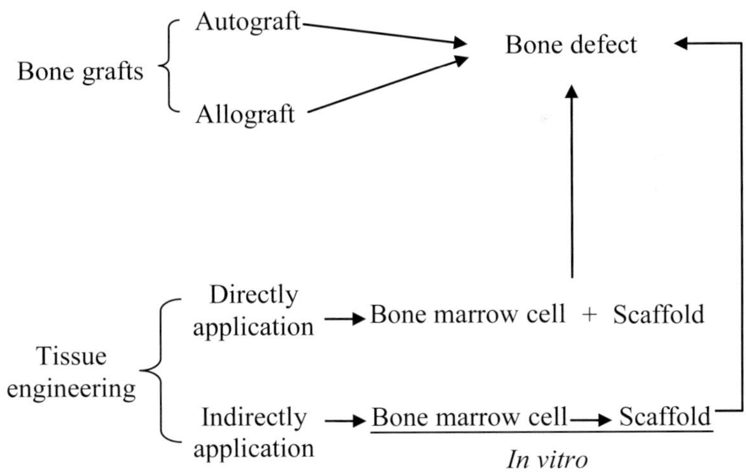

Fig. 18.1. The procedures of routine bone grafts and bone tissue engineering.

A number of innovative therapeutic approaches for tissue engineering have evolved and, typically a three step approach has been proposed for tissue engineering including 1) an expansion phase, 2) initiated tissue differentiation phase and, 3) histotypical differentiation phase [28] (Fig. 18.1). While recent advances in progenitor cells, novel factors, smart materials, and gene therapy offer much hope. However, good clinical practice, consistency, reproducibility, validation and appropriate regulation of these new biological treatments remain significant issues.[18,27]

The symbiosis of bone inductive and conductive strategies is critical for tissue engineering, and is in turn governed by the microenvironment, consisting of the extracellular matrix, which can be duplicated by biomimetic biomaterials such as collagens, hydroxyapatite, calcium phosphates, bioactive glass, proteoglycans, and cell adhesion proteins including fibronectin and bone morphogenetic proteins.[29-34] Advances in scaffold design, bioactive factors and cell culture have improved the prognosis for success in orthopaedic tissue engineering [24,25] and a multidisciplinary approach combining all three key components offers the possibility of clinical success in the long term.[35]

1.3. Osteogenic cells

The potential of using adult & embryonic stem cells for tissue regeneration has generated tremendous scientific and public interest. This follows the discovery of the reasonably well-characterized hematopoietic stem cell (HSC) from bone marrow and peripheral blood, which has found clinical efficacy. The mesenchymal stem cell (MSC) has been shown to be multi potential.[4,36-38] The MSCs is capable of differentiating along multiple lineages under the appropriate culture conditions and to display significant expansion capability (Fig. 18.2).[18,37,39-41] Recent information indicates HSCs and MSCs in general may have more universal differentiation abilities than previously thought.[42,43]

The presence of stem cells for non-hemotapoietic cells in bone marrow was first suggested by the observations of the German pathologist Cohnheim over a century ago.[44] However, the definitive evidence that bone marrow contains multipotential cells came from the pioneering work of Friedenstein.[45] Mesenchymal stem cells are a rare population of undifferentiated cells that have the capacity to differentiate into mesodermal lineages, including bone, fat, muscle, cartilage, tendon, and marrow stroma (Fig. 18.2).[4,36,37,46-50] These stem cells have a number of unique properties: 1) the ability to self-renew, 2) they are multi-potential and, 3) the capacity to regenerate tissue after injury.[36,37,40,43,51,52]

These cell populations may be expanded in culture and subsequently permitted to differentiate into the desired lineage (Fig. 18.2).

The validity of using stromal cells in tissue engineering arises from a number of previous studies: 1) Marrow stromal cells can be isolated from marrow by exploitation of their capacity to adhere to tissue culture plastic and have been shown to have osteogenic activity, [4,44,53] 2) Large quantities of human MSCs can be readily obtained following a simple bone marrow aspiration procedure and subsequent expansion over a million-fold in culture, 3) Culture expanded rat and sheep bone marrow cells have been shown to heal critical-size of bone defects following reimplantation,[3,54-57] 4) Connolly [58] has shown the potential for human bone marrow injection for the treatment of delayed fracture-union and bone defects and, 5) Horwitz et al [59,60] have reported on the therapeutic effects of bone marrow-derived osteoprogenitors transplanted into children with osteogenesis imperfecta.

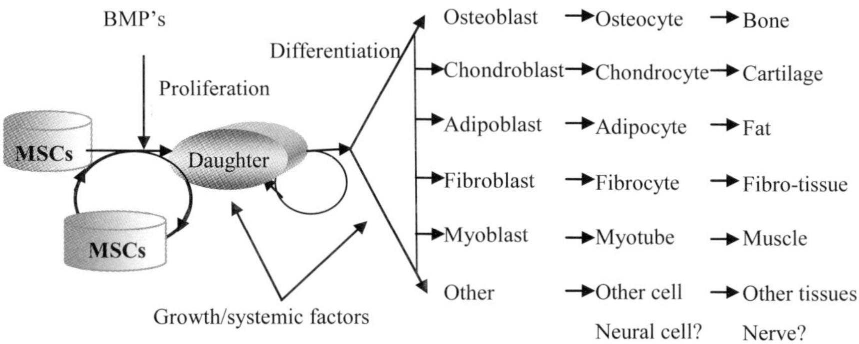

Fig. 18.2. Cell differentiation from the hypothetical, pluripotent mesenchymal stem cell

1.4. Osteoinductive agents

A number of strategies have been developed in recent years in the use of bioactive factors for bone tissue engineering including, extraction and partial purification of growth factors, recombinant protein synthesis and, gene therapy.[61] The capacity of bone for growth, regeneration, and remodelling is largely due to the induction of osteoblasts that are recruited to sites of new bone formation. The process of recruitment remains unclear, though the immediate environment of the cells is likely to play a role via cell-matrix-osteoinductive factor-cell interactions.[25,62] Bone is a physiological storehouse for a multitude of growth factors including insulin-like growth factors I and II (IGF-I, IGF-II), members of transforming growth factor-beta (TGF-β) super family including BMP's, platelet-derived growth factor, and fibroblast growth factors (FGFs). Osteoblasts have been shown to produce many of these growth factors, many of which are incorporated into the extracellular matrix during bone formation. The growth factors are located within the matrix until remodeling or trauma results in their subsequent release.[63] This leeds to modulation of osteoblast and osteoclast metabolism and function during bone remodeling and the initiation and regulation of bone formation after trauma via an autocrine and paracrine mechanisms (Fig. 18.3).[62,64]

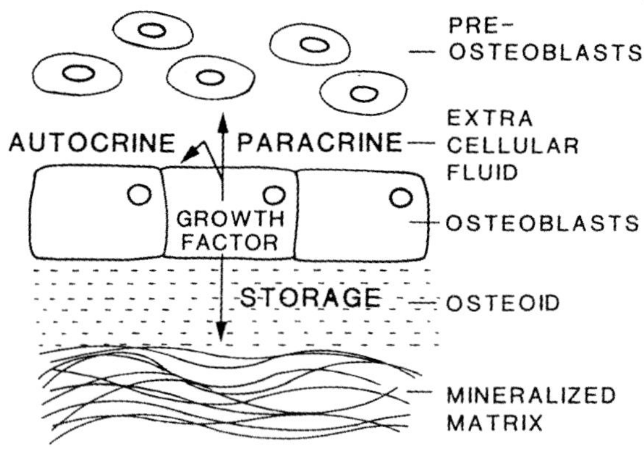

Fig. 18.3. Modulation of bone formation by growth factors. (Reproduced from J Bone Miner Res 1993:8S2:S569 with permission of the American Society for Bone and Mineral Research).

Bone morphogenetic proteins (BMP's)

Bone morphogenetic proteins were first reported by Marshall Urist [65] who described the isolation of a bone inductive extract from adult bone and demonstrated the ability of this extract to induce new endochondral bone formation at ectopic sites in rodents. Since then, a number of osteogenic proteins have been discovered and added to the BMP family.[66] These proteins have been identified as the key signal molecules or stimuli to induce the conversion of mesenchymal stem cells into osteoprogenitor cells and the subsequent development of new bone via endochondral ossification.[47,67] During the differentiation process from mesenchymal progenitors, various hormones and cytokines regulate osteoblast differentiation. Among these, the BMP's are the most potent inducers and stimulators for osteoblast differentiation from bone marrow stromal cells (Table 18.1).[62,68,68-71]

Table 18.1. Action of BMP's on Osteoblasts

Factor	Synonym	KDa	Proliferation	Differentiation
BMP-2	BMP-2A	30-35	↑*	↑
BMP-3	Osteogenin	30-40	↑	↑
BMP-4	BMP-2b	30-35	↑	↑
BMP-5		30-35	↑	?
BMP-6		30-35	↑	↑
BMP-7	Osteogenic protein-1	30-35	↑	↑
BMP-8	Osteogenic protein-2	40-50	?	?
BMP-9	Dorsalin-1	60-63	↑	?

*↑: Increase; ?: not known (Adapted from Baylink et al, 1993; Ebisawa T et al, 1999 and Miller et al, 2000)

The mature BMP monomer consists of about 120 amino acids, with seven cysteine residues. A number of studies have shown the bone inductive potential of BMP-2, 4-7, and 9 in the treatment of fracture repair, segmental bone defects and in the fixation of prosthetic implants and their use in ectopic bone induction.[72-78] To date, BMP's comprise over 30 members, although, BMP-1 is a cysteine-rich peptidase and is not a member of the TGF-β super family.[79]

In vitro experiments have shown that BMP-2 stimulates the formation and mineralisation of bone-like nodules in primary osteoblast cultures [80,81] and promotes the development of an osteoblast-like phenotype in pluripotent mesenchymal stem cell lines.[67] Furthermore, recombinant human BMP-2 (rhBMP-2) has been successfully used to promote a greater degree of osseous and periodontal repair.[82] Following the first report of the healing of large segmental bone defects using BMP-2 implantation, BMP-2 has proven useful in healing critical size defects in rat and sheep femurs.[83] rhBMP-2 can be used to modify the surface chemistry of biomaterial to promot cell attachment and growth on the scaffold to form bone (Fig. 18.4). Anderson and coworkers [84] reported that Saos-2 cells, derived from human osteosarcoma cells, uniquely contain a bone-inducing activity and that components of the Saos-2 cells contain bone morphogenetic proteins (BMP's)-1, -2, - 3, -4, -5, -6, and -7 and the non-collagenous matrix proteins bone sialoprotein, osteonectin (ON), osteopontin (OPN), and osteocalcin (OCN). The combination of

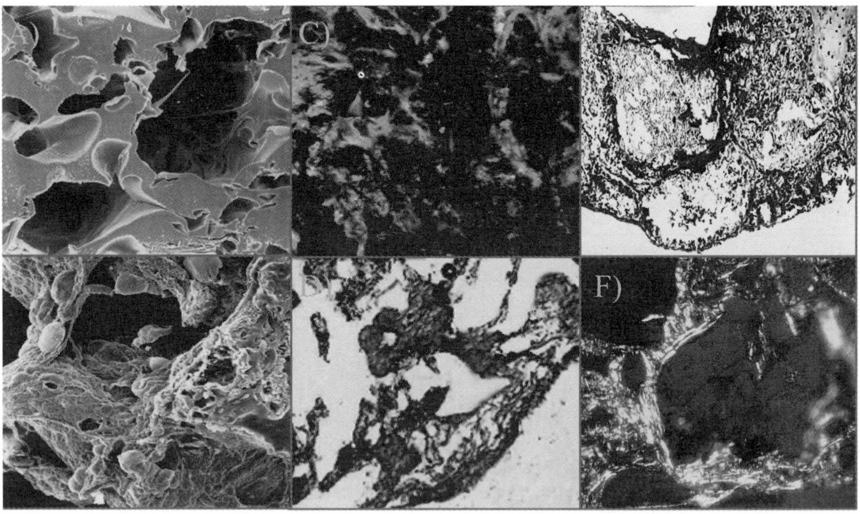

Fig. 18.4. Human bone marrow stromal cell growth on rhBMP-2 adsorbed poly (lactic acid) (PLA) scaffolds. A) Original PLA structure without cells (scanning electron microscopy: SEM). B) SEM. C) Cell adhesion and proliferation on rhBMP-2 adsorbed PLA scaffold as observed by confocal microscopy (viable cells-green and necrotic cells-red). D) Expression of Type I collagen by immunohistochemistry confirmed the maintenance of the osteoblast phenotype. *In vivo* subcutaneous implant: HBM cell in-growth into rhBMP-2 adsorbed PLA scaffolds. New bone matrix formation was observed on/in rhBMP-2 adsorbed PLA scaffolds as detected by toluidine blue staining (E) and birefringence (F). Original magnification: A,C-F) ×100, B) ×500.

BMP-1/tolloid, BMP-3 and BMP-4, and bone sialoprotein was important for the osteoinductive capacity of Saos-2 cells.[85] *In vivo* data has shown the ability of using freeze-dried Saos-2 cells to promote endochondral bone formation.[84] Yang et al (2003) showed that Saos-2 extracted osteoinductive factors significantly stimulated human osteoprogenitors alkaline phosphatase specific activity in basal and osteogenic conditions. Osteoinductive factors present in Saos-2 cell extracts promoted adhesion, expansion, and differentiation of human osteoprogenitor cells on 3-D scaffolds.[86] To date, the optimal mix of BMP's, the appropriate dosage and carrier remain significant challenges.

A number of studies have reported on the clinical application of BMPs. Friedlander and co-workers (2001) reported on the use of type I collagen to deliver rhBMP-7 (OP-1) in the treatment of tibial non-unions in 122 patients.[87] Burkus et al (2002) in a multicenter, prospective, randomized, nonblinded study of 279 patients with degenerative lumbar disc disease found that at 24 months rhBMP-2 on an absorbable collagen sponge demonstrated higher fusion rate (94.5%) than patients (control group) (88.7%) who received autogenous iliac crest bone graft.[88] In a further clinical trial using an LT-CAGE lumbar tapered fusion device, 277 patients had their cages implanted with rhBMP-2 on an absorbable collagen sponge and 402 received autograft transferred from the iliac crest. The patients treated with rhBMP-2 had statistically superior outcomes with regard to length of surgery, blood loss, hospital stay, re-operation rate, median time to return to work, and fusion rates at 6, 12, and 24 months.[89] Govender and colleagues (2002) treated open tibial fractures using rhBMP-2 in a prospective, controlled, randomized study of 450 patients. The study demonstrated that rhBMP-2 implant (1.50 mg/mL) was safe and resulted in significantly superior to accelerated fracture/wound healing, reduced frequency of secondary interventions, the overall invasiveness of the procedures and the infection rate in patients.[90] Boden SD et al (2002) in a prospective randomized clinical study of 25 patients undergoing lumbar arthrodesis demonstrated that rhBMP-2 (with biphasic calcium phosphate granules) induced consistently improved radiographic posterolateral lumbar spine fusion.[91]

Osteoblast Stimulating factor-1/Pleiotrophin (PTN)

Pleiotrophin, also known as heparin-binding growth-associated molecule (HB-GAM), is a 136 amino acid polypeptide which is widely expressed during embryonic life but whose expression is restricted to bone and brain during adulthood.[92,93] PTN was first identified as HB-GAM and shown to processe mitogenic activity in rat and mouse fibroblasts and as a factor that promoted neurite outgrowth in cultures of neonatal rat brain cells.[94,95] In 1990, Tezuka et al [96] detected the same mRNA in calvarial osteoblast–enriched cells and MC3T3-E1 cells by differential hybridisation screening between osteoblastic and fibroblastic cells, and named the factor osteoblast stimulating factor-1 (OSF-1). In bone and cartilage tissues, PTN is expressed in developing and regenerating bone as a matrix-bound form and in culture, it stimulates differentiation of osteoblasts and chondrocytes.[97,98]

PTN is prominently expressed in the cell matrices that act as target substrates for bone formation, probably by mediating chemotactic recruitment and attachment of osteoblasts/osteoblast precursors to the appropriate matrices.[92,99-102] In addition, PTN is thought to play a role in the process of angiogenesis in endochondral ossification.[92,99,101,102] Studies have shown PTN stimulates *in vitro* proliferation and differentiation of osteoblastic cells.[93,97] Furthermore, N-syndecan has been identified as an essential cell surface receptor for PTN, which in turn is immobilized in the extracellular matrices onto which those cells are recruited. The receptor has been found on osteoprogenitors and osteoblasts.[103-106] Recently, Sato and co-workers (2002) demonstrated that PTN had a dose-dependent synergistic or inhibitory effect on BMP-2 induced osteogenesis in endochondral ossification in rat.[107] Yang et al (2003) have shown that PTN has the ability to promote migration, adhesion, expansion and differentiation of human osteoprogenitor cells (Fig. 18.5) and appears to act specifically on late human osteoprogenitor populations.[100] However, Tare RS et al (2002) reported that PTN did not have the osteoinductive potential of bone morphogenetic proteins (BMPs). PTN was fund to have multiple effects on bone formation and the modulation of BMP induced osteoinduction, which were dependent on the concentration of PTN and the timing of its presence.[108]

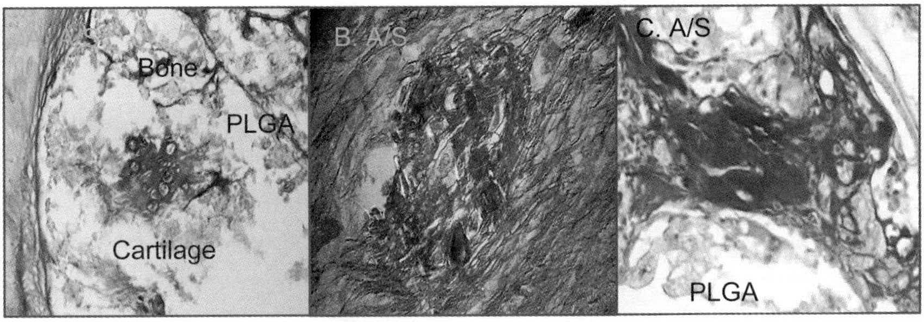

Fig. 18.5. Cartilage and bone formation by human bone marrow cells on PTN adsorbed PLGA scaffolds within diffusion chambers (A, B) and subcutaneously implanted (C) in Nu/Nu mice after 10 weeks as analyzed on paraffin sections. A) Alcian blue and Sirius red staining showed new bone and cartilage matrix formation within chambers; B) Alcian blue/Sirius red staining and birefringence showed new bone formation; and C) Alcian blue/Sirius red staining showed bone matrix deposition within the porous scaffold in subcutaneous implant model. A) ×100, B,C) ×250. * PLGA: poly(lactic-c-÷glycolic acid) scaffold. (Reproduced from J Bone Miner Res 2003:18:47-57 with permission of the American Society for Bone and Mineral Research).[100]

Other factors

There are a number of other important growth/transcription factors involved in the bone formation process.

Runt-related gene 2 (Runx-2): Runx-2 or core-binding factor alpha 1 (cbfa-1) is a helix-loop-helix factor required for the expression of osteoblastic characteristics and bone development.[109] A number of studies have shown that Runx-2 plays a crucial rule in bone cell differentiation, maturation and function in both intramembranous and endochondral ossification (Fig. 18.7).[68,110,111]

Insulin-like growth factors (IGFs): Insulin-like growth factors have been shown to not only enhance bone collagen, matrix synthesis and stimulate the replication of osteoblasts,[112] but also decrease collagen degradation, which plays a central role in the maintenance of bone matrix and bone mass.[113]

Fibroblast growth factors (FGFs): Fibroblast growth factors have angiogenic properties and are considered important in neovascularisation, wound healing and bone repair. Bones treated with FGFs contain a greater number of cells that synthesize bone collagen matrix. However FGFs do not directly affect osteoblast differentiation.[113,114]

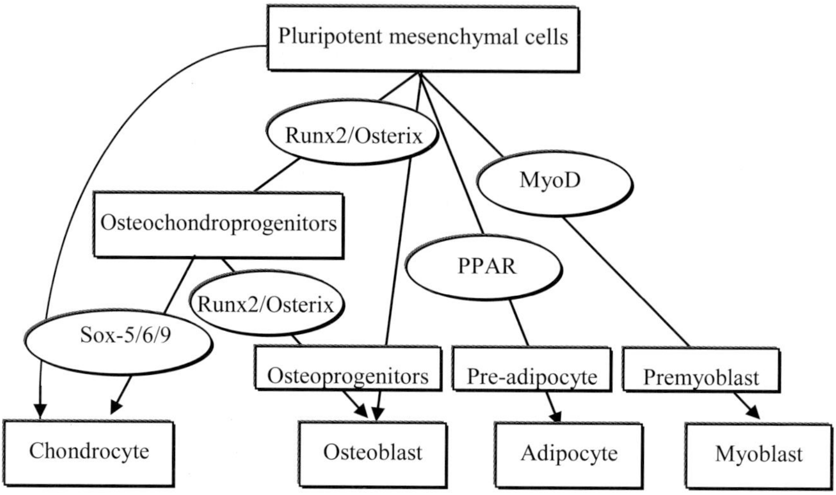

Fig. 18.6. Cell lineage-specific transcription factors. (Adapted from Yamaguchi A et al, 2000; Katagiri T and Takahashi N, 2002).

Parathyroid hormone (PTH): Parathyroid hormone has significant effects on osteoblast activity and has been shown to increase the net release of alkaline phosphatase activity. PTH plays a central role in concert with $1,25(OH)_2D_3$ in maintaining serum calcium and phosphate levels.[115-117]

Growth hormone (GH): Growth hormone is an important regulator of longitudinal bone growth. In vivo and in vitro studies have demonstrated that GH is important in the regulation of both bone formation and bone resorption.[118]

Estrogen: In vitro studies indicate estrogen can modulate osteoblast proliferation, differentiation and the stimulation of other growth factors. It is the major sex steroid regulating the metabolism and maturation of bone and bone turnover in women and men.[119,120] Qu et al [121] demonstrated that estrogen stimulated sequential differentiation of osteoblasts and increased deposition and mineralisation of matrix in mouse bone marrow cultures.

Prostaglandins (PGs): Prostaglandins are important local factors in bone cell metabolism and can stimulate cell proliferation, collagen, non-collagenous protein synthesis and bone formation.[122,123] In addition, PGs have been shown to stimulate osteoclast formation in a variety of cell culture systems. The stimulation of osteoclastic bone resorption may be important in mediating bone loss in response to mechanical forces and inflammation.[124]

Collagen fragment – P-15: Type I collagen has been shown to induce osteoblast-related gene expression of bone marrow cells during osteoblastic differentiation.[125] Bhatnagar et al [126] reported that the potential to construct biomimetic environments by immobilizing a collagen-derived high-affinity cell-binding peptide, P-15, in 3-D templates to promote attachment of human dermal fibroblasts to anorganic bovine bone mineral phase to enhance expression of the osteoblast phenotype.

1.5. Gene therapy

Osteogenic stem cell transplantation has been further developed to incorporate and utilize the principles of gene therapy. The approach is compelling with gene therapy combining endogenous bone stem cells with genes encoding physiological specific osteoinductive growth factors to provide an enhanced and significant bone healing response.[64,127] Boden and co-workers (2000) have suggested that three critical steps are essential in gene therapy for bone formation: 1) An appropriate osteoinductive gene (and effective dose), 2) An appropriate delivery vector (including transduction time/gene transfer method) and, 3) An appropriate carrier material as a scaffold for the new bone formation.[128].

Previous studies have demonstrated the clinical utility of BMP's in spinal fusion, fracture healing and prosthetic joint stabilization.[128-131] Mesenchymal stromal population transfected with bone morphogenetic proteins have, to date, offered the promise of gene therapy for bone tissue engineering and indicate several theoretical advantages over implantation of the recombinant human BMP, including persistent BMP delivery and eliminating the need for a foreign body carrier.[132]

Musgrave et al [132] constructed a replication defective adenoviral vector to carry the rhBMP-2 gene (AdBMP-2) and showed intramuscular bone formation as early as 2 weeks following injection. Alden and co-workers [133] demonstrated that BMP-2 adenoviral vectors could induce striated muscle cells to produce BMP-2, leading to endochondral bone formation in athymic nude rats. In clinically related animal studies, the BMP-2-expressing adenovirus-transfected marrow cells were subsequently injected into critically sized defects in rat femurs and reached higher healing rates compare to controls.[134] Partridge and co-workers (2002) demonstrated the successful delivery of active BMP-2 using human osteoprogenitors on porous biodegradable scaffolds to generate mineralised 3-D structures *in vitro* and *in vivo* (Fig. 18.8).[135] Retroviral BMP-2 gene transfer has also been used effectively in combination with a biodegradable matrix (PLGA-HA scaffold) to stimulate the synthesis of bioactive BMP's and promote bone formation in a mouse model.[21] However, concerns exist regarding clinical safety (immunogenesis *in vivo*, fate of adenovirus) and the long-term complications of injecting genetically altered cells into humans.[43,64,136]

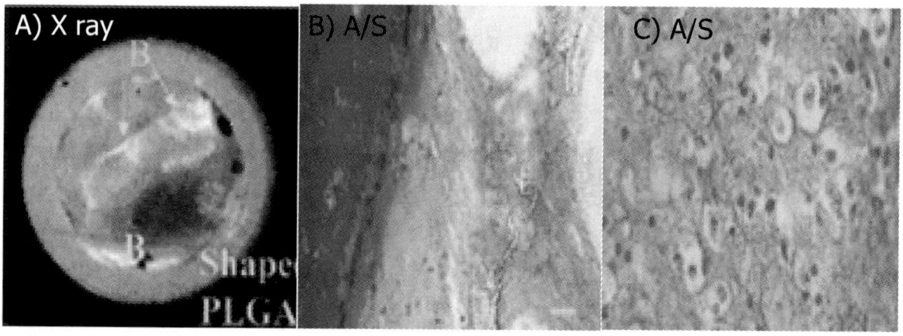

Fig. 18.7. Bone formation by X-ray analysis on PLGA scaffolds within diffusion chambers after 10 weeks by cultured primary human bone marrow cells transduced with AdBMP-2 (A). Bone and cartilage matrix formation as demonstrated by Alcian blue and Siurius red staining (B). Only fibrous tissues was seen in diffusion chambers with human bone marrow cells alone. Original magnification: A) ×5, B,C) ×100. (Reprinted from Biochemical and Biophysical Research Communications, Vol 292, Partridge et al, Adenoviral BMP-2 gene transfer in mesenchymal stem cells, Page 149 (2002), with permission from Elsevier, 2002).[135]

1.6. Osteoconductive materials

Bone formation comprises a complex and temporal sequence of events that begins with the recruitment and proliferation of osteoprogenitors from mesenchymal stem cells.[1-4,49] Central in this process is a material or scaffold with properties compatible to normal bone (Table 18.2) [137-140] for the migration, attachment and growth of stem cells and progenitor cells from the surrounding tissue.[141]

Table 18.2. Mechanical properties of human tissues

	Tensile strength (MPa)	Compressive strength (MPa)
Human femoral head	-	35-86
Human vertebra	-	66±10
Human femoral neck	-	180±28
Cancellous bone	N/A	4-12
Cortical bone	60-160	130-180
Cartilage	3.7-10.5	N/A

(Adapted from Yang S et al, 2001; Ladd et al 1998; Ulrich et al 1997; Bayraktar HH et al 2004)

Table 18.3. Properties of biodegradable polymers

Polymer type*	Melting point (°C)	Glass trans. Temp. (°C)	Degrading time (months)**	Tensile strength (MPa)
PLGA	Amorphous	45-55	Adjustable	41.4-55.2
DL-PLA	Amorphous	55-60	12-16	27.6-41.4
L-PLA	173-178	60-65	>24	55.2-82.7
PGA	225-230	35-40	6-12	>68.9
PCL	58-63	-65	>24	20.7-34.5

*PLGA: poly(l-lactic-co-glycolic acid); DL-PLA: poly(D,L-lactic acid); L-PLA: poly(L-lactic acid); PGA: poly(glycolic acid); PCL: Poly(∊-caprolactone); **Time to complete mass loss. Time also depends in part on geometry. (Adapted from Yang S et al 2001)

The scaffold material should display a number of characteristics including, 1) biocompatibility (surface chemistry for cell attachment, proliferation, and differentiation), 2) bioresorbability (with controllable degradation and resorption rate to match cell/tissue growth *in vitro* and/or *in vivo* (Table 18.3) (i.e., degradable into nontoxic products, leaving the desired living tissue), 3) appropriate porosity/interconnectivity and, 4) mechanical properties to the appropriate tissues at the site of implantation.[137,142-144]

Biodegradable polymers

Biocompatible materials such as metals (stainless steels, titanium-based alloys), ceramics (alumina, coralline hydroxyapatite, porous calcium phosphate, calcium phosphate cements, bioglass) and polymethylmethacrylate (PMMA) have been used extensively for surgical implantations. Coralline hydroxyapatite and porous calcium phosphate have also been used as carriers for osteoinductive factors and as osteoconductive matrices for human bone cells and human bone marrow populations in cell transplantation studies. However, these materials are not themselves osteoinductive and are resorbed relatively slowly *in vivo*. To overcome these limitations, natural or synthetic materials such as poly(lactic acid) (PLA), poly(glycolic acid) (PGA) and copolymers of PLA and PGA as well as biodegradable composite scaffolds based on poly(lactic acid-÷-co-glycolic acid) (PLGA) and polypropylene fumarate have been developed.[5,17,145-147] However, many of the existing three-dimensional scaffolds for tissue engineering are currently less than ideal for clinical applications, due to an absence of mechanical strength, but also a lack of appropriate interconnection porosity critical for cell ingrowth and generation of 3-D tissue.[137,148]

Biomimetic scaffolds

The use of the synthetic polymer materials [137] as well as collagen sponges [149] to generate biomimetic scaffolds for cell transplantation and tissue growth has created significant general interest.[5,150,151] To date, Poly(lactic acid), Poly(glycolic acid), polydioxanone and copolymers are the only synthetic, degradable polymers gained U.S. Food and Drug Administration (FDA) approval and widely used in the formation of resorbable sutures, meshes, scaffolds and drug delivery devices. These materials are biocompatible, processable into a three dimensional structure and, degradable. In recent years, procedures for the surface modification of these materials with biological agents have been developed,[152] leading to wide interest in the use of biomimetic scaffolds, capable of interacting with progenitor to promote differentiation and new bone formation. Furthermore these structures, if coupled with appropriate factors, offer the possibility of positional and environmental information for new tissue growth.

1.7. Biomaterial surface modification

An essential step in successful tissue engineering is the ability of cells to adhere to an extracellular material followed by the ability of the cell to differentiate leading to the production and organisation of an extracellular matrix. The immediate limitation for many polymer materials is the absence of a chemically reactive pendent chain for the easy attachment of cells, drugs, crosslinkers, or biologically active moieties.[137] Generally, cell adhesion is a series of interactive events comprising: 1) initial cell attachment, 2) cell spreading, 3) organisation of an actin cytoskeleton and, 4) formation of focal adhesions.[153] The attachment of the cell to the extracellular matrix (and biomaterials) is known to be controlled by various families of adhesion receptors, including the integrins, selectins, cadherins and immunoglobulins.[22,154,155] Within the integrin receptor family, the binding between integrin receptor and ligand is often mediated through an amino acid recognition sequence Arg-Gly-Asp (RGD), shown to serve as a primary cell attachment cue.[155] In addition, Pierschbacher and Ruoslahti (1984) have shown that synthetic peptides that contain the amino acids RGD, such as GRGDSP, can essentially mimic cell attachment activity of the parental molecule. Thus the RGD peptide cell adhesion ligand provides a simple mechanism of creating polymer surfaces that mimic the extracellular matrix to support osteoblast-like cell adhesion and spreading.[153,156] Fibronectin (FN), vitronectin (VN) and laminin (LN) are believed to play a central role in cellular morphology, migration and the provision of signals that orchestrate cellular proliferation, metabolism, function and differentiation in anchorage dependent cells such as osteoblasts and osteoclasts.[155,157-162] The binding sites of fibronectin include those for fibrin, heparin, collagen, DNA, cells, amyloid P component and fibrin(ogen).[163] Ito et al [164] have demonstrated that the cell adhesion onto RGDS- and FN-immobilised film is based not on physical interactions, but on specific ligand/receptor interactions. Therefore, the adhesion proteins together with their receptors constitute a versatile recognition system providing cells with anchorage, traction for migration, and signals for polarity, position, differentiation and possibly growth (Fig. 18.8).[100,135,154,165]

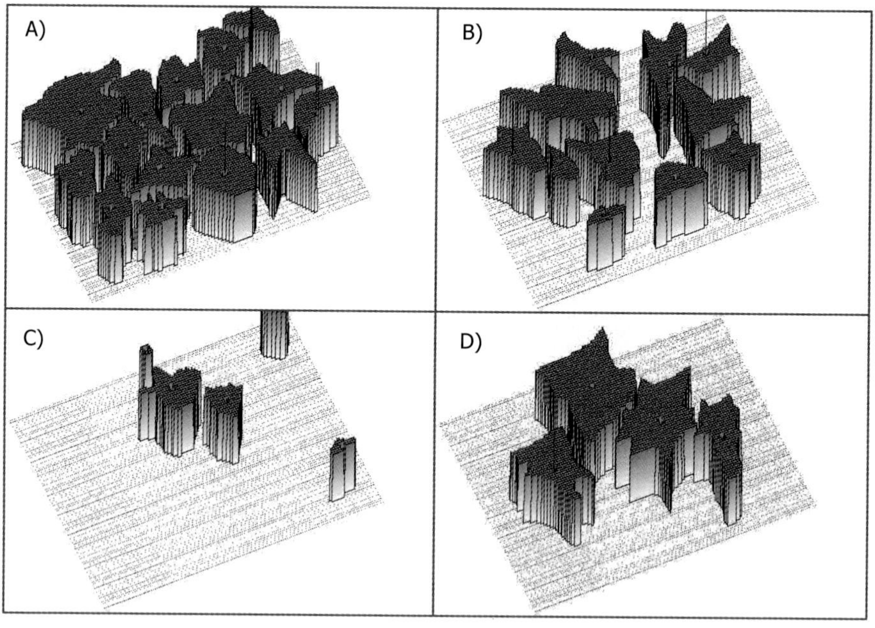

Fig. 18.8. Human bone marrow cell attachment and spreading on PLA films: A) Enhanced cell attachment and spreading were observed in serum free conditions on PLA films coated with FN, or B) coupled with PLL-GRGDS compared to C) PLA film alone. D) tissue culture plastic coated with serum after 5 hours as detected by fluorescence microscopy. Original magnification: ×200

Growth factor and cell encapsulation

Cell encapsulation techniques have been successfully used for the transplantation of pancreatic islets to treat diabetes [166] and for the treatment of Parkinson's disease.[167] In the last decade, drugs and vaccines have also been delivered by encapsulation.[168-170] Gao et al [168] showed that thermoreversible polymers are compatible with rhBMP-2 induced osteogenesis and can serve as novel biomaterials for rhBMP-2 delivery. Thus the encapsulation of osteogenic factors within biodegradable porous polymer scaffold may provide an alternative approach to produce biomimetic osteogenic scaffold for bone regeneration.[171]

2. Conclusion

Tissue engineering has emerged as an important cross-disciplinary approach for the regeneration of traumatic/damaged tissue for a variety of clinical applications including skeletal repair.[17,18,172,173] To achieve this goal, a central strategy has evolved using a source of progenitor cells, appropriate scaffolds and the exploitation of appropriate signaling molecules/growth factors. Internation between researchers in the stem cell research, growth factor biology and biomaterial/biomimetics will aid research in (bone) tissue engineering. Central to such a tissue engineering paradigm will be, however an understanding of how tissues and organs develop and the normal processes of growth and repair.

References

1. S. P. Bruder, N. Jaiswal, N. S. Ricalton, J. D. Mosca, K. H. Kraus, and S. Kadiyala, *Clin. Orthop.,* S247 (1998).
2. M. E. Owen, *J. Cell Sci. Suppl,* 63 (1988).
3. S. Kadiyala, R. G. Young, M. A. Thiede, and S. P. Bruder, *Cell Transplant.,* 125 (1997).
4. A. J. Friedenstein, R. K. Chailakhyan, and U. V. Gerasimov, *Cell Tissue Kinet.,* 263 (1987).
5. C. Chaput, A. Selmani, and C. H. Rivard, *Curr Opin Orthop,* 62 (1996).
6. S. J. Peter, M. J. Miller, A. W. Yasko, M. J. Yaszemski, and A. G. Mikos, *J. Biomed. Mater. Res.,* 422 (1998).
7. R. W. Bucholz, *Clin. Orthop.,* 44 (2002).

8. D. Gazit, Y. Zilberman, G. Turgeman, S. Zhou, and A. Kahn, *J. Cell Biochem.,* 379 (1999).
9. V. M. Goldberg and S. Stevenson, *Semin. Arthroplasty,* 58 (1993).
10. R. R. Betz, *Orthopedics,* s561 (2002).
11. A. J. Weiland, *J. Bone Joint Surg. [Am.],* 166 (1981).
12. H. Burchardt, *Clin. Orthop.,* 28 (1983).
13. S. J. Peter, S. T. Miller, G. Zhu, A. W. Yasko, and A. G. Mikos, *J. Biomed. Mater. Res.,* 1 (1998).
14. G. E. Friedlaender, *Clin. Orthop.,* 58 (1983).
15. C. J. Damien and J. R. Parsons, *J. Appl. Biomater.,* 187 (1991).
16. L. G. Lenke, R. R. Betz, D. Clements, A. Merola, T. Haher, T. Lowe, P. Newton, K. H. Bridwell, and K. Blanke, *Spine,* 604 (2002).
17. R. O. C. Oreffo and J. T. Triffitt, *Bone,* 5S (1999).
18. F. R. Rose and R. O. C. Oreffo, *Biochem. Biophys. Res. Commun.,* 1 (2002).
19. S. N. Khan, E. Tomin, and J. M. Lane, *Orthop. Clin. North Am.,* 389 (2000).
20. D. R. Omstead, L. G. Baird, L. Christenson, M. G. Du, R. Tubo, D. D. Maxted, J. Davis, and F. T. Gentile, *Tissue Eng,* 239 (1998).
21. C. T. Laurencin, M. A. Attawia, L. Q. Lu, M. D. Borden, H. H. Lu, W. J. Gorum, and J. R. Lieberman, *Biomaterials,* 1271 (2001).
22. K. Anselme, *Biomaterials,* 667 (2000).
23. G. F. Muschler and R. J. Midura, *Clin. Orthop.,* 66 (2002).
24. B. D. Boyan, C. H. Lohmann, J. Romero, and Z. Schwartz, *Clin. Plast. Surg.,* 629 (1999).
25. A. H. Reddi, *Tissue Eng,* 351 (2000).
26. D. W. Hutmacher, A. Kirsch, K. L. Ackermann, and M. B. Hurzeler, *Int. J. Periodontics. Restorative. Dent.,* 49 (2001).
27. U. A. Stock and J. P. Vacanti, *Annu. Rev. Med.,* 443 (2001).
28. R. Strehl, K. Schumacher, U. De Vries, and W. W. Minuth, *Tissue Eng,* 37 (2002).
29. J. H. Brekke and J. M. Toth, *J. Biomed. Mater. Res.,* 380 (1998).
30. E. Vogelin, N. F. Jones, J. I. Huang, J. H. Brekke, and J. M. Toth, *Tissue Eng,* 449 (2000).
31. R. Z. LeGeros, *Clin. Orthop.,* 81 (2002).
32. R. S. Yang, C. H. Tang, Q. D. Ling, S. H. Liu, and W. M. Fu, *Mol. Pharmacol.,* 1163 (2002).
33. T. W. Bauer and S. T. Smith, *Clin. Orthop.,* 11 (2002).
34. J. A. Roether, A. R. Boccaccini, L. L. Hench, V. Maquet, S. Gautier, and R. Jerjme, *Biomaterials,* 3871 (2002).
35. J. M. Lane, E. Tomin, and M. P. Bostrom, *Clin. Orthop.,* S107 (1999).
36. M. Owen and A. J. Friedenstein, *Ciba Found. Symp.,* 42 (1988).
37. J. T. Triffitt, in *Principles of bone biology*, Ed. J. P. Bilezikian, L. G. Raisz, and G. A. Rodan (Academic Press, Inc., San Diego, 1996), p. 39.
38. J. N. Beresford, S. E. Graves, and C. A. Smoothy, *Am. J. Med. Genet.,* 163 (1993).
39. I. B. Black and D. Woodbury, *Blood Cells Mol. Dis.,* 632 (2001).
40. A. I. Caplan, *J. Orthop. Res.,* 641 (1991).

41. Y. Jiang, B. N. Jahagirdar, R. L. Reinhardt, R. E. Schwartz, C. D. Keene, X. R. Ortiz-Gonzalez, M. Reyes, T. Lenvik, T. Lund, M. Blackstad, J. Du, S. Aldrich, A. Lisberg, W. C. Low, D. A. Largaespada, and C. M. Verfaillie, *Nature,* 41 (2002).
42. C. B. Ballas, S. P. Zielske, and S. L. Gerson, *J. Cell Biochem. Suppl,* 20 (2002).
43. P. Bianco and P. G. Robey, *Nature,* 118 (2001).
44. D. J. Prockop, *Science,* 71 (1997).
45. A. J. Friedenstein, J. F. Gorskaja, and N. N. Kulagina, *Exp. Hematol.,* 267 (1976).
46. K. H. Wlodarski, *Clin. Orthop.,* 276 (1990).
47. A. Yamaguchi, *Semin. Cell Biol,* 165 (1995).
48. D. P. Lennon, S. E. Haynesworth, R. G. Young, J. E. Dennis, and A. I. Caplan, *Exp Cell Res.,* 211 (1995).
49. J. T. Triffitt and R. O. Oreffo, in *Advances in Organ Biology*, Ed. E. E. Bittar (Jai Press Inc., London, 1998), p. 475.
50. L. G. Raisz, *Clin Chem.,* 1353 (1999).
51. M. E. Owen, in *Marrow stromal cell cuture*, Ed. J. N. a. O. M. E. Beresford (Cambridge University Press, 1998), p.
52. G. Bianchi, A. Muraglia, A. Daga, G. Corte, R. Cancedda, and R. Quarto, *Wound. Repair Regen.,* 460 (2001).
53. P. Charles, L. Mosekilde, L. Risteli, J. Risteli, and E. F. Eriksen, *Bone Miner.,* 81 (1994).
54. J. A. Andrades, B. Han, J. Becerra, N. Sorgente, F. L. Hall, and M. E. Nimni, *Exp. Cell Res.,* 485 (1999).
55. J. E. Dennis and A. I. Caplan, *J. Oral Implantol.,* 106 (1993).
56. S. M. Devine, A. M. Bartholomew, N. Mahmud, M. Nelson, S. Patil, W. Hardy, C. Sturgeon, T. Hewett, T. Chung, W. Stock, D. Sher, S. Weissman, K. Ferrer, J. Mosca, R. Deans, A. Moseley, and R. Hoffman, *Exp Hematol.,* 244 (2001).
57. Q. Qu, P. L. Harkonen, and H. K. Vaananen, *J. Cell Biochem.,* 500 (1999).
58. J. F. Connolly, *Clin. Orthop.,* 8 (1995).
59. E. M. Horwitz, D. J. Prockop, L. A. Fitzpatrick, W. W. Koo, P. L. Gordon, M. Neel, M. Sussman, P. Orchard, J. C. Marx, R. E. Pyeritz, and M. K. Brenner, *Nat. Med.,* 309 (1999).
60. E. M. Horwitz, D. J. Prockop, P. L. Gordon, W. W. Koo, L. A. Fitzpatrick, M. D. Neel, M. E. McCarville, P. J. Orchard, R. E. Pyeritz, and M. K. Brenner, *Blood,* 1227 (2001).
61. S. D. Boden, *Clin. Orthop.,* S84 (1999).
62. D. J. Baylink, R. D. Finkelman, and S. Mohan, *J Bone Miner. Res.,* S565 (1993).
63. G. R. Mundy, in *Primer on the metabolic bone diseases and disorders of mineral metabolism*, Ed. J. F. Murray (Lippincott Williams & Wilkins, Philadelphia, 1999), p. 30.
64. M. Lind and C. Bunger, *Eur Spine J,* S102 (2001).
65. M. R. Urist, *Science,* 893 (1965).
66. J. M. Wozney and V. Rosen, *Clin. Orthop.,* 26 (1998).
67. M. Ahrens, T. Ankenbauer, D. Schroder, A. Hollnagel, H. Mayer, and G. Gross, *DNA Cell Biol.,* 871 (1993).
68. A. Yamaguchi, T. Komori, and T. Suda, *Endocr. Rev.,* 393 (2000).

69. A. F. Miller, S. A. Harvey, R. S. Thies, and M. S. Olson, *J Biol Chem.,* 17937 (2000).
70. R. S. Thies, M. Bauduy, B. A. Ashton, L. Kurtzberg, J. M. Wozney, and V. Rosen, *Endocrinology,* 1318 (1992).
71. T. Ebisawa, K. Tada, I. Kitajima, K. Tojo, T. K. Sampath, M. Kawabata, K. Miyazono, and T. Imamura, *J Cell Sci.,* 3519 (1999).
72. A. J. Celeste, J. J. Song, K. Cox, V. Rosen, and J. M. Wozney, *J. Bone Miner. Res.,* S137 (1994).
73. J. S. D'Alessandro, K. A. Cox, D. I. Israel, P. LaPan, J. K. Moutsatsos, M. C. Ryan, and E. A. Wang, *J. Bone Miner. Res.,* S153 (1991).
74. S. E. Gitelman, M. S. Kobrin, J. Q. Ye, A. R. Lopez, A. Lee, and R. Derynck, *J. Cell Biol.,* 1595 (1994).
75. J. M. Wozney, V. Rosen, A. J. Celeste, L. M. Mitsock, M. J. Whitters, R. W. Kriz, R. M. Hewick, and E. A. Wang, *Science,* 1528 (1988).
76. H. C. Anderson, Curley D.J., Hsu H.H.T., Aguilera X.M., Davis L.S., and Moylan P.F., *Journal of Musculoske1etal Research,* 39 (1999).
77. A. Nifuji and M. Noda, *J. Bone Miner. Res.,* 2057 (1999).
78. J. Rubin, D. Biskobing, X. Fan, C. Rubin, K. McLeod, and W. R. Taylor, *J. Cell Physiol,* 81 (1997).
79. L. C. Yeh, F. Mallein-Gerin, and J. C. Lee, *J Cell Physiol,* 298 (2002).
80. K. A. McCuaig, H. S. Lee, J. C. Clarke, H. Assar, J. Horsford, and J. H. White, *Nucleic Acids Res.,* 1948 (1995).
81. S. E. Harris, L. F. Bonewald, M. A. Harris, M. Sabatini, S. Dallas, J. Q. Feng, N. Ghosh-Choudhury, J. Wozney, and G. R. Mundy, *J. Bone Miner. Res.,* 855 (1994).
82. K. Whang, D. C. Tsai, E. K. Nam, M. Aitken, S. M. Sprague, P. K. Patel, and K. E. Healy, *J. Biomed. Mater. Res.,* 491 (1998).
83. A. W. Yasko, J. M. Lane, E. J. Fellinger, V. Rosen, J. M. Wozney, and E. A. Wang, *J. Bone Joint Surg. Am.,* 659 (1992).
84. H. C. Anderson, K. Sugamoto, D. C. Morris, H. H. Hsu, and T. Hunt, *Bone Miner.,* 49 (1992).
85. H. C. Anderson, P. R. Reynolds, H. H. Hsu, L. Missana, K. Masuhara, P. E. Moylan, and H. I. Roach, *J. Bone Miner. Metab,* 73 (2002).
86. X. B. Yang, D. W. Green, H. I. Roach, N. M. Clarke, H. C. Anderson, S. M. Howdle, K. M. Shakesheff, and R. O. Oreffo, *Connect. Tissue Res.,* 312 (2003).
87. G. E. Friedlaender, C. R. Perry, J. D. Cole, S. D. Cook, G. Cierny, G. F. Muschler, G. A. Zych, J. H. Calhoun, A. J. LaForte, and S. Yin, *J Bone Joint Surg. Am.,* S151 (2001).
88. J. K. Burkus, M. F. Gornet, C. A. Dickman, and T. A. Zdeblick, *J Spinal Disord. Tech.,* 337 (2002).
89. J. K. Burkus, J. D. Dorchak, and D. L. Sanders, *Spine,* 372 (2003).
90. S. Govender, C. Csimma, H. K. Genant, A. Valentin-Opran, Y. Amit, R. Arbel, H. Aro, D. Atar, M. Bishay, M. G. Borner, P. Chiron, P. Choong, J. Cinats, B. Courtenay, R. Feibel, B. Geulette, C. Gravel, N. Haas, M. Raschke, E. Hammacher, d. Van, V, P. Hardy, M. Holt, C. Josten, R. L. Ketterl, B. Lindeque, G. Lob, H. Mathevon, G. McCoy, D. Marsh, R. Miller, E. Munting, S. Oevre, L. Nordsletten,

A. Patel, A. Pohl, W. Rennie, P. Reynders, P. M. Rommens, J. Rondia, W. C. Rossouw, P. J. Daneel, S. Ruff, A. Ruter, S. Santavirta, T. A. Schildhauer, C. Gekle, R. Schnettler, D. Segal, H. Seiler, R. B. Snowdowne, J. Stapert, G. Taglang, R. Verdonk, L. Vogels, A. Weckbach, A. Wentzensen, and T. Wisniewski, *J. Bone Joint Surg. Am.,* 2123 (2002).

91. S. D. Boden, J. Kang, H. Sandhu, and J. G. Heller, *Spine,* 2662 (2002).
92. S. Imai, M. Kaksonen, E. Raulo, T. Kinnunen, C. Fages, X. Meng, M. Lakso, and H. Rauvala, *J. Cell Biol.,* 1113 (1998).
93. H. Masuda, A. Tsujimura, M. Yoshioka, Y. Arai, Y. Kuboki, T. Mukai, T. Nakamura, H. Tsuji, M. Nakagawa, and T. Hashimoto-Gotoh, *Biochem. Biophys. Res. Commun.,* 528 (1997).
94. P. G. Milner, Y. S. Li, R. M. Hoffman, C. M. Kodner, N. R. Siegel, and T. F. Deuel, *Biochem. Biophys. Res. Commun.,* 1096 (1989).
95. H. Rauvala, *EMBO J.,* 2933 (1989).
96. K. Tezuka, S. Takeshita, Y. Hakeda, M. Kumegawa, R. Kikuno, and T. Hashimoto-Gotoh, *Biochem. Biophys. Res. Commun.,* 246 (1990).
97. H. Y. Zhou, Y. Ohnuma, H. Takita, R. Fujisawa, M. Mizuno, and Y. Kuboki, *Biochem. Biophys. Res. Commun.,* 1288 (1992).
98. H. Tapp, D. J. Hernandez, P. J. Neame, and T. J. Koob, *Matrix Biol,* 543 (1999).
99. W. Petersen and M. Rafii, *Arch. Orthop. Trauma Surg,* 414 (2001).
100. X. Yang, R. S. Tare, K. A. Partridge, H. I. Roach, N. M. Clarke, S. M. Howdle, K. M. Shakesheff, and R. O. Oreffo, *J. Bone Miner. Res.,* 47 (2003).
101. B. Souttou, D. Raulais, and M. Vigny, *J. Cell Physiol,* 59 (2001).
102. C. Gieffers, W. Engelhardt, G. Brenzel, T. Matsuishi, and J. Frey, *Eur J Cell Biol,* 352 (1993).
103. J. Merenmies and H. Rauvala, *J. Biol. Chem.,* 16721 (1990).
104. I. Kim, H. Uchiyama, D. Chauhan, and K. C. Anderson, *Br. J. Haematol.,* 483 (1994).
105. E. Szabat and H. Rauvala, *Dev. Biol.,* 77 (1996).
106. I. Kilpelainen, M. Kaksonen, H. Avikainen, M. Fath, R. J. Linhardt, E. Raulo, and H. Rauvala, *J. Biol. Chem.,* 13564 (2000).
107. Y. Sato, H. Takita, N. Ohata, M. Tamura, and Y. Kuboki, *J. Biochem.,* 877 (2002).
108. R. S. Tare, R. O. Oreffo, N. M. Clarke, and H. I. Roach, *J Bone Miner. Res.,* 2009 (2002).
109. P. Ducy, R. Zhang, V. Geoffroy, A. L. Ridall, and G. Karsenty, *Cell,* 747 (1997).
110. T. Komori, H. Yagi, S. Nomura, A. Yamaguchi, K. Sasaki, K. Deguchi, Y. Shimizu, R. T. Bronson, Y. H. Gao, M. Inada, M. Sato, R. Okamoto, Y. Kitamura, S. Yoshiki, and T. Kishimoto, *Cell,* 755 (1997).
111. T. Katagiri and N. Takahashi, *Oral Dis,* 147 (2002).
112. J. M. Hock and I. Gera, *J Bone Miner. Res.,* 65 (1992).
113. J. B. Lian, G. S. Stein, E. Canalis, P. G. Robey, and A. L. Boskey, in *Primer on the metabolic bone diseases and disorders of mineral metabolism*, Ed. M. J. Favus (Lippincott Williams & Wilkins, Philadelphia, 1999), p. 14.
114. T. Nakamura, K. Hanada, M. Tamura, T. Shibanushi, H. Nigi, M. Tagawa, S. Fukumoto, and T. Matsumoto, *Endocrinology,* 1276 (1995).

115. D. W. Dempster, F. Cosman, M. Parisien, V. Shen, and R. Lindsay, *Endocr. Rev.,* 690 (1993).
116. E. Canalis, T. L. McCarthy, and M. Centrella, *J. Endocrinol. Invest,* 577 (1989).
117. D. J. Anh, H. P. Dimai, S. L. Hall, and J. R. Farley, *Calcif. Tissue Int,* 332 (1998).
118. C. Ohlsson, B. A. Bengtsson, O. G. Isaksson, T. T. Andreassen, and M. C. Slootweg, *Endocr. Rev.,* 55 (1998).
119. M. Ernst, J. K. Heath, C. Schmid, R. E. Froesch, and G. A. Rodan, *J. Steroid Biochem.,* 279 (1989).
120. P. B. Saadeh, R. K. Khosla, B. J. Mehrara, D. S. Steinbrech, S. A. McCormick, D. P. DeVore, and M. T. Longaker, *J. Craniofac. Surg.,* 573 (2001).
121. Q. Qu, M. Perala-Heape, A. Kapanen, J. Dahllund, J. Salo, H. K. Vaananen, and P. Harkonen, *Bone,* 201 (1998).
122. H. Kawaguchi, C. C. Pilbeam, J. R. Harrison, and L. G. Raisz, *Clin. Orthop.,* 36 (1995).
123. Y. Dohi, H. Ohgushi, S. Tabata, T. Yoshikawa, K. Dohi, and T. Moriyama, *J. Bone Miner. Res.,* 1173 (1992).
124. L. G. Raisz, C. C. Pilbeam, and P. M. Fall, *Osteoporos. Int,* 136 (1993).
125. M. Mizuno and Y. Kuboki, *J. Biochem.,* 133 (2001).
126. R. S. Bhatnagar, J. J. Qian, A. Wedrychowska, M. Sadeghi, Y. M. Wu, and N. Smith, *Tissue Eng,* 53 (1999).
127. D. A. Oakes and J. R. Lieberman, *Clin. Orthop.,* S101 (2000).
128. S. D. Boden, *Tissue Eng,* 383 (2000).
129. J. S. Fischgrund, S. B. James, M. C. Chabot, R. Hankin, H. N. Herkowitz, J. M. Wozney, and A. Shirkhoda, *J. Spinal Disord.,* 467 (1997).
130. R. D. Welch, A. L. Jones, R. W. Bucholz, C. M. Reinert, J. S. Tjia, W. A. Pierce, J. M. Wozney, and X. J. Li, *J. Bone Miner. Res.,* 1483 (1998).
131. S. L. Cheng, J. Lou, N. M. Wright, C. F. Lai, L. V. Avioli, and K. D. Riew, *Calcif. Tissue Int.,* 87 (2001).
132. D. S. Musgrave, P. Bosch, S. Ghivizzani, P. D. Robbins, C. H. Evans, and J. Huard, *Bone,* 541 (1999).
133. T. D. Alden, D. D. Pittman, G. R. Hankins, E. J. Beres, J. A. Engh, S. Das, S. B. Hudson, K. M. Kerns, D. F. Kallmes, and G. A. Helm, *Hum. Gene Ther.,* 2245 (1999).
134. J. R. Lieberman, A. Daluiski, S. Stevenson, L. Wu, P. McAllister, Y. P. Lee, J. M. Kabo, G. A. Finerman, A. J. Berk, and O. N. Witte, *J. Bone Joint Surg. Am.,* 905 (1999).
135. K. Partridge, X. B. Yang, N. M. P. Clarke, Y. Okubo, K. Bessho, W. Sebald, S. M. Howdle, K. M. Shakesheff, and R. O. C. Oreffo, *Biochem. Biophys. Res. Commun.,* 144 (2002).
136. K. A. Partridge and R. O. Oreffo, *Tissue Eng,* 295 (2004).
137. S. Yang, K. F. Leong, Z. Du, and C. K. Chua, *Tissue Eng,* 679 (2001).
138. A. J. Ladd, J. H. Kinney, D. L. Haupt, and S. A. Goldstein, *J Orthop. Res.,* 622 (1998).
139. D. Ulrich, T. Hildebrand, B. Van Rietbergen, R. Muller, and P. Ruegsegger, *Stud. Health Technol. Inform.,* 97 (1997).

140. H. H. Bayraktar, E. F. Morgan, G. L. Niebur, G. E. Morris, E. K. Wong, and T. M. Keaveny, *J Biomech.,* 27 (2004).
141. J. J. Qian and R. S. Bhatnagar, *J. Biomed. Mater. Res.,* 545 (1996).
142. D. W. Hutmacher, *Biomaterials,* 2529 (2000).
143. L. E. Freed, G. Vunjak-Novakovic, R. J. Biron, D. B. Eagles, D. C. Lesnoy, S. K. Barlow, and R. Langer, *Biotechnology (N. Y.),* 689 (1994).
144. X. Wang, J. Ma, Y. Wang, and B. He, *Biomaterials,* 4167 (2002).
145. R. G. Burwell, in *Bone Grafts, Derivatives and Substitutes*, Ed. M. R. Urist, O'Connor B.T., and Burwell R.G. (Butterworth-Heinemann Ltd., Oxford, 1994), p.
146. W. G. Billotte, G. M. Mehrotra, D. B. Reynolds, and P. K. Bajpai, *Biomed. Sci. Instrum.,* 153 (1995).
147. S. Yang, K. F. Leong, Z. Du, and C. K. Chua, *Tissue Eng,* 1 (2002).
148. S. L. Ishaug, G. M. Crane, M. J. Miller, A. W. Yasko, M. J. Yaszemski, and A. G. Mikos, *J. Biomed. Mater. Res.,* 17 (1997).
149. W. Friess, H. Uludag, S. Foskett, R. Biron, and C. Sargeant, *Int J Pharm.,* 91 (1999).
150. N. Patel, R. Padera, G. H. Sanders, S. M. Cannizzaro, M. C. Davies, R. Langer, C. J. Roberts, S. J. Tendler, P. M. Williams, and K. M. Shakesheff, *FASEB J.,* 1447 (1998).
151. R. A. Quirk, M. C. Davies, S. J. B. Tendler, and K. M. Shakesheff, *Macromolecules,* 158 (2000).
152. K. Shakesheff, S. Cannizzaro, and R. Langer, *J. Biomater. Sci. Polym. Ed,* 507 (1998).
153. R. G. LeBaron and K. A. Athanasiou, *Tissue Eng,* 85 (2000).
154. E. Ruoslahti and M. D. Pierschbacher, *Science,* 491 (1987).
155. R. O. Hynes, *Cell,* 11 (1992).
156. M. D. Pierschbacher and E. Ruoslahti, *Nature,* 30 (1984).
157. A. M. Moursi, R. K. Globus, and C. H. Damsky, *J. Cell Sci.,* 2187 (1997).
158. M. A. Horton, *Int. J. Biochem. Cell Biol.,* 721 (1997).
159. A. J. Garcia, P. Ducheyne, and D. Boettiger, *J. Biomed. Mater. Res.,* 48 (1998).
160. R. K. Globus, S. B. Doty, J. C. Lull, E. Holmuhamedov, M. J. Humphries, and C. H. Damsky, *J. Cell Sci.,* 1385 (1998).
161. A. Rezania and K. E. Healy, *Biotechnol. Prog.,* 19 (1999).
162. I. Villanova, P. A. Townsend, E. Uhlmann, J. Knolle, A. Peyman, M. Amling, R. Baron, M. A. Horton, and A. Teti, *J. Bone Miner. Res.,* 1867 (1999).
163. F. A. Robey, in *Biomedical Applications of Biotechnology*, Ed. W. V. Williams and D. B. Weiner (Technomic Publishing Co., Inc, , Lancaster, PA, 2000), p. 307.
164. Y. Ito, M. Kajihara, and Y. Imanishi, *J. Biomed. Mater. Res.,* 1325 (1991).
165. M. P. Bostrom, X. Yang, M. Kennan, H. Sandhu, E. Dicarlo, and J. M. Lane, *Spine,* 1425 (2001).
166. R. Jaatinen, J. Bondestam, T. Raivio, K. Hilden, L. Dunkel, N. Groome, and O. Ritvos, *J Clin Endocrinol. Metab,* 1254 (2002).
167. M. D. Lindner and D. F. Emerich, *Cell Transplant.,* 165 (1998).
168. T. J. Gao, N. A. Kousinioris, J. M. Wozney, S. Winn, and H. Uludag, *Tissue Eng,* 429 (2002).

169. M. S. Widmer, P. K. Gupta, L. Lu, R. K. Meszlenyi, G. R. Evans, K. Brandt, T. Savel, A. Gurlek, C. W. Patrick, Jr., and A. G. Mikos, *Biomaterials,* 1945 (1998).
170. T. Chandy, G. H. Rao, R. F. Wilson, and G. S. Das, *Drug Deliv.,* 87 (2002).
171. R. H. Li, S. Williams, M. Burkstrand, and E. Roos, *Tissue Eng,* 151 (2000).
172. R. Langer and J. P. Vacanti, *Science,* 920 (1993).
173. S. P. Bruder, N. Jaiswal, and S. E. Haynesworth, *J. Cell Biochem.,* 278 (1997).

CHAPTER 19

BONE GENETIC FACTORS DETERMINED USING MOUSE MODELS

Weikuan Gu, Yan Jiao

Center of Genomics and Bioinformatics, Center of Disease of Connective Tissues, Department of Orthopedic Surgery-Campbell Clinic, Univ. of Tennessee Health Science Center, Memphis, TN

1. Introduction

A complex trait is one that is influenced by multiple genes. A typical example is bone density, for which a number of quantitative trait loci (QTLs) have been identified. Genetic influences have been estimated to account for about 70% of the variance in bone density in young adults (i.e., peak bone density) (1, 2). Decreased peak bone density is a strong determinant of subsequent osteoporotic fractures, which is a prevalent disease affecting millions of people worldwide. Identification of genes that regulate bone density is essential to understanding the molecular basis for the acquisition of peak bone density and, thus, to understanding osteoporosis.

Three approaches have been used to identify genes that regulate bone density: a) association studies using candidate genes. The goal of the association study is to identify the gene allele that is responsible for a complex trait of interest, which, in our case, is bone density. b) genetic Mendelian diseases with a known bone phenotype, mostly diseases. Such a bone disease is caused by the mutation of a single gene or a few genes; and c) quantitative trait loci (QTLs) mapping in humans and mice. Association studies intend to define the role of an allele of a specific known gene in causing a genetic trait, such as bone density, while QTL mapping seeks to identify the chromosomal locations of the unknown genes responsible for bone density. Association studies are usually done in humans, while genetic Mendelian diseases and QTLs are studies in both humans and animal models.

This chapter focuses on the recent progress in the genetic studies of QTLs that influence bone mineral density (BMD) using mouse models.

1.1. Recent view of QTLs by the research community

At the first international meeting of the Complex Trait Consortium (CTC) in Memphis, Tennessee in May of 2002, attendees discussed controversies raised by some publications on the definition of QTLs, and decided that a document should be written, reflecting the research community's view on the definition, mapping and identification of QTLs as a means to identify the molecular players in complex phenotypes.

After ten months of open and wide-ranging discussion through an e-mail exchange, CTC members finally agreed on a "white paper" drafted by a geneticist, Lorraine Flaherty, of the Genomics Institute in New York. This white paper finally was published in Nature Reviews Genetics (3) in an attempt to form a consensus view and to provide the larger scientific community with a realistic set of standards that can be applied to studies involving QTL. According to the white paper, "A quantitative trait locus (QTL) is a genetic locus whose alleles affect this variation. Generally, quantitative traits are multifactorial and influenced by several polymorphic genes and environmental conditions. There can be one or many QTLs influencing a trait or phenotype. Environmental factors also cause variation in the phenotype, independent of genotype, or through gene-environmental interactions. Sometimes a cluster of closely linked polymorphic genes will be responsible for the quantitative variation of a trait. These will be difficult to separate by recombinational events and therefore may be detected as one QTL. However, if distinct QTLs can be separated by genetic or functional means, each should be considered a separate QTL". This is the definition used in this chapter when discussing of QTLs in bone,

Concerning QTL mapping, the white paper stated: "For a given QTL, the mapping resolution will depend on the number of recombinational events in the mapping population. QTLs with smaller effect sizes will require larger mapping populations to ensure that they are detectable". In discussing the significant level of a QTL, the paper indicated that "In regards to QTLs mapped to regions only with a "suggestive" significance, it is generally agreed that these should not be given a locus name. However, reporting such regions is recommended in order to facilitate possible confirmation in future studies". As we will discuss late, the white paper also recognized the advantage of using congenic breeding in the fine mapping of QTL loci.

In regarding to the identification of the genes for QTLs, The paper pointed out difficulties in the identification of genes for QTLs compared to that of trait affected by single major gene. The paper also summarized eight current methodologies in the identification of genes of QTLs. However, as we will discuss late, genes for QTLs of BMD have yet to be identified.

1.2. Genetic factors affecting bone mineral density (BMD) or osteoporosis

Over the past few decades, bone density measurement has been an essential part of the evaluation of patients at risk of osteoporosis. Three methods have been the focus of recent years: pDEXA, QUS, and pQCT.

pDEXA is widely viewed as the preferred method to assess pediatric bone mineral content because of its speed, precision, and minimal radiation exposure, as well as the availability of pediatric reference data (4).

pQCT is a method that allows measurement of trabecular true bone density as well as an analysis of trabecular structure. Peripheral QCT, aimed at measurement of peripheral bones, is also expected to be a sensitive method to monitor therapeutic responses (5).

QUS is a noninvasive method to study bone density and structure in vivo. This technique has the following advantages: it is safe and easy to use, there is no radiation load on the patient, and instruments are relatively cheap and easily transportable compared to traditional osteodensitometry (pDEXA and pQCT). However, some studies indicate that QUS measures something other than the actual mineral content of bone, namely bone quality. In vitro studies have shown that QUS reflects trabecular orientation independently of BMD (6).

BMD is one of the strongest determinants the risk of the subsequent osteoporotic fracture risk (7-8). Hence, very low BMD is considered diagnostic of osteoporosis. Moreover, many women who are osteopenic eventually develop osteoporotic fractures. Therefore, bone density testing has occupied center stage in the diagnosis and treatment of osteoporosis.

Over the last ten years, BMD has been the phenotype of choice for defining heritable markers for osteoporotic fractures. Genetic study of BMD has indicated that: a) Genetic influences account for more than 70% of the variance in BMD (1, 2). b) Bone density is controlled by multiple genetic factors, e.g. QTLs, regulated by many susceptibility genes whose effects on BMD are in turn modified by an interaction with several environmental factors (6-8).

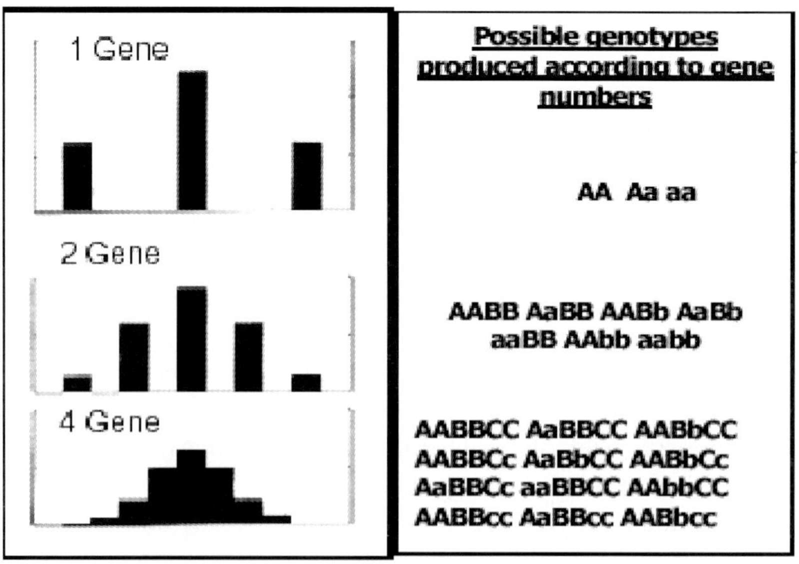

Figure 1. Relationship between gene number and phenotype

However, it is difficult to study the genetic basis of BMD in the human population, because their genetic heterogeneity of the human population is so profound that unreasonably large numbers of subjects would be required for an appropriate assessment of genetic regulation. Figure 1 shows the relationship between the number of genes that control a trait and the phenotype of individuals in a population. The more genes that control a trait, the smaller the differences among the individuals within a population in which the genes are segregating. Because BMD is controlled by multiple genetic loci, the effect of each individual QTL locus on a QTL trait is so small that it can not be distinguished from the environmental influence if the environment is not controlled. Obviously it is very difficult to study a large human population under a controlled environmental condition. Fine mapping of the QTL for the purpose of positional cloning using a human population is even more difficult (9). Alternatively, the mouse model has been widely used to identify the QTL loci of the peak bone density (10).

1.3. The mouse as an excellent animal model for the study of BMD

The application of QTL mapping to humans is, of course, the most relevant for the elucidation of a gene responsible for determining clinical bone density. However, animal models overcome several limitations of humans study subjects, including the high heterozygosity of the genetic background and the non-controllable environmental influences on the phenotype. For these and other reasons, animal models have been developed for the application of linkage studies. Of the animal models, the most common species used for genetic studies is the inbred strain of mice. The mouse model has contributed substantially to genetic studies of complex human diseases (10-12) and has played a unique role in our understanding of many common complex diseases, such as diabetes, obesity, and atherosclerosis (10-12). An example of successful use of mouse model to advance our understanding of human musculoskeletal diseases is a recent study by Ho et al. (13), which identified a spontaneous mutation called progressive ankylosis, which could be relevant to the chronic disease, osteoarthritis.

The mouse model is well-suited for QTL and positional cloning studies because of several important features: 1) the availability of several inbred strains with different phenotypes; 2) the ability to be maintained under a controlled environment to minimize environmental influence; 3) the ability to cross to inbred strains with extreme phenotypes to generate F2 populations and to produce congenic animals for the facilitation of isolating genes relevant to the phenotype under study; 4) short life span; and 5) the ability to test gene function by transgenic and knockout approaches (14). We believe that the mouse model will play a major role in the identification of bone density genes.

1.4. QTLs of BMD identified using the mouse model

In 1996, Beamer and colleagues (15) first reported the diversity of bone mineral density among mouse inbred strains, a finding that has significant impact on the late study of genetic factors of BMD using mouse inbred strains. Beamer et al. made three important contributions. First, they measured BMD using Stratec XCT 960M pQCT, with a specifically modified protocol for small skeletal specimens, particularly mouse femurs and vertebrae. Secondly, they experimentally determined the optimal time for BMD measurement in mice, the peak bone density at 4 months of age. Thirdly, they found the genetic influence on the BMD among 11 mouse inbred strains.

In the past decade, mouse inbred strains have been widely used to determine the genetic factors that influence of bone density. Several groups have been involved in the detection and mapping of QTLs that regulate BMD using strains from these 11. Table 1 lists QTLs that have been reported in the first stage of QTL mapping.

1.5. BMD related QTLs identified using the mouse model

Although bone density is an important factor in determining the bulk bone strength, the bulk bone strength is ultimately determined by both material and geometric properties, which together forming the bone quality. The sensitivity of BMD as a predictor of fracture risk is limited by the fact that this index does not take into account the geometrical and

Table 1. QTL positions of peak bone density identified from mouse model and corresponding human homologous regions

Strains	Chr.	Position (cM)	LOD/P-value*	References	Genomic sequence**	Human homologous***
C57BL/6 X CAST/EiJ	1	82-106	8.8	15, 16, 17	159161492-195694164 bp	1q21-25
C57BL/6 X C3H/HeJ	1	64-82	24	18	123137140 - 159201852 bp	1q13-21 1q25-31-32
AKR/J X SAMP6	2	2.2-23	4.05	19	3374309 - 35049497 bp	10p13-15 9q34-34.1 2q14.2
AKR/J X SAMP6	2	35-45	3.5	19	62721162-67146921 bp	2q23-24
C57BL/6 X DBA/2	2	45-55	p=0.002	20, 16, 21	67146921-104378449 bp (	20q11-q12
C57BL/6 X C3H/HeJ	4	35-54	16.3	18	61124566-113400698 bps	9p21-24 1p31-34
C57BL/6 X DBA/2	4	60-80	9	16, 17	130853280-154299058 bp	1p34-1p36, 1pter
C57BL/6 X CAST/EiJ	5	25-36	3.6	15	42713374-71142400 bps	4p14-16
C57BL/6 X C3H/HeJ	6	51.0	4.56	18	~117348206 bps	7p14-15, 7p21-15, 4q21-22, 2p11-13, 2q11-13, 2pter-qter, 1p31.3-31.2, 5q23.3, 17q23
SAMP6 X SAMR1	7	11-13	P=0.001	19	25522212-29345204 bp	19q13-13.1
C57BL/6 X DBA/2	7	44	P=0.0007	20, 17	~79815461bp	15q24-q26; 11q13/11q21
C57BL/6 X C3H/HeJ	11	31.0	6.76	18	~51857154	5q23-31
AKR/J X SAMP6	11	49-60	5.4	19, 16, 17	87424373-100790282 bp	17q21-23
SAMP6 X SAMP2	11	58-72	10.8	22	97586515-117908416 bp	17q12-25
C57BL/6 X CAST/EiJ	13	0-10	6.1	15	0-21072268 bp	1q42-43, 7p13-15, 6p21-23
C57BL/6 X C3H/HeJ	13	35	7.73	18	~54537729	5q35-qter, 5q22-q32
AKR/J X SAMP6	13	22	5.5	19	~40945185	19q12 6q22
SAMP6 X SAMP2	13	10-30	5.8	22, 23	21072268-63274779 bp	7p14-15; 9q22; 6p21-25, 4q27-28; 17q22-24; 9q12

C57BL/6 X C3H/HeJ	14	40	4.3	18	~63201604	14q21-22, 8pter-21
C57BL/6 X DBA/2	14	2.0	P=0.0007	20	~5168925	3p14
C57BL/6 X CAST/EiJ	15	24-55	3.2	15	59453579 - 100348551 bps	8q24-qter, 12q11-13
C57BL/6 X C3H/HeJ	16	27.6	4.07	13	~38896865	21q22.3
AKR/J X SAMP6	16	9-28.4	P=0.01	14	15488977-38208380 bp	8q11, 21q22 22q11.21, 3q26-29
C57BL/6 X C3H/HeJ	18	24	13.67	13	~45399525	
Total	15	N/A				N/A

* LOD score or P value for each QTL locus is listed according to the availability of information provided from the published reference.
** Mouse genomic sequences are estimated from Ensembl database according to either each molecular markers on both sides for a region or one marker in the QTL position.
*** The human homologous regions corresponding to mouse QTL positions are assigned according to the database of the Human-Mouse Homology Map: www:ncbi.nlm.nih.gov/Homology. The "q" and "p" represent the short and long arms of human chromosomes.

material characteristics of bone that contribute 40% of bone strength. In this context, numerous genetic studies for loci regulating BMD may reveal only a partial description of genetic components determining bone strength (60%). Other important genes contributing to bone strength (40%) remain to be evaluated by alternative approaches.

One of the alternative approaches is the measurement of bulk mechanical prosperities of the bone using three-point bending, which tests measure the breaking strength of the bone. In a recent study of genetic loci that influence femoral-breaking strength as measured by three-point bending, we identified six significant QTLs on chromosomes 1, 2, 8, 9, 10, and 17 (Table 2), which together explained 23% of F2 variance (24). Of those, the QTLs on chromosomes 2, 8, and 10 seem to be unique to bone breaking strength, whereas the remaining three QTLs are concordant with the femoral BMD QTLs. Genetic analysis suggests that, of these six QTLs, three influence BMD, two influence bone quality, and one influences bone size.

Table 2. Significant QTLs for Femur Breaking strength measured by Three-Point Bending

QTL name*	Marker close to the peak of QTL (cM)	% Explained by phenotypic variance	1 Lod interval (cM)	Notes
Fbs17P6.6L8.4	*D17Mit175* (6.6)	8.5	6.6	Shared with BMD QTL in this study
Fbs1P104L4.7	*D1Mit291* (103.8)	3.4	10.8	Shared with BMD QTL in this study
Fbs8P15.7L4.3	*D8Mit125* (15.7)	3.3	11.3	Unique to FBS
Fbs9P47L3.1	*D9Mit270* (41.5)	2.9	21.9	Shared with BMD QTL in this study
Fbs2P54.6L3.6	*D2Mit62* (54.6)	2.6	10.9	Unique to FBS
Fbs10P47L2.7	*D10Mit95* (50.3)	2	8	Unique to FBS
Same QTL after adjustment with periosteal circumference				
Fbs17P6.6L8.3	*D17Mit175* (6.6)	8.5	6.0	
Fbs1P104L6.0	*D1Mit291* (103.8)	4.5	10.2	
Fbs8P15.7L3.5	*D8Mit125* (15.7)	2.4	12	
Fbs9P47L2.7	*D9Mit270* (41.5)	2.1	20	
Fbs2P54.6L1.1	*D2Mit62* (54.6)	0.6	32	
Fbs10P47L2.3	*D10Mit95* (50.3)	1.6	8	

(Li et al., 2002)

Using the same set of bones, QTLs for BMD were also obtained (Table 3). By comparing the QTLs in table 2 and 3, it is clear that some of QTLs detected for breaking strength and for BMD are overlapped while others are unique to either breaking strength or BMD. This is experimental evidence that bone quality is determined not only by BMD but also by its structure and components.

The second type of QTLs related to BMD is an indirect indicator of the BMD QTL, the serum insulin-like growth factor-1 (IGF-1), which is assumed to be a regulator of BMD. The study is done with two mouse inbred strains, C3H/HeJ (C3H) and C57BL/6J (B6), which have been used for detection of BMD. Bone density of C3H is 30% higher than that of B6. Similarly, skeletal IGF-1 content, bone formation, mineral apposition, and marrow stromal cell numbers are higher in C3H than in B6 mice. Because IGF-1 and several bone parameters cosegregate,

Table 3. Significant QTL for BMD delectated using the same population from table 2

QTL name	Marker close to the peak of QTL (cM)	% Explained by phenotypic variance	1 Lod interval (cM)	Notes
Bmd17P6.6L6	*D17Mit175* (6.6)	7.4	6.6	Unique BMD QTL identified in this study
Bmd1P80.4L5.1	*D1Mit33* (82)	5.5	6.6	Identified in previous BMD study
Bmd1P108.8L5.2	*D1Mit291* (103.8)	4.4	15	Unique BMD QTL identified in this study
Bmd14P10L2.7	*D14Mit132* (0)	4.4	17	Identified in previous BMD study
Bmd9P46.5L3.6	*D9Mit270* (41.5)	3	14.8	Identified in previous BMD study
Bmd4P55.3L2.8	*D4Mit204* (61.2)	2.6	11.1	Identified in previous BMD study
Bmd19P37.8L2.8	*D19Mit53* (32.8)	2.6	12.6	Unique BMD QTL identified in this study
Bmd18P25.1L2.7	*D18Mit185* (27.3)	2.1	9.6	Identified in previous BMD study
Bmd12P23L2.9	*D12Mit156* (28.4)	2	6.1	Unique BMD QTL identified in this study

(Li et al., 2002)

Rosen et al (25) hypothesized that the serum IGF-1 phenotype has a strong heritable component and that genetic determinants for serum IGF-1 are involved in the regulation of BMD.

In an attempt to identify genetic factors that regulate the level of IGF-I between these two strains, Rosen et al. (25) conducted a study on genetic mapping of the QTLs that regulate levels of IGF-I. They intercrossed (B6 X C3H) F1 hybrids and analyzed by radioimmunoassay 682 F2 female offspring at 4 months of age for serum IGF-I. Using PCR (polymerase chain reaction), they generated114 genetic markers at average distances of 14 centimorgans along each mouse chromosome. They then genotyped every individual in the F2 population using these markers. Using the data from serum IGF-I and genetic markers, they conducted a genome screening of QTL, meaning analysis of association between polymorphism of molecular markers and phenotypes on every chromosome. As the result, they identified three major QTLs on mouse chromosomes 6, 10, and 15 (Figure 3) and several potential QTLs on 1, 3, 4, and 17.

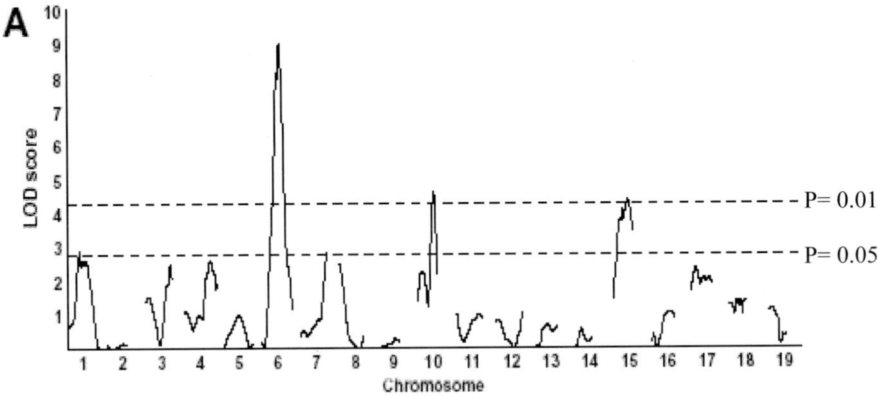

Figure 2A. Genome wide scans for correlations of IGF-I level and molecular markers on each of the 19 autosomes. Association of serum IGF-I for each marker tested on the 19 autosomes. Y-axis represents the LOD score while the X-axis represents the chromosome from 1 to 19, each from left to right (Courtesy of Rosen et al. 2000).

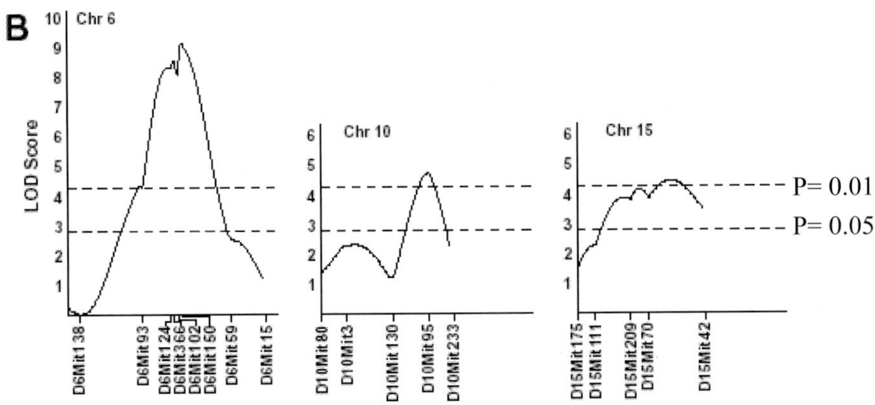

Figure 2B. Representative interval maps for three major QTLs (Chrs 6, 10, and 15) affecting levels of serum IGF-I. Intermarker intervals and chromosome lengths in recombination distances are scaled. Statistical analyses are presented as LOD scores calculated for molecular markers beginning with the centromeric end of each chromosome on the left and extending toward the telomeric end. The horizontal dashed line indicates the critical LOD score of 3 and 4 for each QTL (Courtesy of Rosen et al. 2000).

Figure 2B shows the detailed information of the three major QTLs, each of which covers a large chromosome region. Among them, the QTL locus on chromosome 6 appears to have the highest LOD score. Interestingly, this QTL locus is overlap with that of BMD detected from the same cross (table 1). While the QTL on chromosome 10 is on the similar chromosome region of the QTL of bone strength identified from three-point bending (table 2). The QTL on chromosome 15 has some overlap with a QTL of BMD detected from F2 population derived from a cross B6 X CAST/EiJ. These data seems support their hypothesis that genetic determinants for serum IGF-1 are involved in the regulation of BMD.

1.6. New technologies for identifying QTLs that regulate bone quality

While it has been shown that BMD and bulk mechanical properties are highly heritable, bone mineral distribution and architecture at the microstructural level also are under strong genetic influence (26-27). As our geneticists make the progress on the QTLs of BMD and bone breaking strength, more investigation is needed into bone structure. However, in the past, the difficulty of measuring of these properties has been one of the major obstacles in identification of QTLs. Fortunately, some new techniques have been developed in the past few years.

One of these techniques is the nanoindentation technique, which has been used to measure the bone lamellar properties of human and animals. Its reproducibility and accuracy have been shown by several research groups (28-31). A group of researchers at the University of Memphis recently applied nanoindentation techniques to determine the bone matrix properties in mice. The model of nanoindenter that they currently use has been designed not only for precision but also for high throughput (32-35).

At present, the Oliver-Pharr method is commonly used for determining the indentation modulus and hardness of bone. Measurements of load and displacement were used to determine the contact stiffness. The reduced modulus E_r is determined from the contact stiffness. The equations used to calculate the hardness (H) and the reduced modulus (E_r) are:

$$H = \frac{P_{max}}{A_c} \quad \textbf{(1)}$$

and

$$E_r = \frac{\sqrt{\pi}}{2\sqrt{A_c}} S \quad \textbf{(2)}$$

where P_{max} is the maximum force, A_c is the contact area, and S the contact stiffness.

The elastic modulus of specimen was derived from:

$$\frac{1}{E_r} = \frac{1-\nu_{specimen}^2}{E_{specimen}} + \frac{1-\nu_{tip}^2}{E_{tip}} \quad \textbf{(3)}$$

where ν is Poisson's ratio. The Poisson's ratio and elastic modulus of the diamond indenter tip, ν_{tip} and E_{tip}, were 0.07 and 1140 GPa, respectively.

To establish the validity of nanoindention, the University of Memphis researchers [34-35] evaluated bone cross-sectional geometry at one femoral site at the mid shaft. An approximately 1 mm thick transversal section was cut from each femur using an Isomet 1000 (Bulhler, Lake Bluff, IL). Cross sectional pictures were obtained using a digitizing optical microscope. A custom-written Visual C++ program was used to measure the geometric parameters by selecting the periosteal boundary points and endocortical points. Endocortical and periosteal diameters, cortical area, average cortical thickness, and principal moment inertia were also measured by digitizing.

At present, at least two research groups are investigating the genetic factors that determinant of bone nanoindentation properties. Publication of the QTLs determined by nanoindentation is expected in the near future.

Interestingly, our recent study shows there is no relationship between bone matrix properties measured by nanoindentation and bulk bone mechanical properties measured by three-point bending (Gu et al, unpublished data). It has been suspected that the elastic modulus evaluated by three-point bending may not represent the true material properties because of several simplifying assumptions. The cross-sectional geometry of a long bone varies along the length of the bone and also is poorly represented by a hollow circular cross section. Deriving elastic modulus (material properties) obtained from whole bone tests (structural tests) generally assumes both constant geometric and material properties (31, 35). The three-point bending test has been used extensively in investigating bulk mechanical properties of bone in mice. Although BMD and bulk mechanical properties are important parameters for investigating QTLs that affect bone, neither bone mass nor bulk mechanical properties provide a complete picture of characteristics of bone properties. Deriving bulk mechanical properties generally assumes both constant geometric and material properties. The deterioration of the bone material itself or increased heterogeneities of bone matrix properties have been shown to substantially affect bulk mechanical properties. One of the great advantages of nanoindentation is its ability to probe a surface and map its properties with high resolution. We believe that bulk mechanical properties and nanoindentation parameters will provide complimentary information.

1.7. Confirmation and fine mapping of QTLs

The difficulty in studying of QTLs is how to evaluate the effect of an individual QTL locus. This is because that the effect of each of individual QTL locus on a QTL trait is so small that it can not be distinguished from the environmental influence if the environment is not controlled.

There are several ways to confirm the effect of a QTL locus. First, additional and independent crosses can be performed. In table 1, several QTLs have multiple references because they were mapped by different research groups and/or with different inbred strains.

Another method is to make a congenic strain containing the QTL interval or critical region. Congenic strains are the animals in which a genetic locus containing the QTL has been moved from one strain/line (donor) to the background of another strain/line (recipient) by backcrossing. In other words, a congenic strain for one QTL locus of peak bone density contains only the targeted QTL region from the donor

while the rest of the genome is from the recipient. Polymorphism of molecular markers is used to detect the source strain of the genome components of a congenic strain. Analysis of the genotype of the molecular markers can determine the size of a QTL region that is transferred from a donor to a congenic strain. The most important advantage of using a congenic strain is that subcongenic strains can be used in the fine mapping of QTL. To examine the effect of each of three major QTLs on the level of IGF-I, Rosen and colleagues (36) developed three congenic strains. Each congenic strain was developed by transfer of a specific chromosomal region containing a IGF-I level QTL from the high IGF-I level strain, C3H , to the low IGF-I level, B6, strain. Transfer of the donor region was accomplished by first producing (B6 × C3H)N1F1 offspring, and then backcrossing a N1F1 mouse to a recipient B6 strain mouse to obtain N2F1 progeny. Tail tip DNA samples, made from female and male offspring by standard NaOH digestion method, were genotyped to find carriers of the desired chromosomal regions. These carriers were mated to new B6 mice to generate N3F1 progeny for genotyping. This backcross mating system, followed by genotyping for carriers, was conducted for nine cycles. This procedure is shown in figure 3.

As the result, three congenic strains were developed. Following in figure 4 shows the first congenic strain as an ample. It contains a Chr 6 fragment between *D6Mit93* and *D6Mit150* from C3H, which is from 26.3 to 51.0 cM on genetic map and contains approximately 60 million nucleotides (from 53.5 to 116.9 millions).

Because the congenic strain contains only chromosomal fragment hosting QTL locus from C3H while rest of genome is from B3. The effect of the QTL locus can be analyzed by comparing the phenotype of the congenic strain with that of B6. In fact the authors in this work compared the level of IGF-I of three congenic strains wth that of B6 and C3H.

As shown in figure 5, at the age of 4 months, the B6.C3H-6T has a serum IGF-I level lower than B6. The B6.C3H-10 has an IGF-I level that is higher than B6 but similar to that of C3H. The B6.C3H-15 has a similar level of IGF-I with B6. The variation on the IGF-I levels under

the same B6 genomic background of these congenic strains provide us an unique opportunity to examine the effect of IGF-I QTLs on the longevity under natural polymorphic condition.

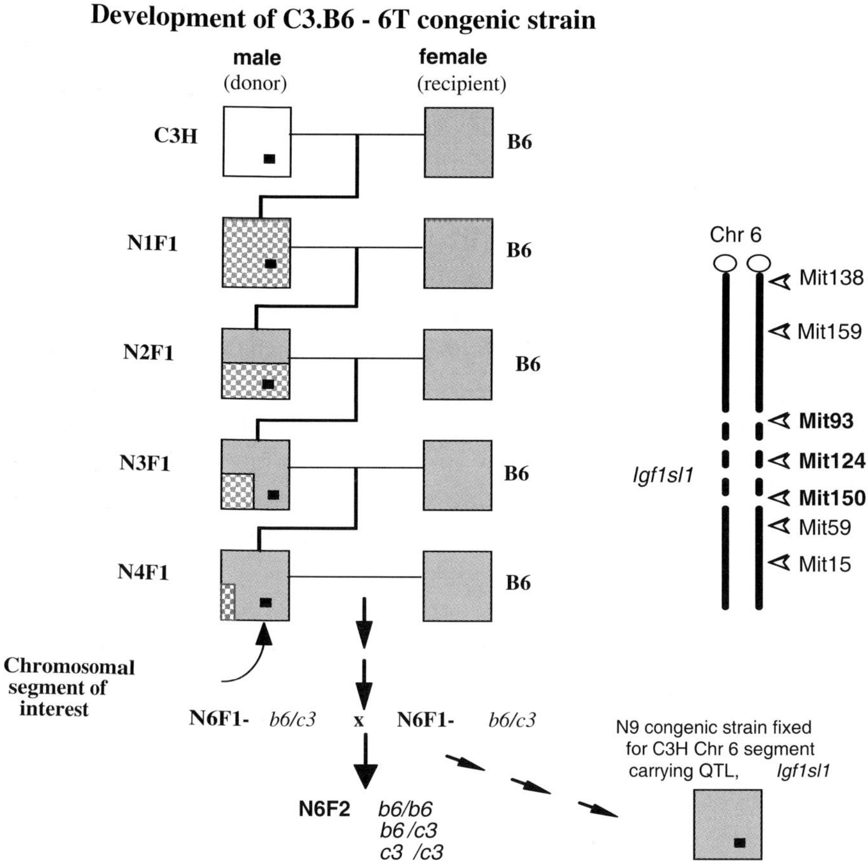

Figure 3. The protocol for congenic breeding to transfer QTL locus from C3H to B6 genome background. QTL on chr 6 is shown as an example for the breeding procedure. The N9 congenic strain with fixed QTL segment theoretically contains more than 99% genome from B6. (Courtesy of Bouxsein et al. 2002).

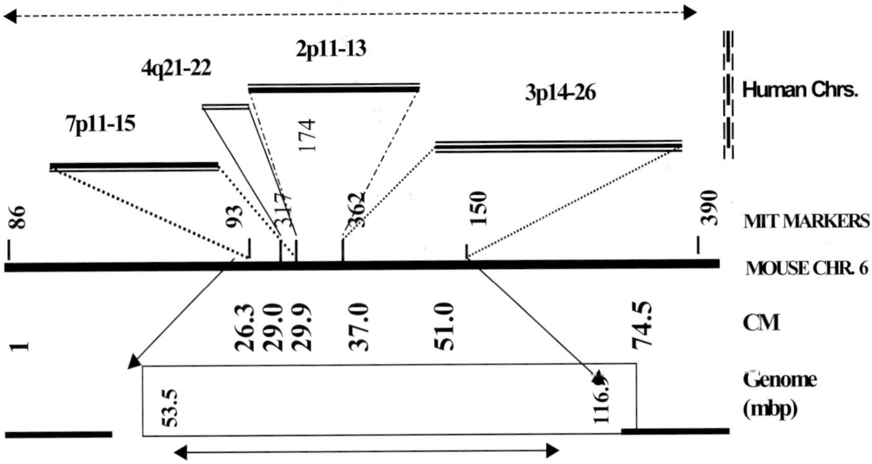

Figure 4. Transferred chromosomal region in the first congenic strain. (a) At the top are chromosomal regions that homologues to the mouse region. (b) In the middle is the map of mouse chromosome 6. The solid bar represents the original region of the Chr 6 from C3H within the congenic strain B6.C3H-6T. Numbers above the solid bar represent the position and names of D6mit microsatellite markers while the number under the solid bar represents the cM distance on the mouse Chr 6. (c) At the bottom is the genomic region that is located to the genetic region flanked by the two makers, D6mit93 and D6mit150.

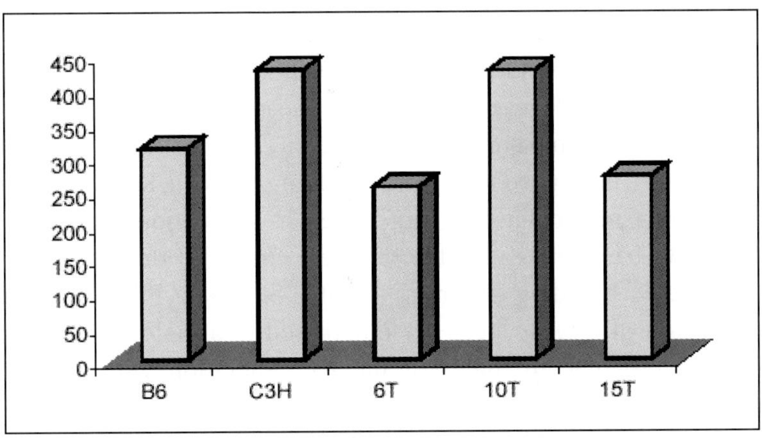

Figure 5. This schematic figure shows the difference of IGF-I levels among congentic and two progenitor strains at the age of 4 months. Serum IGF-I measured by use of RIA from ALPCO Diagnostics (Salem, NH). (Error bars not shown, they have been mentioned in the above three paragraphs in this section). (Courtesy of Bouxsein et al. 2002).

A third method, which has not been used in the mapping of QTLs for BMD in mice, is to perform selection studies where a short-term selective breeding study is performed in which mice are selected for three to five generations for the phenotypic trait (27, 28). At every generation, DNA markers are also scored. If the DNA markers show a concomitant selection for the QTL regions, then this would be considered as additional evidence that the QTL was mapped to the correct general location.

Speed congenic breeding

As shown in figure 3, the classical protocol of congenic breeding needs about 10 generations, at which 99.9% of the genetic background of the progeny is of recipient origin while still retaining heterozygosity at the region of interest. However, the availability of dense genetic maps of the mouse genome has allowed the development of marker-assisted breeding strategies (27), which reduces the number of generations required to eliminate donor strain-derived alleles outside the genetic region of interest. Using marker-based selection for genetically optimal breeders at each generation, a >99.9% of recipient strain genome in the congenic strain can be reached after five backcrosses (generation N5). The time frame for derivation of such a congenic strain is approximately 100 days per generation or 1.3 years to complete the project, as opposed to 2.5 years for a conventional congenic.

The key in the speed congenic breeding is to select the recipient genome using molecular markers in each generation and use such individuals to back cross with the recipient strain. At the beginning of a speed congenic breeding program, about 2-3 female mice carrying the genetic region of a QTL locus are outcrossed to a male of the recipient strain to assure that all males of the F1 generation will carry the Y chromosome from the recipient strain. Two to three male F1 animals carrying the genetic region of interest are then backcrossed to recipient-strain females to produce N2 progeny. By doing this step, all N2 males will carry both sex chromosomes from the recipient strain.

Molecular screening begins at the N2 generation. First, male mice of two to three littermates (~20) are genotyped to identify carriers of the

QTL region of interest. The identified carriers are then subjected to autosomal genome scans using microsatellite or SNP markers that are polymorphic between the donor and recipient strains. Two to four males with the highest percentage of recipient strain contribution are selected as breeders for the next backcross. This process is repeated at each generation out to N5. At this point, male and female mice with >99.9% recipient genome content can be intercrossed to produce homozygous congenics.

As the recently rapid development of tools for genotyping, most of the congenic breeding programs are using molecular markers in their selection procedure now.

Fine mapping

Fine mapping is more difficult since it requires more recombinational events to separate the genes regulating the quantitative trait from closely linked markers. Sometimes, it also require more sensitive phenotyping procedures because of several linked QTLs in one QTL region, each individual QTL will likely have a smaller effect on the phenotype. One can make crosses that involve high numbers of recombinational events, however, the production of subcongenics has been the most efficient and successful way of accomplishing this task in the study of QTLs of BMD. Subcongenic strains are ones that have a shorter differential chromosomal segment than their congenic parent. A set of these subcongenic strains can be made that subdivide this critical interval into a number of segments that can be individually tested for the QTL. Thus, these subcongenics are powerful tools for fine mapping since they allow multiple tests for phenotype on genetically identical mice. If other types of crosses are used for narrowing the critical region, progeny testing is often necessary to confirm the phenotype of the recombinant mice.

As shown in table 1, two QTL loci on mouse chromosome 1 were identified from two crosses with a common strain B6. Those two QTLs almost overlapped each other. To further dissect the QTL locus from B6 and C3H, Shultz et al. (38) analyzed eight B6.C3H Chr 1 subcongenic lines, which not only sharpened the mapping for the *Bmd5* QTL map on distal Chr 1 but also revealed a second BMD QTL, *Bmd19*, in the proximal end of the B6.C3H-1T congenic segment. The identification of multiple QTL loci from originally one locus emphasizes the complexity of QTL study. As we will discuss in the next topic, interaction among QTLs is much like that of genes.

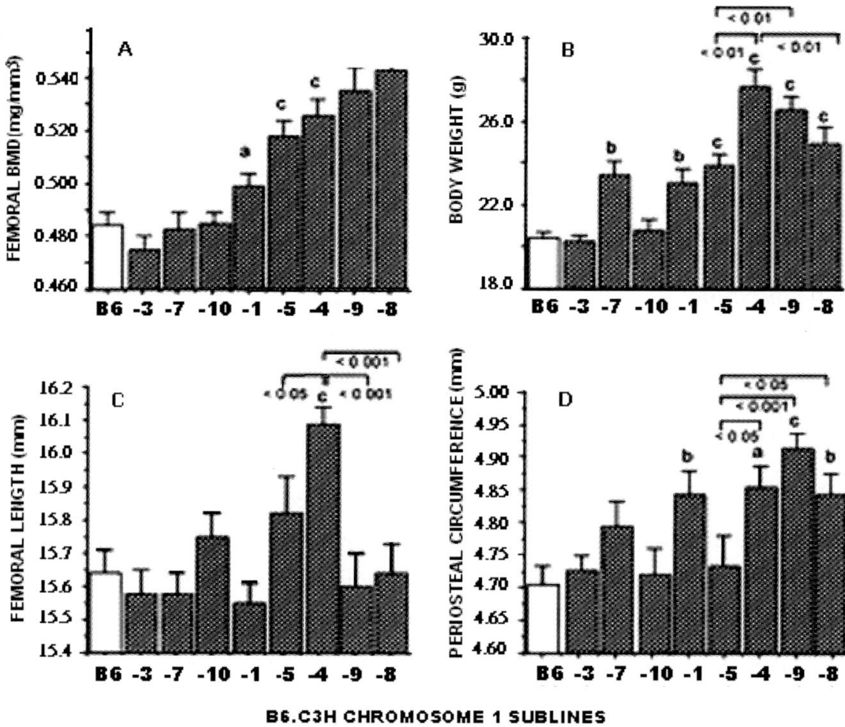

Figure 6. (Courtesy of Shultz et al., 2003) Femoral and body weight data from subcongenic lines obtained by genetic recombination that partitioned the initial BMD QTL region carried in the B6.C3H-1T congenic strain into eight overlapping segments (designated Chr 1-1, -3, -4, -5, -7, -8, -9, and -10). Data (means ± SE) were gathered from groups of 11-22 females at 4 months of age. (A) Femoral total BMD, (B) body weight, (C) femoral length, and (D) mid-diaphyseal periosteal circumference. In each panel, the open bar represents the B6 progenitor controls mice and the striped bars represent the experimental subline data. Significant differences between B6 control and subline means are indicated by lowercase letters (a, $p = 0.05$; b, $p = 0.005$; c, $p = 0.0005$). Significant differences among the four B6.C3H-1 sublines that share the *Bmd5* QTL region are indicated by brackets, along with the appropriate p values.

1.8. Complex and interactions of QTLs detected from mouse model

Genetic study has revealed complicated interactions among genes. The main question is, because the QTLs represent function of a gene or a group of genes along a chromosome region, will the same complexity exist among QTLs? The answer is YES. Several types of interactions have been reported. They include epistatic effect, pleiotropic effect, and gender differences

The first example is the gender difference. For the convenience of producing females, QTLs have been produced from the F2 population of females. However, there are obvious gender differences in the skeletal phenotype. Generally, men have larger bone size and greater cortical mass than that of women. Sex steroid action has been the major factors that influence skeletal development and determine these gender differences. For example, androgens have been reported to increase bone formation (39-40), particularly in cortical areas, and the absence of androgen action has been related to lower bone mass and smaller size (41). Androgens also may influence weight and muscle mass, thereby indirectly affecting skeletal development.

However, gender has been found to exert effects on other phenotypes via genetic mechanisms apparently unrelated to the control of androgen levels. It has been speculated that the marked gender differences in peak bone size/mass were in part the result of genetic influences. QTL analysis of males in inbred strains derived from a cross between B6 and DBA/2 progenitors (BXD) revealed evidence of chromosome locations of a number of QTLs affecting BMD in male mice (42).

In analysis of a F2 population, performed by interval mapping on both male and female $B6D2F_2$ data for the chromosome 2 and 7 QTLs with Map Manager QT, Orwoll et al. (42) found that the chromosome 2 and 7 QTLs clearly exhibited gender specificity. Shown in the figure 6, a significant phenotype-genotype relationship existed for female mice at the midportion of chromosome 2 and for male mice at the proximal portion of chromosome 7, whereas in the opposite gender these same QTLs did not appear to exist.

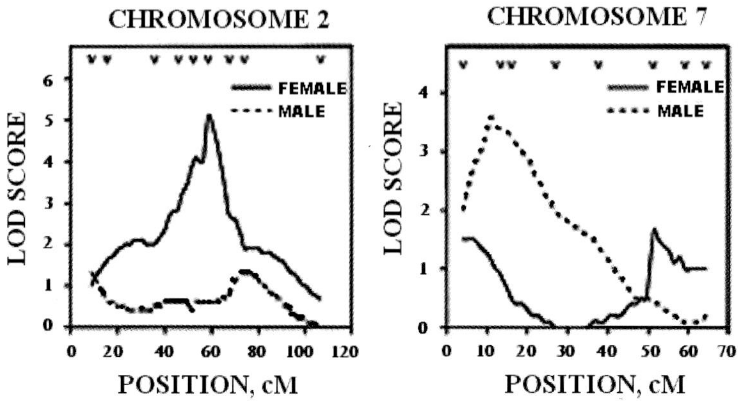

Figure 7. LOD plots of chromosomes 2 and 7 for peak whole body WC-BMD as determined by an interval mapping approach (Map Manager QT). In both instances, the LOD curves exceeded significance threshold of 3.3 (df = 2) for F_2 data. Because the results for the "additive" and "free" QTL models were virtually identical at each chromosomal region, only the "free" models are shown. The chromosomal positions of the individual markers used for mapping are indicated at the top of each graph. (Courtesy of Orwoll et al., 2001).

The same group also reported the evidence of gender-specific genetic influences on femoral geometry at three other chromosomal sites (chromosomes 2, 7, and 12) in a following publication (43).

Epistatic

By a broad definition, an allele (or particular genotype) at one locus (the epistatic gene) renders the genotype at a second locus (the "hypostatic" gene) irrelevant: the phenotype will be dictated by the genotype of the epistatic gene alone. In the study of bone breaking strength and BMD from the same F2 population, we found epistatic interaction among QTLs. By Appling Two-way ANOVA between QTLs of femur breaking strength and BMD in the same study listed on table 1 and 2, we found three significant loci interactions contributing to 14.6% F2 variance (24). In consideration of the complex network of gene action in regulating quantitative traits, the observation is more likely to represent a common genetic mechanism than a special case for FBS. Indeed, we also identified substantial epistatic interactions for BMD, which explained

17.4% of F2 variance compared with 34% for single QTL effects. These multiple interaction loci may contain the genes that encode transcription factors, activation factors, or co-activators that are common to multiple pathways

Table 4. Interactions between QTLs that determine FBS and BMD.

Locus 1 x Locus 2	FBS		Femur BMD	
	P<	% F2 variance	*P*<	% F2 variance
D3mit217 x *D12mit182*	0.001	5.7	ns	ns
D1mit33 x *D11mit36*	ns	ns	0.000	14.4
D10mit95 x *D9mit270*	0.006	4.1	ns	
D1mit291 x *D7mit246*	0.006	4.8	ns	
D11mit36 x *D4mit204*	ns		0.01	3.4
Total % of F2 variance		14.6		17.8

(Courtesy of Li et al., 2002)

Similarly, four epistatic interactions were identified which accounted for 37.6% of phenotype variance (44) by analysis of bone size. The loci interaction identified here emphasizes the importance of epistasis in regulating complex traits and the need to perform such an analysis in genetic dissection of complex traits. The analysis of these interactions may point to biological relationships among the genes. Identification of the loci involved in multiple interactions is critical during this analysis because they could have a more important role than single QTL.

Pleiotropic effect

Pleiotropy is the phenomenon whereby a single gene affects several different aspects of the phenotype. In evolution, pleiotropy is commonly used to refer to phenotypic effects of a gene other than those originally favored by natural selection. In the same study we conducted for bone breaking strength and BMD, one of the most interesting findings was that the largest FBS QTL, Fbs17P6.6L8.4, has a pleiotropic effect on other bone-related phenotypes. As shown in figure 8, these phenotypes include femur BMD, femur length, femur mid-shaft periosteal circumference, and possible work required to break the femur. This is strong evidence that suggests that the gene underlying this QTL may have a general regulatory role on multiple bone metabolism pathways. However, this study did not rule out the possibility of multiple genes/QTLs within this QTL region. As we showed in figure 6 when we discussed the fine mapping of QTL locus, subcongenic strains may reveal more than one QTL gene within the QTL that was originally mapped by a F2 population. This again emphasizes the fact that QTLs are not genes. The exact molecular mechanism of a QTL will not be understood till the gene that causes the QTL is identified.

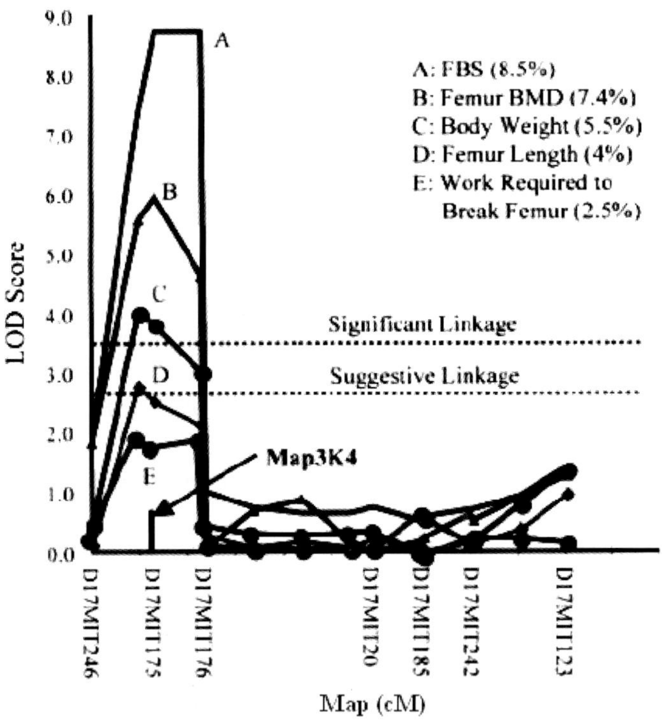

Figure. 8. A novel QTL, Fbs17P6.6L8.4, has pleiotropic effects on multiple bonerelated phenotypes and body weight. Dashed horizontal lines represent significant and suggestive lod thresholds determined by permutation test. Numbers in parentheses represent the percentage of F2 variance explained by this QTL. (Courtesy of Li et al., 2002).

1.9. Bioinformatic approaches to identify QTL candidate genes using mouse model

Up to now, we have talked about the identification, fine mapping, and analysis of function of QTLs. However, the ultimate goal of studying QTLs is to identify the genes that regulate the phenotypes of those QTLs. Therefore, we will discuss how to identify QT genes in the next two sections.

It is much more difficult to identify QT genes than to the single major mutation. The identification of QT genes face two major obstacles. The first one is the large chromosomal regions that QTLs cover. The strategy of congenic and subcongenic breeding all aim to narrow down the QTL regions. The second is the effect of the QT gene is small, which causes difficulty in the linking of gene to phenotype. Analyzing hundreds of genes for their very small effect on the trait requires a tremendous amount of work. In this regard, it is clear that strategies for identifying genes in the past has been limited by the availability of genetic-based data and technology. Today, this limitation still exists, but tremendous progress has been made in the technology and genome resource. One important technology that will assist us in the identification of candidate genes for QT genes is microarray.

Microarray has been widely used in the analysis of gene profiles and will be a powerful addition to the study of QTL genes. A microarray refers to a microscopic array of immobilized nucleic acids. In other words, microarray is the collection of cDNAs or oligonucleotides immobilized on a surface, such as a glass slide or silica wafer. A common glass slide of the size 20X30 mm can hold thousands of PCR products or oligos of genes or ESTs. Currently, the most common application of microarray technology is to analyze the gene expression profile by determination of differences in abundance between two mRNA samples.

In a study of QTL locus on mouse chromosome 1 identified from a cross between B6 and CAST, we generated a congenic strain, B6.CAST-1T, in which the chromosomal fragment containing this QTL had been transferred from CAST to the B6 background. The congenic mice have a significantly higher bone density than the B6 mice. We then performed cDNA microarray analysis to evaluate the gene expression profile that might yield insights into the mechanisms controlling the high bone density by this QTL (45).

In this microarray study, we initially analyzed the expression level of 8734 gene accessions GEM I chips (Incyte Genomics, Inc.). To accomplish this, total RNA was isolated from mouse femurs as previously described [45] and then applied to the OligoTex mRNA isolation column (Qiagen Inc., Valencia, CA). The eluted polyA mRNA was repurified by a second pass through of this column prior to ethanol precipitation. After being dissolved in TE buffer in 1.5-ml Ambion siliconized RNase-free microfuge tubes, the concentration of the purified mRNA was measured with the Spectronic Genesys spectrophotometer. The purified mRNA was used at a concentration of 50ng/ml to hybridize to the mouse GEM I microarray chip after being labeled with Cy3 and Cy5 dyes. The level of EST-specific gene expression was measured by relative intensity of the fluorescence of the dyes from the hybridized cDNA. When the expression profiles were analyzed, we surprisingly found that approximately 60 % of 8734 ESTs and genes were expressed in the femur of C57BL/6J mice. Since the function of two-third of these expressed accessions (ESTs) had not been documented previously, a blast search of these ESTs vs the Genbank nr database was preformed to identify these sequences, which could be the candidates for bone QTL genes. This search identified a total of 207 ESTs with the following criteria: match (i.e. >= 80% identity to a known sequence of >= 300bp or >= 90% identity to a sequence >= 70 bp) to a cDNA newly deposited to the Genbank without a known function; to a known gene; and /or to a genomic sequence (46). The search of the NCBI database revealed the genome sequences of 83 out of 207 of these ESTs by their homologs to genome sequences in either the human or mouse genome. At present, the locations on human chromosomes of 34 of these genomic sequences are known. Then a search of UCSC database located the chromosomal locations of 131 out of the 207 ESTs. To determine if some of these ESTs were located in the identified QTLs for bone density, we compared the locations of these genomic sequences to the QTL chromosomal regions listed in table **2**. Surprisingly, 39 (marked with **) ESTs (10 from NCBI and additional 29 from UCSC) were located in the QTL regions, indicating that nearly one third of these ESTs are candidate genes for the QTL of bone density. If this ratio is extrapolated to the 207 ESTs, at least 60 ESTs out of the 207 could possibly be the new candidate genes in bone density QTL regions, as listed in Table

However, the application of microarray technology has just begun. Another new application in the QTL study is called transcriptome-QTL mapping, which has developed a powerful new bioinformatic method to analyze complex transcriptional infrastructure.

The transcriptome-QTL mapping is developed by a group of investigators at the University of Tennessee Health Science Center. The method exploits isogenic lines of mice that can be studied using a battery of computational, statistical, molecular, and even morphometric methods. Databases on variation in gene expression are coupled to image and brain behavioral databases. A researcher interested in a particular behavioral phenotype in mice (e.g. BMD) can search for transcriptional regulators that may influence that trait, thus identifying major causes that underlie the variation.

In the transcriptome-QTL mapping analysis, the microarray data (expression levels of genes) will be treated as the phenotype in instead of the disease phenotype. By its principal, the traditional linkage analysis is an association between the genotype of a particular location of the chromosome/genome and a particular phenotype, such as disease. In the transcriptome-QTL mapping, the association is sought between genotypes (markers on the chromosome) and the levels of expression of genes. In this case, the procedure of transcriptome-QTL mapping is essentially the same to that of genetic QTL mapping. However, unlike the disease phenotype, there will be thousands of genes having different levels of expression. Thus, the computation in the transcriptome-QTL mapping will repeat thousands of more cycles than that of disease mapping. Practically, the transcriptome-QTL mapping has much more statistic task than that of genetic QTL mapping. But the QTL mapping program can be easily used for the transcriptome-QTL mapping . This group has successfully accomplished transcriptome-QTL maps using the adult brain mRNA levels, adult hematopoietic stem cell mRNA levels, and other published phenotypes of the mouse model. The data are posted on the webpage **http://headmaster.utmem.edu/search.html** as well as on **http://webqtl.roswellpark.org/search.html**. It is expected that the similar analysis on the bone study in the near future.

Following figure 8 shows part of a picture of transcriptome-QTL mapping using Affymetrix mouse U74Av2 chip. In this case, left are the LOD score while the bottom is the microsatellite markers along the chromosome. The peaks of the QTL mapping reflect the degree of the association of expression levels with the microsatellite markers underline of the figure. Click any QTL point will give the information of gene that mapped to that point. Readers are encouraged to view their webpage using the address mentioned above.

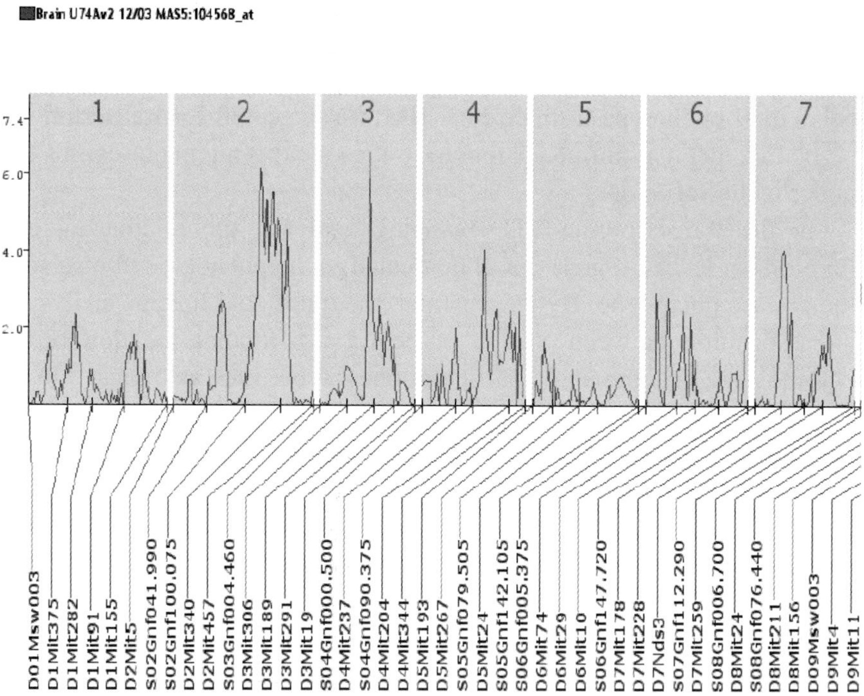

Figure 9. Partial picture of a transcriptom QTL mapping using U74Av2 chip. The figure shows the result of transcriptom mapping of chromosome 1-7 (designed 1 through 7) using microarray data from U74Av2 chip. Left bar indicates the LOD score. Bottom part of the figure shows the molecular markers along each chromosome. Peaks in red color indicate the significance of genes for trait of interest and their position along chromosome.

1.10. Genome resource for positional cloning of genes regulation of BMD

While microarray provides a powerful tool in the identification of candidate genes for QT genes, positional cloning has been the major tool for the identification of QT genes. Because each QTL may contribute to only a small portion of the phenotype, which is also influenced by environmental factors, many strategies, such as differential display, have not been successful for identifying QTL genes. At present, map-based positional cloning is the primary strategy that has been adopted to isolate QTL genes. Positional cloning, a technique whereby a gene is identified and isolated by chromosomal location, with no prior hypothesis in terms of its biology or function, has been used widely to identify disease-related genes from many species. The overall strategy of positional cloning, which has been used to successfully identify, e.g., genes that cause Huntington's disease and cystic fibrosis, is to map the location of a disease gene by linkage analysis and then to use the mapped location on the chromosome to clone the gene. The same strategy is applicable in the finding of the gene relevant to bone density phenotype.

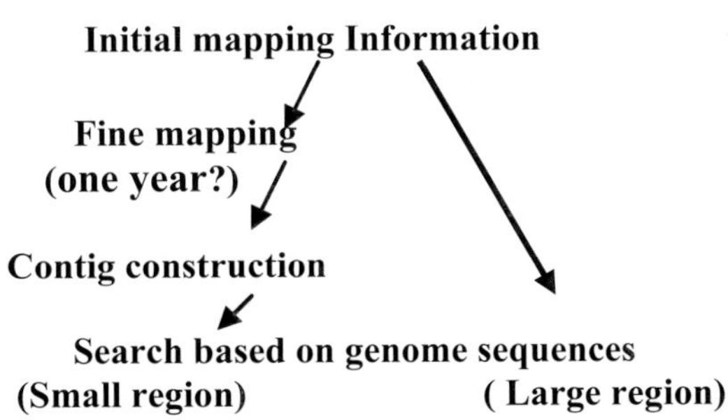

Figure 10. Strategy for cloning of QT genes

Most procedures in traditional positional cloning have been labor intensive and time consuming. As shown in figure 10, the first step is to define a candidate region as precisely as possible, including initial linkage analysis and fine mapping. Linkage analysis requires production of a large pedigree and PCR-based analysis of microsatellite markers for a whole genome search for linkage. Fine mapping is the particularly difficult task of breaking the linkage and identifying useful markers in the targeted region. The next step is to construct a contig connecting the targeted region, which entails identification of a large insert genomic library, either BAC (bacterial artificial chromosome) or YAC (yeast artificial chromosome) with known markers. Finally, the contig will be sequenced and analyzed, using a technique termed chromosomal walking. Because all of these complicated procedures are a tremendous amount of work, positional cloning has needed team effort spanning a number of years.

A very important point is that, the recent completion of the Human and Mouse Genome Projects, along with other new technology, such as mutation analysis and microarrays, allows for dramatically faster progress in the positional cloning of genes from mutated models. The technique of positional cloning has changed in that: i. We do not need to make BAC clones that contain the genomic sequences within the targeted/fine mapped region. ii. We do not need to sequence the entire region –usually 10Mbp of the genomic sequences. Their sequences are available through public (Ensembl) and private databases (Celera). iii. We now know that the majority of the mouse genome are repetitive sequences such as transposons, which are easy to identify and therefore can be eliminated from further analysis. iv. On average, a 10 cM region may contain 300 genes. Assuming each gene has five exons, we need to analyze 1,500 coding regions, which is feasible using current high throughput technology either to sequence or to analyze the expression profiles in less than half a year. v. Recently, technologies for gene expression profiling (e.g. microarray) and SNP identification (e.g. SpectruMedix, which we use in our procedure) will allow us to analyze large numbers of genes in a shorter time (less than six months as in our current work).

It is interesting to read a paragraph from a publication entitled "Initial Sequencing and Comparative Analysis of the Mouse Genome" by the "Mouse Genome Sequencing Consortium". It says "The availability of an annotated mouse genome sequence now provides the most efficient tool yet in the gene hunter's toolkit. One can move directly from genetic mapping to identification of candidate genes, and the experimental process is reduced to PCR amplification and sequencing of exons and other conserved elements in the candidate interval. With this streamlined protocol, it is anticipated that many decades-old mouse mutants will be understood precisely at the DNA level in the near future."

While this paragraph illustrates a workable protocol for the single gene mutation, it is practically feasible to screen every gene for a narrowed QTL locus of a 10 to 20 cM region. The real difficulty in analysis of the QT gene is to correctly identify the QT gene from many of these genes. While the mutation is the search for the single major gene mutation among a large number of genes within the targeted region, the QT gene appears as polymorphic variation, which occurs in many genes and multiple places on the single gene.

1.11. Identification and confirmation of QT gene

Although several genes have been identified and confirmed as underlying QTLs, the confirmation of these genes still requires a great deal of effort. At present, only QTLs with strong effect (or large effect size) on the phenotypes have been readily identified by positional cloning. With the rapid progress on genome assembly and development of biotechnology, we believe that QT genes with minor effect will be eventually identified and confirmed. The CTC consortium listed some of the conditions that generally used in the formation of the QT genes. Generally more than one of these conditions should be applied and some are more important than others. One should keep in mind that there is no standard criterion for the confirmation of QT genes so far. Below are the criteria cited from the CTC publication. For detailed information, readers are encouraged to review the paper (3). 1). Polymorphisms in the coding or regulatory regions. DNA sequence differences, leading to changes in either the structure or regulation (or quantity) of a gene product, should be detected between the strains used for mapping and known to differ in the quantitative trait. However, one should notice that here the "changes" are based on the current knowledge on the biology function of a gene structure. 2). Gene function. It is obviously that some evidence should support a link between the function of the gene and the expression of the QTL being analyzed, thus, it makes sense biologically. 3). In vitro functional studies showing differences in activity of the two alleles. Often in vitro tests can substitute for in vivo ones. If an in vitro functional test can be designed, then transfection experiments can be used to test the effects of the alternative alleles on relevant cellular phenotypes. This may be particularly tricky for the trait of a small effect. 4). Transgenesis. Transgenesis with bacterial artificial chromosomes (BACs) (or other large chromosomal segments) can also be used to confirm the identity of the candidate gene. 5). Knockins. Knockins can also be used to confirm candidate genes. 6). Deficiency Complementation Test. If a knockout (or a null allele) of the candidate QTL is available, then complementation tests between the knockout (or mutant strain) and the strains containing the QTL variant alleles could be used as evidence of identity. 7). Mutational analysis. With the advent of new convenient mutational techniques, it is now possible to perform gene-specific mutational analyses, i.e. collect a series of mutations in a specific gene. 8). Homology searches. The mouse and human genomes are strikingly similar where high homology occurs between more functionally important regions.

It seems that with the accumulation of data of gene expression profiling, we are able to predict function of more and more genes and therefore construct more molecular pathways. The confirmation of QT genes should be much easier when we have better understanding of the function of a majority of genes within a QTL region. We feel confident that this may be achieved in the near future.

1.12. Future directions on the study of QTL genes using mouse model

Strategies to identify genes in the past have been based on the availability of limited gene-based data and technologies. Obviously, the recent completion of human genome working draft sequence and pending completion of both human and mouse genome sequence, along with the other new technologies such as high through put polymorphic screening and microarrays, will allow faster progress on identifying QTL genes that regulate bone density and other phenotypes. It is, therefore, likely that future identification of QTL genes will rely heavily on an integrated strategy involving many genomic resources, existing technologies, and new technologies to elucidate gene expression and gene product function. Such an integrated strategy for identification of bone density genes may involve a number of steps including: (i) the collection of genomic sequences and ESTs within a QTL region; (ii) the identification of important gene regions, such as the coding region, regulation, and *intron-exon* conjunction regions; (iii) the expression analysis of all the possible genes within the QTL region by comparing, for example, the gene expression f a high and a low bone density QTL; (iv) the polymorphic analysis of genes in the high and low bone density loci; (v) the association analysis between any detected polymorphism and bone density; and (vi) confirmation by additional *in vitro* and *in vivo* functional studies.

References

1. Jouanny P, Guillemin F, Kuntz C, Jeandel C, and Pourel J. (1995) Environmental and genetic factors affecting bone mass. Similarity of bone density among members of healthy families. *Arthritis & Rheumatism* 38(1);61-7.
2. Smith DM. Nance WE. Kang KW. Christian JC. and Johnston CC Jr. (1973) Genetic factors in determining bone mass. *Journal of Clinical Investigation.* 52(11):2800-8.
3. Biola O, Angel JM, Avner P, Bachmanov AA, Belknap JK, Bennett B, Blankenhorn EP, Blizard DA, Bolivar V, Brockmann GA, Buck KJ, Bureau JF, Casley WL, Chesler EJ, Cheverud JM, Churchill GA, Cook M, Crabbe JC, Crusio WE, Darvasi A, de Haan G, Dermant P, Doerge RW, Elliot RW, Farber CR, Flaherty L, Flint J, Gershenfeld H, Gibson JP, Gu J, Gu W, Himmelbauer H, Hitzemann R, Hsu HC, Hunter K, Iraqi FF, Jansen RC, Johnson TE, Jones BC, Kempermann G, Lammert F, Lu L, Manly KF, Matthews DB, Medrano JF, Mehrabian M, Mittlemann G, Mock BA, Mogil JS, Montagutelli X, Morahan G, Mountz JD, Nagase H, Nowakowski RS, O'Hara BF, Osadchuk AV, Paigen B, Palmer AA, Peirce JL, Pomp D, Rosemann M, Rosen GD, Schalkwyk LC, Seltzer Z, Settle S, Shimomura K, Shou S, Sikela JM, Siracusa LD, Spearow JL, Teuscher C, Threadgill DW, Toth LA, Toye AA, Vadasz C, Van Zant G, Wakeland E, Williams RW, Zhang HG, Zou F; Complex Trait Consortium. The nature and identification of quantitative trait loci: a community's view. Nat Rev Genet. 2003 Nov;4(11):911-6.
4. Blake GM, Fogelman I. Bone densitometry and the diagnosis of osteoporosis. Semin Nucl Med. 2001 Jan;31(1):69-81.
5. Eysel P, Schwitalle M, Oberstein A, Rompe JD, Hopf C, Kullmer K. Preoperative estimation of screw fixation strength in vertebral bodies. Spine. 1998 Jan 15;23(2):174-80.
6. Wuster C, Heilmann P, Pereira-Lima J, Schlegel J, Anstatt K, Soballa T. Quantitative ultrasonometry (QUS) for the evaluation of osteoporosis risk: reference

data for various measurement sites, limitations and application possibilities. Exp Clin Endocrinol Diabetes. 1998;106(4):277-88.
7. Dequeker J. Nijs J. Verstraeten A. Geusens P. and Gevers G. (1987) Genetic determinants of bone mineral content at the spine and radius: a twin study. *Bone.* 8(4):207-9.
8. Seeman E. Hopper JL. Bach LA. Cooper ME. Parkinson E. McKay J. and Jerums G. (1989) Reduced bone mass in daughters of women with osteoporosis. *New England Journal of Medicine.* 320(9):554-8.
9. Rosen CJ, Beamer WG, Donahue LR. Defining the genetics of osteoporosis: using the mouse to understand man. Osteoporos Int. 2001;12(10):803-10.
10. Moore KJ Utilization of mouse models in the discovery of human disease genes. Drug Discov Today. 1999 Mar;4(3):123-128.
11. Moore KJ, Nagle DL Complex trait analysis in the mouse: The strengths, the limitations and the promise yet to come. Annu Rev Genet. 2000;34:653-686. Review.
12. McPeek MS From mouse to human: fine mapping of quantitative trait loci in a model organism. Proc Natl Acad Sci U S A. 2000 Nov 7;97(23):12389-90.
13. Ho NC, Jia L, Driscoll CC, Gutter EM, Francomano CA. A skeletal gene database J Bone Miner Res. 2000 Nov;15(11):2095-122.
14. Bedell MA, Jenkins NA, Copeland NG. Mouse models of human disease. Part I: techniques and resources for genetic analysis in mice. Genes Dev. 1997 Jan 1;11(1):1-10. Review.
15. Beamer WG, Donahue LR, Rosen CJ, Baylink DJ. Genetic variability in adult bone density among inbred strains of mice. Bone. 1996 May;18(5):397-403.
16. Beamer WG, Shultz KL, Churchill GA, Frankel WN, Baylink DJ, Rosen CJ, Donahue LR Quantitative trait loci for bone density in C57BL/6J and CAST/EiJ inbred mice. Mamm Genome. 1999 Nov;10(11):1043-9.
17. Shultz KL, Donahue LR, Bouxsein ML, Baylink DJ, Rosen CJ, Beamer WG. Congenic strains of mice for verification and genetic decomposition of quantitative trait loci for femoral bone mineral density. J Bone Miner Res. 2003 Feb;18(2):175-85.
18. Beamer WG, Shultz KL, Donahue LR, Churchill GA, Sen S, Wergedal JR, Baylink DJ, Rosen CJ. Quantitative trait loci for femoral and lumbar vertebral bone mineral density in C57BL/6J and C3H/HeJ inbred strains of mice. J Bone Miner Res. 2001 Jul;16(7):1195-206.
19. Benes H, Weinstein RS, Zheng W, Thaden JJ, Jilka RL, Manolagas SC, Shmookler Reis RJ Chromosomal mapping of osteopenia-associated quantitative trait loci using closely related mouse strains. J Bone Miner Res. 2000 Apr;15(4):626-33.
20. Klein RF, Mitchell SR, Phillips TJ, Belknap JK, Orwoll ES Quantitative trait loci affecting peak bone mineral density in mice. J Bone Miner Res. 1998 Nov;13(11):1648-56.
21. Klein OF, Carlos AS, Vartanian KA, Chambers VK, Turner EJ, Phillips TJ, Belknap JK, Orwoll ES. Confirmation and fine mapping of chromosomal regions influencing peak bone mass in mice. J Bone Miner Res. 2001 Nov;16(11):1953-61.

22. Shimizu M, Higuchi K, Bennett B, Xia C, Tsuboyama T, Kasai S, Chiba T, Fujisawa H, Kogishi K, Kitado H, Kimoto M, Takeda N, Matsushita M, Okumura H, Serikawa T, Nakamura T, Johnson TE, Hosokawa M Identification of peak bone mass QTL in a spontaneously osteoporotic mouse strain. Mamm Genome. 1999 Feb;10(2):81-7.
23. Shimizu M, Higuchi K, Kasai S, Tsuboyama T, Matsushita M, Matsumura T, Okudaira S, Mori M, Koizumi A, Nakamura T, Hosokawa M. A congenic mouse and candidate gene at the Chromosome 13 locus regulating bone density. Mamm Genome. 2002 Jul;13(7):335-40.
24. Li X, Masinde G, Gu W, Wergedal J, Mohan S, Baylink DJ. Genetic dissection of femur breaking strength in a large population (MRL/MpJ x SJL/J) of F2 Mice: single QTL effects, epistasis, and pleiotropy. Genomics. 2002 May;79(5):734-40.
25. Rosen CJ, Churchill GA, Donahue LR, Shultz KL, Burgess JK, Powell DR, Ackert C, Beamer WG. Mapping quantitative trait loci for serum insulin-like growth factor-1 levels in mice. Bone. 2000 Oct;27(4):521-8. Erratum in: Bone 2000 Dec;27(6):877.
26. Turner CH, Hsieh YF, Muller R, Bouxsein ML, Baylink DJ, Rosen CJ, Grynpas MD, Donahue LR, Beamer WG. Genetic regulation of cortical and trabecular bone strength and microstructure in inbred strains of mice. J Bone Miner Res. 2000 Jun;15(6):1126-31.
27. Blake GM, Fogelman I. Bone densitometry and the diagnosis of osteoporosis. Semin Nucl Med. 2001 Jan;31(1):69-81.
28. Rho JY, Tsui TY, Pharr GM. Elastic properties of human cortical and trabecular lamellar bone measured by nanoindentation. Biomaterials. 1997 Oct;18(20):1325-30.
29. Hoffler CE, Moore KE, Kozloff K, Zysset PK, Brown MB, Goldstein SA Heterogeneity of bone lamellar-level elastic moduli. Bone. 2000 Jun;26(6):603-9.
30. Zysset PK, Guo XE, Hoffler CE, Moore KE, Goldstein SA Elastic modulus and hardness of cortical and trabecular bone lamellae measured by nanoindentation in the human femur. J Biomech. 1999 Oct;32(10):1005-12.
31. Rho JY, Zioupos P, Currey JD, Pharr GM. Variations in the individual thick lamellar properties within osteons by nanoindentation. Bone. 1999 Sep;25(3):295-300.
32. Zysset PK, Guo XE, Hoffler CE, Moore KE, Goldstein SA. Mechanical properties of human trabecular bone lamellae quantified by nanoindentation. Technol Health Care. 1998 Dec;6(5-6):429-32.
33. Rho JY, Mishra SR, Chung K, Bai J, Pharr GM. Relationship between ultrastructure and the nanoindentation properties of intramuscular herring bones. Ann Biomed Eng. 2001 Dec;29(12):1082-8.
34. Rho JY, Currey JD, Zioupos P, Pharr GM. The anisotropic Young's modulus of equine secondary osteones and interstitial bone determined by nanoindentation. J Exp Biol. 2001 May;204(Pt 10):1775-81.
35. Rho JY, Roy ME 2nd, Tsui TY, Pharr GM. Elastic properties of microstructural components of human bone tissue as measured by nanoindentation. J Biomed Mater Res. 1999 Apr;45(1):48-54.

36. Bouxsein ML, Rosen CJ, Turner CH, Ackert CL, Shultz KL, Donahue LR, Churchill G, Adamo ML, Powell DR, Turner RT, Muller R, Beamer WG. Generation of a new congenic mouse strain to test the relationships among serum insulin-like growth factor I, bone mineral density, and skeletal morphology in vivo. J Bone Miner Res. 2002 Apr;17(4):570-9.
37. Markel P, Shu P, Ebeling C, Carlson GA, Nagle DL, Smutko JS, Moore KJ. Theoretical and empirical issues for marker-assisted breeding of congenic mouse strains. Nat Genet. 1997 Nov;17(3):280-4.
38. Shultz KL, Donahue LR, Bouxsein ML, Baylink DJ, Rosen CJ, Beamer WG. Congenic strains of mice for verification and genetic decomposition of quantitative trait loci for femoral bone mineral density. J Bone Miner Res. 2003 Feb;18(2):175-85.
39. Szulc P, Claustrat B, Marchand F, Delmas PD. Increased risk of falls and increased bone resorption in elderly men with partial androgen deficiency: the MINOS study.J Clin Endocrinol Metab. 2003 Nov;88(11):5240-7.
40. Wickman S, Kajantie E, Dunkel L. Effects of suppression of estrogen action by the p450 aromatase inhibitor letrozole on bone mineral density and bone turnover in pubertal boys. J Clin Endocrinol Metab. 2003 Aug;88(8):3785-93.
41. Kung AW. Androgen and bone mass in men. Asian J Androl. 2003 Jun;5(2):148-54. Review.
42. Orwoll ES, Belknap JK, Klein RF. Gender specificity in the genetic determinants of peak bone mass. J Bone Miner Res. 2001 Nov;16(11):1962-71.
43. Klein RF, Turner RJ, Skinner LD, Vartanian KA, Serang M, Carlos AS, Shea M, Belknap JK, Orwoll ES. Mapping quantitative trait loci that influence femoral cross-sectional area in mice. J Bone Miner Res. 2002 Oct;17(10):1752-60.
44. Masinde GL, Wergedal J, Davidson H, Mohan S, Li R, Li X, Baylink DJ. Quantitative trait loci for periosteal circumference (PC): identification of single loci and epistatic effects in F2 MRL/SJL mice. Bone. 2003 May;32(5):554-60.
45. Gu W, Li X, Lau KH, Edderkaoui B, Donahae LR, Rosen CJ, Beamer WG, Shultz KL, Srivastava A, Mohan S, Baylink DJ. Gene expression between a congenic strain that contains a quantitative trait locus of high bone density from CAST/EiJ and its wild-type strain C57BL/6J. Funct Integr Genomics. 2002 Apr;1(6):375-86.
46. Gu W-K., X-M. Li, B. A. Roe, K-H. William Lau, B. Edderkaoui, S. Mohan and D. J. Baylink Application of Genomic Resources and Gene Expression Profiles to Identify Genes That Regulate Bone Density. Current Genomics 2003, *4,* 75-102.

CHAPTER 20

RECENT ADVANCES IN BONE BIOLOGY RESEARCH

Di Chen[1,2], Ying Yan[1], Mo Chen[1], Hui Shen[3], Hong-Wen Deng[3,4]

[1] *Center for Musculoskeletal Research, University of Rochester Medical Center, Rochester, New York 14642, USA*

[2] *Medical College, Nankai University, Tianjin 300071, P.R. China*

[3] *Osteoporosis Research Center and Department of Biomedical Sciences, Creighton University Medical Center, Omaha, NE 68131, USA*

[4] *Laboratory of Molecular and Statistical Genetics, College of Life Sciences, Hunan Normal University, Changsha, Hunan 410081, P.R. China*

1. Introduction

In the past decade, a tremendous advancement has been achieved in understanding the molecular mechanisms of osteoblast and osteoclast biology. The key signaling pathways which control osteoblastic bone formation and osteoclastic bone resorption have been identified. Using the molecular and genetic approaches, irrefutable data have been obtained demonstrating that Runx2, Osterix and LRP5 are critical proteins for bone development during embryogenesis and bone formation in postnatal and adult life. Additionally, it has been established that the RANKL-RANK pathway is absolutely required for osteoclast formation and bone resorption. These findings provide us opportunities to further investigate the molecular mechanisms by which bone remodeling is controlled and regulated. They also provide important insight into the pathlogical mechanisms of numerous diseases, including osteoporosis.

2. Bone Remodeling

In most physiological and pathological circumstances, the coupling of bone formation to previous bone resorption occurs faithfully. Packets of bone which are removed during resorption will be replaced by site-specific bone formation. The cellular and molecular mechanisms which are responsible for mediating the coupling process or disrupting it during disease are the focus of ongoing effort in the field. A number of hypotheses have been proposed for the coupling mechanism with the most favorable notion predicting that coupling is mediated locally by multiple growth factors released from bone matrix during the process of osteoclast bone resorption. Factors that participate in this process undoubtedly include IGFs, TGFβs, FGFs and BMPs, all of which are known to stimulate osteoblast activity and new bone formation[1]. More recently, it has been suggested that coupling, or at least the bone formation phase of bone remodeling, may also be systemically mediated by the actions of leptin on the hypothalamus which leads to inhibited bone formation following β2-adrenergic stimulation of cells in the osteoblast lineage. A full understanding of this sequence of cellular events will lead to clarification of the mechanisms responsible for the decreased osteoblast activity which occurs in age-related bone loss. Additionally, it will expand our grasp of the pathophysiology seen in osteoporosis, as well as the specific defects in osteoblast function which occur in malignancies such as myeloma and breast cancer.

All of the diseases of bone are superimposed on the normal bone remodeling sequence. In diseases where osteoclasts are activated, such as osteoporosis, primary hyperparathyroidism, hyperthyroidism, and Paget's disease, in which osteoclasts are activated, there is a compensatory and relatively balanced increase in the formation of new bone. However, there are also a number of well-described conditions in which osteoblast activity does not completely repair and replace the defect left by previous resorptive activity. One example is myeloma, usually characterized by punched-out osteolytic bone lesions with little new bone formation[2]. In myeloma, there appears to be a specific defect in osteoblast maturation[3]. There are probably increased numbers of osteoblasts around the edges of the osteolytic lesions, but the osteoblasts fail (in the great majority of patients) to synthesize more than thin osteoid seams. In solid tumors associated with malignancy, there is also a failure of bone formation to repair resorptive defects[4]. Although bone formation usually occurs at sites of previous osteoclastic resorption in normal adult humans, there are special situations in which osteoblasts may lay down new bone on surfaces not previously resorbed. The examples include osteoblastic metastases associated with tumors such as prostate and breast cancer metastasis. In elderly patients with osteoporosis, there is a decrease in mean wall thickness, presumably reflecting the inability of osteoblasts to repair adequately the resorptive defects made during osteoclastic resorption[5]. In recent years, several key transcription factors and receptors for osteoblast and osteoclast development have been identified and signaling pathways mediated by these factors may be critical for bone remodeling as well.

3. Runx2

Runx2 (also named Cbfa1, PEBP2αA, AML-3 and Osf2) is a bone-selective transcription factor that belongs to the runt-domain gene family. DNA-binding sites for Runx2 have been identified in the promoter regions of many osteoblast-specific genes[6] such as osteopontin[7], type I collagen[8], osteocalcin[9,10], and bone sialoprotein[11]. Runx2 binds response elements in these promoters and transcriptionally activates or suppresses these genes. In fact, over-expression of Runx2 in non-osteoblastic cells leads to expression of osteoblast-specific genes such as osteocalcin and bone sialoprotein[6]. Targeted disruption of Runx2 in mice reveals that Runx2 expression is absolutely required for bone development *in vivo*. Homozygous Runx2-deficient mice die soon after birth due to an inability to breathe. The most pronounced effect is a complete lack of both endochondral and intramembranous ossification[12,13], with an absence of mature osteoblasts throughout the body. Heterozygous mutant mice have skeletal abnormalities, similar to those seen in a human mutation called cleidocranial dysplasia (CCD) syndrome[14,15], including hypoplasia of the clavicle and delayed development of membranous bones[12,13].

To investigate the function of Runx2 in bone formation in postnatal mice, Ducy et al.[16] generated transgenic mice over-expressing the Runx2 DNA-binding domain (ΔRunx2) driven by the osteocalcin gene 2 (OG2) promoter. ΔRunx2 was expressed in differentiated osteoblasts only postnatally and acted in a dominant-negative fashion due to its higher affinity for binding to DNA than Runx2 itself[16]. Skeletons of the ΔRunx2-transgenic mice were normal at birth, but the mice suffered from osteopenia due to decreases in bone volume and bone formation rates, evident 3 weeks after birth[16]. These results indicate that Runx2 plays a crucial role not only in osteoblast differentiation and bone development but also osteoblast function and postnatal bone formation.

Bone morphogenetic proteins up-regulate Runx2 mRNA expression *in vitro*[6,17]. Recent reports show that Smad1 and 5 interact with Runx1, 2 and 3[18] and a truncated Runx2 identified in a CCD patient fails to interact with Smad1 and 5[19]. Also, Runx2 cooperates with Smad1 and 5 to induce osteoblast differentiation in C2C12 cells[19,20]. These lines of evidence suggest that Runx2 interacts tightly with BMP signaling protein Smad1 and 5 and regulates osteoblast differentiation.

PTH is a potent modulator of bone metabolism. In experimental animals and patients with osteoporosis, intermittent administration of PTH increases bone mass by stimulating *de novo* bone formation[21-24]. A recent report showed that PTH induces bone formation by enhancing Runx2 expression and this effect is predominantly mediated by the protein kinase A (PKA) signaling pathway[25]. In contrast, sustained administration of PTH induces bone resorption and inhibits bone formation. Recent reports show that continuous over-expression of Runx2 in osteoblasts in transgenic mice (the Runx2 gene is driven by the 2.3 kb Col1a1 promoter) induces progressive osteopenia and high cortical bone turnover during adulthood and aging[26,27]. This is likely occurs for two reasons: a) continuous over-expression of the Runx2 induces expression of RANKL[27,28] in osteoblasts, which stimulates osteoclast formation and bone resorption; and b) continuous over-expression of Runx2 inhibits osteoblast maturation since in these transgenic mice, numbers of osteopontin-positive cells are increased and osteocalcin-positive cells and osteocytes are decreased[26]. Taken together, these findings suggest that the dual effects of PTH on bone formation and bone resorption are determined by the temporal expression patterns of Runx2 induced by PTH and also suggest that Runx2 may play a critical role in bone remodeling. An animal model in which Runx2 (or ΔRunx2) expression is inducible in osteoblasts in adult animals will be required to further investigate the role of Runx2 in bone remodeling in adult life.

So far, only one study has examined the relationship between polymorphisms in Runx2 gene and bone mineral density (BMD) variation[29]. In 495 randomly selected healthy women and 800 female fracture patients, two common polymorphisms within exon 1 of Runx2 gene were identified, an 18-bp deletion and a synonymous alanine codon polymorphism with alleles GCA and GCG. The former was not significantly associated with BMD variation, whereas the GCA allele of the latter variant was related to significantly greater BMD at all measured bone sites, including spine (L2-L4), femoral neck, trochanter, ultra-distal forearm, whole body, etc. In addition, the GCA allele was associated with approximately threefold protection against Colle's fracture. These results suggested that Runx2 variants might be related to genetic effects on BMD and osteoporosis.

4. Osterix

Osterix is a novel zinc finger-containing transcription factor that is specifically expressed in all developing bones. In osterix null mutant mice, bone formation is completely absent. In endochondral skeletal elements of these mice, mesenchymal cells, together with osteoclasts and blood vessels, invade the mineralized cartilage matrix. However, the mesenchymal cells do not deposit bone matrix. Similarly, cells in the periosteum and in the condensed mesenchyme of membranous skeletal elements cannot differentiate into osteoblasts. The mesenchymal cells in osterix null mutant mice express Runx2 while osterix is not expressed in Runx2 null mutant mice, suggesting that osterix acts downstream of Runx2. Additionally, the preosteoblasts in osterix null mutant mice express chondrocyte marker genes, suggesting that osterix is a transcription factor which specifically regulates osteoblast differentiation[30]. The finding that Osx-null cells acquire a chondrocyte phenotype implies that Osterix is a negative regulator of Sox9 and of the chondrocyte phenotype overall[31].. The role of osterix on BMD variation is still waiting for exploration.

5. LRP5

Osteoporosis pseudoglioma (OPPG) syndrome is an autosomal recessive disorder involving both skeletal and eye abnormalities that have been mapped to the human chromosome 11q12-13 locus[32]. While OPPG patients posess normal bone growth, they show severe osteopenia without abnormal collagen synthesis or hormonal defects. Patients also show congenital or juvenile-onset blindness due primarily to hyperplasia of the primary vitreous. OPPG patients harbor inactivating mutations in the low-density-lipoprotein (LDL)-receptor like protein 5 (LRP5) gene[33], and heterozygous mouse carriers of the mutations have reduced bone mass. Mice deficient for the LRP5 gene show a similar phenotype to OPPG syndrome in humans[34]. Interestingly, the high bone mass (HBM) syndrome has also been mapped to this chromosomal region[35] and it has been reported that an activating mutation (Gly171Val) in LRP5 is responsible for the HBM syndrome[36]. Carriers of the autosomal dominant HBM trait show very high spinal bone mineral density without other clinical features[35]. Given the heightened interest of LRP5, the genomic region harboring this gene (chromosome 11q12-13) has been investigated for its linkage to normal BMD variation by different groups[37-39]. However, the results are largely inconsistent, owing partially to the limited power of the conventional linkage approach and the high false positive rate with a whole-genome scan (Also see Chapter 26 in this book)[39]. An interesting finding on the relationship between LRP5 variants and osteoporosis-related phenotypes has been reported recently by Ferrari et al.[40]. In a cohort of 889 healthy whites, they perform a cross-sectional association study for 13 polymorphisms in LRP5 gene with bone mineral content (BMC), areal BMD, and bone size at lumbar spine, and stature. Based on the allele frequency and LD pattern, 5 variants among the 13 polymorphisms were selected as "informative' and eventually analyzed. Significant associations were found for a missense SNP in exon 9 with spine BMC, bone size, and stature, respectively. The associations were observed mainly in adult men. Haplotype analyses of those 5 polymorphisms suggest that additional genetic variation within the locus might also contribute to bone mass and bone size variation. Although similar cross-sectional studies in children did not produce significant result as observed in adults, a 1-year longitudinal study of 386 children from the original cohort supported the hypothesis that LRP5 gene variants influence the bone phenotypes during growth, in which significant associations were observed between LRP5 haplotypes and changes in BMC and bone area in males but not females. Taken together with this evidence, LRP5 variants may be important determinants for vertebral bone mass and size, mostly in white males. The underlying physiological mechanisms and such effects in other races remain uncertain and deserve further examination.

At the surface of cells, two receptor proteins are involved in receiving the Wnt signal: Frizzled and LRP-5/6. There are many genes encoding Frizzled proteins (ten in the human genome), and different Frizzled proteins probably have different affinities for various Wnt family members. Wnt proteins can form a complex with the cysteine-rich domain (CRD) of Frizzled and with LRP-5/6, leading to the formation of a dual-receptor complex. LRP5 is a single pass membrane receptor whose extracellular domain contains four modules consisting of six YWTD repeats followed by an epidermal growth factor (EGF)-like motif and a LDLR-like ligand-binding domain[41]. A recent study has shown that LRP5 is involved in the Wnt canonical signaling pathway[42]. Wnt proteins are secreted factors playing critical roles in early during development, for instance controlling mesoderm induction, patterning, cell fate determination and morphogenesis[43]. There is also some limited information about the roles of Wnt proteins beyond development. Wnt proteins trigger signaling pathways inside cells that proceed through several protein complexes. One protein in these pathways is the β-catenin molecule. Normally, β-catenin forms a large complex with several proteins that includes Disheveled, casein kinase I, glycogen synthase kinase 3β (GSK-3β) and the scaffolding protein axin. This complex promotes the addition of phosphate groups to β-catenin by GSK-3β, enabling it to be detected by the cellular protein degradation machinery. The 1st and 2nd most N-terminal modules of LRP5 mediate its interaction with the Wnt-frizzled ligand-receptor complex. The intracellular tail of LRP-5/6 binds axin which controls β-catenin in a Wnt-dependent manner. This association results in inhibition of β-catenin phosphorylation by GSK-3β. Therefore, signaling from Wnt releases β-catenin from its binding protein axin, allowing it to move to the nucleus, where it combines with a protein called TCF to activate the expression of target genes. LRP-5/6 has a second function. It binds to a molecule that counteracts Wnt. That molecule is called Dickkopf 1 (dkk1), which blocks Wnt function. Binding of dkk1 to LRP-5/6 might alter the confirmation of LRP-5/6, so that it can no longer interact with Wnt and Frizzled and thus halting the intracellular signaling.

LRP5 is expressed in osteoblasts, although at a very low level. LRP5 and LRP6 expression is stimulated by the treatment with BMP-2 in the ST2 marrow stromal cells[33]. In the same cells, Wnt1, Wnt2 and Wnt3a but not Wnt4 and Wnt5a can induce the expression of alkaline phosphatase (ALP). Over-expression of LRP5 cannot enhance Wnt3a induction of ALP but a gain-of-function mutation in this gene does induce the HBM phenotype in humans. This suggests that other Wnts, or even other ligands, may be involved in signaling through LRP5 and that Wnt-induced ALP stimulation may reflect only one aspect of their activity.

6. OPG-RANKL-RANK

It has become clear in the last few years that the tumor necrosis factor (TNF) ligand family member, Receptor Activator of NF-κB (RANK) ligand and its two known receptors RANK and osteoprotegerin (OPG), are the key local regulators of osteoclastic bone resorption *in vivo*.

6.1. Osteoprotegerin (OPG)

OPG is a TNF receptor (TNFR) superfamily member that lacks a transmembrane domain and is thus secreted. When expressed, recombinant OPG inhibits both physiological and pathological bone resorption. Hepatic over-expression of the *OPG* gene in mice results in severe osteopetrosis[44]. OPG has only two known ligands, RANK ligand

(RANKL) and TRAIL, both of which are type II membrane-bound TNF homologs[45,46]. In contrast to most other TNF receptor family members, OPG is secreted and circulates *in vivo*[44]. Cumulative evidence shows that OPG acts as a non-signaling decoy receptor for RANKL, and thereby regulates bone turnover[47-49]. Although OPG can also binds to TRAIL, the significance of this is unknown. It is unlikely that the interaction between TRAIL and OPG interfers bone remodeling since *Trail*-deficient mice have no skeletal abnormalities[50]. The relationship between circulating OPG levels and bone turnover remains unclear. The effects of OPG on bone have been best shown in rodents where *OPG*-deficient mice exhibit profound osteoporosis from birth caused by enhanced osteoclast formation and function as well as prolonged osteoclast survival. Histology reveals a destruction of growth plates, lack of trabeculae and histomorphometric analyses demonstrate an increase in bone resorption in long bones of adult null mutant mice. *OPG*-deficient mice also develop calcification of the aorta and renal arteries[51,52]. These findings indicate that OPG is a physiological regulator of osteoclast-mediated bone resorption during postnatal bone growth. It also suggests that OPG might play a role in preventing calcification of larger arteries. In vivo, parenteral administration of OPG results in a marked increase in bone mineral density and bone volume associated with a decrease in the number of active osteoclasts both in normal and ovariectomized rats[53]. Serum calcium concentration also decreases rapidly upon parenteral administration of OPG, independent of any changes in urinary calcium excretion, in thyroparathyroidectomized rats whose serum calcium levels were raised acutely by PTH infusion[54]. This suggests that OPG, in addition to its effect on osteoclastogenesis, also affects the function and/or survival of mature osteoclasts.

In vitro, in the presence of M-CSF, RANKL induces osteoclast formation in the absence of osteoblasts/stromal cells, and addition of OPG abrogates this[47,48]. OPG also strongly inhibits osteoclast formation induced by a range of osteotropic hormones and cytokines including 1,25-dihydroxyvitamin D3, PTH, PGE_2, IL-1 and IL-11 in co-cultures of osteoblasts/stromal cells and hemapoietic osteoclast progenitors. Interestingly, almost all of the factors that stimulate RANKL expression conversely inhibit OPG production[47-49].

Compared with other genes described above, the relationship between variants in the OPG gene and normal BMD variation or osteoporotic fractures has been tested extensively. Several groups have conducted linkage and/or association studies for various polymorphisms inside the OPG gene with BMD variation and/or osteoporotic fractures[55-60]. Langdahl *et al.*[55] identified 12 polymorphisms in the promoter region and the five exons of the human OPG gene. Although none of them were associated with BMD variation or biochemical markers of bone turnover, two polymorphisms (A163G and T245G) in the promoter region were significantly associated with vertebral fractures. Interestingly, Arko *et al.*[56] reported a weak association between the T245G polymorphism with lumbar spine BMD. Recently, a case-control study in postmenopausal Danish women also showed significant association between the A163G polymorphism and forearm BMD, heel BUA, heel SOS, as well as fractures[59]. Therefore, polymorphisms in the promoter of the OPG gene may affect the BMD variation and risk of osteoporotic fractures. The molecular mechanisms underlying the effects of these variants deserve further investigation.

6.2. RANKL and its signaling receptor, RANK

The existence of a cell surface-associated factor for osteoclast formation and differentiation (termed osteoclast differentiation factor, ODF) has been postulated for many years. This factor is now known to be a TNF ligand superfamily member, which is cloned by four independent groups and designated as RANKL[61], TRANCE[62], OPG ligand[63] and ODF[64]. The expression of this molecule is obligatory for osteoclastic bone resorption and normal bone modeling and remodeling. As proposed by American Society for Bone and Mineral Research, this cytokine will be referred to hereafter as RANKL. Although the existence of a secreted form of RANKL encompassing representing the extracellular C-terminal domain has been described in a number of studies[63,65,66], there is to date no unequivocal evidence that a soluble form of RANKL exists *in vivo* or is generated by proteolytic cleavage in the bone microenvironment.

In recent years, our understanding of the role of RANKL in bone resorption *in vivo* has increased tremendously with generation of

RANKL knockout mice. In these mice, typical osteopetrosis with total occlusion of bone marrow space with endosteal bone was found. The bones of the RANKL null mutant mice lack osteoclasts, although they contain osteoclast progenitors which differentiate into functionally active osteoclasts when co-cultured with normal osteoblasts/stromal cells from wild-type littermates[67]. Administration of recombinant RANKL to mice induces osteoclast formation and increases blood ionized calcium[63,68]. These results suggest that RANKL is absolutly required for osteoclast development.

In the presence of M-CSF, RANKL induces osteoclast formation in all model systems presently available to study osteoclast ontogeny[47,49]. Treatment of stromal/osteoblastic cells with known stimulators of osteoclast formation, 1,25-dihydroxyvitamin D_3, PTH, PGE_2, IL-1β, TNF-α, IL-11 and IL-6 induces or enhances RANKL mRNA expression[47,49]. A recombinant soluble form of RANKL stimulates bone resorption in organ cultures that is completely inhibitable by OPG. Polyclonal antibodies against RANKL inhibit bone resorption in organ cultures induced by not only soluble RANKL but also by a variety of unrelated hormones and cytokines, clearly indicating that bone resorption induced by these osteotropic factors is mediated by RANKL. In cultures of isolated rat osteoclasts, devoid of stromal or osteoblastic cells and where there is no new osteoclast formation, recombinant RANKL markedly increases the bone-resorbing activity of the osteoclasts as well as prolonged their survival[47,48].

The TNFR super family member RANK is the only known signaling receptor for RANKL. RANK, a type I transmembrane protein, mediates all of the signals essential for osteoclast differentiation from hematopoietic progenitors as well as activation of mature osteoclasts[47,69,70]. Interestingly, over-expression of RANK in human embryonic kidney fibroblasts (293) cells induces ligand-independent NF-κB activation[69], suggesting that pathological conditions associated with RANK over-expression or mis-expression may result in increased osteoclast formation independent of RANKL. In accord with this notion, mutations in exon 1 of the RANK gene have been detected in Familial Expansile Osteolysis (FEO), a condition characterized by osteolytic lesions and generalized osteopenia. The resulting mutant RANK proteins are constitutively active or exhibit increased NF-κB activation *in vitro* consistent with a gain-of-function mutation[71,72]. *Rank*-deficient mice have been generated[70,73], and as expected they exhibit severe osteopetrosis due to complete absence of osteoclasts and lack of bone resorption. Although these RANK null mutant mice form incisors, there is complete failure of teeth eruption, thus confirming the absolute requirement of an intact RANKL-RANK pathway for osteoclastogenesis *in vitro* and *in vivo*.

Genetic studies on variants of the RANK and the RANKL genes are very rare. No linkage evidence has been observed at regions around the two genes with BMD variation in 164 English families[74]. One association study identified 13 polymorphisms in the RANK gene and found that one of the polymorphism in intron 6 is marginally associated with risk of osteoporotic fractures and BMD variation[75]. Further analyses for these two genes are certainly desired.

7. Alox15

Through combined quantitative trait loci (QTL) mapping and gene expression analysis, Klein and his colleagues identified the lipoxygenase gene Alox15 as a negative regulator of peak BMD in mice[76]. Analysis of genomic DNA identified 15 polymorphisms in the Alox15 gene that distinguished the D2 (low peak BMD and femoral shaft strength) and B6 (high peak BMD and femoral shaft strength) strains[76]. The Alox15 gene encodes 12/15-lipoxygenase (12/15-LO), an enzyme that converts arachidonic and linoleic acids into endogenous ligands for the peroxisome proliferator-activated receptor-γ (PPARγ)[77]. Activation of this pathway in marrow-derived mesenchymal progenitors stimulates adipogenesis and inhibits osteblastogenesis[78]. Transient over-expression of 12/15-LO in murine bone marrow stromal cells reduced alkaline phosphatase activity and secretion of osteocalcin, which reflected restrict osteoblast differentiation[76]. To explore the role of 12/15-LO in skeletal development *in vivo*, Klein et al. examined the skeletal phenotype of 12/15-LO knockout mice. Compared with the B6 mice, the 12/15-LO knockout mice have similar body weight and whole-body BMD, but with increased femoral BMD and biomechanical indices of femoral shaft strength[76]. In addition, cross-breeding experiments between the D2 mice and 12/15-LO knockout mice confirmed the Alox15 plays a role in skeletal development[76]. Moreover, pharmacological inhibitors of this enzyme improved bone density and strength in two rodent models of osteoporosis[76]. These findings demonstrate that genetic variation within the 12/15-LO locus contributes to variation in peak bone mass and identify a novel pathway that regulates skeletal development. Further studies in both animal models and human populations will be required to further understand the role of 12/15-LO pathway in bone development and pinpoint the causal polymorphism(s) within this gene for BMD variation.

References

1. Howard GA, Bottemiller BL, Turner RT, et al. *Proc Natl Acad Sci* USA 78:3204-3208 (1981).
2. Snapper I, Kahn A. Myelomatosis. (Karger, Basel), 1971.
3. Valentin-Opran A, Charhon SA, Meunier PJ, et al. *Brit J Haematol* 52:601-610 (1982).
4. Stewart AF, Vignery A, Silvergate A, et al. *J Clin Endocrinol Metab* 55:219-227 (1982).
5. Darby AJ, Meunier PJ. *Calcif Tiss Intl* 33:199-204 (1981).
6. Ducy P, Zhang R, Geoffroy V, et al. *Cell* 89:747-754 (1997).
7. Sato M, Morii E, Komori T, et al. *Oncogene* 17:1517-1525 (1998).
8. Kern B, Shen J, Starbuck M, et al. *J Biol Chem* 276:7101-7107 (2001).
9. Ducy P, and Karsenty G. *Mol Cell Biol* 15:1858-1869 (1995).
10. Geoffroy V, Ducy P, Karsenty G. *J Biol Chem* 270:30973-30979 (1995).
11. Javed A, Barnes GL, Jasanya BO, et al. *Mol Cell Biol* 21:2891-2905 (2001).
12. Komori T, Yagi H, Nomura S, et al. *Cell* 89:755-764 (1997).
13. Otto F, Thornell AP, Crompton T, et al. *Cell* 89:765-771 (1997).
14. Mundlos S, Mulliken JB, Abramson DL, et al. *Hum Mol Genet* 4:71-75 (1995).
15. Mundlos S, Otto F, Mundlos C, et al. *Cell* 89:773-779 (1997).
16. Ducy P, Starbuck M, Priemel M, et al. *Genes Dev* 13:1025-1036 (1999).
17. Chen D, Ji X, Harris MA, et al. *J Cell Biol* 142:295-305 (1998).
18. Hanai J, Chen LF, Kanno T, et al. *J Biol Chem* 274:31577-31582 (1999).
19. Zhang YW, Yasui N, Ito K, et al. *Proc Natl Acad Sci USA* 97:10549-10554 (2000).
20. Lee KS, Kim HJ, Li QL, et al. *Mol Cell Biol* 20:8783-8792 (2000).
21. Hock JM, Fonseca GJ, and Raisz LG. *Endocrinology* 122:2899-2904 (1988).
22. Hock JM and Gera I. *J Bone Miner Res* 7:65-72 (1992).
23. Dempster DW, Cosman F, Parisien M, et al. *Endocr Rev* 14:690-708 (1993).
24. Neer RM, Arnaud CD, Zanchetta JR, et al. *N Engl J Med* 344:1434-1441 (2001).
25. Krishnan V, Moore TL, Ma YL, et al. *Mol Endocrinol* 17:423-435 (2003).

26. Liu W, Toyosawa S, Furuichi T, et al. *J Cell Biol* 155:157-166 (2001).
27. Geoffroy V, Kneissel M, Fournier B, et al. *Mol Cell Biol* 22:6222-6233 (2002).
28. Enomoto H, Shiojiri S, Hoshi K, et al. *J Biol Chem* 278:23971-23977 (2003)
29. Vaughan T, Pasco JA, Kotowicz MA, et al. *J Bone Miner Res* 17:1527-1534 (2002).
30. Nakashima K, Zhou X, Kunkel G, et al. *Cell* 108:17-29 (2002).
31. Nakashima K, de Crombrugghe B. *Trends Genet* 19:458-466 (2003).
32. Gong Y, Vikkula M, Boon L, et al. *Am J Hum Genet* 59:146-151 (1996).
33. Gong Y, Slee RB, Fukai N, et al. *Cell* 107:513-523 (2001).
34. Kato M, Patel MS, Levasseur R, et al. *J Cell Biol* 157:303-314 (2002).
35. Johnson ML, Gong G, Kimberling W, et al. *Am J Hum Genet* 60:1326-1332 (1997).
36. Little RD, Carulli JP, Del Mastro RG, et al. *Am J Hum. Genet* 70:11-19 (2002).
37. Koller DL, Econs MJ, Morin PA, et al. *J Clin Endocrinol Metab* 85:3116-20 (2000)
38. Koller DL, Liu G, Econs MJ, et al. *J Bone Miner Res* 15 (Suppl 1):1094 (2000).
39. Deng HW, Xu FH, Conway T, et al. *J Clin Endocrinol Metab* 86:3735-41 (2001).
40. Ferrari SL, Deutsch S, Choudhury U et, et al. Am J Hum Genet 74:866-875 (2004).
41. Bejsovec A. *Curr Biol* 10:R919-922 (2000).
42. Mao J, Wang J, Liu B, et al. *Mol Cell* 7:801-809 (2001).
43. Huelsken J, Birchmeier W. *Curr Opin Genet Dev* 11:547-53 (2001).
44. Simonet WS, Lacey DL, Dunstan CR et al. *Cell* 89:309-310 (1997)
45. Simonet WS, Lacey DL, Timms E, et al. *Cell* 93:165-176 (1998).
46. Emery JG, McDonnell P, Burke MB, et al. *J Biol Chem* 273:14363-14367 (1998).
47. Hofbauer LC, Heufelder AE. *J Mol Med* 79:243-253 (2001).
48. Theill LE, Boyle WJ, Penninger JM. *Annu Rev Immunol* 20:795-823 (2002).
49. Sordillo EM, Pearse RN. *Cancer*; 97(3 Suppl):802-812 (2003).
50. Sedger LM, Glaccum MB, JC Schuh JC, et al. *Eur J Immunol* 32:2246-2254 (2002).
51. Bucay N, Sarosi I, CR Dunstan CR, et al. *Genes Develop* 12:1260-1268 (1998).
52. Mizuno A, N Amizuka, K Irie, et al. *Biochem Biophys Res Commun* 247:610-615 (1998).
53. Akatsu T, Murakami T, Kono M, et al. *Bone* 23:495-498 (1998).
54. Yamamoto M, Murakami T, Nishikawa M, et al. *Endocrinology* 139:4012-4015 (1998).
55. Langdahl BL, Carstens M, Stenkjaer L, et al. *J Bone Miner Res* 17:1245-1255 (2002).
56. Arko B, Prezelj J, Komel R, et al. *J Clin Endocrinol Metab* 87:4080-4084 (2002).
57. Wynne F, Drummond F, O'Sullivan K, et al. *Calcif Tissue Int* 71:26-35 (2002).
58. Ohmori H, Makita Y, Funamizu M, et al. *J Hum Genet* 47:400-406 (2002).
59. Jorgensen HL, Kusk P, Madsen B, et al. *J Bone Miner Metab* 22:132-8 (2004).
60. Brandstrom H, Gerdhem P, Stiger F, et al. *Calcif Tissue Int* 74:18-24 (2004).
61. Anderson DM, Maraskovsky E, Billingsley WL, et al. *Nature* 390:175-179 (1997).
62. Wong BR, Rho J, Arron J, et al. *J Biol Chem* 272:25190-25194 (1997).
63. Simonet WS, Lacey DL, Timms E, et al. *Cell* 93:165-176 (1998).
64. Yasuda H, Shima N, Nakagawa N, et al. *Endocrinology* 139:1329-1337 (1998).
65. Lum L, Wong BR, Josien R, et al. *J Biol Chem* 274:13613-13618 (1999).
66. Wong BR, Josien R, Choi Y. *J Leukocyte Biol* 65:715-724 (1999).
67. Kong Y-Y, Yoshida H, Sarosi I, et al. *Nature* 397:315-323 (1999).

68. Burgess TL, Qian Y-X, Kaufman S, et al. *J Cell Biol* 145:527-538 (1999).
69. Hsu H, DL Lacey, CR Dunstan, et al. *Proc Natl Acad Sci USA* 96:3540-3545 (1999).
70. Li J, Sarosi I, Yan X-Q, et al. *Proc Natl Acad Sci USA* 97:1566-1571 (1999).
71. Hughes AE, Ralston SH, Marken J, et al. *Nature Genet* 24:45-48 (2000).
72. Johnson-Pais T, Singer F, Bone HG, et al. *J Bone Miner Res* 18:376-380 (2003).
73. Dougall B, Glaccum M, Charrier K, et al. *Genes Develop* 13:2412-2424 (1999).
74. Miles LJ, Duncan EL, Crane AM, et al. *J Bone Miner Res* 17(Suppl 1):S322 (2002)
75. Langdahl BL, Husted LB, Hocking LJ, et al. *J Bone Miner Res* 18 (Suppl2):S210 (2003).
76. Klein RF, Allard J, Avnur Z et al. *Science* 303:229-232 (2004).
77. Huang JT, Welch JS, Ricote M, et al. *Nature* 400:378-82 (1999).
78. Khan E and Abu-Amer Y. *J Lab Clin Med* 142:29-34 (2003).

APPENDIX

AN INTRODUCTION TO HOLOGIC TECHNOLOGY

- Introducing:
- Discover the Difference 10 Seconds Makes:
- The new Standard in point-of-care fracture risk assessment.
- Now with Express BMD.
- The revolutionary Discovery – the latest addition to the Hologic QDR Series –
- Combine the proven clinical value of bone mineral density (BMD) measurement with Instant Vertebral Assessment (IVA), enabling the fastest point-of-care evaluation of the two most definitive factors associated with osteoporotic fracture: low BMD and the presence of vertebral fracture.
- Utilizing OnePass single-sweep scanning technology, Discovery raises the standard of bone densitometry with features that include:
- **10-second Express BMD** – Regional spine and hip BMD results in just 10 seconds with better than 1% precision.
- **10-second Instant Vertebral Assessment** – single-energy imaging of the spine (L4-T4) with exceptional image quality in only 10 seconds

- **e-Reporting** – Advanced remote interpretation and reporting software, including speech recognition compatibility

- **CADfx** – Computer-aided fracture assessment tool quantifies degree of vertebral compression and simplifies IVA interpretation

- **QDR Mobility** – An exclusive package of mobile reporting tools, enabling physicians to receive BMD studies and generate reports using a wireless network.

- See specifications data sheet for available features on specific Discovery models.

INDEX